Budin Michov
**Electrophoresis Fundamentals**

## Also of Interest

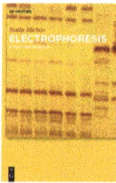

*Electrophoresis.*
*Theory and Practice*
Budin Michov, 2020
ISBN 978-3-11-033071-7, e-ISBN 978-3-11-033075-5

*Praktische Labordiagnostik.*
*Lehrbuch zur Laboratoriumsmedizin, klinischen Chemie*
*und Hämatologie*
Harald Renz (Ed.), 2018
ISBN 978-3-11-047376-6, e-ISBN 978-3-11-047407-7

*Organic Trace Analysis*
Reinhard Nießner and Andreas Schäffer, 2017
ISBN 978-3-11-044114-7, e-ISBN 978-3-11-044115-4

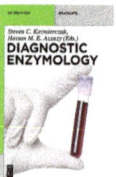

*Diagnostic Enzymology*
Steven Kazmierczak and Hassan M. E. Azzazy (Eds.), 2014
ISBN 978-3-11-020724-8, e-ISBN 978-3-11-022780-2

Budin Michov

# Electrophoresis Fundamentals

Essential Theory and Practice

DE GRUYTER

**Author**
Prof. Dr. Budin Michov, MD, PhD, DSc
bmichov@gmail.com

ISBN 978-3-11-076162-7
e-ISBN (PDF) 978-3-11-076164-1
e-ISBN (EPUB) 978-3-11-076167-2

**Library of Congress Control Number: 2021948482**

**Bibliographic information published by the Deutsche Nationalbibliothek**
The Deutsche Nationalbibliothek lists this publication in the Deutsche Nationalbibliografie;
detailed bibliographic data are available on the Internet at http://dnb.dnb.de.

To people who desire to study new techniques for exploring
proteins and nucleic acids in our organisms.

# Preface

Electrophoresis is referred to as moving of charged dissolved particles in an electric field, which causes their resolution as a result of velocities and interaction with the separation medium. The resolving is performed by charge, size, or binding affinity.

Numerous electrophoresis techniques are known. They are used for resolving proteins and nucleic acids, chromosomes, viruses, cell organelles (mitochondria, ribosomes), cells (red cells, tissue cells, and parasites), *etc.*

The electrophoresis methods are used in medicine, molecular biology, biochemistry, proteomic and genomic studies, microbiology, and forensics.

The proposed book *Electrophoresis Fundamentals* is based on my recent book *Electrophoresis. Theory and Practice*, published in 2020 by Walter de Gruyter in Berlin and Boston. While the published book is intended for specialists, the present one is a book for laboratory technicians, students, young scientists, biochemists, medical doctors, and more.

It was a challenge and a hard work for me to write the book *Electrophoresis Fundamentals* after the great progress in protein and nucleic acid electrophoresis in the last years. I will be happy, if it will help people who want to understand the basis of electrophoresis and desire to use its methods in the practice.

Nuremberg, Sofia, 2022
Budin Michov

https://doi.org/10.1515/9783110761641-202

# About the Author

**Prof. Budin M. Michov, MD, PhD, DSc,** studied medicine and specialized in biochemistry. He worked as a chemist and biochemist at Technical University in Munich and as a scientific adviser at SERVA Electrophoresis in Heidelberg, Germany. On the basis of his patented method, he founded a company in Germany for electrophoretic methods and products. Later, he became the chairman of the Department of Biochemistry at the Medical Faculty of Sofia University, Bulgaria.

Prof. Michov published articles on a new complex compound in TRIS-borate buffers, mobility calculation of composed ions, calculating the TRIS-borate ion mobility, ionic mobility parameter, simplifying the Henry's function, electrophoresis in one buffer at two pH values, electrophoresis in expanded stationary boundary, TRIS-formate-taurinate buffer system for SDS electrophoresis in homogeneous polyacrylamide gels, electrophoresis of CSF proteins in agarose gels, geometric and electrokinetic radii and electric potentials of charged and dissolved particles, and more. He invented a new device for concentration gradient system, and a buffer recirculator for electrophoresis.

Prof. Michov has published several books: *Electrophorese*, Walter de Gruyter, Berlin and New York (1996), in German; *Molecular Mechanisms of Inflammation and Allergy* (1977); *Electrophoresis. Theory and Practice* (2020), Walter de Gruyter, Berlin and Boston; and more.

https://doi.org/10.1515/9783110761641-203

# Contents

Preface —— VII

About the Author —— IX

Abbreviations —— XXIX

1          Fundamentals of electrophoresis —— 1
                Overview on electrophoresis —— 1
                Separation media —— 3
                Electrophoresis resolution and sharpness —— 3
                Detection of resolved bands —— 3
                References —— 4

1.1        Electric double layer of a charged particle —— 5
1.1.1          Electric charges of polyions —— 5
1.1.2          Model of Helmholtz —— 5
1.1.3          Model of Gouy–Chapman —— 6
1.1.4          Theory of Debye–Hückel —— 7
1.1.5          Model of Stern —— 7
1.1.6          Two radii and two electric potentials of a charged particle —— 9
                References —— 11

1.2        Proteins and nucleic acids form polyions in solution —— 13
1.2.1          Structure and conformation of proteins —— 13
1.2.1.1        Masses and electric charges of proteins —— 13
1.2.1.2        Isoelectric points of proteins —— 14
1.2.1.3        Native and denatured proteins —— 15
1.2.2          Structure and conformation of nucleic acids —— 15
1.2.2.1        Masses and electric charges of nucleic acids —— 15
1.2.2.2        Native and denatured nucleic acids —— 16
                References —— 17

1.3        Electrophoresis is running in buffers —— 19
1.3.1          Buffers —— 19
1.3.1.1        Buffer capacity —— 20
1.3.2          Buffers used in electrophoresis —— 22
1.3.3          Biological buffers —— 23
                References —— 26

1.4        **Polyions are moving in electric field** —— 27
1.4.1          Ionic and polyionic mobility —— 27
1.4.1.1        Effective mobility —— 28
1.4.2          Equations of polyionic mobility —— 29
1.4.2.1        Equation of Smoluchowski —— 29
1.4.2.2        Equation of Hückel —— 29
1.4.2.3        Henry's function —— 29
1.4.2.4        New expression of Henry's function —— 30
1.4.2.5        Equation of Onsager —— 31
1.4.2.6        Equation of Robinson–Stokes —— 32
1.4.2.7        Parametric equation —— 33
1.4.2.8        Quadratic equation —— 33
1.4.3          Mobilities of ions used in electrophoretic methods —— 33
1.4.3.1        Calculating the mobilities of composed ions —— 35
               References —— 37

1.5        **Electrophoresis is carried out in different solid media** —— 39
1.5.1          Cellulose acetate —— 39
1.5.2          Agarose gel —— 39
1.5.3          Polyacrylamide gel —— 41
1.5.3.1        Acrylamide —— 42
1.5.3.2        Bisacrylamide —— 42
1.5.3.3        The magnitudes $T$ and $C$ —— 42
1.5.3.4        Alternative cross-linkers —— 42
1.5.3.5        Initiator-catalyst systems —— 43
1.5.3.6        Copolymerization of acrylamide and BIS —— 45
1.5.3.7        Homogeneous polyacrylamide gels —— 46
1.5.3.8        Thin and ultrathin polyacrylamide gels —— 46
1.5.3.9        Gradient polyacrylamide gels —— 47
               References —— 47

1.6        **General theory of electrophoresis** —— 49
1.6.1          What is the polyionic mobility depending on? —— 50
1.6.1.1        Influence of polyionic nature —— 50
1.6.1.2        Influence of buffer —— 50
1.6.1.3        Influence of medium —— 52
1.6.1.4        Electroosmosis —— 52
1.6.2          Ionic boundaries —— 52
1.6.2.1        Moving boundary —— 52
1.6.2.2        Stationary boundary —— 53
1.6.3          Regulating function —— 53
1.6.4          Diffusion —— 53

1.6.5        Joule heating —— **53**
             References —— **54**

**1.7        Electrophoresis instrumentation —— 55**
1.7.1        Electrophoresis cells —— **56**
1.7.2        Power supplies —— **56**
1.7.3        Thermostats —— **56**
1.7.4        Scanners and densitometers —— **56**
1.7.5        Gel casting cassettes —— **57**
1.7.6        Gradient makers —— **57**
1.7.7        Buffer mixers —— **58**
1.7.8        Blotters —— **59**
1.7.9        Equipment for semi-automatic electrophoresis —— **59**
             References —— **60**

**1.8        Classification of electrophoretic methods —— 61**
1.8.1        Zone electrophoresis —— **62**
1.8.1.1      Tiselius electrophoresis —— **62**
1.8.1.2      Capillary electrophoresis —— **62**
1.8.1.3      Free-flow electrophoresis —— **62**
1.8.1.4      Solid media electrophoresis —— **62**
1.8.1.5      Cellulose acetate electrophoresis —— **63**
1.8.1.6      Agarose gel electrophoresis —— **63**
1.8.1.7      Pulsed-field electrophoresis —— **63**
1.8.1.8      Immunoelectrophoresis and immunofixation —— **63**
1.8.1.9      Affinity electrophoresis —— **64**
1.8.1.10     Polyacrylamide gel electrophoresis —— **64**
1.8.1.11     Iontophoresis —— **64**
1.8.2        Isotachophoresis —— **64**
1.8.2.1      Disc-electrophoresis —— **64**
1.8.2.2      Native disc-electrophoresis —— **65**
1.8.2.3      SDS disc-electrophoresis —— **65**
1.8.3        Isoelectric focusing —— **65**
1.8.3.1      Isoelectric focusing with carrier ampholytes —— **65**
1.8.3.2      Isoelectric focusing in immobilized pH gradients —— **65**
1.8.3.3      Two-dimensional electrophoresis —— **66**
1.8.4        Dielectrophoresis —— **66**
1.8.4.1      Dielectrophoretic force —— **66**
1.8.4.2      DEP technology —— **67**
1.8.4.3      Applications of dielectrophoresis —— **68**
             References —— **68**

| | | |
|---|---|---|
| 2 | **Electrophoresis of proteins** —— 71 | |
| | References —— 71 | |
| | | |
| **2.1** | **Cellulose acetate electrophoresis of proteins** —— 73 | |
| 2.1.1 | Theory of cellulose acetate electrophoresis of proteins —— 73 | |
| 2.1.2 | Practice of cellulose acetate electrophoresis of proteins —— 73 | |
| 2.1.3 | Protocols —— 74 | |
| | Cellulose Acetate Electrophoresis of Serum Proteins —— 74 | |
| | References —— 75 | |
| | | |
| **2.2** | **Agarose gel electrophoresis of proteins** —— 77 | |
| 2.2.1 | Theory of agarose gel electrophoresis of proteins —— 78 | |
| 2.2.2 | Agarose gel electrophoresis of serum proteins —— 78 | |
| 2.2.2.1 | Albumin —— 80 | |
| | Intermediate zone of albumin-$\alpha_1$-globulins —— 80 | |
| 2.2.2.2 | Alpha-1 globulins —— 81 | |
| | Intermediate zone of $\alpha_1$-$\alpha_2$-globulins —— 81 | |
| 2.2.2.3 | Alpha-2 globulins —— 82 | |
| | Intermediate zone of $\alpha_2$-$\beta$-globulins —— 82 | |
| 2.2.2.4 | Beta-globulins —— 83 | |
| | Intermediate zone of $\beta$-$\gamma$-globulins —— 84 | |
| 2.2.2.5 | Gamma-globulins —— 84 | |
| 2.2.3 | Agarose gel electrophoresis of lipoproteins —— 85 | |
| 2.2.3.1 | High-density lipoproteins —— 86 | |
| 2.2.3.2 | Low-density lipoproteins —— 86 | |
| 2.2.3.3 | Intermediate-density lipoproteins —— 86 | |
| 2.2.3.4 | Very-low-density lipoproteins —— 86 | |
| 2.2.3.5 | Chylomicrons —— 86 | |
| 2.2.3.6 | Hyperlipoproteinemias —— 87 | |
| 2.2.4 | Agarose gel electrophoresis of hemoglobins —— 89 | |
| 2.2.4.1 | Normal hemoglobins —— 89 | |
| 2.2.4.2 | Abnormal and pathological hemoglobins —— 90 | |
| | Sickle cell disease —— 91 | |
| | Thalassemias —— 91 | |
| 2.2.4.3 | Running hemoglobin electrophoresis —— 93 | |
| 2.2.5 | Agarose gel electrophoresis of cerebrospinal fluid proteins —— 94 | |
| 2.2.6 | Electrophoresis of creatine kinase isoenzymes —— 95 | |
| 2.2.7 | Electrophoresis of lactate dehydrogenase isoenzymes —— 96 | |
| 2.2.8 | Agarose gel electrophoresis of urinary proteins —— 97 | |
| 2.2.9 | Protocols —— 98 | |
| | Agarose Gel Electrophoresis of Serum Proteins —— 98 | |
| | Agarose Gel Electrophoresis of Lipoproteins —— 100 | |

Agarose Gel Electrophoresis of Hemoglobins —— 102
Alkaline Agarose Gel Electrophoresis of Hemoglobins —— 102
Acidic Agarose Gel Electrophoresis of Hemoglobins —— 103
Agarose Gel Electrophoresis of Cerebrospinal Fluid Proteins —— 104
Agarose Gel Electrophoresis of Urinary Proteins —— 105
References —— 107

2.3       Immunoelectrophoresis —— 109
2.3.1     Immunodiffusion electrophoresis according to Grabar and
          Williams —— 110
2.3.2     Rocket immunoelectrophoresis according to Laurell —— 110
2.3.3     Immunofixation and immunoprinting —— 111
2.3.4     Protocols —— 112
          Immunofixation of Serum Proteins —— 112
          Immunofixation of Bence Jones Proteins —— 115
          Immunoprobing with Avidin-biotin Coupling to Secondary
          Antibody —— 116
          References —— 117

2.4       Affinity electrophoresis —— 119
2.4.1     Theory of affinity electrophoresis —— 119
2.4.2     Lectin affinity electrophoresis —— 119
2.4.2.1   Electrophoresis of alkaline phosphatase isoenzymes on lectin
          agarose gels —— 120
2.4.3     Saccharide affinity electrophoresis —— 122
2.4.4     Affinity supported molecular matrix electrophoresis —— 123
2.4.5     Phosphate affinity electrophoresis —— 123
2.4.6     Capillary affinity electrophoresis —— 125
2.4.7     Affinity-trap electrophoresis —— 125
2.4.8     Charge shift electrophoresis —— 126
2.4.9     Mobility shift electrophoresis —— 126
2.4.10    Protocols —— 127
          Electrophoresis of ALP Isoenzymes on Lectin Agarose Gels —— 127
          Staining the ALP Isoenzymes —— 128
          References —— 129

2.5       Polyacrylamide gel zone electrophoresis of proteins —— 131
2.5.1     Homogeneous gel zone electrophoresis of proteins —— 131
2.5.1.1   Theory of polyacrylamide gel zone electrophoresis of
          proteins —— 132
2.5.1.2   McLellan buffers —— 133
2.5.1.3   Running electrophoresis —— 134

2.5.2       Gradient gel zone electrophoresis of proteins —— 134
2.5.2.1     Theory of gradient gel zone electrophoresis —— 135
2.5.2.2     Ferguson plots —— 137
2.5.2.3     Determination of Stokes radii and masses of native proteins —— 138
2.5.3       Blue native polyacrylamide gel electrophoresis —— 139
2.5.4       Protocols —— 140
            Horizontal Electrophoresis of Proteins on Gradient Gels —— 140
            Blue Native Polyacrylamide Gel Electrophoresis —— 142
            Preparation of Dialyzed Cell Lysate —— 142
            Casting Gradient BN Gels —— 143
            Running BN Electrophoresis —— 145
            References —— 145

2.6         **Isotachophoresis of proteins** —— 147
2.6.1       Theory of isotachophoresis of proteins —— 147
2.6.1.1     Kohlrausch regulating function —— 147
            References —— 150

2.7         **Disc-electrophoresis of proteins** —— 151
2.7.1       Theory of disc-electrophoresis —— 152
2.7.2       Native disc-electrophoresis —— 153
2.7.2.1     Disc-electrophoresis according to Ornstein and Davis —— 153
2.7.2.2     Buffer-gel systems for disc-electrophoresis according to Ornstein
            and Davis —— 155
2.7.2.3     Disc-electrophoresis according to Allen *et al* —— 156
2.7.2.4     Effect of Hjerten —— 156
2.7.2.5     Buffer-gel systems for disc-electrophoresis according to
            Allen *et al* —— 156
2.7.2.6     Disc-electrophoresis in one buffer at two pH values according to
            Michov —— 157
2.7.2.7     Theory of disc-electrophoresis in one buffer at two pH values —— 157
2.7.2.8     Buffer-gel systems for disc-electrophoresis in one buffer at two
            pH values —— 159
2.7.3       Denatured SDS disc-electrophoresis —— 161
2.7.3.1     Detergents —— 161
            Sodium dodecyl sulfate —— 163
            Sample preparation for SDS polyacrylamide gel
            electrophoresis —— 165
            Nonreducing sample preparation —— 165
            Reducing sample preparation —— 165
            Reducing sample preparation with alkylation —— 166
2.7.3.2     Practice of SDS disc-electrophoresis —— 167

SDS disc-electrophoresis in TRIS-chloride-glycinate buffer
system according to Laemmli — 167
SDS disc-electrophoresis in TRIS-acetate-TRICINEate buffer
system according to Schägger–Jagow — 168
SDS disc-electrophoresis in TRIS-formate-taurinate buffer system
according to Michov — 169
SDS disc-electrophoresis in one buffer at two pH values
according to Michov — 170
SDS disc-electrophoresis in gradient gels — 171
2.7.3.3 Staining of SDS protein bands — 171
Determination of molecular masses of SDS-denatured
proteins — 171
2.7.4 Protocols — 173
Disc-electrophoresis in Alkaline Buffer-gel System According to
Ornstein–Davis — 173
Disc-electrophoresis in One Buffer at Two pH Values According to
Michov — 176
SDS Disc-electrophoresis According to Laemmli — 178
SDS Electrophoresis in TRIS-formate-taurinate Buffer System
According to Michov — 180
Coomassie Brilliant Blue R-250 Staining of Proteins Resolved in
SDS Gels — 183
References — 184

2.8 Isoelectric focusing of proteins — 187
2.8.1 Theory of isoelectric focusing — 188
2.8.2 Isoelectric focusing with carrier ampholytes — 189
2.8.2.1 Properties of carrier ampholytes — 189
2.8.2.2 Formation of a pH gradient by carrier ampholytes — 190
Separator electrofocusing — 191
2.8.2.3 IEF with carrier ampholytes on polyacrylamide gels — 191
Thin and ultrathin polyacrylamide gels for IEF with carrier
ampholytes — 191
Rehydratable polyacrylamide gels for IEF with carrier
ampholytes — 191
2.8.2.4 Sample preparation and application — 192
2.8.2.5 Electrode solutions — 192
2.8.2.6 Running isoelectric focusing with carrier ampholytes — 193
2.8.3 Isoelectric focusing in immobilized pH gradients — 195
2.8.3.1 Properties of immobilines — 196
2.8.3.2 Casting IPG gels — 197
Rehydratable IPG gels — 198

2.8.3.3        Running isoelectric focusing on IPG gels —— **198**
2.8.4          Isoelectric focusing in the clinical laboratory —— **199**
2.8.5          Protocols —— **199**
               Casting Gels for Isoelectric Focusing with Carrier Ampholytes —— **199**
               Running Isoelectric Focusing with Carrier Ampholytes —— **200**
               Casting Immobiline Gels —— **202**
               IEF with IPG Gel Strips —— **202**
               References —— **206**

2.9      **Free-flow electrophoresis of proteins** —— **209**
2.9.1          Theory of free-flow electrophoresis —— **209**
2.9.2          Types of free-flow electrophoresis —— **210**
2.9.2.1        Free-flow zone electrophoresis —— **211**
2.9.2.2        Free-flow isotachophoresis —— **211**
2.9.2.3        Free-flow isoelectric focusing —— **212**
2.9.3          Device technology of free-flow electrophoresis —— **213**
2.9.4          Detection system of free-flow electrophoresis —— **213**
2.9.5          Applications of free-flow electrophoresis —— **214**
2.9.6          Protocols —— **214**
               Free-flow Zone Electrophoresis of Human T and B
               Lymphocytes —— **214**
               References —— **215**

2.10     **Capillary electrophoresis of proteins** —— **217**
2.10.1         Theory of capillary electrophoresis of proteins —— **217**
2.10.2         Instrumentation —— **218**
2.10.2.1       Coating the capillaries —— **219**
2.10.2.2       Sieving matrix in capillary electrophoresis —— **220**
2.10.3         Practice of capillary electrophoresis —— **222**
2.10.3.1       Injection —— **222**
2.10.3.2       Separation —— **222**
2.10.3.3       Detection —— **223**
2.10.4         Types of capillary electrophoresis —— **224**
2.10.5         Applications of capillary electrophoresis —— **224**
2.10.5.1       Serum protein analysis —— **224**
2.10.5.2       Hemoglobins —— **225**
2.10.5.3       Isoenzymes —— **226**
2.10.5.4       Immune complexes —— **226**
2.10.5.5       Single cell analysis —— **227**
2.10.6         Protocols —— **227**
               Capillary IEF of Proteins —— **227**
               References —— **228**

2.11        Two-dimensional electrophoresis ── 231
2.11.1          Theory of 2D electrophoresis ── 231
2.11.2          Isoelectric focusing in the first dimension ── 232
2.11.2.1        Sample preparation ── 232
2.11.2.2        ISO-DALT and IPG-DALT ── 233
2.11.3          SDS disc-electrophoresis in the second dimension ── 234
2.11.4          Detection and evaluation of proteins in 2D pherograms ── 236
2.11.4.1        Autoradiography and fluorography ── 236
2.11.4.2        Two-dimensional gel image analysis ── 236
2.11.5          Protocols ── 237
                Two-dimensional Gel Electrophoresis Using the O'Farrell
                System ── 237
                Sample preparation ── 237
                First-dimensional gels (isoelectric focusing) ── 237
                Second-dimensional gels (SDS electrophoresis) ── 238
                References ── 239

2.12        Preparative electrophoresis of proteins ── 241
2.12.1          Preparative disc-electrophoresis ── 241
2.12.1.1        Elution of proteins during electrophoresis ── 241
2.12.1.2        Elution of proteins after electrophoresis ── 242
                Elution by diffusion ── 243
                Elution by gel dissolving ── 243
                Electroelution ── 243
2.12.2          Preparative isoelectric focusing ── 245
2.12.2.1        Preparative IEF with carrier ampholytes in granulated gels ── 245
2.12.2.2        Preparative IEF in immobilized pH gradients ── 247
2.12.2.3        Recycling isoelectric focusing ── 248
2.12.3          QPNC-PAGE ── 248
2.12.4          Protocols ── 249
                QPNC-PAGE ── 249
                References ── 250

2.13        Microchip electrophoresis of proteins ── 253
2.13.1          Microchip materials ── 254
2.13.1.1        PDMS ── 254
2.13.1.2        PMMA ── 255
2.13.1.3        PC ── 255
2.13.2          Microchip fabrication ── 256
2.13.2.1        Fabrication of channel plate ── 257
2.13.2.2        Fabrication of cover plate ── 258
2.13.2.3        Wall coating ── 258

Dynamic wall coating — 258
Permanent wall coating — 259
2.13.2.4 Bonding the plates — 259
2.13.3 Zone electrophoresis on microchip — 261
2.13.3.1 Free-flow electrophoresis on microchip — 261
2.13.3.2 Affinity- and immunoelectrophoresis on microchip — 262
2.13.4 Isotachophoresis on microchip — 262
2.13.4.1 Disc-electrophoresis on microchip — 262
2.13.5 Isoelectric focusing on microchip — 262
2.13.6 Two-dimensional electrophoresis on microchip — 263
2.13.7 Protein separation technique on microchip — 263
2.13.7.1 Concentrating the protein samples prior to microchip
electrophoresis — 263
2.13.7.2 Running protein electrophoresis — 264
2.13.7.3 Detecting the proteins — 264
Staining the proteins — 264
Fluorescence detection — 264
Chemiluminescence detection — 265
Mass spectrometry detection — 265
2.13.8 Microchips in clinical diagnostics — 265
References — 266

2.14 Blotting of proteins — 271
2.14.1 Theory of protein blotting — 271
2.14.2 Blot membranes — 271
2.14.3 Transfer of proteins — 272
2.14.3.1 Electrotransfer of proteins — 272
Tank blotting of proteins — 273
Semidry blotting of proteins — 273
2.14.3.2 Capillary transfer of proteins — 274
2.14.4 Blocking — 275
2.14.5 Detection — 276
2.14.5.1 Detection by dyes — 276
2.14.5.2 Detection by probes — 276
2.14.5.3 Immunoblotting — 277
2.14.6 Making the blot membranes transparent — 277
2.14.7 Blotting techniques — 278
2.14.7.1 Western blotting — 278
2.14.7.2 Far-Western blotting — 279
2.14.7.3 Southwestern blotting — 279
2.14.7.4 Northwestern blotting — 279
2.14.7.5 Eastern blotting — 280

2.14.8       Protocols —— 280
             Western Blotting —— 280
             Semidry Blotting —— 282
             Capillary Blotting —— 283
             India Ink Staining of Proteins on Membrane —— 284
             Detecting Proteins by Immunoblotting —— 285
             Immunoprobing with Avidin–biotin Coupling to Secondary
             Antibody —— 286
             References —— 287

2.15    **Evaluation of protein pherograms —— 289**
2.15.1       Qualitative evaluation of protein pherograms —— 289
2.15.1.1     Fixing of proteins —— 290
2.15.1.2     Staining of proteins —— 290
             Anionic dye staining of proteins —— 291
             Ponceau S staining of proteins —— 291
             Amido black staining of proteins —— 291
             Coomassie brilliant blue staining of proteins —— 292
             Conventional Coomassie brilliant blue staining of proteins —— 292
             Colloidal Coomassie brilliant blue staining of proteins —— 293
             India ink staining of proteins —— 293
             Counter-ionic dye staining of proteins —— 293
             Metal staining of proteins —— 294
             Silver staining of proteins —— 294
             Chelate dye staining of proteins —— 295
             Staining of glycoproteins —— 295
             Staining of lipoproteins —— 296
             Staining of enzymes —— 296
             Autoradiography of proteins —— 297
             Fluorography of proteins —— 297
             SYPRO Ruby staining of proteins —— 298
             Destaining of gel background —— 298
             Drying of gels —— 299
2.15.2       Quantitative evaluation of a protein pherogram —— 299
2.15.2.1     Densitometry —— 299
             Densitometers —— 301
2.15.2.2     Scanning —— 302
2.15.3       Protocols —— 302
             Coomassie Brilliant Blue Staining of Proteins in Polyacrylamide
             Gels —— 302
             Colloidal Coomassie Staining —— 303
             Rapid Silver Staining —— 304

SYPRO Ruby Staining of Proteins —— 305
Staining of Lipoproteins with Sudan Black B —— 305
References —— 306

2.16      Precast gels for protein electrophoresis. Rehydratable gels —— 309
2.16.1        Precast agarose gels —— 309
2.16.2        Precast polyacrylamide gels —— 309
2.16.3        Rehydratable polyacrylamide gels —— 311
2.16.4        Protocols —— 311
              Silanization of Glass Plates —— 311
              Casting Agarose Gels on Support Film by Capillary Technique —— 311
              Casting Thin PAA Gels by Cassette Technique —— 312
              Casting Ultrathin PAA Gels by Flap Technique —— 314
              References —— 315

3         Electrophoresis of nucleic acids —— 317
              Buffers for electrophoresis of nucleic acids —— 317
              Polymerase chain reaction —— 318
              Procedure —— 318
              Applications —— 320
              Surface electrophoresis —— 321
              References —— 323

3.1       Agarose gel electrophoresis of nucleic acids —— 325
3.1.1         Zone polyacrylamide gel electrophoresis of native nucleic
              acids —— 326
3.1.1.1       Theory of sieving migration —— 326
              Factors affecting migration of nucleic acids —— 326
3.1.1.2       Practice of zone agarose gel electrophoresis of native nucleic
              acids —— 327
              Submarine electrophoresis of nucleic acids —— 327
              Restriction fragment length polymorphism —— 328
              Variable number tandem repeat —— 329
              DNA profiling —— 329
              Paternity or maternity testing for child —— 330
              Interpretation of DNA test results —— 331
3.1.2         Zone agarose gel electrophoresis of denatured nucleic acids —— 332
3.1.3         DNA sequencing —— 332
3.1.3.1       Maxam–Gilbert sequencing method —— 332
3.1.3.2       Sanger sequencing method —— 333
3.1.3.3       Single-strand conformation polymorphism method —— 334
3.1.3.4       DNase footprinting assay —— 336

3.1.3.5      Nuclease protection assay —— 337
             S$_1$-nuclease protection assay —— 337
3.1.4        RNA separation —— 338
3.1.4.1      Primer extension assay —— 339
3.1.5        Protocols —— 339
             Native DNA Electrophoresis on Agarose Gels —— 339
             Sanger Sequencing Reactions Using *Taq* DNA Polymerase —— 340
             References —— 341

3.2      **Pulsed-field gel electrophoresis of nucleic acids —— 343**
3.2.1        Theory of pulsed-field gel electrophoresis of nucleic acids —— 343
3.2.2        Types of pulsed-field gel electrophoresis —— 344
3.2.3        Mapping the human genome —— 345
3.2.3.1      STR analysis —— 345
3.2.3.2      AmpFLP —— 345
3.2.3.3      DNA family relationship analysis —— 346
3.2.3.4      Y-chromosome analysis —— 346
3.2.3.5      Mitochondrial analysis —— 346
3.2.4        Protocols —— 346
             Pulsed-field Gel Electrophoresis of Chromosomal DNA —— 346
             References —— 347

3.3      **Capillary electrophoresis of nucleic acids —— 349**
3.3.1        Theory of capillary electrophoresis of nucleic acids —— 349
3.3.2        Instrumentation for capillary electrophoresis —— 350
3.3.2.1      Capillaries —— 350
3.3.2.2      Coating polymers —— 350
3.3.2.3      Gels —— 351
3.3.3        Running capillary electrophoresis —— 352
3.3.3.1      Injection —— 352
3.3.3.2      Separation —— 352
3.3.3.3      Detection —— 352
3.3.4        Pulsed-field capillary electrophoresis —— 353
3.3.5        Applications of capillary electrophoresis —— 353
3.3.6        Protocols —— 354
             Quantitative PCR Analysis —— 354
             References —— 354

3.4      **Polyacrylamide gel electrophoresis of nucleic acids —— 357**
3.4.1        Zone polyacrylamide gel electrophoresis of native nucleic
             acids —— 357

3.4.1.1        Practice of zone polyacrylamide gel electrophoresis of native
               nucleic acids —— 358
3.4.2          Disc-electrophoresis of native nucleic acids —— 360
3.4.2.1        Theory of disc-electrophoresis of native nucleic acids —— 360
3.4.2.2        Running disc-electrophoresis of native nucleic acids —— 360
3.4.2.3        Disc-electrophoresis of double-stranded PCR products —— 361
3.4.3          Electrophoretic mobility shift assay —— 363
3.4.4          Clinical applications —— 364
3.4.5          Protocols —— 364
               DNA Disc-electrophoresis in TRIS-taurinate Buffer at Two pH
               Values According to Michov —— 364
               Separation of PCR Products by Polyacrylamide Gel Disc-
               Electrophoresis According to Michov —— 366
               Electrophoretic Mobility Shift Assay —— 367
               Electroelution of Small DNA Fragments from Polyacrylamide
               Gel —— 368
               References —— 368

3.5       Microchip electrophoresis of nucleic acids —— 371
3.5.1          Theory of microchip electrophoresis of nucleic acids —— 371
3.5.2          Construction of a microchip for electrophoresis of nucleic acids —— 371
3.5.2.1        Polymers —— 372
3.5.3          Running microchip electrophoresis of nucleic acids —— 373
3.5.4          Applications of DNA microchip electrophoresis —— 373
               References —— 373

3.6       Blotting of nucleic acids —— 375
3.6.1          Blotting principles —— 375
3.6.1.1        Transfer —— 375
               Diffusion transfer —— 376
               Capillary transfer —— 376
               Vacuum transfer —— 376
               Electrotransfer —— 377
3.6.1.2        Blocking —— 378
3.6.1.3        Detection by probes, dyes, and autoradiography —— 378
3.6.2          Southern blotting —— 379
3.6.3          Northern blotting —— 380
3.6.4          Middle-Eastern blotting —— 381
3.6.5          Protocols —— 382
               Southern Blotting by Downward Capillary Transfer —— 382
               Electroblotting of Polyacrylamide Gel onto Nylon Membrane —— 383
               References —— 384

3.7        Evaluation of nucleic acid pherograms —— 385
3.7.1          Counter-ion dye staining of nucleic acids —— 385
3.7.2          Silver staining of nucleic acids —— 385
3.7.3          Fluorescence methods for detecting nucleic acids —— 386
3.7.4          Autoradiography of nucleic acids —— 389
3.7.5          Labeling of nucleic acids with proteins —— 390
3.7.6          Absorption spectroscopy of nucleic acids —— 391
3.7.7          Protocols —— 391
               Detection of DNA and RNA in Gel Using Ethidium Bromide —— 391
               Fast and Sensitive Silver Staining of DNA Bands According to Han
               et al. [34] —— 392
               Autoradiography of Radiolabeled DNA in Gels and Blots —— 392
               References —— 394

3.8        Precast gels for nucleic acid electrophoresis —— 395
3.8.1          Precast agarose gels —— 395
3.8.2          Precast polyacrylamide gels —— 395
3.8.3          Protocols —— 396
               Casting Mini- and Midi-agarose Gels —— 396

4          Iontophoresis —— 397
4.1            Theory of iontophoresis —— 397
4.2            Factors affecting iontophoresis —— 398
4.2.1          Physicochemical factors —— 398
4.2.2          Biological factors —— 399
4.3            Calculating the iontophoretic current —— 399
4.4            Iontophoresis device —— 399
4.5            Diagnostic iontophoresis —— 400
4.6            Therapeutic iontophoresis —— 400
4.6.1          Transdermal iontophoresis —— 400
               Hyperhidrosis iontophoresis —— 401
               Other transdermal applications —— 401
4.6.2          Ocular iontophoresis —— 402
               References —— 403

5          History of electrophoresis and iontophoresis —— 405
5.1            History of electrophoresis —— 405
5.1.1          Discovery of electrophoresis —— 405
5.1.2          Zone electrophoresis of Tiselius —— 406
5.1.3          Paper and cellulose acetate electrophoresis —— 406
5.1.4          Gel electrophoresis —— 406
               Agarose gel electrophoresis —— 406

Polyacryalamide gel electrophoresis — 407
5.1.5 Isoelectric focusing — 407
5.1.6 Two-dimensional electrophoresis — 407
5.1.7 Blotting — 408
Blotting of proteins — 408
Blotting of nucleic acids — 409
5.1.8 Staining methods — 409
5.1.9 Outline history of electrophoresis — 409
5.2 History of iontophoresis — 413
5.2.1 Outline history of iontophoresis — 414
References — 415

6 Troubleshooting — 421
6.1 Protein electrophoresis troubleshooting — 421
6.2 IEF troubleshooting — 421
6.3 Nucleic acid electrophoresis troubleshooting — 427
6.4 Blotting troubleshooting — 431
6.5 Iontophoresis troubleshooting — 433

Problems — 435
1 Fundamentals of electrophoresis — 435
2 Electrophoresis of proteins — 436
3 Electrophoresis of nucleic acids — 438

Solution of problems — 441
1 Fundamentals of electrophoresis — 441
2 Electrophoresis of proteins — 442
3 Electrophoresis of nucleic acids — 445

Reagents for electrophoresis — 447

Recipes for electrophoresis solutions — 459
Buffers — 459
Solutions for agarose gel electrophoresis — 463
Solutions for affinity electrophoresis — 465
Solutions for native disc-electrophoresis — 466
Solutions for SDS disc-electrophoresis — 467
Solutions for IEF — 468
Blotting solutions — 469
Fixative solutions — 470
Staining solutions — 471
Destaining solutions — 472

Silver staining solutions —— **472**
Other solutions —— **473**

**SI units and physical constants used in electrophoresis** —— **475**
References —— **479**

**Electrophoresis terms** —— **481**

**Index** —— **485**

# Abbreviations

| | |
|---|---|
| $\eta$ | Dynamic viscosity |
| $\zeta$ | Electrokinetic potential of an ion, in V |
| $\alpha$ | Ionization degree of an electrolyte, dimensionless |
| $\mu_{\infty i}$ | Absolute mobility of ion $i$, in m$^2$/(s V) |
| $\mu_i$ | Mobility of ion $i$, in m$^2$/(s V). [No difference exists between the terms ionic mobility and electrophoretic mobility.] |
| $\mu_i'$ | Effective mobility of ion $i$, in m$^2$/(s V) |
| [B] | Equilibrium concentration of substance B, in mol/dm$^3$ (mol/L) or kmol/m$^3$ |
| [H$^+$] | Equilibrium concentration of the proton, in mol/dm$^3$ (mol/L) or kmol/m$^3$ |
| **[H$^+$]** | Dimensionless proton concentration. **[H$^+$]** = [H$^+$]/[H$^+$]$_0$, where [H$^+$]$_0$ is the standard proton concentration of 1 mol/dm$^3$ |
| [OH$^-$] | Equilibrium concentration of hydroxide ion, in mol/dm$^3$ (mol/L) or kmol/m$^3$ |
| **[OH$^-$]** | Dimensionless concentration of hydroxide ion. [OH$^-$] = [OH$^-$]/[OH$^-$]$_0$, where [OH$^-$]$_0$ is the standard hydroxide concentration of 1 mol/dm$^3$ |
| A | Adenine |
| ACES | N-(2-**ace**tamido)-2-aminoethanesulfonic acid |
| $a_{i(pi)}$ | Electrokinetic radius of ion $i$ (polyion $pi$) |
| ALP | **Al**kaline **p**hosphatase |
| Ammediol | 2-**Am**ino-2-**me**thyl-1,3-propane**diol** |
| APS | **A**mmonium **p**eroxydi**s**ulphate |
| BCIP | 5-**B**romo-4-**c**hloro-3-**i**ndolyl **p**hosphate |
| BES | N,N-**b**is(2-hydroxyethyl)-2-amino-**e**thane**s**ulfonic acid |
| BICINE | N,N-**b**is(2-hydroxyethyl)-gly**cine** |
| BIS | N,N'-methylene**bis**acrylamide, N,N'-methylenediacrylamide |
| BISTRIS | **Bis**(2-hydroxyethyl)-amino-**tris**-(hydroxymethyl)-methane |
| bp | **B**ase **p**airs |
| BSA | **B**ovine **s**erum **a**lbumin |
| C | **C**ytosine |
| $C$ | **C**rosslinking |
| CA | **C**ellulose **a**cetate |
| CHAPS | 3-[(3-**c**holamidopropyl)dimethyl**a**mmonio]-**p**ropane**s**ulfonate |
| CHES | 2-(**c**yclo**h**exylamino)**e**thane**s**ulfonic acid |
| $c_i$ | Volume concentration of ion $i$, in mol/dm$^3$ (mol/L) or kmol/m$^3$ |
| DEAE | **D**i**e**thyl**a**mino**e**thyl |
| DMS | **D**i**m**ethyl **s**ulfate |
| DMSO | **D**i**m**ethyl **s**ulf**o**xide |
| DNA | **D**eoxyribo**n**ucleic **a**cid(s) |
| dNTP | **D**eoxy**n**ucleoside **t**ri**p**hosphate |
| EDTA | **E**thylene**d**initrilo**t**etra**a**cetate (ethylenediaminetetraacetate) |
| Eq. | **Eq**uation(s) |
| G | **G**uanine |
| GABA | **γ**-**A**mino**b**utyric **a**cid |
| GlyGly | **Gly**cyl**gly**cine |
| [H$^+$] | Dimensionless concentration of hydrogen ion. $[\mathbf{H}^+] = [H^+]/[H_0^+]$, where $[H_0^+]$ is the standard hydrogen ion concentration of 1 mol/dm$^3$. |
| HDL | **H**igh **d**ensity **l**ipoproteins |
| HEPES | 4-(2-**h**ydroxy**e**thyl)**p**iperazine-1-**e**thane**s**ulfonic acid |

https://doi.org/10.1515/9783110761641-205

| | |
|---|---|
| HRP | Horseradish peroxidase |
| $I$ | Ionic strength, in mol/kg or mol/dm$^3$ (mol/L) |
| $\boldsymbol{I}$ | Dimensionless ionic strength. $\boldsymbol{I} = I/I_0$, where $I_0$ is the standard ionic strength of 1 mol/dm$^3$ |
| $I_c$ | Volume ionic strength in mol/dm$^3$ (mol/L) |
| IEF | Isoelectric focusing |
| Ig | Immunoglobulin |
| $I_m$ | Mass ionic strength in mol/kg |
| IPG | Immobilized pH gradient |
| $K$ | Thermodynamic equilibrium constant; different dimensions |
| $\boldsymbol{K}$ | Dimensionless thermodynamic equilibrium constant. $\boldsymbol{K} = K/K_0$, where $K_0$ is the standard thermodynamic equilibrium constant, which has the dimensions of $K$. |
| $K_c$ | Concentration equilibrium constant; different dimensions |
| $\boldsymbol{K_c}$ | Dimensionless concentration equilibrium constant. $\boldsymbol{K_c} = K_c/K_0$, where $K_0$ is the concentration equilibrium constant with dimensions of $K_c$. |
| kb | Kilobases |
| kbp | Kilobases pairs |
| $K_R$ | Retardation coefficient |
| $K_w$ | Ionic product of the water, in (mol/dm$^3$)$^2$, (mol/L)$^2$ |
| LDL | Low density lipoproteins |
| Mbp | Megabases pairs |
| MES | 2-(N-morpholino)ethanesulfonic acid |
| $m_i$ | Mass concentration (molality) of ion $i$, in mol/kg |
| MOPS | 3-Morpholino-propanesulfonic acid |
| $M_r$ | Relative molecular (ionic, polyionic) mass. It represents the ratio between the mass of a particle, and 1/12th of the mass of carbon isotope $^{12}$C, dimensionless. |
| NBT | Nitro blue tetrazolium |
| Nonidet | Non-ionic detergent |
| OD | Optical density |
| PAGE | Polyacrylamide gel electrophoresis |
| PAS | Periodacid-Schiff's reagent |
| pH | Hydrogen exponent (pondus, potentia hydrogenii) – the negative logarithm of [H$^+$], dimensionless |
| pH(I), pI | Isoelectric point (pondus isoelectricus), dimensionless |
| p$K$ | Negative logarithm of $K$, dimensionless |
| p$K'$ | Negative logarithm of $K'$, dimensionless |
| p$K_c$ | Negative logarithm of $K_c$, dimensionless |
| PBS | Phosphate-buffered saline |
| PCR | Polymerize chain reaction |
| PVC | Polyvinyl chloride |
| PVDF | Polyvinylidene difluoride |
| $r$ | Molecular radius |
| $R_\mu$ | Relative mobility |
| $R_f$ | Ratio to front |
| $r_{i(pi)}$ | Geometric radius of ion $i$ (polyion $pi$) |
| RNA | Ribonucleic acid(s) |
| s | Number of ionic species in a solution, dimensionless |
| SDS | Sodium dodecyl sulfate |

| | |
|---|---|
| SDS-PAGE | **SDS p**olyacrylamide **g**el **e**lectrophoresis |
| *T* | Total monomeric concentration (polyacrylamide concentration) |
| T | **T**hymine |
| *TAE* | *TRIS-acetate-EDTA* |
| TBE | TRIS-**b**orate-EDTA |
| TBS | TRIS-**b**uffered **s**aline |
| TCA | **T**richloroacetic acid |
| TEA | **T**ri**e**thanol**a**mine |
| TES | N-[tris(hydroxymethyl)methyl]-2-amino**e**thane**s**ulfonic acid |
| TMEDA (TEMED) | N,N,N',N'-**te**tra**m**ethyl**e**thylen**e**di**a**mine, 1,2-bis(dimethylamino)ethane |
| TRICINE | N-[**tri**s(hydroxymelhyl)methyl]-gly**cine** |
| TRIS | **Tris**(hydroxymethyl)-aminomethane |
| UV | **U**ltra**v**iolet light |
| VLDL | **V**ery **l**ow **d**ensity **l**ipoproteins |
| $z_i$ | Charge number (electrovalence) of ion $i$, dimensionless |

NB: To derive logarithmic values of ionization constants, use dimensionless ionization constants. They can be considered as ratios between the ionization constants and a standard constant of 1 mol/dm$^3$ (1 mol/L). Dimensionless concentrations of H$^+$ (or other ions) can also be defined as ratios between the concentrations of H$^+$ (or other ions) and a standard concentration of H$^+$ (or other ions), which should also be 1 mol/dm$^3$ (1 mol/L). The same is valid for the ionic strengths.

# 1 Fundamentals of electrophoresis

Overview on electrophoresis —— 1
Separation media —— 3
Electrophoresis resolution and sharpness —— 3
Detection of resolved bands —— 3
References —— 4

The term *electrophoresis* means moving of charged dissolved particles in an electric field, which causes their resolution depending on their velocities and interaction with the separation medium [1–3]. In its current form, the electrophoresis is connected with the studies of Tiselius [4] in the 1930s.

The electrophoresis of positively charged particles (cations) is called *cataphoresis*; the electrophoresis of negatively charged particles (anions) is called *anaphoresis*.

The electrophoresis is used for resolving of proteins and nucleic acids, chromosomes, viruses, cell membranes (plasma, lysosomal, nuclear, and other), cell organelles (mitochondria, ribosomes), cells (red cells, tissue cells, and parasites), *etc.*, it takes place in biochemistry, proteomic and genomic studies, forensics, molecular biology, and microbiology. By electrophoresis, more than the half of all separations and almost all separations of blood proteins and DNA are carried out.

## Overview on electrophoresis

Proteins and nucleic acids form polyions. In an electric field, the positively charged polyions move toward the negative pole, while the negatively charged polyions move toward the positive pole (Figure 1.1).

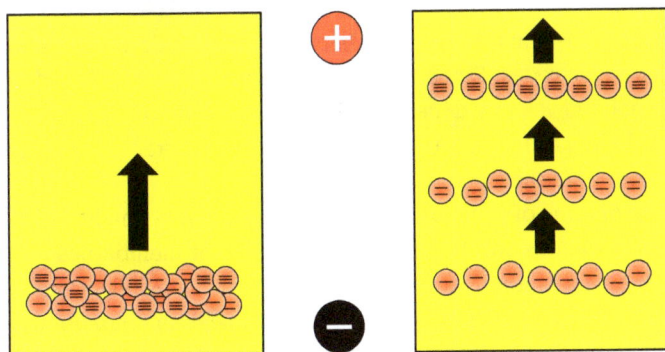

**Figure 1.1:** Electrophoresis of unipolar particles carrying different electric charges. *Left* – start of electrophoresis; *Right* – end of electrophoresis.

https://doi.org/10.1515/9783110761641-001

Commonly, polyions that are to be resolved are applied onto a separation medium, which in its turn is placed into an electrophoresis cell that is connected to a power supply. When electric current is turned on, the larger and less charged polyions move slower through the medium, while the smaller and more charged polyions move faster.

Mostly, electrophoresis is carried out in agarose or polyacrylamide gels, soaked with buffers. Other separation media are: starch gel, cellulose acetate, and paper. Today they have lost their actuality. Electrophoresis can also be carried out in buffers only. The buffers are electrolyte solutions, which maintain constant pH values, e.g. TRIS-borate buffer, TRIS-histidinate buffer, Goods buffers, and more.

Since proteins and nucleic acids are mostly colorless, their movement through the gel cannot be followed during electrophoresis. Therefore, tracking dyes are usually included in the sample buffer. At electrophoresis of negatively charged proteins, Bromophenol blue, Xylene cyanol, which runs slower than Bromophenol blue, or Orange G are used; at electrophoresis of positively charged proteins Bromocresol green or Methylene blue are used.

Proteins can be separated also in pH gradients, where they stop at their isoelectric (pI) points. This electrophoresis, called *isoelectric focusing*, can be carried out in two types of pH gradients: produced by carrier ampholytes, or by immobilines.

The electrophoresis takes place in horizontal or vertical separation cells. The electrodes can be placed on the gel or in separate electrode tanks.

Beside electrophoresis cells and power supplies, thermostats for reserving the resolving medium temperature, and densitometers or scanners for analyzing the *pherograms* (the resolved polyions in the medium) are used. Semi-automated electrophoresis devices are also offered on the market.

The electrophoretic velocity $v$ of a polyion is proportional to its effective mobility $\mu'$ and the electric field strength $E$. In its turn, the effective mobility depends on the total electric charge of the polyion and is inversely proportional to the viscosity of the separation medium. The total electric charge is determined by the buffer pH value; and the viscosity of the resolving medium depends mainly on the medium structure and temperature.

The electric field strength is equal to the ratio between the electric voltage and distance between the two electrodes. Since the distance remains constant during the electrophoresis, the polyion velocity depends only on the electric voltage.

Electrophoresis should be carried out at voltage and electric current, when the heating could be drawn out from the electrophoresis cell.

After electrophoresis, *blotting* of proteins and nucleic acids can be made. Using this technique, the resolved bands can be immobilized onto blot membranes and treated afterward. The blotting methods are carried out in four steps: electrophoretic separation of proteins or nucleic acid fragments; their transfer and immobilization onto blot membranes; binding of analytical probes to the blotted substances; and visualization of the blotted bands.

The blotting of DNA bands is called Southern blotting, the blotting of RNA bands is called Northern blotting, and the blotting of protein bands is called Western blotting. The blotting membrane consists usually of nitrocellulose, nylon or **polyvinylidene difluoride** (PVDF).

## Separation media

The electrophoresis can be carried out in a solution, but the diffusion there is too strong. In order to limit the diffusion, solid media are used. The earliest solid medium was cellulose contained in the filter paper. The paper electrophoresis was invented by Kunkel and Tiselius [5] in 1951. Cellulose acetate membranes were the next step in the electrophoresis progress. Today most common are agarose and polyacrylamide gel electrophoresis.

## Electrophoresis resolution and sharpness

The *resolution* of electrophoresis is referred to the ability of electrophoresis to separate two sample components from each other. It depends on the Gaussian profiles of the bands and is calculated by dividing the distance between the centers of adjacent bands by their average bandwidths.

The *sharpness* of electrophoresis is referred to as the reciprocal value of the bandwidth; narrow the bands, the higher the sharpness.

## Detection of resolved bands

The majority of polyion bands are not visible to the naked eye, with a few exceptions. Direct optical detection of resolved bands can be applied, for example, to hemoglobins (which are red colored). Therefore, a couple of methods are created for detection, localization, and quantitation of separated bands.

The detection of resolved polyions is carried out directly or indirectly. The direct detection is performed in the resolving medium by nonspecific or specific staining, enzyme-substrate reactions, immune precipitation, autoradiography, and fluorography. The indirect detection is performed by immune printing or blotting.

The proteins can be stained in gels. Common dye is Coomassie brilliant blue, which can detect 0.3 µg of protein in a spot. DNA is detected usually by fluorescent intercalating of ethidium bromide. Both proteins and DNA can react with silver ions to form black bands. The silver methods are 100 times more sensitive than the other staining methods and can detect 2 ng of protein in a spot.

The process of staining is followed by destaining of the gel background to remove unbound dye. In some cases, placing the gel on an **u**ltraviolet (UV) lightbox or under a UV lamp can reveal UV absorbing bands, or UV fluorescence bands.

If radioactive atoms have been incorporated in the polyions prior to electrophoresis, their position in the gel or membrane can be determined by autoradiography. For this purpose, the gel or membrane is placed on a photographic film or overlaid with it and let to expose in the dark and cold to show the positions of the radioactive bands.

The most modern methods of both detection and characterization of resolved proteins involve mass spectrometry.

The electrophoresis results can be recorded with a computer operated camera, and the intensity of a band or spot of interest can be compared against markers in the same gel, using specialized software.

After electrophoresis, the gels with bands can be saved in dry forms.

# References

[1]   Lyklema J. Fund Interface Colloid Sci, 1995, 2, 3, 208.
[2]   Hunter RJ. Foundations of Colloid Science, 2nd edn, Oxford University Press, Oxford, 2001.
[3]   Russel WB, Saville DA, Schowalter WR. Colloidal Dispersions. Cambridge University Press, Cambridge, 1989.
[4]   Tiselius A. Trans Faraday Soc, 1937, 33, 524–531.
[5]   Kunkel HG, Tiselius A. Electrophoresis of proteins on filter paper, J Gen Physiol, 1951, 35, 89–118.

# 1.1 Electric double layer of a charged particle

1.1.1    Electric charges of polyions —— 5
1.1.2    Model of Helmholtz —— 5
1.1.3    Model of Gouy–Chapman —— 6
1.1.4    Theory of Debye–Hückel —— 7
1.1.5    Model of Stern —— 7
1.1.6    Two radii and two electric potentials of a charged particle —— 9
         References —— 11

## 1.1.1 Electric charges of polyions

Polyions carry electric charges. In a solution, two types of electric charges exist: total and net charge. The *total charge* of a polyion is the sum of all electric charges, which the polyion carries: the charges of the amino acid residues, metal ions, cofactors, *etc.* These charges can be observed in an infinitely dilute solution.

The magnitude of a total electric charge is given by the equation

$$Q_i = z_i e \tag{1.1.1}$$

where $z_i$ is the total number of elementary charges (electrovalence) of the particle $i$ and the fixed part of its ionic atmosphere, and $e$ is the electric charge of the proton, equal to $1.602\ 1892 \times 10^{-19}$ C. Since $z_i$ has a positive or negative value, the value of $Q_i$ can have different signs.

The *net charge* is less (in absolute units) than the total charge. The oppositely charged ions (counter-ion) in the solution, which form ionic atmosphere are responsible for this. The surface of a charged and solvated (hydrated) ion or polyion, and its ionic atmosphere build an electric double layer [1, 2]. The electric field strength inside the electric double layer can be from 0 to over $10^9$ V/m.

Several models describe the electric double layer. Most common are the models of Helmholtz, Gouy–Chapman, Debye–Hückel, and Stern.

## 1.1.2 Model of Helmholtz

The first model of the electric double layer was developed by Helmholtz [3] and Perrin [4]. According to it, the counter-ions are located at a certain distance from the charged surface. As a result, the charged particle can be considered as a ball capacitor. It consists of two concentric spheres similarly to a capacitor electric double layer, and the electric potential is decreased linearly by the distance from the charged surface (Figure 1.1.1).

https://doi.org/10.1515/9783110761641-002

**6** — 1.1 Electric double layer of a charged particle

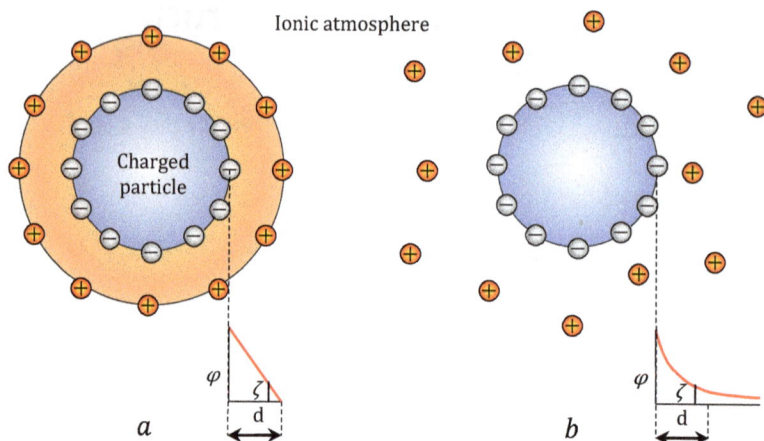

Figure 1.1.1: Models of the electric double layer according to Helmholtz (*a*) and Gouy–Chapman (*b*).

The electric potential of the surface of a charged particle (the inner electric potential of Helmholtz)

$$\varphi_{in} = \frac{Q}{4\pi\varepsilon a} \tag{1.1.2}$$

where $Q$ is the electric charge of the particle, in C; $\varepsilon$ is the (di)electric permeability of the solvent (water), which is equal to the product of the relative (di)electric permeability $\varepsilon_r$ and the (di)electric-constant $\varepsilon_0$ (8.854 187 817 × 10$^{-12}$ F/m), and $a$ is the radius of the particle in m. The electric potential of the ionic atmosphere of the particle (the outer electric potential of Helmholtz)

$$\varphi_{ex} = \frac{Q}{4\pi\varepsilon(a+d)} \tag{1.1.3}$$

where $d$ is the thickness of the electric double layer. It follows from eqs. (1.1.2) and (1.1.3) that the electrokinetic potential $\zeta$ of a charged particle and its ionic atmosphere may be described by the expression

$$\zeta = \frac{Q}{4\pi\varepsilon a} - \frac{Q}{4\pi\varepsilon(a+d)} = \frac{Qd}{4\pi\varepsilon a(a+d)} = \frac{Q}{4\pi\varepsilon\left(1+\frac{a}{d}\right)} \tag{1.1.4}$$

## 1.1.3 Model of Gouy–Chapman

Gouy [5] and Chapman [6] have shown that the fixed arrangement of the counterions in the model of Helmholtz does not exist, because they are moving to and from the charged particle. According to the model of Gouy–Chapman, the number of counter-ions, and thereby the electric potential of the double layer, decreases exponentially with the distance from the charged surface (Figure 1.1.1).

The relationship between the counter-ion concentration at the charged surface, $c$, and the counter-ion concentration in the solution, $c_0$, is

$$c = c_0 e^{-\left(\frac{ze\varphi}{kT}\right)} \tag{1.1.5}$$

where $z$ is the number of elementary charges on the counter-ion, $e$ is the elementary charge, $k$ is the Boltzmann constant, and $\varphi$ is the potential of the charged surface.

## 1.1.4 Theory of Debye–Hückel

Later Debye and Hückel [7] developed a similar theory of the electric double layer. According to it, the thickness of the ionic atmosphere

$$\frac{1}{\kappa} = \frac{1}{F}\left(\frac{\varepsilon RT}{2I}\right)^{1/2} \tag{1.1.6}$$

where $\kappa$ is the parameter of Debye–Hückel in $m^{-1}$, $F$ is the Faraday constant (96,484,554.61 C/mol), $R$ is the molar gas constant [8,318.41 J/(kmol K)], $T$ is the thermodynamic temperature in K, and $I$ is the ionic strength of the solution in $mol/dm^3$ (mol/L).

It follows from eq. (1.1.6) that the thickness of ionic atmosphere depends exclusively on the ionic strength, i.e. on the concentration and electrovalence of all ions in the solution. Moreover, the electric double layer, according to Gouy–Chapman, can be considered as a spherical capacitor, if $d$ in the Helmholtz eq. (1.1.4) is replaced by the magnitude $1/\kappa$. Then, the following equation is obtained:

$$\zeta = \frac{Q}{4\pi\varepsilon a(1 + \kappa a)} \tag{1.1.7}$$

## 1.1.5 Model of Stern

In 1924, Stern [8] suggested a combination of the Helmholtz model and the Gouy–Chapman model. According to his model, the ionic atmosphere consists of a fixed (adsorption, inner) part, known as adsorption layer, and a diffuse (free, outer) part, known as diffuse layer. The thickness of the fixed layer is equal to $\Delta$, and the thickness of the diffused layer is equal to $1/\kappa$. The fixed atmosphere is composed of hydrated counter-ions, which are adsorbed on the hydrated charged surface. The diffuse layer is composed of the remaining counter-ions, which are also hydrated. The border between the fixed and diffuse parts is named *slipping plane*. The ions of the diffuse layer are moving freely and its concentration decreases exponentially with the distance from the charged surface (Figure 1.1.2).

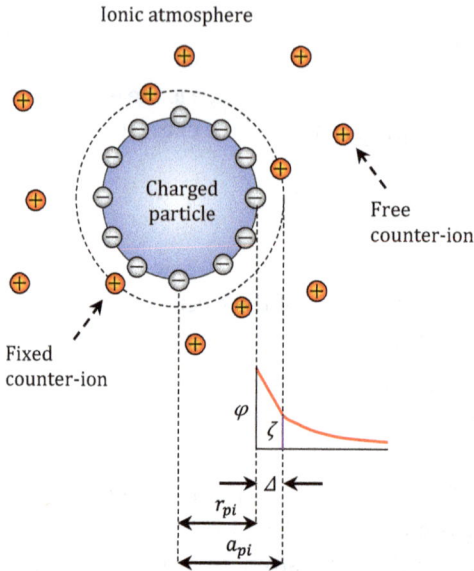

**Figure 1.1.2:** Spherical protein polyion of radius $r_{pi}$ possessing a negative total charge of $-12e$ is covered by the fixed part of its ionic atmosphere of $\Delta$ thickness. The fixed part consists of 3 counter-ions with a positive net charge of $3e$. The diffuse part, which has a thickness of $(1/\kappa - \Delta)$, contains the remaining counter-ions, which have a positive total charge of $9e$.

The fixed part of the ionic atmosphere is moving in an electric field together with the charged particles, while the counter-ions of the diffuse part separate from the fixed part and are moving to the opposite pole. As a result, the electric potential of the charged particle decreases at the beginning linearly up to the boundary between the fixed and diffuse part of the ionic atmosphere, and then decreases exponentially [9–11].

The model of Stern explains best the polyion properties. It is known that the $pK_c$ values of some chemical groups of proteins differ by up to 1.5 pH units from the $pK_c$ values of the same groups in corresponding amino acids. This can be explained so: If a protein polyion has a positive total (sum) charge, its protons are repelled and hydroxide ions from the solution are attracted. As a result, the $pK_c$ values of the acidic groups reduce. If a protein polyion has a negative total charge, its hydroxide ions are repelled and protons from the solution are attracted. In this case, the $pK_c$ values of the alkaline groups grow up.

When the positive and negative charges of a polyion are equal, the polyion is located in its *isoelectric point*. Then the $\zeta$-potential of the polyion is equal to 0. The isoelectric point depends on the pH value of the solution, which influences the ionization of the polyion [12, 13].

The $\zeta$-potential cannot be measured directly but can be calculated using theoretical models and the ionic (polyionic) electrophoretic mobility. The widely used theory for calculating the $\zeta$-potential is that of Smoluchowski [14]. Smoluchowski theory is valid for dispersed particles of any shape and concentration. However, it is valid only for a thin double layer, when the Debye length, $1/\kappa$, is much smaller than the particle radius $a$, namely

$$\kappa a \gg 1 \tag{1.1.8}$$

When the Debye length is much larger (for example, in some nanocolloids), the following expression is obtained:

$$\kappa a < 1 \tag{1.1.9}$$

## 1.1.6 Two radii and two electric potentials of a charged particle

Recently, we have proposed that every charged and dissolved particle has two radii and two electric potentials [15].

We assume that the counter-ion is in continuous movement to and from the central ion due to the electrostatic forces of attraction and the repulsion forces of diffusion. Therefore, it can be suggested that in every infinite small interval of time a part of the counter-ion is linked with the central ion, forming the **ad**sorption layer (*ad*) of the ionic atmosphere, while the remaining part of the counter-ion forms the **di**ffusion layer (*di*) of the ionic atmosphere.

In accordance with these considerations we assume that the radius of the central ion, which we will call *geometric radius R*, increases in the presence of ionic atmosphere to another radius, which we will call *electrokinetic radius A*. These radii characterize different ions that should be called *geometric* and *electrokinetic ion*, the electrokinetic ion being, of course, the larger particle. Hence, the electrokinetic radius is not constant, but varies to $A + dA$, as $R \leq A \leq R + r$, where $r$ is the geometric radius of the counter-ion, and $A - R = \Delta$ is the thickness of the adsorption layer.

In addition, the geometric ion has a specific electric potential, which we will call *geometric (thermodynamic) potential*; and the electrokinetic ion has another electric potential, which we will call *electrokinetic potential* (Figure 1.1.3). The electrokinetic potential $\varphi$ is a function not only of the distance from it, but also of the ionic strength of the solution.

On the surface of the electrokinetic ion, the electric potential of the electrokinetic ion and its ionic atmosphere, known as $\zeta$-potential, is

$$\varphi_A = \varphi_R \frac{1 - \kappa(A - R)}{1 + \kappa R} \tag{1.1.10}$$

where $\varphi_R$ is the electric potential of the geometric ion and its ionic atmosphere. The electric field of the ionic atmosphere (the counter-ions) is located around the central ion, whose electric charges create an internal electric field. Therefore, the electric potential on the surface of the geometric ion and its ionic atmosphere is

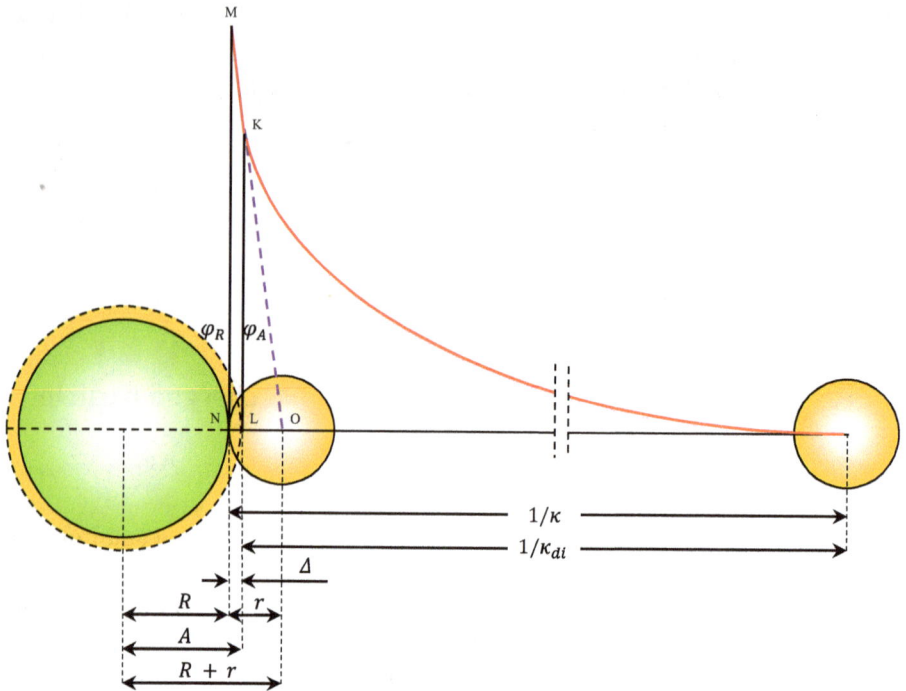

**Figure 1.1.3:** Electric potentials and radii of the central ion and its ionic atmosphere. $\varphi_R$ – electric potential of the geometric ion and its ionic atmosphere; $\varphi_A$ – electric potential of the electrokinetic ion and its ionic atmosphere; $R$ and $r$ – geometric radii of the central ion and its counter-ion, respectively; $A$ and $R+r$ – electrokinetic and the maximum electrokinetic radius of the central ion; $\Delta$ – thickness of the adsorption layer; $1/\kappa$ – thickness of the ionic atmosphere; $1/\kappa_{di}$ – thickness of the diffuse layer of the ionic atmosphere.

$$\varphi_R = \varphi_{R_0} + \varphi_{1/\kappa} \qquad (1.1.11)$$

where $\varphi_{R_0}$ is the electric potential of the geometric ion.

When $1/\kappa = \Delta$, i.e. on the surface of the electrokinetic ion, the electric potential of the electrokinetic ion and its ionic atmosphere ($\zeta$-potential) is

$$\varphi_A = \varphi_R(1 - \kappa r) \qquad (1.1.12)$$

Using the above equations, we proved that

$$A = R + \kappa r^2 \qquad (1.1.13)$$

where $r \leq \kappa^{-1} \leq \infty$, i.e. $0 \leq \kappa r \leq 1$.

The last equation shows that the electrokinetic radius $A$ is a function of the parameter $\kappa$, namely, of the ionic strength $I$ of the solution, at a slope of $r^2$. According to it, the thickness of the adsorption layer is

$$\Delta = A - R = \kappa r^2 \qquad (1.1.14)$$

Assuming that $R = nr$, it follows that

$$A = nr + \kappa r^2 = r(n + \kappa r) \qquad (1.1.15)$$

If $n = 1$, the last equation is simplified to give

$$a = r(1 + \kappa r) \qquad (1.1.16)$$

where $a$ is the electrokinetic radius of the central ion, and $r$ is the geometric radius of the central ion and its counter-ion.

Having in mind the existence of two radii for each charged particle, we [16] integrated the differential equations of Henry's function [17, 18] in the interval $r, 2r$, and obtained the new function

$$f(\kappa r) = \frac{1 + \kappa a}{1 + \kappa r} = \frac{1 + \kappa r + (\kappa r)^2}{1 + \kappa r} \qquad (1.1.17)$$

which helps us to define the electrokinetic potential as

$$\zeta = \frac{1.5\mu\eta}{\varepsilon} \frac{1 + \kappa r}{1 + \kappa a} \qquad (1.1.18)$$

The deduced equations show that the electrokinetic radius $A$ of a central ion is a function of the parameter of the Debye–Hückel $\kappa$, hence, a function of the ionic strength $I$ of the solution, while the geometric radius $R$ of the central ion is constant. The slope of the curve $dA/d\kappa$ of the displayed equations is equal to $r^2$, i.e. to the square of the geometric radius $r$ of the counter-ion. When the thickness of the ionic atmosphere $1/\kappa \to \infty$, i.e. when $\kappa \to 0$, then $A \to R$; and when $1/\kappa \to 0$, i.e. when $\kappa \to 1/r$, then $A \to R + r$.

# References

[1]   Quincke G. Pogg Ann, 1859, 107, 1–47.
[2]   Quincke G. Pogg Ann, 1861, 113, 513–598.
[3]   Helmholtz HL. Wiedemann's Ann Phys Chem, 1879, 7, 337–382.
[4]   Perrin J. J Chim Phys, 1904, 2, 601–651.
[5]   Gouy MG. J Phys France, 1910, 9, 457–468.
[6]   Chapman DL. Phil Mag, 1913, 25, 475–481.
[7]   Debye P, Hückel E. Phys Z, 1923, 24, 185–206.
[8]   Stern O. Z Elektrochem, 1924, 30, 508–516.
[9]   Morrison ID, Ross S. Colloidal Dispersions: Suspensions, Emulsions, and Foams, 2nd ed. Wiley, New York, 2002.
[10]  Jiang J, Oberdörster G, Biswas P. J Nanopart Res, 2008, 11, 77–89.
[11]  McNaught AND, Wilkinson A. Definition of Electrokinetic Potential. IUPAC. Compendium of Chemical Terminology, 2nd ed. Blackwell Scientific Publications, Oxford, 1997.
[12]  Kirby BJ. Micro- and Nanoscale Fluid Mechanics: Transport in Microfluidic Devices. Cambridge University Press, Cambridge, 2010.

[13]  Lyklema J. Fundamentals of Interface and Colloid Science. Academic Press, London, 1995, Vol. **2**, 3, 208.

[14]  Smoluchowski M. Bull Int Acad Sci Cracovie. 1903, 184–199.

[15]  Michov BM. Electrochim Acta, 2013, 108, 79–85.

[16]  Michov BM. Electrophoresis, 1988, 9, 199–200.

[17]  Henry DC. Proc Roy Soc A, 1931, 133, 106–129.

[18]  Henry DC. Trans Faraday Soc, 1948, 44, 1021–1026.

# 1.2 Proteins and nucleic acids form polyions in solution

1.2.1     Structure and conformation of proteins —— 13
1.2.1.1   Masses and electric charges of proteins —— 13
1.2.1.2   Isoelectric points of proteins —— 14
1.2.1.3   Native and denatured proteins —— 15
1.2.2     Structure and conformation of nucleic acids —— 15
1.2.2.1   Masses and electric charges of nucleic acids —— 15
1.2.2.2   Native and denatured nucleic acids —— 16
          References —— 17

The electrophoresis is used for separation of proteins and nucleic acids. These compounds split protons in a solution (the acidic proteins and nucleic acids) or bind protons (the alkaline proteins). As a result, they convert themselves into charged particles (polyions). The proteins and nucleic acids can be resolved electrophoretically in native or denatured state.

## 1.2.1 Structure and conformation of proteins

Proteins are made up of amino acid residues and can in addition contain non-amino acid compounds. To the first group belong the globular proteins (soluble proteins with oval-shaped molecules, such as albumins and globulins), and some fibrous proteins (scleroproteins), such as collagens and keratins, which are water-insoluble. The second group includes protein complexes that are composed of a protein part and external compounds, such as hemoglobins.

### 1.2.1.1 Masses and electric charges of proteins

The electrophoretic mobility of proteins is determined by their masses and electric charges. The masses depend on the number of amino acid residues. The amino acid number in most proteins extends from less than a hundred to many thousands. Since the relative molecular mass $M_r$ of an amino acid residue is approximately 110, the relative molecular mass of the proteins is approximately 10,000 to 100,000 and more.

The proteins are amphoteric polyelectrolytes that form in an aqueous solution polyions with both positive and negative electric charges. The charges are carried by ionizable groups in the amino acid residues. The negatively charged proteins are known as *proteinates*, and the positively charged proteins as *protein polycations*.

https://doi.org/10.1515/9783110761641-003

In a neutral aqueous solution, the carboxyl groups of the aspartic and glutamic acid residues split protons and obtain negative charges. On the contrary, the imidazole group of the histidine residue, the amino group of the lysine residue, and the amidino group of the arginine residue (its imino group) bind protons and become positively charged.

The splitting or binding of a proton depends on the solution pH value: the first case is predominating at a high pH value; the second case is prevailing at a low pH value.

The total charge (the number of elementary charges) of a protein polyion is calculated as a sum of the negative and positive charges. For example, if a proteate has 200 negative and 180 positive charges, its total number of elementary charges is

$$z = -200 + 180 = -20$$

pH ranges, suitable for calculating the charges of a protein polyion and close to the neutral pH range, are 8.5 to 9.0, and 5.0 to 5.5. At pH = 8.5 to 9.0, the groups $\alpha$-COOH, $\beta$-COOH, and $\gamma$-COOH are deprotonated, and the groups $\varepsilon$-NH$_2$ and $-C(NH)NH_2$ are protonated. The groups $-C_3N_2H_3$, $\alpha$-NH$_2$, $-SH$, and $-C_6H_4OH$ are not ionized. This pH value is most commonly used for electrophoresis of acidic proteins.

### 1.2.1.2 Isoelectric points of proteins

The proteins have amphoteric molecules. That is why they can be considered as zwitterions and are denoted by the formula $H_3^+N-Pr-COO^-$. Their charges vary when the proton concentration in the solution is changed. The carboxyl groups at low pH values and the amino, imidazole and amidino groups at higher pH values are neutral:

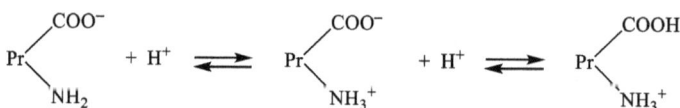

The pH value, at which the negative charges of an amphoteric polyion neutralize its positive charges and the total charge becomes equal to zero, is defined as *isoelectric point*. Typically, the isoelectric point is noticed as pI, however this term should be replaced by the term pH(I) point, because the isoelectric point is a pH value [1].

The absence of a total electric charge in the isoelectric point destroys the electric double layer of a protein and makes it defenseless against precipitants. Then, methanol and ethanol can bind water dipoles and disrupt the hydration shell of the protein polyion, and, as a result, the protein precipitates. Precipitation can also be caused by salts, which also destroy the hydration shell.

The pH(I) values of 95% of the known proteins are located in the pH range 3 to 10 [2]. The acidic glycoprotein of chimpanzee, whose pH(I) = 1.8, seems to have the lowest isoelectric point; and the lysozyme from human placenta, whose pH(I) = 11.7, seems to have the highest isoelectric point.

### 1.2.1.3 Native and denatured proteins

The secondary, tertiary, and quaternary structures of proteins determine their biological activity (enzyme and hormone reactions, antigen-antibody reactions, oxygen transport, *etc.*). This is their native state. Changes in the quaternary, tertiary, and secondary structure of proteins is referred to as *denaturation*.

The denaturation of proteins may be caused by organic solvents (alcohols), decreasing pH (acid denaturation), high temperature (heat denaturation), ionic or non-ionic detergents, and high-energy radiation (UV, X-rays, β-rays).

When denaturing is carried out, the interactions, which stabilize the native protein structure, are broken. The polypeptide chains are unfolded and new interactions take place between them. The hydrophobic groups, previously hidden in the protein internal, come outside and the protein becomes insoluble.

## 1.2.2 Structure and conformation of nucleic acids

Nucleic acids are the largest macromolecules in the biosphere. They consist of numerous mononucleotide residues, which are connected to each other by phosphodiester bonds in polynucleotide chains. The phosphodiester bond represents a phosphoric acid residue, which connects a 3′-hydroxyl group of a mononucleotide residue with a 5′-hydroxyl group of an adjacent mononucleotide residue.

A mononucleotide residue consists of a base, a sugar (deoxyribose or ribose), and a phosphate. The bases contain purine or pyrimidine. Purine bases are adenine (A) and guanine (G), pyrimidine bases are cytosine (C), uracil (U), and thymine (T). Uracil is not presented in the deoxyribonucleic acids (DNA) and thymine is not presented in the ribonucleic acids (RNA).

DNA has linear or circular molecules, which are double-stranded or single-stranded. The chromosomes of eukaryotes contain double-stranded linear DNA. RNA may also be linear or circular. mRNA, rRNA, and tRNA are single-stranded linear RNA.

The conformation of the DNA molecule exerts influence on DNA movement. For example, supercoiled DNA migrates faster than relaxed DNA because supercoiled DNA is more compact. The velocity of DNA movement changes under various electrophoresis conditions [3].

### 1.2.2.1 Masses and electric charges of nucleic acids

The relative molecular mass $M_r$ of a mononucleotide residue is approximately 330. This means that three mononucleotide residues in DNA or RNA have $M_r$ of about 1,000. Each mononucleotide residue has a negative charge in neutral or alkaline solutions. Therefore, the number of elementary charges can be calculated when the relative nucleate mass is divided by −330.

In a solution the nucleic acids are polyanions (nucleates), which migrate in an electric field in the anodic direction. In contrast to the proteins, the number of elementary charges of the nucleic acids does not change in neutral or alkaline buffers. Consequently, the ratio of the electric charge to the mass of DNA and RNA polyions is constant; therefore, their mobilities in a solution are equal [4].

Nucleate mobilities depend only on the ionic strength and the nativity of the polyions. For example, the mobility of the native DNA from calf thymus is $-15.1 \times 10^{-9}$ m$^2$/(s V) at an ionic strength of 0.1 mol/L and $-21.7 \times 10^{-9}$ m$^2$/(s V) at an ionic strength of 0.01 mol/L (for denatured DNA from calf thymus, these data are $-13.3 \times 10^{-9}$ and $-19.0 \times 10^{-9}$ m$^2$/(s V), respectively). The increased temperature also diminishes the nucleate mobilities because the denaturation increases the nucleate volume.

### 1.2.2.2 Native and denatured nucleic acids

The nucleic acids, as the proteins, can be in native or denatured state (Figure 1.2.1). The native nucleic acids contain base pairs; the denatured nucleic acids, however, contain no base pairs. This means that double-stranded nucleic acids exist only under native conditions, whereas single-stranded nucleic acids can occur under native or denaturing conditions.

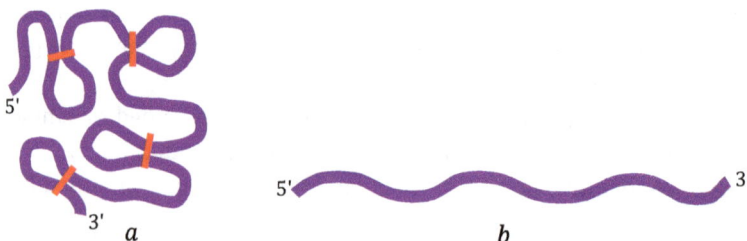

**Figure 1.2.1:** Native (*a*) and denatured (*b*) nucleic acid (rRNA). The base pairs are shown in red.

The nativity of a double-stranded nucleic acid or a hybrid depends on the number of hydrogen bonds between the two strands and the repulsion between the negatively charged phosphate groups of the polynucleotide chains. If the strength of hydrogen bonds is greater than the strength of repulsion between the negatively charged phosphate groups of nucleic acids, they remain natural. If the binding forces of base pairing are not sufficient, the double helix denatures. If the repulsion between the negative charges of the polynucleotide chains is reduced (for example, by increasing the ionic strength), the double helix is stabilized.

The native DNA molecules may be present in globular or linear conformation. The globular molecules are either supercoiled or circled.

Many factors are known, which disrupt the inter- or intramolecular base pairing, and cause denaturing of nucleic acids: high temperature, high pH values, methylmercury hydroxide, glyoxal or formaldehyde, and urea and formamide.

*Temperature.* The building of base pairs releases energy. Therefore, a supply of thermal energy allows the paired regions to separate from each other – a process named *thermal denaturing*. The thermal denaturing of DNA can be compared with the melting of the crystal lattice of DNA; therefore, is called *DNA melting*. The DNA melting occurs at 70–90 °C in a saline solution and destroys the DNA double helix into single strands. As a result, the physical properties of DNA (viscosity, light absorption, and optical rotation) change.

The melting of nucleic acids depends on the base pairing: GC-rich DNA have a higher melting point, as three hydrogen bonds connect guanine and cytosine; in contrast, TA-rich DNA have a lower melting point, as two hydrogen bonds connect adenine and thymine.

*Alkaline pH values.* Another possibility to break hydrogen bonds in nucleic acids is to increase the pH value. Increasing their concentration, the negative charged hydroxide ions (OH⁻) interact with the protons in the hydrogen bonds and form with them own hydrogen bonds, which denature the nucleates into single strands.

*Methylmercury hydroxide, glyoxal, or formaldehyde.* These chemicals are volatile and toxic, especially methylmercury hydroxide; therefore, they should be carefully handled. Methylmercury hydroxide, glyoxal, or formaldehyde denature nucleates because they react with the nitrogen atoms in the purine and pyrimidine rings, which are involved in the base pairing.

*Urea and formamide.* Highly concentrated urea ($H_2C-CO-NH_2$), alone or in combination with formamide ($OHC-NH_2$), forms hydrogen bonds with the nucleic acid bases and, as a result, destroys the base pairs. So, it breaks the secondary structure of the nucleic acids, namely the inter- and intramolecular base pairing in DNA or RNA.

# References

[1]  Ferard G. JIFCC, 1992, 4, 122–126.
[2]  Gianazza E, Righetti PB. J Chromatogr, 1980, 193, 1–8.
[3]  Sambrook JF, Russell DW, eds. A Laboratory Manual, 3rd ed. Cold Spring Harbor Laboratory Press, 2001.
[4]  Olivera BM, Baine P, Davidson N. Biopolymers, 1964, 2, 245–257.

# 1.3 Electrophoresis is running in buffers

1.3.1     Buffers —— 19
1.3.1.1   Buffer capacity —— 20
1.3.2     Buffers used in electrophoresis —— 22
1.3.3     Biological buffers —— 23
         References —— 26

To keep the concentration of $H^+$ and $OH^-$, hence, the pH values constant, all electrophoretic methods, with the exception of isoelectric focusing, are carried out in buffers.

## 1.3.1 Buffers

According to Brønsted [1], buffers are systems of weak bases and their conjugated acids. Their titration curves show a plateau region (buffer zone), which includes up to 1.5 pH units (Figure 1.3.1). In the buffer zone, the buffer pH value changes insignificantly when base or acid is added or the buffer solution is diluted.

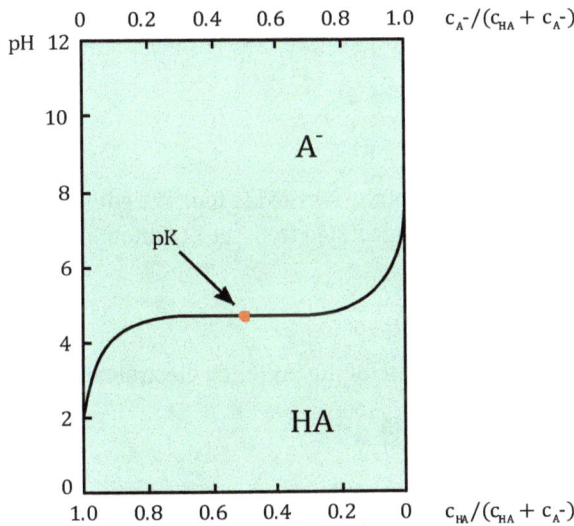

**Figure 1.3.1:** Titration of acetic acid (HA). The buffer zone of acetate buffer is between pH = 4 and pH = 5.4.

https://doi.org/10.1515/9783110761641-004

The buffers obey the equation of Henderson–Hasselbalch

$$pH = pK_c + \log \frac{[Base]}{[Acid]} \qquad (1.3.1)$$

With the aid of this equation, the buffer pH value can be calculated, if $pK_c$ of the buffering acid and the equilibrium concentrations of the base and its conjugated acid are known. When $pH = pK_c$, the concentrations of the base and acid are equal, i.e. half of the protolyte is ionized.

### 1.3.1.1 Buffer capacity

Each buffer has a specific buffer capacity $\beta$. The buffer capacity is equal to the changed concentration of the buffer base or acid, causing a change of the buffer pH value by one pH unit, i.e.

$$\beta = \frac{dc_A}{dpH} \qquad (1.3.2)$$

where $c_A$ is the base concentration.

Let us derive the expression of buffer capacity. For the equilibrium reaction

$$HA \leftrightarrow H^+ + A^-$$

the total base concentration in the buffer solution is

$$c_A = [A^-] + [OH^-] - [H^+] \qquad (1.3.3)$$

where $[OH^-]$ is the equilibrium concentration of the hydroxide ion. The equilibrium concentration of the buffer anion $[A^-]$ can be calculated from the equation

$$K_c = \frac{[H^+][A^-]}{[HA]} \qquad (1.3.4)$$

Since the total concentration of the acid and base of the buffering electrolyte is

$$c = [HA] + [A^-] \qquad (1.3.5)$$

it follows from eqs. (1.3.4) and (1.3.5) that

$$[A^-] = \frac{cK_c}{K_c + [H^+]} \qquad (1.3.6)$$

It is known that

$$[OH^-] = \frac{K_w}{[H^+]} \qquad (1.3.7)$$

where $K_w$ is the ionic product of water, equal to about $10^{-14}$ $(mol/L)^2$ at 25 °C. Hence, eqs. (1.3.2), (1.3.6), and (1.3.7) can be united into

$$\beta = \frac{d\left(c\frac{K_c}{K_c+[H^+]} + \frac{K_w}{[H^+]} - [H^+]\right)}{d(-log[H^+])} = \ln10\left(c\frac{K_c[H^+]}{(K_c+[H^+])^2} + \frac{K_w}{[H^+]} + [H^+]\right) \quad (1.3.8)$$

In neutral, weakly acidic, or weakly basic buffers, the concentrations of the proton and hydroxide ion are relatively low. Therefore, they can be neglected in eq. (1.3.8) that gives

$$\beta = \ln10c\frac{K_c[H^+]}{(K_c+[H^+])^2} = \ln10c\frac{10^{-pK_c-pH}}{\left(10^{-pK_c}+10^{-pH}\right)^2} \quad (1.3.9)$$

According to eqs. (1.3.8) and (1.3.9), the buffer capacity depends on the concentration and pH value of the buffer. When $pH = pK_c$, it follows from eq. (1.3.8) that

$$\beta = \ln10\left(\frac{c}{4} + \frac{K_w}{[H^+]} + [H^+]\right) \quad (1.3.10)$$

and from eq. (1.3.9) that

$$\beta = \ln10\frac{c}{4} \quad (1.3.11)$$

How a buffer counteracts a pH change? Let us assume that a buffer is created after neutralization of a weak acid of concentration $c$ with a strong base of concentration $x$. Because all basic molecules react stoichiometrically with the weak acid, the Henderson–Hasselbalch equation (eq. (1.3.1)) can be presented as

$$pH = pK_c + log\frac{x}{c-x} \quad (1.3.12)$$

To calculate the pH sensitivity of a buffer, the first derivative of eq. (1.3.12) should be derived. Using the relationship between the natural and decimal logarithms

$$log\,n = \frac{\ln n}{\ln 10} \quad (1.3.13)$$

eq. (1.3.12) gives

$$\frac{dpH}{dx} = \frac{1}{\ln10}\left[\frac{1}{x} - \frac{1}{c-x}(-1)\right] = \frac{c}{\ln10x(c-x)} \quad (1.3.14)$$

It can be determined from this expression that a buffer is more effective, if its first derivative is in its minimum, i.e. when its reciprocal value, named buffer capacity (see above), has a maximum value. To find when the quotient $dpH/dx$ has its minimum, the second derivative of eq. (1.3.12) should be derived:

$$\frac{d^2\mathrm{pH}}{dx^2} = \frac{c(2x-c)}{\ln 10 x^2 (c-x)^2} \tag{1.3.15}$$

According to eq. (1.3.15), the second derivative is equal to 0 when $x = \frac{c}{2}$, hence, when the added base neutralizes half of the weak acid. If the $x$ value is substituted in the Henderson–Hasselbalch equation, then

$$\mathrm{pH} = \mathrm{p}K_c + \frac{1}{\ln 10}\ln\frac{2c}{2c} = \mathrm{p}K_c \tag{1.3.16}$$

## 1.3.2 Buffers used in electrophoresis

In electrophoresis, acidic, neutral, or alkaline buffers are used (Table 1.3.1). The alkaline buffers are most widespread because the most proteins and all nucleic acids are acidic.

**Table 1.3.1:** Buffers used in electrophoresis.

| Buffer and pH | Concentrations | Preparation |
|---|---|---|
| **Acidic buffers** | | |
| Acetate buffer pH = 5.3, 10x | 0.60 mol/L sodium acetate $CH_3COOH$ | Adjust 49.22 g anhydrous sodium acetate with 1 mol/L acetic acid to pH = 5.3. Add deionized water to 1,000.0 mL. |
| **Neutral buffers** | | |
| TRIS-citrate-EDTA buffer pH = 7.2, 10x | 0.25 mol/L TRIS Citric acid 0.01 mol/L $Na_2EDTA$ | Adjust 30.29 g TRIS with 2 mol/L citric acid to pH = 7.2 and add 3.72 g $Na_2EDTA \cdot H_2O$ Add deionized water to 1,000.0 mL. |
| Hydrogen phosphate buffer pH = 7.3, 10x | 0.0034 mol/L $KH_2PO_4$ 0.0071 mol/L $Na_2HPO_4$ | Mix 1.36 g/L $KH_2PO_4$ and 1.78 g/L $Na_2HPO_4$ in ratio 435:565 (V/V). |
| **Alkaline buffers** | | |
| TRIS-hydrogen phosphate-EDTA buffer pH = 7.7, 10x | 0.35 mol/L TRIS 0.35 mol/L $NaH_2PO_4$ 0.01 mol/L $Na_2EDTA$ | 42.40 g TRIS 49.69 g $Na_2HPO_4$ 3.72 g $Na_2EDTA \cdot H_2O$ Add deionized water to 1,000.0 mL. |
| TRIS-formate buffer pH = 7.8, 10x | 1.29 mol/L TRIS 1.00 mol/L formic acid | 156.27 g TRIS 38.36 mL formic acid Add deionized water to 1,000.0 mL. |

**Table 1.3.1** (continued)

| Buffer and pH | Concentrations | Preparation |
|---|---|---|
| TRIS-borate-EDTA buffer pH = 8.3, 10x | 0.89 mol/L TRIS 0.89 mol/L boric acid 0.02 mol/L Na$_2$EDTA | 107.81 g TRIS 55.03 g boric acid 7.44 g Na$_2$EDTA·H$_2$O Add deionized water to 1,000.0 mL. |
| TRIS-taurinate-EDTA buffer pH = 8.5, 2x | 0.564 mol/L TRIS 0.782 mol/L taurine 0.002 mol/L Na$_2$EDTA | 68.32 g TRIS 97.86 g taurine 0.74 g Na$_2$EDTA·H$_2$O Add deionized water to 1,000.0 mL. |
| TRIS-TRICINEate buffer pH = 8.6 | 0.324 mol/L TRIS 0.096 mol/L TRICINE | 39.25 g TRIS 17.20 g TRICINE Add deionized water to 1,000.0 mL. |
| Barbitalate buffer pH = 8.6 | 0.05 mol/L sodium barbitalate 0.01 mol/L barbital | 10.31 g sodium barbitalate 1.84 g barbital Add deionized water to 1,000.0 mL. |
| TRIS-chloride buffer pH = 8.8, 4x | 1.5 mol/L TRIS HCl | Solve 181.71 g TRIS in 800 mL deionized water and titrate with HCl to pH = 8.8. Add deionized water to 1,000.0 mL. |
| TRIS-taurinate buffer pH = 9.0, 4x | 1.96 mol/L TRIS 0.54 mol/L taurine | 237.43 g TRIS 67.58 g taurine Add deionized water to 1,000.0 mL. |
| Glycinate buffer pH = 10.5, 10x | 0.20 mol/L glycine NaOH | Adjust 15.01 g glycine with 0.1 mol/L NaOH to pH = 10.5. Add deionized water to 1,000.0 mL. |

We proved that the TRIS-borate buffer contains a complex compound formed by a condensation reaction between boric acid and TRIS [2]. The complex compound, which we called *TRIS-boric acid*, has a zwitterionic structure. It dissociates a hydrogen ion and a *TRIS-borate* ion whose mobility was calculated [3]. The existence of TRIS-boric acid in the TRIS-borate buffer was reproved 30 years later by a large scientific group in Paris [4].

The formation of TRIS-boric acid and the dissociation of its protonated amino group take place according to the following scheme (Figure 1.3.2)

## 1.3.3 Biological buffers

The major intracellular buffer is the hydrogen phosphate buffer ($HPO_4^{2-}/H_2PO_4^-$), often incorrectly referred to as phosphate buffer. Its maximum buffer capacity is at pH = 6.7. The fluids around the cells of living organisms have also constant pH. To

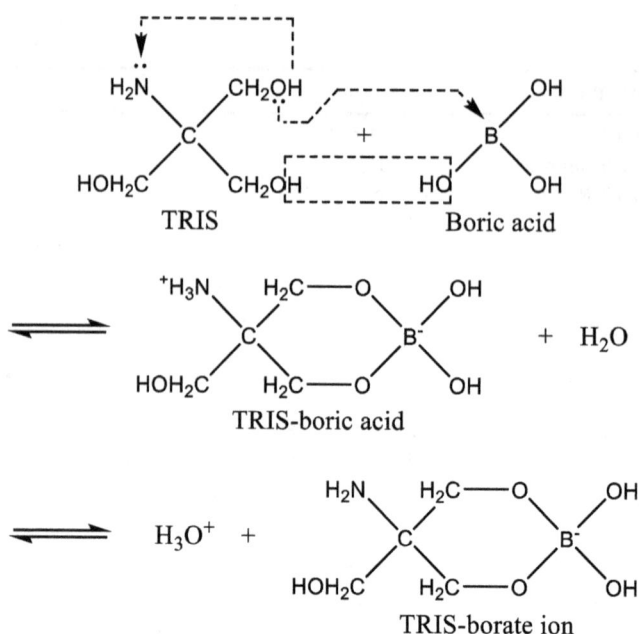

**Figure 1.3.2:** Formation and dissociation of TRIS-boric acid.

study biological processes in the laboratory, scientists use *biological buffers*, which cover the pH range of 2 to 11. Many of them were described by Norman Good and colleagues [5–7]. Most of Good buffers are solutions of zwitterionic compounds.

The specifications of biological buffers used in the electrophoresis techniques are given in Table 1.3.2. Their $pK_a$ values are temperature and concentration dependent.

**Table 1.3.2:** Biological buffers used in electrophoresis.

| Buffer | Full compound name | Chemical formulas | $M_r$ | $pK_a$ at 25 °C | Buffering range |
|---|---|---|---|---|---|
| MES | 2-(*N*-morpholino)-ethanesulfonic acid | | 195.24 | 6.10 | 5.4–6.8 |
| BISTRIS | 2,2-Bis(hydroxymethyl)-2,2′,2″-nitrilotriethanol | | 209.24 | 6.50 | 5.8–7.2 |

**Table 1.3.2** (continued)

| Buffer | Full compound name | Chemical formulas | $M_r$ | pK$_a$ at 25 °C | Buffering range |
|---|---|---|---|---|---|
| PIPES | 1,4-Piperazinediethane-sulfonic acid | | 302.37 | 6.76 | 6.1–7.5 |
| ACES | 2-(Carbamoylmethylamino)-ethanesulfonic acid | | 182.20 | 6.78 | 6.1–7.5 |
| BES | N,N-Bis(2-hydroxyethyl)-2-aminoethanesulfonic acid | | 213.25 | 7.09 | 6.4–7.8 |
| MOPS | 3-Morpholinopropane-1-sulfonic acid | | 209.26 | 7.20 | 6.5–7.9 |
| TES | 2-{[Tris(hydroxymethyl)methyl]amino}ethanesulfonic acid | | 229.25 | 7.40 | 6.7–8.1 |
| HEPES | 2-[4-(2-Hydroxyethyl)-piperazin-1-yl]ethanesulfonic acid | | 238.30 | 7.48 | 6.8–8.2 |
| TRICINE | N-(2-Hydroxy-1,1-bis-(hydroxymethyl)ethyl)glycine | | 179.17 | 8.05 | 7.4–8.8 |
| TRIS | 2-Amino-2-hydroxymethyl-propane-1,3-diol | | 121.14 | 8.06 | 7.4–8.8 |
| GLY-GLY | 2-[(2-Aminoacetyl)amino]-acetic acid | | 132.12 | 8.20 | 7.5–8.9 |

**Table 1.3.2** (continued)

| Buffer | Full compound name | Chemical formulas | $M_r$ | pK$_a$ at 25 °C | Buffering range |
|--------|--------------------|--------------------|-------|------------------|-----------------|
| BICINE | 2-(Bis(2-hydroxyethyl)-amino) acetic acid | | 163.17 | 8.26 | 7.6–9.0 |
| CHES | 2-(Cyclohexylamino)-ethanesulfonic acid | | 207.29 | 9.49 | 8.8–10.2 |

# References

[1]    Brønsted JN. Rec Trav Chim, 1923, 42, 718–728.
[2]    Michov BM. J Appl Biochem, 1982, 4, 436–440.
[3]    Michov BM. Electrophoresis, 1984, 5, 171.
[4]    Tournie A, Majerus O, Lefevre G, Rager MN, Walme S, Caurant D, Barboux P. J Colloid Interface Sci, 2013, 400, 161–167.
[5]    Good NE, Winget GD, Winter W, Connolly TN, Izawa S, Singh RMM. Biochemistry, 1966, 5, 467–477.
[6]    Good NE, Izawa S. Methods Enzymol, 1972, 24, 53–68.
[7]    Ferguson W, Braunschweiger KI, Braunschweiger WR, Smith JR, McCormick JJ, Wasmann CC, Jarvis NP, Bell DH, Good NE. Anal Biochem, 1980, 104, 300–310.

# 1.4 Polyions are moving in electric field

1.4.1     Ionic and polyionic mobility —— 27
1.4.1.1   Effective mobility —— 28
1.4.2     Equations of polyionic mobility —— 29
1.4.2.1   Equation of Smoluchowski —— 29
1.4.2.2   Equation of Hückel —— 29
1.4.2.3   Henry's function —— 29
1.4.2.4   New expression of Henry's function —— 30
1.4.2.5   Equation of Onsager —— 31
1.4.2.6   Equation of Robinson–Stokes —— 32
1.4.2.7   Parametric equation —— 33
1.4.2.8   Quadratic equation —— 33
1.4.3     Mobilities of ions used in electrophoretic methods —— 33
1.4.3.1   Calculating the mobilities of composed ions —— 35
          References —— 37

The mobilities play important role in electrophoresis of proteins and nucleic acids.

## 1.4.1 Ionic and polyionic mobility

In an electric field, a charged particle (ion or polyion) is moved by the electrophoretic force $F_e$ (in N) to the counter-pole. This force is directed toward the frictional force $F_f$ (in N). The frictional force is caused by the dipole molecules of the solvent (water), which stay on the path of the moving particle. If the charged particle moves with a constant speed, the two forces are equal to each other:

$$F_e = F_f \tag{1.4.1}$$

It is known that

$$F_e = QE = z_\infty eE \tag{1.4.2}$$

where $Q$ is the electric charge of the particle, in C, which in its turn is equal to the product of the elementary charge number $z_\infty$ and the elementary charge $e$; and $E$ is the strength of the electric field, in V/m. Simultaneously, it is known that

$$F_f = fv_\infty \tag{1.4.3}$$

where $f$ is the friction coefficient (in Js/m$^2$), and $v_\infty$ is the velocity of a charged particle (in m/s) in an infinitely dilute solution. For spherical particles, according to Stokes' law,

$$f = 6\pi\eta r \tag{1.4.4}$$

where $\eta$ is the dynamic viscosity of the solvent (in Pa s), and $r$ is the radius of the charged particle (in m). From eqs. (1.4.1)–(1.4.4), the equation

https://doi.org/10.1515/9783110761641-005

$$z_\infty e = 6\pi\eta r \frac{v_\infty}{E} \qquad (1.4.5)$$

is obtained.

It is known also that

$$\frac{v_\infty}{E} = \mu_\infty \qquad (1.4.6)$$

where $\mu_\infty$ is the absolute mobility of the particle, in $m^2/(s\ V)$, i.e. in an infinitely dilute solution. Then, it follows in from eqs. (1.4.5) and (1.4.6) that

$$\mu_\infty = \frac{z_\infty e}{6\pi\eta r} \qquad (1.4.7)$$

In the reality, a charged particle moves together with a part of its ionic atmosphere. Therefore, the electrokinetic potential $\zeta$, measured on the charged particle, is less than its absolute potential $\varphi_0$. According to most scientists, $\zeta$-potential is measured on the slipping plane between the particle (together with its adsorbed counter-ions), and the diffuse part of its ionic atmosphere. When the ionic strength is increased, the mobility of the charged particle decreases:

$$\mu = \frac{v}{E} = \frac{ze}{6\pi\eta r} \qquad (1.4.8)$$

At the isoelectric point, the electric charge of a protein and its mobility are equal to zero.

Regardless of their masses, nucleates have same mobility because their charge density is equal. However, if the electrophoretic separation takes place in a gel, the larger nucleates meet greater resistance, which decreases their mobility.

### 1.4.1.1 Effective mobility

In contrast to strong electrolyte solutions, effective mobility $\mu'$ is measured in weak electrolyte solutions. This is due to the fact that not all molecules are dissociated, but only a part of them, namely $\alpha c$, where $\alpha$ is the dissociation degree (dimensionless), and $c$ is the total electrolyte concentration, in $mol/dm^3$ [1, 2]:

$$\mu' = \alpha\mu \qquad (1.4.9)$$

Consequently, the effective mobility of an ion in a solution of weak acid or weak base depends on the pH value of the solution.

## 1.4.2 Equations of polyionic mobility

In the scientific literature, a few equations for polyionic mobility are known: equation of Smoluchowski, Hückel, Onsager, Robinson–Stokes, parametric equation, and quadratic equation.

### 1.4.2.1 Equation of Smoluchowski

Smoluchowski [3, 4] found that the mobility $\mu$ of a charged particle can be expressed by the following equation

$$\mu = \frac{\zeta \varepsilon}{\eta} \qquad (1.4.10)$$

where $\varepsilon$ is the (di)electric permittivity of the solvent (water), which equals the product of the relative (di)electric permittivity $\varepsilon_r$ and the (di)electric constant $\varepsilon_0$ (8.854 187 817 × $10^{-12}$ F/m); and $\eta$ is the dynamic viscosity of the solution (in Pa s).

### 1.4.2.2 Equation of Hückel

Hückel [5] showed that the ionic atmosphere reduces the absolute mobility $(1 + \kappa a)$ times, according to the equation

$$\mu = \frac{\mu_\infty}{1 + \kappa a} = \frac{\zeta \varepsilon}{1.5 \eta} \qquad (1.4.11)$$

where $\kappa$ is the reciprocal value of the ionic atmosphere thickness $(1/\kappa)$, and $a$ is the radius of a charged particle. When the ionic concentration in a solution (buffer) increases, the thickness of the ionic atmosphere decreases, i.e. the sum $(1 + \kappa a)$ increases. This means that the $\zeta$-potential depends on the charge of the particle, as well as on the concentration of the solution (buffer).

### 1.4.2.3 Henry's function

Comparing the equations of Smoluchowski (1.4.10) and Hückel (1.4.11), we could establish that the mobility, according to the first equation, is 1.5 times higher than the mobility, according to the second equation. In order to unite both equations, Henry [6, 7] added to the Hückel equation his complex function $f(\kappa a)$ and obtained that

$$\mu = \frac{\zeta\varepsilon}{1.5\eta} f(\kappa a) \qquad (1.4.12)$$

where

$$f(\kappa a) = 1 + \frac{(\kappa r)^2}{16} - \frac{5(\kappa r)^3}{48} - \frac{(\kappa r)^4}{96} + \frac{(\kappa r)^5}{96} - \frac{11}{96} e^{\kappa r} \int\limits_\infty^{\kappa r} \frac{e^{-t}}{t} dt \qquad (1.4.13)$$

According to Henry's function, the mobility of an ion or polyion depends not only on the ionic strength of the solution, which takes place in the parameter $\kappa$, but also on the radius of the charged particle $a$. When $\kappa a \to \infty$, then $f(\kappa a) \to 1.5$ and eq. (1.4.12) is transformed into the equation of Smoluchowski (1.4.10); when $\kappa a \to 0$, then $f(\kappa a) \to 1.0$ and eq. (1.4.12) gives the Hückel eq. (1.4.11).

### 1.4.2.4 New expression of Henry's function

We proved that each ion (polyion) has two radii: a geometric $r$ and an electrokinetic $a$ [8]. In addition, we showed that the radius $a$ is a function of the Debye–Hückel parameter, according to the equation

$$a = r(1 + \kappa r) \qquad (1.4.14)$$

where $0 \le \kappa r \le 1$. In accordance with this, we established that every ion and its ionic atmosphere are characterized by four electric potentials: $\varphi_0$, $\varphi_r$, $\zeta$, and $\varphi_a$. Using the theories of Gouy [9], Chapman [10], and Debye–Hückel [11], we proved that the geometric potential

$$\varphi_0 = \frac{ze}{4\eta\varepsilon r} \qquad (1\ 4\ 15)$$

the potential of the geometric ion and its ionic atmosphere

$$\varphi_r = \frac{ze}{4\eta\varepsilon a} = \frac{ze}{4\eta\varepsilon r(1 + \kappa r)} = \frac{\varphi_0}{1 + \kappa r} \qquad (1.4.16)$$

the electrokinetic potential

$$\zeta = \frac{ze}{4\eta\varepsilon r(1 + \kappa a)} = \frac{\varphi_0}{1 + \kappa a} = \varphi_r \frac{1 + \kappa r}{1 + \kappa a} \qquad (1.4.17)$$

and the potential of the electrokinetic ion and its ionic atmosphere

$$\varphi_a = \frac{ze\,exp[-\kappa(a - r)]}{4\pi\varepsilon a(1 + \kappa r)} = \frac{\varphi_r}{1 + \kappa a} = \frac{\zeta}{1 + \kappa r} = \zeta\frac{r}{a} \qquad (1.4.18)$$

If we introduce the two ionic radii and the four potentials in the Henry's expressions, we obtain a new function, which we refer to as $f(\kappa r)$ [12]:

$$f(\kappa r) = \frac{1 + \kappa a}{1 + \kappa r} = \frac{1 + \kappa r + (\kappa r)^2}{1 + \kappa r} \tag{1.4.19}$$

Now eq. (1.4.12) could be converted into

$$\mu = \frac{\zeta \varepsilon}{1.5\eta} f(\kappa r) = \frac{\zeta \varepsilon}{1.5\eta} \frac{1 + \kappa a}{1 + \kappa r} \tag{1.4.20}$$

When $\kappa r \to 0$, then $f(\kappa r) \to 1$, and

$$\mu = \frac{\zeta \varepsilon}{1.5\eta} \tag{1.4.21}$$

i.e., the equation of Hückel occurs; and when $\kappa r \to 1$, then $f(\kappa r) \to 1.5$, and

$$\mu = \frac{\zeta \varepsilon}{\eta} \tag{1.4.22}$$

i.e., the equation of Smoluchowski occurs.

The graphics of $f(\kappa r)$ as a function of $\kappa r$ and of its logarithmic value are given on Figure 1.4.1.

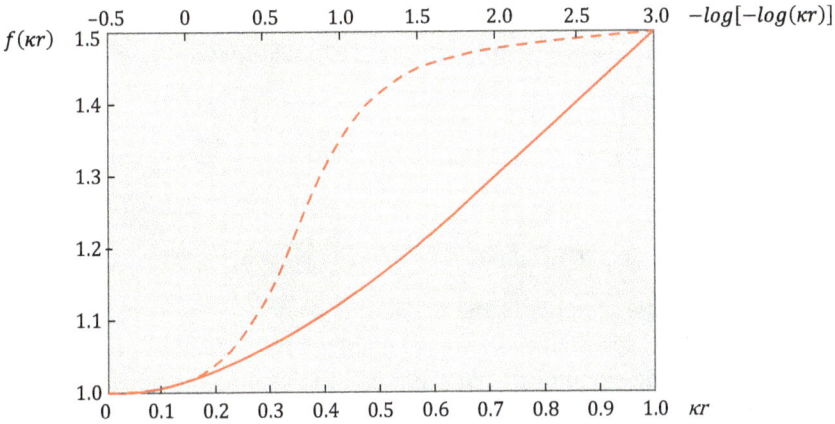

Figure 1.4.1: The dependence of function $f(\kappa r)$ on $\kappa r$ (solid line) and on its logarithmic value $-log[-log(\kappa r)]$ (dashed line).

## 1.4.2.5 Equation of Onsager

Onsager [13, 14] proposed that the polyionic mobility depends on the phenomena cataphoresis and relaxation. *Cataphoresis* causes additional friction force, which reduces the polyionic mobility in an electric field. It occurs because the diffuse part of its ionic atmosphere is moving toward the polyion. *Relaxation* is referred to the restructuring of the complex polyion – ionic atmosphere in an electric field: both the polyion and its ionic atmosphere move in opposite directions in an electric field, which results in polarization of the complex and additional reduction of the polyionic mobility (Figure 1.4.2).

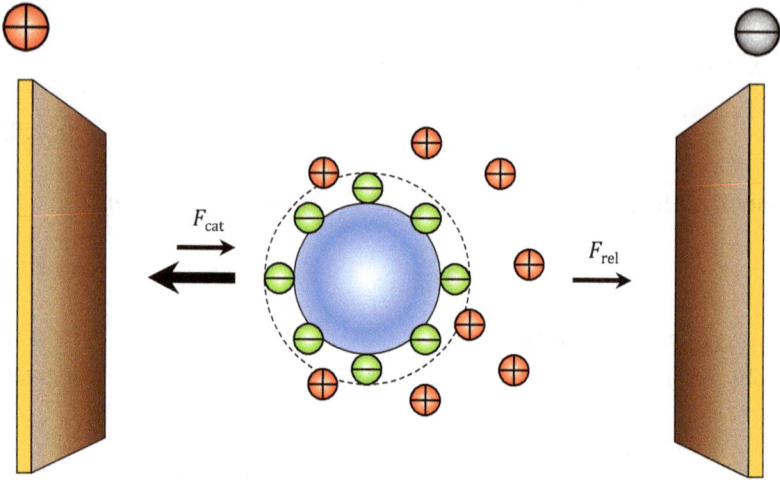

**Figure 1.4.2:** Influence of the cataphoretic strength ($F_{cat}$) and relaxation strength ($F_{rel}$) on the polyionic mobility.

The effect of cataphoresis and relaxation on the particle mobility is described by the Onsager equation

$$\mu = \mu_\infty - \left(\mu_{cat} + \mu_{rel}\right) = \mu_\infty - zsI^{1/2} \tag{1.4.23}$$

where $\mu_\infty$ is the absolute mobility of the polyion; $\mu_{cat}$ and $\mu_{rel}$ are its cataphoresis and relaxation mobility, respectively; $z$ is the number of its elementary charges (the electrovalence); $s$ is the slope of the Onsager calibration line, and $I$ is the ionic strength of the solution (buffer).

### 1.4.2.6 Equation of Robinson–Stokes

The Onsager equation can be used only for solutions with low ionic strengths. At higher ionic strength (up to 0.1 mol/L), the equation of Robinson–Stokes [15] can be applied in a good agreement with the experiment:

$$\mu = \mu_\infty - \frac{sI^{1/2}}{1 + \kappa a} \tag{1.4.24}$$

According to this equation, the ions and polyions are not considered as point charges, as in Onsager theory, but as charged particles. Therefore, $I^{1/2}$ is divided by the sum $(1 + \kappa a)$ where the radius $a$ of the charged particle could have different values, depending on the experimental conditions.

### 1.4.2.7 Parametric equation

The ionic (polyionic) mobility can be calculated also according to the linear parametric equation

$$\mu = \mu_\infty - p \tag{1.4.25}$$

where $p$ is the parameter of the ionic (polyionic) mobility [16] expressed as

$$p = \frac{ze\kappa}{6\pi\eta} \tag{1.4.26}$$

For weak electrolytes, the above equation should be transformed into

$$\mu' = \alpha(\mu_\infty - p) \tag{1.4.27}$$

### 1.4.2.8 Quadratic equation

If eq. (1.4.14), which represents our concept of existing of two radii of a charged particle (geometric and electrokinetic radius), is inserted in the equation of Robinson–Stokes (1.4.24) and the resulting expression is simplified, the following quadratic equation is obtained:

$$\mu = \mu_\infty - \frac{sI^{1/2}}{1 + \kappa r + \kappa^2 r^2} \tag{1.4.28}$$

The experimental values of the ionic mobilities can be well expressed by the parametric and quadratic equation as well as by the equation of Robinson–Stokes. However, while in the Robinson–Stokes equation the context-dependent value of $a$ is used, the quadratic equation contains the well-defined geometric radius $r$ of the ion. The experimental data lie closest to the results obtained by the quadratic equation.

## 1.4.3 Mobilities of ions used in electrophoretic methods

In Table 1.4.1, the absolute mobilities of ions, which are used in the electrophoretic methods, are listed [17–19].

**Table 1.4.1:** Absolute ionic mobilities, in $m^2/(s\,V)$, of ions, at two temperatures.

| Ions | Chemical formulas | $\mu_\infty \times 10^9\,(0\ °C)$ | $\mu_\infty \times 10^9\,(25\ °C)$ |
|---|---|---|---|
| Hydronium ion | $H_3O^+$ | 248.74 | 362.39 |
| Potassium ion | $K^+$ | 39.57 | 76.18 |
| Sodium ion | $Na^+$ | 27.40 | 51.95 |
| Imidazolium ion | | 26.65 | 43.74 |
| Alaninium ion | $^+H_3NCH(CH_3)COOH$ | 17.46 | 32.67 |
| 1/2 Calcium ion | $Ca^{2+}$ | 15.55 | 30.84 |
| β-Alaninium ion | $^+H_3NCH_2CH_2COOH$ | 17.22 | 29.30 |
| Ammediolium ion | $HOCH_2(^+H_3N)C(CH_3)CH_2OH$ | 13.37 | 27.86 |
| TRIS ion | $(HOCH_2)_3CNH_3^+$ | 12.75 | 27.86 |
| BISTRIS ion | $(HOCH_2CH_2)_2NH^+C(CH_2OH)_3$ | 12.00 | 24.01 |
| HEPESate ion | $HOCH_2CH_2N\quad NCH_2CH_2SO_{3-}$ | −9.52 | −22.85 |
| TRICINEate ion | $(HOCH_2)_3CNHCH_2COO^-$ | −10.51 | −24.01 |
| Barbitalate (veronalate) ion | | −13.99 | −24.97 |
| TESate ion | $(HOCH_2)_3NHCH_2CH_2SO_3^-$ | −12.75 | −26.90 |
| ACESate ion | $H_2NCOCH_2NHCH_2CH_2SO_3^-$ | −14.24 | −27.86 |
| MESate ion | $NCH_2CH_2SO_{3-}$ | −13.49 | −28.19 |
| 1/2 Hydrogen phosphate ion | $HPO_4^{2-}$ | −14.15 | −29.71 |
| Glycylglycinate ion | $H_2NCH_2CONHCH_2COO^-$ | −16.97 | −30.26 |
| 4-Aminobutirate ion (GABAate ion) | $H_2NCH_2CH_2CH_2COO^-$ | −16.22 | −32.19 |
| Borate ion | $B(OH)_4^-$ | −18.21 | −33.63 |
| Taurinate ion | $H_2NCH_2CH_2SO_3^-$ | −17.71 | −34.11 |
| Dihydrogen phosphate ion | $H_2PO_4^-$ | −17.46 | −34.28 |

**Table 1.4.1** (continued)

| Ions | Chemical formulas | $\mu_\infty \times 10^9 (0\ °C)$ | $\mu_\infty \times 10^9 (25\ °C)$ |
|---|---|---|---|
| BICINEate ion | $(HOCH_2CH_2)_2NCH_2COO^-$ | −19.70 | −36.04 |
| Glycinate ion | $H_2NCH_2COO^-$ | −20.94 | −38.45 |
| Lactate ion | $CH_3CHOHCOO^-$ | −18.95 | −40.22 |
| 1/2 Sulfate ion | $SO_4{}^{2-}$ | −21.25 | −41.47 |
| Acetate ion | $CH_3COO^-$ | −22.18 | −42.36 |
| Formate ion | $HCOO^-$ | −30.88 | −56.60 |
| Chloride ion | $Cl^-$ | −42.95 | −79.09 |
| Hydroxide ion | $OH^-$ | −108.83 | −204.92 |

## 1.4.3.1 Calculating the mobilities of composed ions

We propose equations for calculating the mobilities of ions that could be considered as composed of ions with known mobilities [20]. Different ions have different shapes but in order to simplify the equations we suppose that all ions are spherical.

Let us accept that ion $x$ is composed of ions $a$ and $b$ (all ions under discussion are meant to be hydrated) after splitting off small ions or molecules such as $H^+$, $HO^-$, $H_2O$, $H_2$, etc. Let us also admit that the volume of ion $x$ is approximately equal to the sum of the volumes of forming ions. If we express the volumes of all ionic spheres with the formula $V = \frac{4}{3}\pi r^3$ and simplifying the equation, the following expression is obtained:

$$r_x^3 = r_a^3 + r_b^3 \tag{1.4.29}$$

where $r_x$, $r_a$, and $r_b$ are the radii of the composed and forming ions, respectively.

It is known that the electric field acts on a moving ion with a force, which is in equilibrium with the friction force defined by Stokes' equation [21]. Transforming this equilibrium, we obtain the equation

$$r = \frac{ze}{6\pi\eta\mu_\infty}\ [m] \tag{1.4.30}$$

where $z$ is the ionic electrovalency, $e$ is the electron charge ($1.6021892 \times 10^{-19}$ C), $\eta$ is the dynamic viscosity of the water, and $\mu_\infty$ is the absolute ionic mobility, i.e. the mobility in an extremely diluted solution. Hence, eq. (1.4.29) could be represented as

$$\left(\frac{z_x}{\mu_{\infty x}}\right)^3 = \left(\frac{z_a}{\mu_{\infty a}}\right)^3 + \left(\frac{z_b}{\mu_{\infty b}}\right)^3 \tag{1.4.31}$$

which gives that

$$\mu_{\infty x} = z_x \left[\left(\frac{z_a}{\mu_{\infty a}}\right)^3 + \left(\frac{z_b}{\mu_{\infty b}}\right)^3\right]^{-1/3} \ [\mathrm{m}^2/(\mathrm{s\,V})] \tag{1.4.32}$$

If ion $x$ is formed by two equal ions with absolute mobility $\mu_{\infty a}$ and if the composed ion and the forming ions have the same electrovalency, eq. (1.4.32) could be transformed into

$$\mu_{\infty x} = 2^{-1/3} \mu_{\infty a} \ [\mathrm{m}^2/(\mathrm{s\,V})] \tag{1.4.33}$$

Eqs. (1.4.32) and (1.4.33) could be applied for extremely diluted solutions. However, every ion is surrounded by counter-ionic atmosphere whose radius

$$\kappa^{-1} = 1.9885 \times 10^{-12} (\varepsilon_r T/I)^{1/2} \ [\mathrm{m}] \tag{1.4.34}$$

where $\varepsilon_r$ is the relative electric permittivity, $T$ is the thermodynamic temperature, and $I$ is the ionic strength of the solution [22]. This atmosphere diminishes the absolute mobility of an ion $(1+\kappa r)$ times, i.e.

$$\mu = \frac{\mu_\infty}{1+\kappa r} \ [\mathrm{m}^2/(\mathrm{s\,V})] \tag{1.4.35}$$

From eqs. (1.4.30) and (1.4.35) follows that

$$\mu_w = \frac{\mu}{2} \pm \left[\frac{\mu}{2}\left(\frac{\mu}{2} + \frac{ze\kappa}{3\pi\eta}\right)\right]^{1/2} \ [\mathrm{m}^2/(\mathrm{s\,V})] \tag{1.4.36}$$

The last equation can be shortened, if the square root is expanded into a power series and if we take only the first correction term. Then, eq. (1.4.36) could be transformed into

$$\mu_\infty = \mu + \frac{ze\kappa}{6\pi\eta} \ [\mathrm{m}/(\mathrm{s\,V})] \tag{1.4.37}$$

We gave evidences for the existence of a complex ion in TRIS-borate buffers, which we called TRIS-borate ion [23]. TRIS-borate ion [(HO)$_2$B$^-$(OCH$_2$)$_2$C(CH$_2$OH)NH$_2$, TB$^-$] could be considered as composed of a TRIS ion [(HOCH$_2$)$_3$CN$^+$H$_3$, HT$^+$] and a borate ion [B(OH)$_4^-$, B$^-$] when a proton and two water molecules are split off.

Let us calculate the mobility of TRIS-borate ion at 25 °C and $I = 0.01$ mol/L [24]. It is known that at 25 °C (298.15 K) and $I = 0.01$ mol/L $\mu_{HT^+} = 24.06 \times 10^{-9}$ and $\mu_{B^-} = -29.84 \times 10^{-9}$ m$^2$/(s V) [4]. Taking into account that at this temperature $\varepsilon_r = 78.54$ and

$\eta = 0.8904 \times 10^{-3}$ Pa s, we could calculate that $ze\kappa/(6\pi\eta) = (\pm)3.14 \times 10^{-9}$ m²/(s V). Hence, it follows from eqs. (1.4.37) and (1.4.32) that at 25 °C $\mu_{\infty HT^+} = 27.20 \times 10^{-9}$ m²/(s V), $\mu_{\infty B^-} = -32.98 \times 10^{-9}$ m²/(s V), and $\mu_{\infty HB^-} = -23.45 \times 10^{-9}$ m²/(s V), and when $I = 0.01$ mol/L

$$\mu_{TB^-} = -23.45 \times 10^{-9} + 3.14 \times 10^{-9} = -20.31 \times 10^{-9} \text{ m}^2/(\text{s V})$$

## References

[1]   Longsworth LG. Bier M (ed.), Electrophoresis. Theory, Methods and Applications. Academic Press, New York, 1959, 92–136.
[2]   Alberty RA. J Am Chem Soc, 1950, 72, 2361–2367.
[3]   Smoluchowski M. Bull Acad Sci Cracovie, 1903, A8, 182–200.
[4]   Smoluchowski M. Graetz L (ed.), Handbuch Der Elektrizität Und Des Magnetismus. Leipzig, Barth, Vol. **2**, 1921, 366–428.
[5]   Hückel E. Phys Z, 1924, 25, 204–210.
[6]   Henry DC. Proc Roy Soc, 1931, A133, 106–129.
[7]   Henry DC. Trans Faraday Soc, 1948, 44, 1021–1026.
[8]   Michov BM. Electrochim Acta, 2013, 108, 79–85.
[9]   Gouy MG. J Phys, 1910, 9, 457–468.
[10]  Chapman DL. Phil Mag, 1913, 25, 475–481.
[11]  Debye P, Hückel E. Phys Z, 1923, 24, 185–206.
[12]  Michov BM. Electrophoresis, 1988, 9, 199–200.
[13]  Onsager L. Phys Z, 1926, 27, 388–392.
[14]  Onsager L. Chem Rev, 1933, 13, 73–89.
[15]  Robinson RA, Stokes RH. Electrolyte Solutions, 2nd. Butterworths, London, 1965.
[16]  Michov BM. Electrophoresis, 1985, 6, 471–474.
[17]  Jovin TM. Ann N Y Acad Sci, 1973, 209, 477–496.
[18]  Marcus Y. Ion Properties. Marcel Dekker, New York, 1997.
[19]  Dean JA. Lange's Handbook of Chemistry, 15th ed. McGraw-Hill, New York, 1999.
[20]  Michov BM. Electrophoresis, 1983, 4, 312–313.
[21]  Stokes GG. Trans Camb Phyl Soc, Part 2, 1856, 9, 8–150.
[22]  Debye P, Hückel E. Phys Z, 1923, 24, 185–206.
[23]  Michov BM. J Appl Biochem, 1982, 4, 436–440.
[24]  Michov BM. Electrophoresis, 1984, 5, 171.

# 1.5 Electrophoresis is carried out in different solid media

1.5.1    Cellulose acetate —— 39
1.5.2    Agarose gel —— 39
1.5.3    Polyacrylamide gel —— 41
1.5.3.1  Acrylamide —— 42
1.5.3.2  Bisacrylamide —— 42
1.5.3.3  The magnitudes *T* and *C* —— 42
1.5.3.4  Alternative cross-linkers —— 42
1.5.3.5  Initiator-catalyst systems —— 43
1.5.3.6  Copolymerization of acrylamide and BIS —— 45
1.5.3.7  Homogeneous polyacrylamide gels —— 46
1.5.3.8  Thin and ultrathin polyacrylamide gels —— 46
1.5.3.9  Gradient polyacrylamide gels —— 47
         References —— 47

Electrophoresis is carried out in buffers or solid media containing buffers. The solid media act as sieves and separate the polyions according to their volumes.

Different solid media were used for electrophoresis. Nowadays, cellulose acetate and first of all polyacrylamide (PAA) and agarose gels are used.

## 1.5.1 Cellulose acetate

Cellulose acetate (CA) is composed of glucose residues that are esterified by acetic acid. Each glucose residue is β-glycosidically connected with the hydroxyl group at the C-4 atom of the next glucose residue. In this way, long fibrous molecules are formed, which are linked each with other by hydrogen bonds. CA possesses large pores, which have no sieving effect on the migrating polyions; therefore, the electrophoretic separation is fulfilled only according to the electric charges of proteins.

The CA films were used first by Joachim Kohn [1]. They do not counteract the diffusion, as a result of which the resolution of CA electrophoresis is of low grade.

## 1.5.2 Agarose gel

Agarose is one of the two main components of agar, which is extracted from red seaweed [2]. It consists of linear polysaccharide chains, which are built of approximately 400 residues of agarobiose. Agarobiose contains an α-(1→3)-linked β-D-galactopyranose and a β-(1→4)-linked 3,6-anhydro-L-α-galactopyranose [3]. The 3,6-anhydro-L-α-galactopyranose represents L-galactose that possess an anhydrobridge

https://doi.org/10.1515/9783110761641-006

between the third and sixth positions. Some D- and L-galactose units can be methyl-ated; ionizable groups are also found in small amounts [4] (Figure 1.5.1).

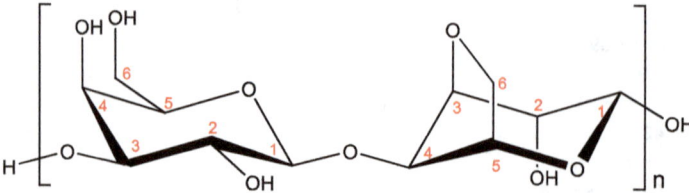

**Figure 1.5.1:** Structure of the residue of an agarobiose monomer.

Each polysaccharide chain contains about 800 galactose residues, and has $M_r$ between 100,000 and 200,000 [5]. Many chains form helical fibers, which build supercoiled structures of a radius of 20–30 nm [6]. The fibers are quasi-rigid and have different length that depends on the agarose concentration [7]. They form pores of different di-ameter. This structure is held together by hydrogen bonds and can be disrupted by heating that melts the gel to give a liquid.

The agarose concentration in the gel is usually 0.7–2.0 g/dL. Low-concentration gels (0.1–0.2 g/dL) are fragile and are not easily to handle; high concentration gels are brittle. In a 0.16 g/dL agarose gel, the pores have a diameter of about 500 nm; and in a 0.075 g/dL agarose gel, they have a diameter of about 800 nm [8]. For protein and DNA separations, 1 g/dL agarose gel is usually used whose pores have a diameter of about 150 nm.

The agarose gels were introduced in the electrophoresis by Hjerten [9]. They are optimal for electrophoresis of proteins with $M_r$ larger than 200,000 [10]. Gels with a concentration of 0.8–1.0 g/dL are suitable for electrophoresis of 5–10 kb DNA frag-ments; gels with a concentration of 2 g/dL are suitable for electrophoresis of 0.2–1 kb DNA fragments.

The agarose gels are usually cast on support films. The support films are made of 0.18–0.20 mm polyester film that binds to the melt agarose. The supported gels can be cut in different sizes, are stable at temperatures up to 110 °C, and are trans-parent to UV light above 310 nm.

The gelling and melting temperatures of agarose gels vary depending on the agarose type. Agaroses derived from *Gelidium* have a gelling temperature of 34–38 °C and a melting temperature of 90–95 °C, while agaroses derived from *Gracilaria* have a gelling temperature of 40–52 °C and a melting temperature of 85–90 °C. The gelling temperature is a function of the concentration of the methyl group in agarose: in-creasing methylation lowers the gelling temperature [11].

*Low-melting agarose.* The standard agarose melts at 80–90 °C, when DNA dena-tures. To hinder this, low-melting agarose was invented, which melts at 30–35–40 °C. The low-melting agarose is chemically modified agarose that possess hydroxyethyl groups in its polysaccharide chains and fewer sulfate groups [12]. It has a lower resolution and lower mechanical strength than the standard agarose [13]. The low-melting agarose allows the gel slices with DNA to be melt after electrophoresis and placed at disposal to polymerases, restriction endonucleases, and ligases.

*High-strength agarose.* If very large molecules have to be resolved, large pores in low-concentrated agarose gel are needed. However, when the agarose concentration falls, the gels degrade easily. The mechanical strength of an agarose gel is expressed in $g/cm^2$, i.e. is equal to the mass, which can be carried by 1 $cm^2$ agarose gel. The standard gels have mechanical strength of about 1,000 to 2,000 $g/cm^2$, the low melting gels of about 200 $g/cm^2$. To overcome this disadvantage, an agarose type was discovered that has higher mechanical strength at low concentration – up to 6,000 $g/cm^2$. This agarose is used for pulsed-field electrophoresis of chromosomes.

*Charge free agarose.* As mentioned, the agarose gels contain charged groups, which cause electroosmosis. To prevent this phenomenon, chemically modified agarose was invented that has no acidic groups [14, 15].

*Electroosmosis.* Electroosmosis is referred to as movement of the solvent (water) when located in an electric field [16]. It is caused in the following way: Residual chemical groups of sulfuric, pyruvic, or carbonic acid on an agarose molecule split hydrogen ions in neutral and alkaline buffers, and transform themselves into the negatively charged sulfate, pyruvate, or carboxylate groups. The negatively charged agarose gel does not migrate to the anode (stationary phase); however, its positively charged counter-ions migrate, together with their own hydration envelopes, to the cathode (mobile phase). As a result, the cathodic gel pole binds water (hydrates) while the anodic gel pole loses water (dehydrates).

## 1.5.3 Polyacrylamide gel

The polyacrylamide (PAA, poly(2-propenamide)) gel, introduced in 1959 by Raymond and Weintraub [17], is a copolymer of the monomer acrylamide and the comonomer bisacrylamide. It is the best medium for electrophoretic separations [18, 19]. The PAA gel is relatively nontoxic, because it contains minute residual amounts of acrylamide after its production [20].

The PAA gel has some important advantages in comparison with other gels: it is hydrophilic and electrical neutral; it does not show electroosmosis, since it carries almost no electric charges; it is transparent to light of wavelengths above 250 nm; it is chemically inert and thermo-stable; it can be prepared with desired pore size, if the concentration of acrylamide and bisacrylamide is changed; and it does not interact with polyions to be resolved, or with dyes.

### 1.5.3.1 Acrylamide

Acrylamide (prop-2-enamide, $M_r = 71.08$) (Figure 1.5.5) crystallizes into white crystals, which are soluble in water, methanol, glycerol, acetone, trichloromethane, and other solvents, and has a boiling point of $84.5 \pm 0.3$ °C. It is highly neurotoxic and accumulates in the body. Acrylamide polymerizes spontaneously during prolonged storage.

### 1.5.3.2 Bisacrylamide

**Bis**acrylamide (BIS), ($N,N'$-methylene-bis-acrylamide, $N,N'$-methylene-diacrylamide, $M_r = 154.17$) (Figure 1.5.5) is a white less toxic cross-linker, also soluble in water. It melts at 185 °C and also polymerizes spontaneously during prolonged storage.

### 1.5.3.3 The magnitudes *T* and *C*

Acrylamide (the monomer) and BIS (the comonomer) are often referred to as *monomers* and their solution as *monomeric solution*. Hjerten [21] improved formulas for the *total monomeric concentration T* and the *cross-linking degree C* of the PAA gel, which we have transformed into:

$$T = a + b \,(g/dL)$$ (1.5.1)

and

$$C = \frac{b}{a+b} \,(\text{dimensionless})$$ (1.5.2)

where *a* and *b* are the concentrations of acrylamide and BIS, respectively, in g/dL.

The total monomeric concentration *T* and the cross-linking *C* determine the pore size.

### 1.5.3.4 Alternative cross-linkers

The BIS cross-linked PAA gels have high optical transparency, good mechanical properties at low concentrations, are electric neutral, and may possess variable porosities. However, their relatively low stability in alkaline solutions, and their reduced optical transparency at high concentrations limit their application. Therefore, besides BIS, alternative cross-linkers are developed: DATD [22], BAP (PDA) [23], DHEBA [24], BAC [25], AcrylAide (*FMC*, Rockland, Me.), *etc.* (Table 1.5.1).

Some of the cross-linkers are used for liquefaction of PAA gel to free the separated polyions after electrophoresis. If a gel contains DHEBA [24], the liquefaction is carried out by oxidation with periodic acid; if it contains BAC [25], the BAC disulfide bonds should be split with thiols.

**Table 1.5.1:** Some alternative cross-linkers.

| Cross-linkers | Chemical formulas | $M_r$ |
|---|---|---|
| DATD (N,N'-diallyltartardiamide) |  | 228.25 |
| BAP (1,4-bis-acryloyl-piperazine), or PDA (piperazine-diacrylamide) |  | 194.23 |
| DHEBA (N,N'-(1,2-dihydroxyethylen)-bis-acrylamide) |  | 200.20 |
| BAC (N,N'-bis-acryloylcystamine) |  | 260.38 |

Irrespective of this, the alternative cross-linkers have many disadvantages. Therefore, BIS is still the most used comonomer.

### 1.5.3.5 Initiator-catalyst systems

The copolymerization of acrylamide and BIS requires free radicals, which are generated by initiator-catalyst systems [26]. Among them, the APS-TMEDA system is optimal (Figure 1.5.2).

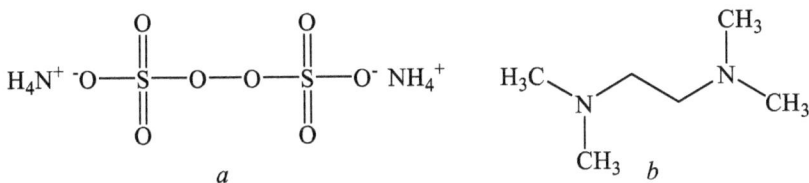

**Figure 1.5.2:** Chemical formulas of APS (*a*) and TMEDA (*b*).

The *initiator* APS (**ammonium peroxydisulfate**, ammonium persulfate, $M_r = 228.18$) dissociates in solution giving two ammonium cations and a peroxydisulfate anion. The peroxydisulfate anion forms two sulfate free radicals [27] (Figure 1.5.3):

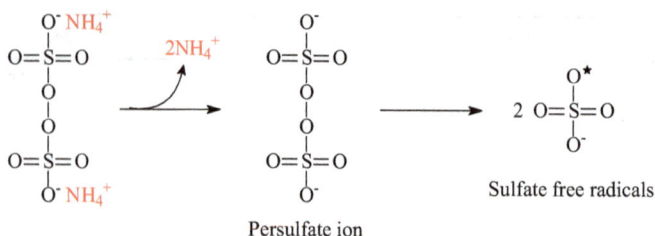

Persulfate ion

Sulfate free radicals

**Figure 1.5.3:** Building of sulfate free radicals from APS.

APS is used in concentrations of 1–4 mmol/L (0.02–0.10 g/dL). Compared to other oxidizing agents, it is preferred because it releases no molecular oxygen, which inhibits the copolymerization.

The velocity of copolymerization process increases with the increase of concentrations of the monomers and APS. We propose that TMEDA concentration should remain constant, while the APS concentration should be changed, according to the formula:

$$c_{APS} = 0.32 \ (g/dL)^2 / T \ [g/dL] \qquad (1.5.3)$$

This means that at $T = 5$ g/dL $c_{APS}$ should be 0.06 g/dL; at $T = 11$ g/dL it should be 0.03 g/dL; and at $T = 16$ g/dL it should be 0.02 g/dL.

The tertiary amine TMEDA [$N,N,N',N'$-tetramethylethylenediamine, TEMED, 1,2-bis(dimethylamino)-ethane, $M_r = 116.21$] is the most common used *catalyst*, in concentrations of 2–4 mmol/L (0.02–0.06 mL/dL). It accelerates the decomposition of peroxydisulfate ions into sulfate free radicals.

APS and TMEDA are used at approximately equimolar concentrations in the range of 1–10 mmol/L.

Riboflavin (riboflavin-5′-phosphate) also generates free radicals in a photochemical reaction, often in combination with TMEDA. When irradiated with blue to ultraviolet light in the presence of oxygen, it converts into a leuco (colorless) form, which initiates the copolymerization (Figure 1.5.4). This is referred to as *photochemical copolymerization*.

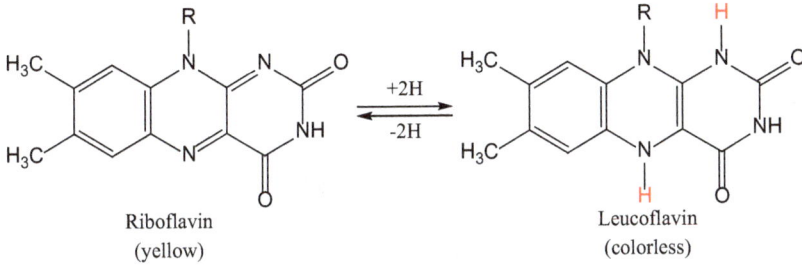

Figure 1.5.4: Riboflavin and its leuco form.

## 1.5.3.6 Copolymerization of acrylamide and BIS

The copolymerization of acrylamide and BIS proceeds in two steps. At first, the sulfate radicals bind to the alkene (the vinyl groups of acrylamide) molecules forming sulfate ester radicals, which take part in the polymerization producing long polymeric chains [28]. Then, the polymeric chains are cross-linked by BIS molecules. So, the spatial structure of the PAA gel is formed, which can be compared to a sponge [29, 30] (Figure 1.5.5).

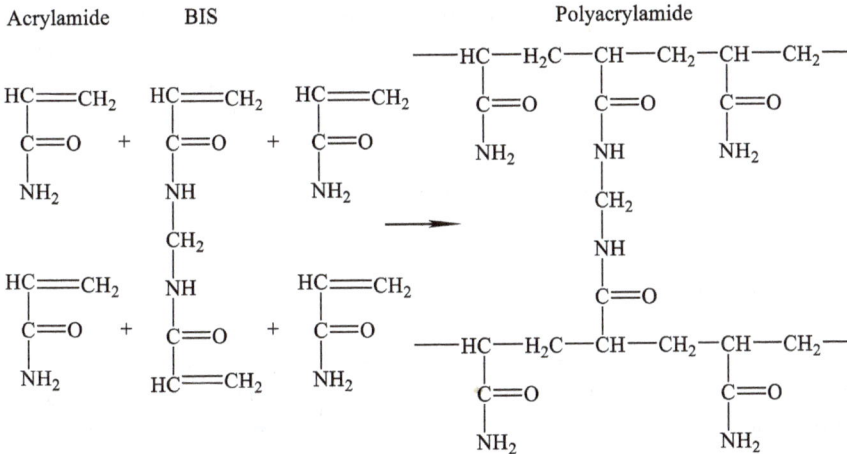

Figure 1.5.5: Copolymerization of acrylamide and BIS. A little part of the neutral carboxamide groups in the polyacrylamide gel can hydrolyze giving negatively charged carboxylate groups.

The viscosity, elasticity, and strength of a PAA gel depend on the ratio between the length of PAA chains and the frequency of the cross-linkages, i.e. on the ratio between the molar concentrations of acrylamide and BIS. If the ratio between the molar concentrations of acrylamide and BIS is 200:1 (between the mass concentrations 100:1), the PAA gel is elastic, soft, and transparent, because it has long chains. If the ratio between the molar concentrations of acrylamide and BIS is less than 20:1

(between the mass concentrations 10:1), the gel is fragile, brittle, and dull, since it is made of short PAA chains.

When $T$ is increased at fixed $C$, the number of PAA chains increases and the pore size decreases in a nearly linear relationship. The relationship of $C$ to the pore size is more complex. When $T$ is held constant and $C$ is increased, the pore size decreases to a minimum at about $C = 0.05$. Gels with low $T$ (e.g., 7.5 g/dL) are used for separation of large proteins, while gels with high $T$ (e.g., 15 g/dL) are used for small proteins.

For most proteins and SDS-proteins the value of $C$ should be about 0.0026 (37.5:1); for most native DNA and RNA value of $C$ should be about 0.033 (29:1); for most denaturing DNA and RNA the value of $C$ should be about 0.05 (19:1 acrylamide/BIS). Gel with $T = 4$ g/dL and $C = 0.03$ have no sieving properties. It is important that total acrylamide is polymerized. If this is not the case, the nonpolymerized acrylamide binds covalently to the polyions during electrophoresis [31].

Besides the initiator–catalyst system, the copolymerization between acrylamide and BIS depends on the pH value, temperature, purity of the chemicals, and inhibitors.

The hydroxide ions accelerate the copolymerization process. Therefore, it should take place at basic pH values. On the contrary, the rate of the copolymerization process decreases at low pH values. The temperature accelerates the copolymerization process; at temperatures below 20 °C, it slows down.

Inhibitors of the copolymerization are the oxygen and peroxides, the last building oxygen at their cleavage. The oxygen slows down the process, because it catches free radicals. Therefore, the copolymerization should be carried out with exclusion of air.

Similar to the PAA gel is the gel, formed by the monomer (**N**-acryloyl-tris(hydroxymethyl)aminomethane [32, 33].

Two types of PAA gels are known: homogeneous and gradient gels.

### 1.5.3.7 Homogeneous polyacrylamide gels

The homogeneous PAA gels have different thickness. Thin and ultrathin PAA gels are preferable. They are cast on support film or support fabric and are used for horizontal electrophoresis. The support film and fabric for PAA gel are usually prepared from pretreated polyester, which binds chemically to the PAA gel.

### 1.5.3.8 Thin and ultrathin polyacrylamide gels

The thin gels are 0.5–1.0 mm thick; the ultrathin gels are thinner than 0.5 mm. Both gels have the following advantages against the vertical gels [34]: the temperature gradient between the gel surfaces is smaller since the heat dissipates easier; the samples applied on the gel surface have smaller volumes; electrode strips, soaked with electrode solutions, could be placed on the gel; the staining and destaining procedures are quickly performed.

The thin and ultrathin horizontal gels [35] on support films can be cut into strips and can be stored dry. However, they are not suitable for electroblotting. Alternatives for this purpose are the fabric-supported gels [36]. The fabric is commonly made from polyester fibers, which form pores with a diameter of 10–60 µm [37].

### 1.5.3.9 Gradient polyacrylamide gels

The gradient PAA gels (pore gradient PAA gels) were introduced by Margolis and Kenrick [38]. Their concentration varies continuously from one gel end to the other gel end, resulting in a continuous changing of the pore diameters [39, 40]. The $T$-value of the gradient gels is usually of 4–28 g/dL.

There are two types of gradient gels: linear and exponential. In the gradient gels with linearly increased total concentration $T$, at constant cross-linking degree $C$, the average pore radius decreases exponentially. The maximum pore radius $r_{max} = a - T^{-b}$, where $a$ and $b$ are empirical constants [41].

## References

[1]   Rocco RM. Clin Chem, 2005, 51, 1896.
[2]   Griess GA, Guiseley KB, Serwer P. Biophys J, 1993, 65, 138–148.
[3]   Araki C. Wolfram MC (ed.). Carbohydrate chemistry of substances of biological interest, Proc 4th Int Congress Biochem, 1958, 1, 15.
[4]   Armisen R, Galatas F, McHugh DJ (ed.). Production and utilization of products from commercial seaweeds, FAO Fish Tech Pap, 1987, 1–57.
[5]   Kirkpatrick FH, Dumais MM, White HW, Guiseley KB. Electrophoresis, 1993, 14, 349–354.
[6]   Maniatis T, Fritsch EF, Sambrook J. Chapter 5, Protocol 1. In Molecular Cloning – A Laboratory Manual, 3rd ed., 1, 5.4.
[7]   Stephen AM, Phillips GO (eds.). Food Polysaccharides and Their Applications. CRC Press, 2006, 226.
[8]   Serwer P. Biochemistry, 1980, 19, 3001–3005.
[9]   Hjerten S. Biochem Biophys Acta, 1961, 53, 514–517.
[10]  Smisek DL, Hoagland DA. Macromolecules, 1989, 22, 2270–2277.
[11]  Workshop on Marine Algae Biotechnology: Summary Report. National Academy Press, 1986, 25.
[12]  Maniatis T, Fritsch EF, Sambrook J. Chapter 5, Protocol 6. In Molecular Cloning – A Laboratory Manual, 1, 5, 29.
[13]  Fotadar UI, Shapiro LE, Surks MI. Biotechniques, 1991, 10, 171–172.
[14]  Saravis CA, Zamcheck N. J Immunol Methods, 1979, 29, 91–96.
[15]  Rosen A, Ek K, Aman P. J Immunogenet, 1979, 28, 1–11.
[16]  Lonza Group. Appendix B. Agarose Physical Chemistry.
[17]  Raymond S, Weintraub L. Science, 1959, 130, 711.

[18] Hoffman AS. DeRossi D, Kajiwara K, Osada V, Yamauchi A (eds.). Polymer Gels: Fundamentals Und Biomedical Applications. Plenum Press, New York, 1991, 289–297.

[19] Fawcett JS, Morris CJOR. Sep Sci, 1966, 1, 9–26.

[20] Woodrow JE, Seiber JN, Miller GC. J Agr Food Chem, 2008, 56, 2773–2779.

[21] Hjerten S. Arch Biochem Biophys, 1962, Suppl 1, 147–151.

[22] Anderson LE, McClure WO. Anal Biochem, 1973, 51, 173–179.

[23] Hochstrasser DF, Patchornik A, Merril CR. Anal Biochem, 1988, 173, 412–423.

[24] O´Connell PBH, Brady CJ. Anal Biochem, 1976, 76, 63–73.

[25] Hanson JN. Anal Biochem, 1976, 76, 37–48.

[26] Gordon AH (Work TS, Work E (eds.). Laboratory Techniques in Biochemistry and Molecular Biology. North Holland Publ Co, Amsterdam – London, 1975, Vol. 1, part 1.

[27] Jakob H, Leininger S, Lehmann T, Jacobi S, Gutewort S. Ullmann's Encyclopedia of Industrial Chemistry. Wiley-VCH, Weinheim, 2005.

[28] Shi Q, Jackowski G. Hames BD (ed.). Gel Electrophoresis of Proteins: A Practical Approach. Oxford University Press, New York, 1998, 1.

[29] Rüchel R, Steere RL, Erbe EF. J Chromatogr, 1978, 166, 563–575.

[30] Chrambach A. The Practice of Quantitative Gel Electrophoresis. VCH Publishers. Florida, 1985, 265.

[31] Ornstein L. Ann N Y Acad Sci, 1964, 121, 321–349.

[32] Kozulic M, Kozulic B, Mosbach K. Anal Biochem, 1987, 163, 506–512.

[33] Kozulic B, Mosbach K, Pietrzak M. Anal Biochem, 1988, 170, 478–484.

[34] Görg A, Postel W, Westermeier R, Gianazza E, Righetti PG. J Biochem Biophys Methods, 1980, 3, 273–284.

[35] Görg A, Postel W, Westermeier R. GIT Labor Med, 1979, 1, 32–40.

[36] Nishizawa H, Murakami A, Hayashi N, Iida M, Abe Y. Electrophoresis, 1985, 6, 349–350.

[37] Kinzkofer-Peresch A, Patestos NP, Fauth M, Kögel F, Zok R, Radola BJ. Electrophoresis, 1988, 9, 497–511.

[38] Margolis J, Kenrick KG. Biochem Biophys Res Commun, 1967, 27, 68–73.

[39] Wright GL, Farrell KB, Roberts DB. Biochim Biophys Acta, 1973, 295, 396–411.

[40] Mahadik SP, Korenovsky A, Rapport MM. Anal Biochem, 1976, 76, 615–633.

[41] Rothe GM. Electrophoresis, 1988, 9, 307–316.

# 1.6 General theory of electrophoresis

1.6.1    What is the polyionic mobility depending on? —— 50
1.6.1.1  Influence of polyionic nature —— 50
1.6.1.2  Influence of buffer —— 50
1.6.1.3  Influence of medium —— 52
1.6.1.4  Electroosmosis —— 52
1.6.2    Ionic boundaries —— 52
1.6.2.1  Moving boundary —— 52
1.6.2.2  Stationary boundary —— 53
1.6.3    Regulating function —— 53
1.6.4    Diffusion —— 53
1.6.5    Joule heating —— 53
         References —— 54

The most widely used theory of electrophoresis was developed in 1903 by Smolu-chowski [1, 2]. According to it

$$\mu = \frac{\zeta\varepsilon}{\eta} = \frac{\zeta\varepsilon_r\varepsilon_0}{\eta} \tag{1.6.1}$$

where $\mu$ (in m$^2$/(s V)) is the mobility of the polyion $pi$, $\zeta$ (in V) is its zeta potential (i.e., the electrokinetic potential at the slipping plane of the polyionic double elec-tric layer), $\varepsilon$ (in F/m) is the (di)electric permittivity of the medium, $\varepsilon_r$ (dimension-less) is the relative (di)electric permittivity of the medium, $\varepsilon_0$ is the (di)electric constant (8.854 187 818 × 10$^{-12}$ F/m), and $\eta$ (in Pa s) is the dynamic viscosity of the medium.

The electrophoretic velocity $v$ (in m/s) of a polyion can be calculated from the equation

$$v = \mu E \tag{1.6.2}$$

hence, it is a product of the polyionic mobility and the strength (intensity) of the electric field $E$ (in V/m). The last equation shows that the electrophoretic velocity depends on factors, which affect the polyionic mobility and field strength [3].

If direct voltage is applied between two electrodes, the electric field

$$E = \frac{U}{l} = \frac{J}{\gamma} \tag{1.6.3}$$

where $U$ is the voltage (in V), $l$ is the distance between the electrodes (in m), $J$ is the electric current density (in A/m$^2$), and $\gamma$ is the specific conductivity of the buffer (in S/m). The distance remains unchanged during electrophoresis; there-fore, the strength of the electric field depends only on the voltage.

https://doi.org/10.1515/9783110761641-007

## 1.6.1 What is the polyionic mobility depending on?

The polyionic mobility depends on the polyionic nature, buffer and medium.

### 1.6.1.1 Influence of polyionic nature

The polyionic nature determines the geometric potential $\varphi$ and the geometric radius of the polyion. The $\varphi$-potential depends on both the electric charge of the polyion and its geometric radius.

According to our ideas [4, 5], the relation

$$A = R + \kappa r^2 \tag{1.6.4}$$

exists, where $A$ (in m) is the electrokinetic radius, $R$ (in m) is the geometric radius of a polyion, $\kappa$ (in m$^{-1}$) is the parameter of Debye–Hückel [6], and $r$ (in m) is the geometric radius of the counter-ion.

Besides, we [7] have proved that the polyionic mobility

$$\mu = \frac{\zeta\varepsilon}{1.5\eta} f(\kappa r) \tag{1.6.5}$$

The function $f(\kappa r)$ is described by the equation

$$f(\kappa r) = \frac{1 + \kappa a}{1 + \kappa r} = \frac{1 + \kappa r + (\kappa r)^2}{1 + \kappa r} \tag{1.6.6}$$

where $a$ and $r$ are the electrokinetic and geometric radii of the counter-ion, respectively. The function $f(\kappa r)$ is similar to the Henry's function $f(\kappa a)$ [8, 9], however, in contrast to the Henry's function $f(\kappa a)$, it has a real and definite value.

If we introduce eq. (1.6.6) in (1.6.5), the following expression is obtained:

$$\mu = \frac{\zeta\varepsilon}{1.5\eta} \frac{1 + \kappa a}{1 + \kappa r} \tag{1.6.7}$$

### 1.6.1.2 Influence of buffer

The buffer influences the polyionic mobility mainly by its pH value and its ionic strength, as well as by its (di)electric permittivity and temperature.

*Buffer pH.* A buffer keeps the pH value, i.e. the proton concentration of a solution, unchanged. For strong electrolytes, which are completely ionized, the pH value does not influence the ionic mobility. However, in weak-acid and weak-base buffers, as well as in ampholyte buffers, the ionic mobility is a function of the dissociation degree of the buffer electrolyte, according to the equation

$$\mu_{i(pi)}' = \alpha\mu_{i(pi)} \tag{1.6.8}$$

In this equation, $\mu_{i(pi)}'$ is the effective mobility of the ion $i$ (polyion $pi$), in $m^2/(s\ V)$; $\alpha$ is the dissociation degree; and $\mu_{i(pi)}$ is the mobility of the ion (polyion) of the weak protolyte in an indefinitely diluted solution.

*Ionic strength of buffer.* The polyionic mobility decreases when the ionic strength of a buffer increases [10]. This can be explained with the concept of the existing of geometric and electrokinetic radius (see above). When the ionic strength increases, the product $\kappa r_{pi}$ grows, the electrokinetic radius of the polyion $a_{pi}$ is enlarged, and the $\zeta$-potential is decreased. The reverse takes place when the ionic strength decreases. Then, the electrokinetic radius of the polyion decreases and its $\zeta$-potential increases (Figure 1.6.1).

$\mu_{pi}\times10^9, m^2/(s\ V)$

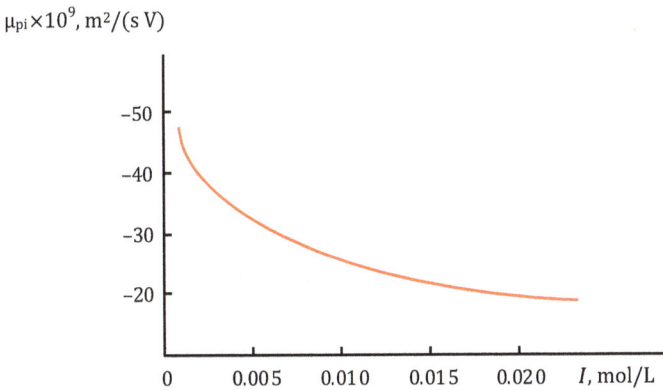

**Figure 1.6.1:** Dependence of polyionic mobility on buffer ionic strength.

*(Di)electric permittivity of buffer.* The relative (di)electric permittivity $\varepsilon_r$ of the buffer solvent (water) remains constant during the electrophoresis. It depends on the temperature: at 0 °C, $\varepsilon_r$ is equal to 88.00; and at 25 °C, it is equal to 78.64.

*Temperature of buffer.* The temperature decreases the (di)electric permittivity $\varepsilon$ of the solvent and its dynamic viscosity $\eta$. It has been found that the temperature decreases the dynamic viscosity according to the equation

$$\eta = Ae^{b/T} \tag{1.6.9}$$

where $A$ and $b$ are empirical constants. Studies indicate that a temperature increase of 1 K increases the ionic mobility of about 3%.

### 1.6.1.3 Influence of medium

The medium influences the electrophoresis because of its dynamic viscosity and electroosmosis.

### 1.6.1.4 Electroosmosis

Electroosmosis (electroendosmosis, **e**lectro**o**smotic flow, EOF) is an essential component of the electrophoresis techniques. It is referred to as moving of the solvent (water) across a porous material, a capillary tube, a microchannel, or any conduit filled with electrolyte solution, when electric field is applied. Electroosmosis was first described in 1809 by the Russian scientist Reuss [11].

Electroosmosis is an important characteristic of the agarose gel. The agarose gel has acidic groups, which carry negative electric charges in neutral and alkaline buffers. It is the immobile medium (stationary phase) but its counter-ions, which are positively charged, move, together with their hydration envelopes (mobile phase), to the cathode. As a result, the gel cathode stacks water (hydrates), whereas the gel anode loses water (dehydrates).

## 1.6.2 Ionic boundaries

Two ionic boundaries in electrophoresis are known: moving and stationary boundary.

### 1.6.2.1 Moving boundary

The moving ionic boundary [12–14] across a buffer system is characterized by the magnitude $W$, which expresses its volume $V$ swept by the passed electric charge $Q$ [15]:

$$W = \frac{V}{Q} \ (\mathrm{m^3/C}) \tag{1.6.10}$$

The relationship between the magnitude $W$ and the effective velocity $v'$ is described by the equation

$$v' = W\frac{I}{S} \tag{1.6.11}$$

where $I$ (in A) is the total electric current, and $S$ (in $\mathrm{m^2}$) is the cross section.

### 1.6.2.2 Stationary boundary

The stationary ionic boundary [16] does not move in a buffer system. It can also be obtained in a same buffer if the buffer parts have different pH values [17].

## 1.6.3 Regulating function

The regulating function of Kohlrausch [18], valid for strong and weak mono- and multivalent electrolytes, can be expressed by the relationship [19]

$$\omega(x) = \sum_i \frac{c_i'(x)z_i}{\mu_i} = \text{const}(x) \tag{1.6.12}$$

where $c_i'$ is the effective ionic concentration of electrolyte $i$, present at a given point $x$ along the migration path. The function $\omega(x)$ has a constant value independent of the time of the electric current passage. Hence, the composition of the moving boundary is regulated automatically.

## 1.6.4 Diffusion

With the time, the resolved bands are broadened by diffusion. After a certain time $t$, a Gaussian concentration profile is observed as the result of diffusion. It may be characterized [20] by the standard deviation $\sigma$ given by the relationship

$$\sigma = \sqrt{2Dt} \ \ [m] \tag{1.6.13}$$

where $D$ is the *diffusivity* or diffusion coefficient (in $m^2/s$).

## 1.6.5 Joule heating

The Joule heating (the electric power) that is contained in the equation

$$P = IU = I^2R \ \ [W, \text{watts}] \tag{1.6.14}$$

is also important for the electrophoresis. The heating causes convection currents, diffusional broadening, evaporation, viscosity and pH changes, thermal denaturing of polyions, especially proteins, gel drying, and even fire. The temperature difference between the center and the outer parts of the resolving medium can reach more than 10 °C. This can be diminished by cooling systems, or using dilute buffers and low voltages.

## References

[1] Smoluchowski M. Bull Int Acad Sci Cracovie, 1903, 3, 184–199.
[2] Smoluchowski M. In Graetz L (ed.). Handbuch Der Elektrizität Und Des Magnetismus. Barth, Leipzig, 1921, Vol. 2, 366–428.
[3] Michov BM. Elektrophoresis. Theory and Practice. Walter de Gruyter, Berlin – Boston, 2020.
[4] Michov BM. Electrophoresis, 1989, 10, 16–19.
[5] Michov BM. Electrochim Acta, 2013, 108, 79–85.
[6] Debye P, Hückel E. Phys Z, 1923, 24, 185–206.
[7] Michov BM. Electrophoresis, 1988, 9, 199–200.
[8] Henry DC. Proc Roy Soc, 1931, A133, 106–129.
[9] Henry DC. Trans Faraday Soc, 1948, 44, 1021–1026.
[10] Bahga SS, Bercovici M, Santiago JG. Electrophoresis, 2010, 31, 910–919.
[11] Reuss FF. Mem Soc Imperiale Naturalistes De Moscow, 1809, 2, 327–337.
[12] Longsworth LG. J Am Chem Soc, 1945, 67, 1109–1119.
[13] Svensson H. Acta Chem Scand, 1948, 2, 841–855.
[14] Brausten I, Svensson H. Acta Chem Scand, 1949, 2, 359–373.
[15] Dole VV. J Am Chem Soc, 1945, 67, 1119–1126.
[16] Alberty RA. J Am Chem Soc, 1950, 72, 2361–2367.
[17] Michov BM. Electrophoresis, 1989, 10, 686–689.
[18] Kohlrausch F. Ann Phys Chem, 1897, 62, 209–239.
[19] Dismukes EB, Alberty RA. J Am Chem Soc, 1954, 76, 191–197.
[20] Giddings JC. Sep Sci, 1969, 4, 181–189.

# 1.7 Electrophoresis instrumentation

1.7.1    Electrophoresis cells —— 56
1.7.2    Power supplies —— 56
1.7.3    Thermostats —— 56
1.7.4    Scanners and densitometers —— 56
1.7.5    Gel casting cassettes —— 57
1.7.6    Gradient makers —— 57
1.7.7    Buffer mixers —— 58
1.7.8    Blotters —— 59
1.7.9    Equipment for semi-automatic electrophoresis —— 59
         References —— 60

A modern electrophoresis labor has at its disposal: electrophoretic cell, DC power supply, thermostat, scanner or densitometer, and computer, casting cassettes, gradient maker, blotter, and other devices (Figure 1.7.1).

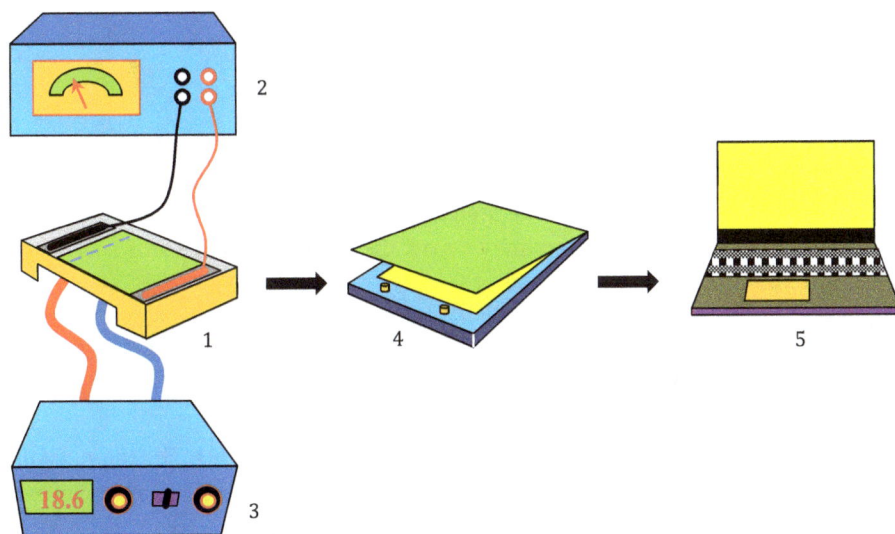

**Figure 1.7.1:** Apparatuses used for electrophoresis.
1. Electrophoresis cell; 2. power supply; 3. thermostat; 4. scanner; 5. computer.

Other laboratory devices and equipment that are used for electrophoresis research are: Becher glasses, Erlenmeyer flasks, graduated cylinders, test tubes, glass rods, pipettes, automatic pipettes, pump, heating plate, magnetic stirrer, magnetic rods, height-adjustable platforms (laboratory boy), shaker, centrifuge, scissors, spatula, microwave oven, dryer or incubator, UV lamp, fan, water pump, paper cutting machine, and more.

https://doi.org/10.1515/9783110761641-008

### 1.7.1 Electrophoresis cells

The electrophoretic separation is carried out in electrophoretic cells. The electrophoretic cells are prepared by isolate materials such as ceramic, polycarbonate, Plexiglas, PVC, Piacryl, and glass. Platinum wires or graphite rods are used as electrodes. For protection, the electric current is automatically stopped in most cases when opening the cell.

The gels have a slab or cylindrical form. The slab gels, either vertical or horizontal, are much more common. They are cast between two glass plates separated by spacer strips. The electrophoretic cells for slab gels are preferable if compared with the electrophoretic cells for cylindrical gels, because the slab gels are easier to be analyzed by a densitometer or scanner and are suitable for immunoelectrophoresis and autoradiography. In addition, in slab gels, the separated polyions are faster stained and the slab gels can be dried as evidence.

There are two types of slab-gels electrophoresis cells: for horizontal and vertical electrophoresis.

### 1.7.2 Power supplies

There are different types of power supplies. A power supply suitable for all types of electrophoresis should produce a voltage of 50–5,000 V, and a direct electric current of 5–100 mA. For the pulsed-field electrophoresis, a control unit is connected to the power supply, which alternately drives the electrodes in a north/south and east/west direction.

### 1.7.3 Thermostats

The cooling of an electrophoretic cell is carried out by thermostats. The thermostats control the desired temperature of a liquid and pump it through the electrophoresis cell. Thus, the temperature inside the electrophoretic cell remains constant.

### 1.7.4 Scanners and densitometers

In many cases, it is sufficient to detect the presence or absence of stained bands or spots of the separate polyions with the naked eye. However, it is impossible to determine with the naked eye the exact differences between the intensities (concentrations) of the bands or spots. This can be carried out with the aid of densitometers.

In the analytical chemistry, the term **o**ptical **d**ensity (OD, dimensionless) is used. It is different from the extinction [1]. OD shows that the absorbance of a solution increases linearly with the concentration of the solution. The extinction values are calculated from either a reflected or transmitted radiation, depending on whether the measurement is carried out before or behind the medium.

Today, scanners are mostly used for analyzing the polyionic fractions obtained by electrophoresis. The pherograms are scanned and the scans are "read" using appropriate computer programs.

## 1.7.5 Gel casting cassettes

A casting cassette is constructed by two glass plates and a U-shaped spacer (Figure 1.7.2), which are held together by brackets.

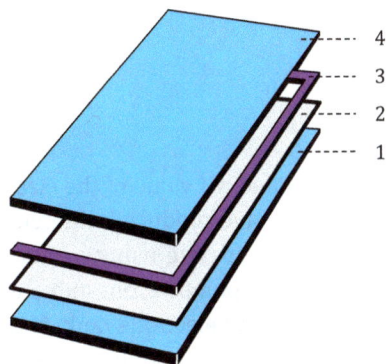

**Figure 1.7.2:** Assembly of a cassette for casting thin and ultrathin gels.
1. Lower glass plate; 2. support film; 3. U-shaped spacer; 4. upper glass plate.

Gel thickness can be varied by spacers inserted between the glass plates prior to the gel formation. The spacers are made of rubber or silicone. They can be glued to one of the glass plates. The distance between the glass plates can also be determined by Parafilm layers (0.12 mm thick each). There are also casting cassettes with mounted 0.5 mm thick gaskets (for example, of Desaga, Heidelberg).

## 1.7.6 Gradient makers

A simple and widespread gradient maker consists of two communicating tubes: a mixing chamber and a reservoir, located at the same level (Figure 1.7.3).

The mixing chamber contains a magnetic bar; the reservoir contains a compensation bar. The compensation bar in the reservoir compensates the volume of the magnetic bar in the mixing chamber and the increase of its solution volume created during the twisting of the magnetic bar in the mixing chamber. Thus, the two solutions remain on the same level during the casting, and the solution in the mixing chamber does not flow back into the reservoir. The magnetic bar is driven by a motor.

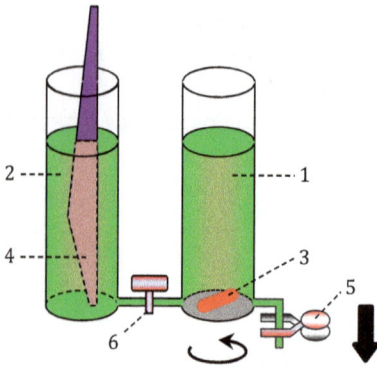

**Figure 1.7.3:** Gradient maker.
1. Mixing chamber; 2. reservoir; 3. magnetic bar;
4. compesation bar; 5. outlet clamp; 6. pinchcock.

A simple gradient maker, constructed by tubes, was described [2]. It is used to generate linear, exponential, and other concentration gradients.

## 1.7.7 Buffer mixers

During electrophoresis, changes in the ionic composition and pH value of the electrode buffers occur. To avoid this, numerous companies have produced different electrode buffer mixers (recirculators). Most recirculators operate using a same principle: a pump sucks buffer from one electrode tank and pumps it out into another electrode tank; after mixing the buffer returns to the first electrode tank, and so on [3, 4]. These apparatuses depend on electric power and are too sophisticated.

We have proposed a water-driven buffer recirculator [5]. Its construction is presented on Figure 1.7.4.

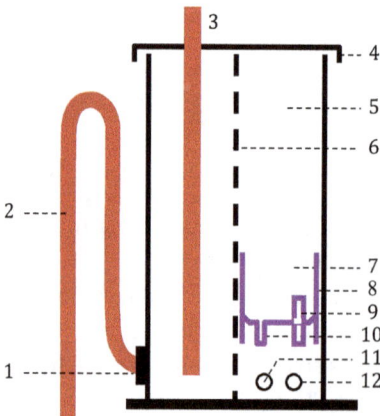

**Figure 1.7.4:** Scheme of recirculator.
1. Big aperture; 2. syphon; 3. hose; 4. recirculator lid; 5. recirculator vessel; 6. fin; 7. float; 8. rib; 9. inlet tube; 10. outlet tube; 11 and 12. small apertures.

The water fills up this recirculator through the hose and flows out through its syphon. This makes the float move up and down. When the float is in the lower part of the recirculator it is filled through the inlet tubing with electrode buffer. When it is in the upper part of the recirculator, it feeds the buffer through the outlet tubing into the electrophoretic device.

If the vessels of the electrophoretic device are one above the other, as in the vertical electrophoretic devices, the recirculating buffer mixes with the buffer in the upper vessel, spills over into the lower vessel and passes into the recirculator float. If the two electrode vessels are on the same level, as in the horizontal electrophoretic devices, the circulating buffer is poured out into one of the vessels, then it passes through tubing into the second vessel and afterwards in the recirculator float.

The quantity of the recirculating buffer and the rate of recirculation can be regulated by changing the water flow.

## 1.7.8 Blotters

There are different blotters that are used for various blotting techniques, such as capillary, vacuum, tank, or semidry blotters (see *Blotting*).

## 1.7.9 Equipment for semi-automatic electrophoresis

In addition to the classic electrophoresis devices, semi-automatic electrophoresis systems have been developed. The main advantages of a semi-automatic system are the diminished work, the fast and reproducible determination of the results, and their easy documentation.

The first semi-automated electrophoresis system was PhastSystem [6] (Pharmacia Biotech, Uppsala). It consists of a horizontal separation cell with an integrated thermostat, a power supply, and a staining chamber. The electric current, the temperature in the cell, and the staining can be programmed. The staining chamber can be warmed up to 50 °C and filled with staining solutions using a diaphragm pump. In addition, the system contains a device for rotating the gels in the solutions. It contains also a blotting unit with graphite electrodes for electrophoretic transfer.

Devices for preparative electrophoresis have also been developed [7]. Using them, the electrophoretic fractions are collected in separate vessels. Thus, the biomolecules can be separately analyzed.

## References

[1]   IUPAC. Compendium of Chemical Terminology, 2nd edn, 1997, Online corrected version, 2006.

[2]   Michov BM. Anal Biochem, 1978, 86, 432–442.

[3]   Pharmacia Gel Electrophoresis System. Pharmacia Fine Chemicals AB, Uppsala, Sweden, 1978.

[4]   LKB Systems and Methods for Biochemical Research. LKB-Produkter AB, Bromma, Sweden, 1976.

[5]   Michov BM. Biotech Bioeng, 1979, 21, 2147–2148.

[6]   Olsson I, Axiö-Fredriksson UB, Degerman M, Olsson B. Electrophoresis, 1988, 9, 16–22.

[7]   Chrambach A. The Practice of Quantitative Gel Electrophoresis. VCH, Weinheim – Deerfield Beach, 1985.

# 1.8 Classification of electrophoretic methods

1.8.1     Zone electrophoresis —— 62
1.8.1.1   Tiselius electrophoresis —— 62
1.8.1.2   Capillary electrophoresis —— 62
1.8.1.3   Free-flow electrophoresis —— 62
1.8.1.4   Solid media electrophoresis —— 62
1.8.1.5   Cellulose acetate electrophoresis —— 63
1.8.1.6   Agarose gel electrophoresis —— 63
1.8.1.7   Pulsed-field electrophoresis —— 63
1.8.1.8   Immunoelectrophoresis and immunofixation —— 63
1.8.1.9   Affinity electrophoresis —— 64
1.8.1.10  Polyacrylamide gel electrophoresis —— 64
1.8.1.11  Iontophoresis —— 64
1.8.2     Isotachophoresis —— 64
1.8.2.1   Disc-electrophoresis —— 64
1.8.2.2   Native disc-electrophoresis —— 65
1.8.2.3   SDS disc-electrophoresis —— 65
1.8.3     Isoelectric focusing —— 65
1.8.3.1   Isoelectric focusing with carrier ampholytes —— 65
1.8.3.2   Isoelectric focusing in immobilized pH gradients —— 65
1.8.3.3   Two-dimensional electrophoresis —— 66
1.8.4     Dielectrophoresis —— 66
1.8.4.1   Dielectrophoretic force —— 66
1.8.4.2   DEP technology —— 67
1.8.4.3   Applications of dielectrophoresis —— 68
          References —— 68

Depending on the gel position, vertical and horizontal electrophoresis can be distinguished; depending on the volume of the sample to be resolved, analytical and preparative electrophoresis exist.

*Vertical and horizontal electrophoresis.* The vertical electrophoresis can be performed in slab or cylindrical gels, while the horizontal electrophoresis can be carried out only in slab gels. The horizontal electrophoresis offers numerous advantages over the vertical electrophoresis: easier handling; usage of electrode strips soaked with buffer instead of tank buffers; selection of application points, which is important for native electrophoresis and isoelectric focusing (IEF); free access to the gel surface during electrophoresis, *etc.*

*Analytical and preparative electrophoresis.* Most electrophoresis methods are analytical. However, there are also methods that allow a preparative separation of polyions. The IEF can also be performed analytically or preparative.

According to the buffer, all electrophoresis methods can be classified as: zone electrophoresis, isotachophoresis (ITP), and IEF. There are also combinations between these methods.

https://doi.org/10.1515/9783110761641-009

## 1.8.1 Zone electrophoresis

The zone electrophoresis is known also as continuous or conventional electrophoresis. It is carried out in one buffer, i.e. in an electric field of continuous strength, whose pH value remains constant during the electrophoresis. The zone electrophoresis can take place free (only in buffer) or in a solid medium soaked with buffer.

### 1.8.1.1 Tiselius electrophoresis

The classic Tiselius electrophoresis [1] is a free zone electrophoresis. The polyions are mixed with a buffer, placed in a U-shaped glass tube, and overlaid with the same buffer. When an electric voltage is applied, they separate from each other forming zones (layers) in the two legs of the tube.

The separated colorless proteins are detected using a Schlieren-scanning system. It shows the streaks refraction shadows formed by the analytes in the electrophoresis tube, which in turn are focused onto a photographic plate. Today Tiselius electrophoresis has little practical usage.

### 1.8.1.2 Capillary electrophoresis

The capillary electrophoresis (CE), named also high performance capillary electrophoresis, is carried out in capillaries or micro- and nanofluidic channels [2, 3]. Commonly, CE refers to capillary zone electrophoresis, but capillary gel electrophoresis, capillary IEF, capillary isotachophoresis, and micellar electrokinetic chromatography also belong to it [4]. In the capillary isotachophoresis the polyions are separated from each other by spacers.

### 1.8.1.3 Free-flow electrophoresis

In the free-flow electrophoresis (FFE) [5], a buffer flows perpendicular to an electric field, and the electrophoretic fractions are collected separately. FFE is used for separation of polyions, organelles, and cells. Free-flow isotachophoresis and free-flow isoelectric focusing are also known.

### 1.8.1.4 Solid media electrophoresis

The electrophoresis can be run also in solid media that contain buffers in their pores. Diverse media are known: paper, starch, cellulose acetate, agarose, and polyacrylamide. Today, the agarose and polyacrylamide gels are the most widespread electrophoresis media. Their pores can act as a sieve. So, the polyions are separated according to their electric charges, as well as to their volumes. As a result, the electrophoresis resolution increases.

### 1.8.1.5 Cellulose acetate electrophoresis

The cellulose acetate electrophoresis [6, 7] is the simplest electrophoretic method. The cellulose acetate membranes have large pores; therefore, the electrophoretic separation is only charge dependent. Nowadays, it is almost totally replaced by the agarose and polyacrylamide gels electrophoresis.

### 1.8.1.6 Agarose gel electrophoresis

The **a**garose **g**el **e**lectrophoresis (AGE) [8, 9] is widespread. Depending on the polyion nativity, two types of AGE are known: electrophoresis of native polyions, i.e., of polyions with intact structure and properties; and electrophoresis of denatured polyions, i.e., of polyions with altered structure and properties.

### 1.8.1.7 Pulsed-field electrophoresis

The AGE is the preferred method for resolving DNA fragments of about 1,000–23,000 bp. For larger fragments, for example, chromosomes, the **p**ulsed-**f**ield **g**el **e**lectrophoresis (PFGE) is used [10]. In the pulsed-field gel electrophoresis, DNA fragments greater than 23 kbp are forced by a pulsing electric field so that they relax and expand. The separation takes place many hours or even days.

### 1.8.1.8 Immunoelectrophoresis and immunofixation

The **immuno**electrophoresis (IE) is a combination between a zone electrophoresis in an agarose gel and immune reactions. The immunoglobulins are located in the agarose gel. The **immuno**fixation (IF) is an immunoelectrophoretic method, during which immunoglobulins are applied on the gel surface after the electrophoresis.

### 1.8.1.9 Affinity electrophoresis

The **affinity electrophoresis** (AE) is a zone electrophoresis carried out in a ligands containing agarose gel [11, 12]. It is based on the interactions between chelating compounds and their ligands.

### 1.8.1.10 Polyacrylamide gel electrophoresis

The polyacrylamide gel, in contrast to the agarose gel, has no electric charges and, as a result, shows no electroosmosis. The **polyacrylamide gel electrophoresis** (PAGE) [13, 14] is a zone electrophoresis that is carried out for separation of native proteins and nucleic acids.

### 1.8.1.11 Iontophoresis

The **iontophoresis** (IP) is a type of zone electrophoresis, during which ions are introduced into the human body. This chemical flux is measured commonly in $\mu mol/cm^2 h$. The reverse IP is a technique, which helps to remove ions from the human body.

## 1.8.2 Isotachophoresis

The term **isotachophoresis** (ITP) originates from the Greek words *isos* (same), *tachos* (speed), and *phorein* (to carry). It means moving of ions at the same velocity.

ITP takes place in a buffer system consisting of leading and trailing buffers, which form the moving ionic boundary (function) of Kohlrausch [15, 16]. The poly-ions arrange themselves in the moving boundary according to their effective mobilities [17, 18]. In contrast to the continuous electrophoresis, the electric field strength is distributed discontinuously and the pH value changes during the electrophoresis.

### 1.8.2.1 Disc-electrophoresis

There are native and denatured disc-electrophoresis methods.

## 1.8.2.2 Native disc-electrophoresis

The native disc-electrophoresis is a combination between ITP and zone electrophoresis. In the beginning of electrophoresis, the polyions are concentrated into a very thin stack, which is then resolved in fractions. It is based on the Ornstein theory [19].

## 1.8.2.3 SDS disc-electrophoresis

Typical denatured disc-electrophoresis is the **s**odium **d**odecyl **s**ulfate (SDS) electrophoresis. It resembles a native disc-electrophoresis carried out in a SDS containing gel buffer system. SDS denatures the proteins and binds to them so that they become strongly negatively charged. Since the ratio of charge/mass is similar for all proteins, the proteins separate from each other only according to their masses.

As the native PAGE, SDS disc-electrophoresis can be also performed in gradient gels. The gradient gels sharpen the protein bands because they separate from each other according to their volume and conformation.

# 1.8.3 Isoelectric focusing

The **i**soelectric **f**ocusing (IEF) is based on the migration of proteins through a pH gradient until they reach their isoelectric points. The pH gradients may be formed by mobile zwitterion ampholytes (carrier ampholytes) or immobile zwitterion ampholytes (immobilines).

## 1.8.3.1 Isoelectric focusing with carrier ampholytes

The carrier ampholytes are moving in an electric field until they lose their electric charges at their isoelectric points and stop. As a result, stationary pH gradients are formed, in which analyzed substances can be separated and focused into narrow bands. Usually IEF with carrier ampholytes is carried out in polyacrylamide gels. Agarose gels are also used for IEF, because they have large pores [20, 21].

## 1.8.3.2 Isoelectric focusing in immobilized pH gradients

The immobilines are acrylamide derivatives, which contain buffering groups. They can be copolymerized into polyacrylamide gels, so that **i**mmobilized **pH g**radients (IPG) are produced. In these gels IEF can also be run.

### 1.8.3.3 Two-dimensional electrophoresis

The two-dimensional electrophoresis (2D-electrophoresis) represents a combination between IEF and SDS disc-electrophoresis. They are carried out sequentially in two mutually perpendicular directions [22, 23]. Depending on the pH gradient, there are two types of 2D-electrophoresis: ampholyte 2D-electrophoresis, and immobiline 2D-electrophoresis.

## 1.8.4 Dielectrophoresis

The term *dielectrophoresis* (DEP) was first adopted by Pohl [24]. It contains the Greek words *di* (two), *electron* (amber, hence electricity), and *phorein* (to carry). Nowadays DEP is referred to as a method for separating dielectric particles in a non-uniform electric field [25–27]. The particles are not required to be charged. Many articles are dedicated on the theory, technology, and applications of DEP. Prominent among them is the review article by Ronald Pethig [28].

DEP is based on the fact that an inhomogeneous electric field polarizes all particles into *dipoles*, according to Maxwell [29, 30] and Hatfield [31], creating dipole moments [32]. The magnitude of the force exerted on the dielectric particles depends on their electric properties, shape, and volume; on the medium; and on the frequency of the electric field.

If the particles are moving in the direction of an increasing electric field, the DEP is referred to as positive DEP; if the particles are moving in the direction of a decreasing electric field, it is referred to as negative DEP.

### 1.8.4.1 Dielectrophoretic force

According to most scientists [33, 34], the time-average DEP force $F_{DEP}$, which acts on a spherical particle, is

$$F_{DEP} = 2\pi\varepsilon r^3 F_{CM}\nabla E^2 \tag{1.8.1}$$

where $\varepsilon$ is the permittivity of the surrounding medium (equal to the product of the relative permittivity $\varepsilon_r$ and the (di)electric constant $\varepsilon_0$); $r$ is the particle radius; $F_{CM}$ is the Clausius–Mossotti factor [35, 36] related to the polarizability of the particle; $\nabla$ represents the gradient operator; and $E$ is the electric field strength. As seen, $F_{DEP}$ depends on the square of the applied electric field strength, which indicates that this process can be observed using either direct-current or alternative-current field.

Most particles, especially biological ones, are not homogeneous. Bacteria and cells have so-called multishell structure [37]. For example, erythrocytes, which have about 7 μm diameter, can be represented as a cytoplasm (first shell) surrounded by a thin spherical membrane (second shell). Other cells, which have nuclei, such as leukocytes, have a three-shell model (the first shell is the nucleus; the second shell is the cytoplasm; and the third shell is the plasma membrane).

In accordance with this theory, the permittivity of a cell (e.g., human erythrocyte) is given by the expression

$$\varepsilon_{cell} = \varepsilon_{mem} \frac{\left(\frac{r_{mem}}{r_{cyt}}\right)^3 + 2\frac{\varepsilon_{cyt} - \varepsilon_{mem}}{\varepsilon_{cyt} + 2\varepsilon_{mem}}}{\left(\frac{r_{mem}}{r_{cyt}}\right)^3 - \frac{\varepsilon_{cyt} - \varepsilon_{mem}}{\varepsilon_{cyt} + 2\varepsilon_{mem}}} \tag{1.8.2}$$

where cyt is the index for the cytoplasm, mem is the index for the plasma membrane, $r_{cyt}$ is the radius of the cytoplasm, and $r_{mem}$ is the radius of the plasma membrane.

### 1.8.4.2 DEP technology

The developing of the DEP technology includes the fabrication of microelectrodes, introduction of silicone polymers, fabrication of microfluidics devices for DEP, and developing "funnel" electrodes.

The *microelectrodes* are fabricated by photolithography and metallic vapor. They have interdigitated and castellated geometry, and use modest voltage [38]. In older techniques (Figure 1.8.1a), the particles are directed into flow paths [39]. Later Yasukawa *et al.* [40] modified this method to separate particles according to their size (Figure 1.8.1b). By coupling acoustic waves into an interdigitated microelectrode system, particles can first be concentrated and then focused into flow channels. These electrodes can be used for both positive and negative DEP of cells [41].

The *introduction of silicone polymers* was a new DEP trend. Cheng *et al.* [42] described a technique they termed molecular DEP, in which a cusp-shaped silica nanocolloid, functionalized with a DNA probe, was used to create a local field gradient for detecting picomolar single-stranded DNA within 1 min.

The *microfluidic* devices for DEP were made by film techniques, photo and electron beam lithography, laser ablation, complementary metal-oxide-semiconductor technology [43, 44], scanning force DEP [45], and assembling the nanoparticles in two or more differing surfaces [46].

"Funnel" electrodes were developed by Fiedler *et al.* [47] (Figure 1.8.2). With their help, the particles were guided by angled electrodes to a small exit gap, at which the particles were concentrated prior to be introduced into the device. Angled electrodes are employed in a DEP microchip design for filtering and sorting of bioparticles [48–50].

*a*

*b*

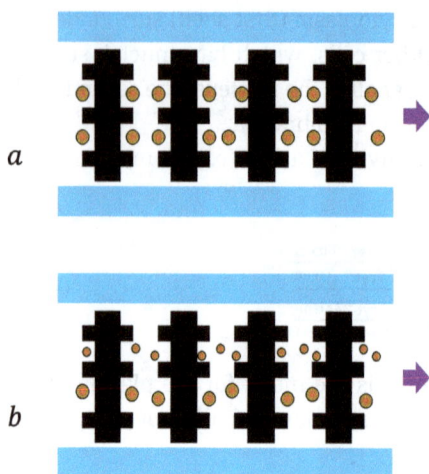

**Figure 1.8.1:** Microelectrodes.
*(a)* Particles focused into narrow bands of flow in an interdigitated and castellated electrode system.
*(b)* A modified interdigitated and castellated design for separating particles according to their size into separate fluid flow streams.

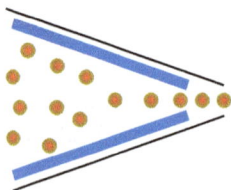

**Figure 1.8.2:** DEP funnel electrode design for concentrating particles in a flowing aqueous suspension.

### 1.8.4.3 Applications of dielectrophoresis

DEP is applied in medical diagnostics, particle filtration, biosensors, nanoassembly, microfluidics, cell therapeutics, and more. It has made possibly the separation of cancer cells [51], proteins [52, 53], DNA [54–56], chromosomes [57], red and white blood cells [58–60], stem cells [61–63], neurons [64, 65], pancreatic β-cells [66], bacteria and yeast [67–69], and viruses [70, 71]. DEP can also be used for drug discovery and deliver [72], and for detecting apoptosis by measuring the changes in the electrophysiological properties of the cells [73].

## References

[1]   Tiselius A. Trans Faraday Soc, 1937, 33, 524–531.
[2]   Hjerten S. J Chromatogr, 1983, 270, 1–6.
[3]   Ewing AG, Wallingford RA, Olefirowicz TM. Anal Chem, 1989, 61, 292A–303A.
[4]   Kemp G. Biotech Appl Biochem, 1998, 27, 9–17.
[5]   Hannig K, Heidrich HG. Free-Flow Electrophoresis. GIT, Darmstadt, 1990.
[6]   Kohn J. Clin Chim Acta, 1957, 2, 297–304.
[7]   Leslie P, Bonner R, Vandevalle I, Olsson R. Electrophoresis, 1983, 4, 318–319.
[8]   Hjerten S. Biochim Biophys Acta, 1961, 53, 514–517.

[9]   Hjerten S. Biochim Biophys Acta, 1962, 62, 445–449.
[10]  Schwartz DC, Cantor CR. Cell, 1984, 37, 67–75.
[11]  Swallow DM. In Isozymes: Curr Top Biol Med Res, 1977, 1, 159.
[12]  Takeo K. Electrophoresis, 1984, 5, 187–195.
[13]  Raymond S, Weintraub L. Science, 1959, 130, 711.
[14]  Hjerten S. Arch Biochem Biophys, Suppl, 1962, 1, 147–151.
[15]  Kohlrausch F. Ann Phys Chem, 1897, 62, 209–239.
[16]  Beckers JL, Bocek P. Electrophoresis, 2000, 21, 2747–2767.
[17]  Everaerts FM, Beckers JL, Verheggen TPEM. Isotachophoresis: Theory, Instrumentation and Applications. Elsevier, Amsterdam, 1976.
[18]  Bocek P, Deml M, Gebauer P, Dolnik V. Analytical Isotachophoresis. VCH Verlagsgesellschaft, Weinheim, 1988.
[19]  Ornstein L. Ann N Y Acad Sci, 1964, 121, 321–349.
[20]  Rosen A, Ek K, Aaman P. J Immunol Methods, 1979, 28, 1–11.
[21]  Saravis CA, O'Brien M, Zamcheck N. J Immunol Methods, 1979, 29, 91–96.
[22]  O'Farrell PH. J Biol Chem, 1975, 250, 4007–4021.
[23]  Klose J. Humangenetik, 1975, 26, 231–143.
[24]  Pohl HA. J Appl Phys, 1951, 22, 869–871.
[25]  Pohl HA. Dielectrophoresis: The Behavior of Neutral Matter in Nonuniform Electric Fields. Cambridge University Press, Cambridge, 1978.
[26]  Jones TB. Electromechanics of Particles. Cambridge University Press, Cambridge, 1995.
[27]  Kirby BJ. Micro- and Nanoscale Fluid Mechanics: Transport in Microfluidic Devices. Cambridge University Press, Cambridge, 2010.
[28]  Pethig R. Biomicrofluidics, 2010, 4, 022811.
[29]  Abraham M. The Classical Theory of Electricity and Magnetism (English Translation of 8th German Edition). Hafner, New York, 1932, 90–91.
[30]  Lo YJ, Lei U. Appl Phys Lett, 2009, 95, 253701–253703.
[31]  Hatfield HS. Trans Inst Min Metall, 1924, 33, 335.
[32]  Jones TB. Electromechanics of Particles. Cambridge University Press, New York, 1995.
[33]  Irimajiri A, Hanai T, Inouye A. J Theoret Biol, 1979, 78, 251–269.
[34]  Jones TB. Electromechanics of Particles. Cambridge University Press, Cambridge, 1995.
[35]  Morgan H, Green N. AC Electrokinetics: Colloids and Nanoparticles. Research Studies Press, Philadelphia, 2002.
[36]  Kirby BJ. Micro- and Nanoscale Fluid Mechanics: Transport in Microfluidic Devices. Cambridge University Press, Cambridge, 2010.
[37]  Irimajiri A, Hanai T, Inouye A. J Theor Biol, 1979, 78, 251–269.
[38]  Price JAR, Burt JPH, Pethig R. Biochim Biophys Acta, 1988, 964, 221–230.
[39]  Becker FF, Wang XB, Huang Y, Pethig R, Vykoukal J, Gascoyne PR. Proc Natl Acad Sci USA, 1995, 92, 860–864.
[40]  Yasukawa T, Suzuki M, Shiku H, Matsue T. Sens Actuators B Chem, 2009, 142, 400–403.
[41]  Pethig R, Huang Y, Wang XB, Burt JPH. J Phys D, 1992, 25, 881–888.
[42]  Cheng IF, Senapati S, Cheng X, Basuray S, Chang HC, Chang HC. Lab Chip, 2010, 10, 828–831.
[43]  Manaresi N, Romani A, Medoro G, Altomare L, Leonardi A, Tartagni M, Guerrieri G. IEEE J Solid-State Circuits, 2003, 38, 2297–2305.
[44]  Castellarnau M, Zine N, Bausells J, Madrid C, Juarez A, Samitier J, Errachid A. Sens Actuators B, 2007, 120, 615–620.
[45]  Lynch BP, Hilton AM, Simpson GJ. Biophys J, 2006, 91, 2678–2686.
[46]  Lumsdon SO, Kaler EW, Velev OD. Langmuir, 2004, 20, 2108–2116.

[47]  Fiedler S, Shirley SG, Schnelle T, Fuhr G. Anal Chem, 1998, 70, 1909–1915.

[48]  Kralj JG, Lis MTW, Schmidt MA, Jensen KF. Anal Chem, 2006, 78, 5019–5025.

[49]  Cheng IF, Chang HC, Hou D, Chang HC. Biomicrofluidics, 2007, 1, 021503.

[50]  Kim U, Qian J, Kenrick SA, Daugherty PS, Soh HT. Anal Chem, 2008, 80, 8656–8661.

[51]  Gascoyne PRC, Wang XB, Huang Y, Becker FF. IEEE Trans Industry Appl, 1997, 33, 670–678.

[52]  Washizu M, Suzuki S, Kurosawa O, Nishizaka T, Shinohara T. IEEE Trans Ind Appl, 1994, 30, 835–843.

[53]  Clarke RW, White SS, Zhou DJ, Ying LM, Klenerman D. Angew Chem, 2005, 117, 3813–3816.

[54]  Asbury CL, Van Den Engh G. Biophys J, 1998, 74, 1024–1030.

[55]  Dalir H, Yanagida Y, Hatsuzawa T. Sens Actuators B, 2009, 136, 472–478.

[56]  Lei KF, Cheng H, Choy KY, Chow LMC. Sens Actuators A, 2009, 156, 381–387.

[57]  Prinz C, Tegenfeldt JO, Austin RH, Cox EC, Sturm JC. Lab Chip, 2002, 2, 207–212.

[58]  Huang Y, Wang XB, Gascoyne PRC, Becker FF. Biochim Biophys Acta, 1999, 1417, 51–62.

[59]  Yang J, Huang Y, Wang XB, Becker FF, Gascoyne PRC. Biophys J, 2000, 78, 2680–2689.

[60]  Cheng IF, Froude VE, Zhu YX, Chang HC, Chang HC. Lab Chip, 2009, 9, 3193–3201.

[61]  Stephens M, Talary MS, Pethig R, Burnett AK, Mills KI. Bone Marrow Transplant, 1996, 18, 777–782.

[62]  Flanagan LA, Lu J, Wang L, Marchenko SA, Jeon NL, Lee AP, Monukii ES. Stem Cells, 2008, 26, 656–665.

[63]  Markx GH, Carney L, Littlefair M, Sebastian A, Buckle AM. Biomed Dev, 2009, 11, 143–150.

[64]  Heida T, Vulto P, Rutten WLC, Marani E. J Neurosci Methods, 2001, 110, 37–44.

[65]  Flanagan LA, Lu J, Wang L, Marchenko SA, Jeon NL, Lee AP, Monukii ES. Stem Cells, 2008, 26, 656–665.

[66]  Pethig R, Jakubek LM, Sanger RH, Heart E, Corson ED, Smith PJS. IEE Proc Nanobiotechnol, 2005, 152, 189–193.

[67]  Li H, Bashir R. Sens Actuators B, 2002, 86, 215–221.

[68]  Bessette PH, Hu XY, Soh HT, Daugherty PS. Anal Chem, 2007, 79, 2174–2178.

[69]  Park S, Beskok A. Anal Chem, 2008, 80, 2832–2841.

[70]  Markx GH, Dyda PA, Pethig R. J Biotechnol, 1996, 51, 175–180.

[71]  Grom F, Kentsch J, Müller T, Schnelle T, Stelzle M. Electrophoresis, 2006, 27, 1386–1393.

[72]  Pethig R. Adv Drug Deliv Rev, 2013, 65, 1589–1599.

[73]  Chin S, Hughes MP, Coley HM, Labeed FH. Int J Nanomed, 2006, 1, 333–337.

# 2 Electrophoresis of proteins

The most commonly used solid media for protein electrophoresis are: paper, starch gel, cellulose acetate, and above all agarose or polyacrylamide gels.

*Paper electrophoresis* [1, 2] has historical meaning. It was replaced by electrophoresis on cellulose acetate membranes. Glass fiber paper can also be applied for electrophoresis, however, it is alkaline and, as a result, causes a strong electroosmotic flow.

*Starch (gel) electrophoresis* was introduced in 1955 by Smithies [3]. Starch gels are prepared from partially hydrolyzed potato starch in a concentration of 8–15 g/dL and in form of 5–10 mm layers. The pore size of the starch gels depends on the starch concentration.

Starch gel electrophoresis is completely detached by the polyacrylamide gel electrophoresis because of the widely varying starch properties, lack of reproducibility, and inconvenience of handling. Nevertheless, it can be used for separations of human enzymes [4], and in population genetics [5, 6] because the starch gel, as a natural product, is an enzyme-friendly medium.

## References

[1]   Wieland T, Fischer E. Naturwissenschaften, 1948, 35, 29–30.
[2]   Cremer HD, Tiselius A. Biochem Z, 1950, 320, 273–283.
[3]   Smithies O. Biochem J, 1955, 61, 629–641.
[4]   Harris H, Hopkinson DA. Handbook of Enzyme Electrophoresis in Human Genetics. North Holland Publishing Co., Amsterdam – Oxford, Elsevier, New York, 1980.
[5]   Ferguson KA. Biochemical Systematics and Evolution. Blackie & Son, Bishopbriggs, Glasgow, 1980.
[6]   Britton-Davidian J. Methods Enzymol, 1993, 224, 98–112.

https://doi.org/10.1515/9783110761641-010

# 2.1 Cellulose acetate electrophoresis of proteins

2.1.1    Theory of cellulose acetate electrophoresis of proteins —— **73**
2.1.2    Practice of cellulose acetate electrophoresis of proteins —— **73**
2.1.3    Protocols —— **74**
         Cellulose Acetate Electrophoresis of Serum Proteins —— **74**
         References —— **75**

Cellulose acetate electrophoresis is zone electrophoresis that is used mainly for analysis of serum proteins and isoenzymes. It takes place in simply designed horizontal cells without refrigeration, and is characterized by simple handling, small sample volumes, short separation times, and rapid staining and destaining.

Cellulose acetate electrophoresis is used for separation of proteins in clinical laboratories.

## 2.1.1 Theory of cellulose acetate electrophoresis of proteins

The cellulose acetate membranes have very large pores and, as a result, no screening effect on proteins to be resolved. The electrophoretic separation depends only on the electric charges of proteins. However, the resultant zones are broader than the start zone due to convection.

The theory of cellulose acetate electrophoresis is the theory of the zone electrophoresis. The electrophoretic separation also depends on the nature of the buffer, the voltage applied, and the distance between the electrodes.

## 2.1.2 Practice of cellulose acetate electrophoresis of proteins

Prior to electrophoresis, the cellulose acetate membrane has to be immersed in a buffer for 2–3 min. The excess buffer on the membrane should be dried between two filter paper sheets and then the CA membrane is spanned on a plastic frame (bridge). The frame is placed into a horizontal electrophoresis cell so that the membrane ends should contact the two electrode buffers. Afterword, the cell should be closed with the lid and undiluted or twice-diluted serum samples are applied onto the cellulose acetate membrane through holes in the lid. The urine and cerebrospinal fluid samples should be first concentrated until the protein concentration reaches 2–3 g/dL. Thereafter, the electrophoresis is started.

Generally, electrophoresis is carried out in the 5,5-diethylbarbiturate buffer of Longworth [1] with a pH value of 8.0 to 9.0, known as barbitalate or veronalate buffer. The barbitalate buffer contains a derivative of barbituric acid (an anesthetic drug).

https://doi.org/10.1515/9783110761641-011

Therefore, it should be replaced by a buffer that contains no barbiturate. Such a buffer is the TRIS-taurinate buffer [2].

To become clear protein bands after the electrophoresis, the background of the CA pherograms should be transparent. This can be achieved when the membrane is soaked in mixtures of dioxane, isobutanol, methanol, or other liquids.

A similar kind of electrophoresis is Cellogel electrophoresis. It is carried out in Cellogel strips. The Cellogel strips contain cellulose, too, but are thicker than the CA membranes. Like the cellulose acetate membranes, the Cellogel strips have very large pores and, therefore, exert no screening effect on the proteins to be resolved.

## 2.1.3 Protocols

### Cellulose Acetate Electrophoresis of Serum Proteins

**Materials and equipment**
Barbitalate buffer (pH = 8.6, $I$ = 0.10 mol/L) or
TRIS-taurinate buffer (pH = 9.0, $I$ = 0.05 mol/L) or
TRIS-glycinate buffer (pH = 9.5, $I$ = 0.025 mol/L)
Cellulose acetate (CA) membranes
Ponceau S
Trichloroacetic acid (TCA)
Acetic acid
Glycerol
Methanol
Dioxane
Isobutanol
Filter paper
Electrophoresis cell
Power supply

***Barbitalate buffer***

| | | |
|---|---|---|
| Sodium barbitalate | 20.62 g | (0.10 mol/L) |
| Barbital | 4.00 g | (0.02 mol/L) |
| Sodium azide | 0.65 g | (0.01 mol/L) |
| Deionized water | 1,000.00 mL | |

***Ponceau S staining solution***

| | |
|---|---|
| Ponceau S | 0.3 g |
| Trichloroacetic acid | 3.0 g |
| Deionized water to | 100.0 mL |

*Clearing solution*

| | |
|---|---|
| Acetic acid | 15.0 mL |
| 87% Glycerol | 0.2 mL |
| Methanol to | 100.0 mL |
| or | |
| Dioxane | 70.0 mL |
| Isobutanol | 30.0 mL |

**Procedure**
- Fill an electrophoresis cell and a Petri dish with an electrophoresis buffer.
- Immerse a CA membrane into the Petri dish for 2–3 min, then dry briefly between filter paper sheets, and place onto the electrophoresis cell bridge.
- Put the bridge with the CA membrane into the electrophoresis cell.
- Cover the electrophoresis cell with its lid.
- Apply the samples through the cell lid, using an applicator onto the cathode side of the membrane.
- Run the electrophoresis at 200–250 V (about 3 mA per membrane) for 25–30 min.
- After the electrophoresis, place the CA membrane into the Ponceau S staining solution for 5 min.
- Destain the membrane three times in 5 mL/dL acetic acid.
- Place the membrane in a Petri with clearing solution for 5 min.
- Place the membrane on a clean glass plate and roll over with a photo roller until the air bubbles under the membrane are removed.
- Clear the membrane at 70 °C for 5 min.
- Evaluate the red-colored protein bands against a control pherogram, or with a densitometer at 530 nm, or using a scanner. Using a computer program, calculate the relative concentrations of the serum proteins.

# References

[1]  Longsworth LG. Chem Rev, 1942, 30, 323–326.
[2]  Michov BM. Protein Separation by SDS Electrophoresis in a Homogeneous Gel Using a TRIS-formate-taurinate Buffer System and Homogeneous Plates Suitable for Electrophoresis. (Proteintrennung durch SDS-Elektrophorese in einem homogenen Gel unter Verwendung eines TRIS-Formiat-Taurinat-Puffersystems und dafür geeignete homogene Elektrophoreseplatten.) German Patent 4127546, 1991.

# 2.2 Agarose gel electrophoresis of proteins

2.2.1    Theory of agarose gel electrophoresis of proteins —— 78
2.2.2    Agarose gel electrophoresis of serum proteins —— 78
2.2.2.1  Albumin —— 80
         Intermediate zone of albumin-$\alpha_1$-globulins —— 80
2.2.2.2  Alpha-1 globulins —— 81
         Intermediate zone of $\alpha_1$-$\alpha_2$-globulins —— 81
2.2.2.3  Alpha-2 globulins —— 82
         Intermediate zone of $\alpha_2$-$\beta$-globulins —— 82
2.2.2.4  Beta-globulins —— 83
         Intermediate zone of $\beta$-$\gamma$-globulins —— 84
2.2.2.5  Gamma-globulins —— 84
2.2.3    Agarose gel electrophoresis of lipoproteins —— 85
2.2.3.1  High-density lipoproteins —— 86
2.2.3.2  Low-density lipoproteins —— 86
2.2.3.3  Intermediate-density lipoproteins —— 86
2.2.3.4  Very-low-density lipoproteins —— 86
2.2.3.5  Chylomicrons —— 86
2.2.3.6  Hyperlipoproteinemias —— 87
2.2.4    Agarose gel electrophoresis of hemoglobins —— 89
2.2.4.1  Normal hemoglobins —— 89
2.2.4.2  Abnormal and pathological hemoglobins —— 90
         Sickle cell disease —— 91
         Thalassemias —— 91
2.2.4.3  Running hemoglobin electrophoresis —— 93
2.2.5    Agarose gel electrophoresis of cerebrospinal fluid proteins —— 94
2.2.6    Electrophoresis of creatine kinase isoenzymes —— 95
2.2.7    Electrophoresis of lactate dehydrogenase isoenzymes —— 96
2.2.8    Agarose gel electrophoresis of urinary proteins —— 97
2.2.9    Protocols —— 98
         Agarose Gel Electrophoresis of Serum Proteins —— 98
         Agarose Gel Electrophoresis of Lipoproteins —— 100
         Agarose Gel Electrophoresis of Hemoglobins —— 102
         Alkaline Agarose Gel Electrophoresis of Hemoglobins —— 102
         Acidic Agarose Gel Electrophoresis of Hemoglobins —— 103
         Agarose Gel Electrophoresis of Cerebrospinal Fluid Proteins —— 104
         Agarose Gel Electrophoresis of Urinary Proteins —— 105
         References —— 107

The agarose gels, compared to the polyacrylamide gels, have larger pores. Therefore, they are suitable for separation of native proteins with very high masses [1]. The agarose gels are simpler in production than the polyacrylamide gels; are nontoxic; and the electrophoresis run is only 20–30 min. The agarose gel electrophoresis is used in research and especially in diagnostics.

https://doi.org/10.1515/9783110761641-012

## 2.2.1 Theory of agarose gel electrophoresis of proteins

The theory of agarose gel electrophoresis of proteins [2–3] does not differ from the general theory of electrophoresis. However, on agarose gels two opposite movements exist: movement of proteins and movement of water (electroosmosis). As a result, $\alpha_1$-, $\alpha_2$-, and $\beta$-globulins are moving to the anode (in front to start line), while $\gamma$-globulins are moving back to the cathode (behind the start line). Nevertheless, electroosmosis helps to separate proteins with small electric charge.

Agarose gels are used generally in concentrations of 0.7–1.2 g/dL. In 1 g/dL agarose gel, particles with $M_r$ up to $50 \times 10^6$ and with radii of up to 30 nm can be separated. This means that they are suitable for resolving globular and membrane proteins with high molecular masses [4], and viruses [5].

The most common electrophoresis on agarose gels is the zone electrophoresis, which is run in various buffers [6–7]. Among them, barbitalate buffers (veronalate buffers) are the most used. However, the usage of barbituric acid derivatives has been restricted by the Medicines Act [8]. Therefore, buffers containing no barbituric acid compounds are preferred. Such a buffer is the TRIS-taurinate buffer [9].

| TRIS-taurinate buffer, pH = 8.5, $I$ = 0.10 mol/L | |
| --- | --- |
| TRIS | 34.16 g |
| Taurine | 48.94 g |
| NaN$_3$ | 0.10 g |
| Deionized water to | 1,000.00 mL |

Horizontal agarose gel electrophoresis is a standard method for separation of proteins in serum, cerebrospinal fluid (CSF), urine, and other fluids.

## 2.2.2 Agarose gel electrophoresis of serum proteins

Serum protein electrophoresis (SPE) is a method used for separating serum proteins. With its help, five main groups (fractions) of serum proteins are obtained: albumin, $\alpha_1$-globulins, $\alpha_2$-globulins, $\beta$-globulins, and $\gamma$-globulins (Figure 2.2.1). The ratio between the mass concentrations of albumin and globulins is 1.5–3:1.

SPE can be used for diagnosing different diseases, for example, multiple myeloma, macroglobulinemia, amyloidosis, hypogammaglobulinemia, and more.

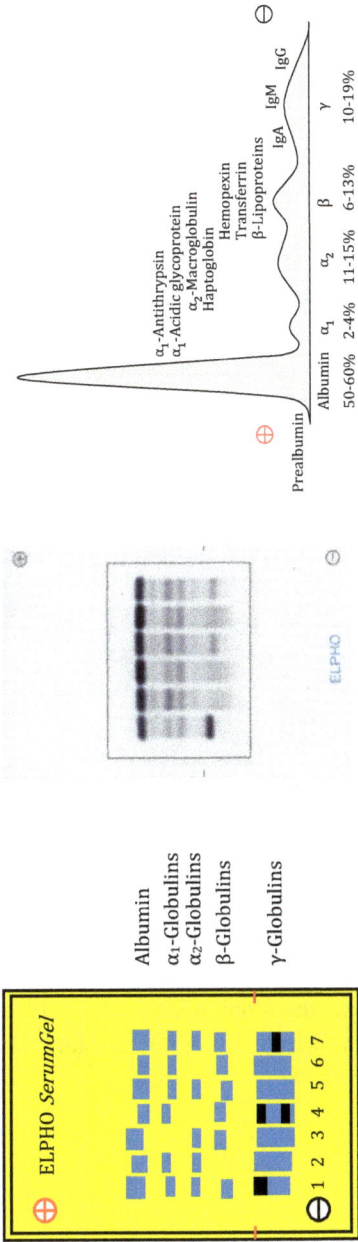

**Figure 2.2.1:** Pherogram of serum proteins obtained by agarose gel electrophoresis.

### 2.2.2.1 Albumin

Albumin, together with the other serum proteins, is synthesized in the liver. Its relative molecular mass is approximately 69,000 and its isoelectric point is 4.9.

Pathological conditions connected with changes in the albumin concentration are: *bisalbuminemia*, characterized with two albumin bands; and *analbuminemia*, characterized with reduction or even absence of the albumin band. Decreased albumin concentration is observed in the nephrotic syndrome (Figure 2.2.2) when albumin passes easily through the damaged glomerular membrane into the urine due to its low molecular mass. The albumin concentration is also low in liver diseases, malnutrition, nutrient absorption disorders, and loss of protein in enteropathy. Increased albumin concentration can be observed in acute alcoholism, during pregnancy, and in puberty.

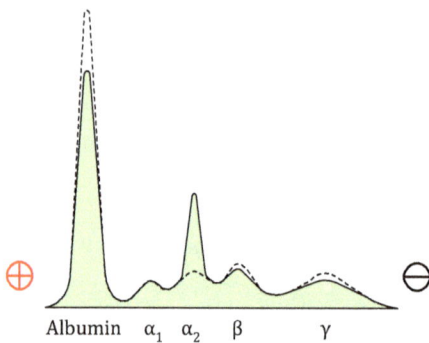

**Figure 2.2.2:** Nephrotic syndrome. The concentration of $\alpha_2$-globulins is increased; the concentration of albumin is decreased.

Before albumin, *transthyretin* (prealbumin) is moving. It transfers the hormone thyroxine. The concentration of transthyretin is too low; therefore, it is usually hidden when using common staining methods. Mutations of transthyretin cause familiar amyloidosis.

### Intermediate zone of albumin-$\alpha_1$-globulins

The intermediate zone of albumin-$\alpha_1$-globulins contains $\alpha_1$-lipoproteins (high-density lipoprotein (HDL)) and $\alpha$-fetoprotein. The concentration of $\alpha_1$-lipoproteins decreases at severe inflammation, acute hepatitis, and liver cirrhosis; the concentration of $\alpha$-fetoprotein increases in hepatocellular carcinoma – then, a sharp band appears between albumin and $\alpha_1$-globulins.

## 2.2.2.2 Alpha-1 globulins

$\alpha_1$-Globulins constitute the lowest main fraction – only 1–4% of the serum proteins. $\alpha_1$-Globulins are: $\alpha_1$-antitrypsin, $\alpha_1$-lipoproteins, acidic $\alpha_1$-glycoprotein, thyroxin-binding globulin, prothrombin, and more.

$\alpha_1$-Antitrypsin ($M_r$ = 54,000) inhibits trypsin and forms the bulk of $\alpha_1$-globulins. It has a sulfhydryl group, which can bind to thiol compounds. The concentration of $\alpha_1$-antitrypsin is too low in juvenile pulmonary emphysema, and decreases in nephrotic syndrome. Its low concentration in lung emphysema is due to the pulmonary tissue destruction by neutrophil elastase. $\alpha_1$-Antitrypsin is a protein of the acute phase of inflammation.

$\alpha_1$-Lipoproteins ($\alpha$-lipoproteins, HDL, $M_r$ = 200,000) transport phospholipids, cholesterol, triacylglycerols, fat soluble vitamins such as A and E, and some hormones.

Acidic $\alpha_1$-glycoprotein ($M_r$ = 44,100), called also orosomucoid, is an acute phase protein, too. Orosomucoid and $\alpha_1$-antitrypsin move together, but orosomucoid is stained worse. Its concentration increases in chronic inflammations as well as in some malignancies.

Thyroxin-binding globulin ($M_r$ = 40,000) transfers thyroxine.

Prothrombin ($M_r$ = 68,500) transforms into thrombin during the clotting process.

The concentration of $\alpha_1$- and $\alpha_2$-globulins increases in malignant tumors while the concentration of albumin decreases (Figure 2.2.3).

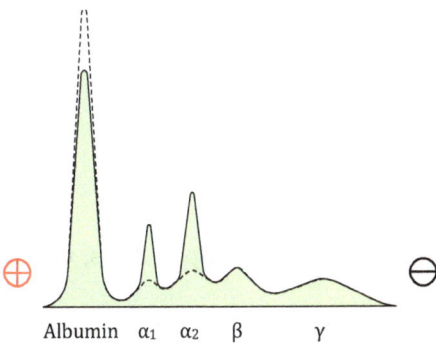

**Figure 2.2.3:** Increased concentrations of $\alpha_1$- and $\alpha_2$-globulins in malignant tumors. The albumin concentration decreases.

## Intermediate zone of $\alpha_1$-$\alpha_2$-globulins

In the intermediate zone of $\alpha_1$-$\alpha_2$-globulins, two pale picks can be seen, which are $\alpha_1$-antichymotrypsin and vitamin-D-binding protein. $\alpha_1$-Antichymotrypsin is an acute phase protein; therefore, its concentration increases in acute inflammation.

### 2.2.2.3 Alpha-2 globulins

$\alpha_2$-Globulins contain 7–12% of total serum proteins. Their most important represen-tatives are: $\alpha_2$-macroglobulin, haptoglobins, $\alpha_2$-lipoproteins, and ceruloplasmin.

$\alpha_2$-Macroglobulin ($M_r$ = 720,000) is an acute phase protein that inhibits plasmin and trypsin, and binds insulin. Its concentration increases in nephrotic syndrome and acute inflammation. The large volume of $\alpha_2$-macroglobulin molecule does not allow passing through the glomerular membrane, whereas the other proteins pass through into the urine. Therefore, its concentration increases significantly in ne-phrotic syndrome.

Haptoglobins ($M_r$ = 100,000) bind hemoglobins after red blood cells are de-stroyed and thus prevent the loss of iron. They represent the largest part of the acute phase proteins, causing an increase in the $\alpha_2$-globulin concentration in acute inflammation (Figure 2.2.4). On the contrary, their concentration decreases in he-molytic anemia.

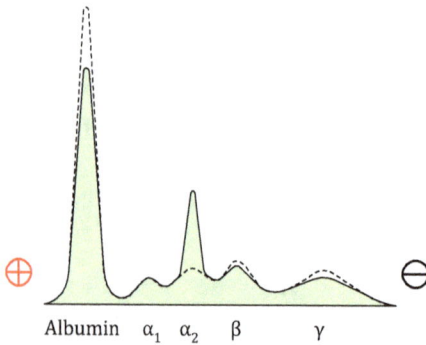

**Figure 2.2.4:** Increased concentration of $\alpha_2$-globulins in acute inflammation.

$\alpha_2$-Lipoproteins (prebeta-lipoproteins, VLDL, $M_r$ = 5–20×10$^6$) transport lipids, mainly triacylglycerols.

Ceruloplasmin ($M_r$ = 132,000) is an oxidase that takes place in copper metabolism. It is also an acute phase protein.

The concentration of $\alpha_2$-zone, combined with that of $\alpha_1$-zone, increases in patients with malignant tumors and cirrhosis, while the albumin concentration decreases.

### Intermediate zone of $\alpha_2$-β-globulins

The intermediate zone of $\alpha_2$-β-globulins contains prebeta-lipoproteins (HDL, VLDL). Their concentration is increased in type II hypercholesterolemia, hypertriglyceridemia, and nephrotic syndrome. Here, haptoglobin-hemoglobin complexes are also present. They move in front of haptoglobins.

### 2.2.2.4 Beta-globulins

The relative concentration of β-globulins is 6–10% of serum proteins. Most important among them are: hemopexin, transferrin, β-lipoproteins, plasminogen, complement components, and fibrinogen (in plasma).

Hemopexin ($M_r$ = 80,000) binds the heme of hemoglobins.

Transferrin ($M_r$ = 80,000) transfers iron. It is an acute phase protein.

β-Lipoproteins (LDL, $M_r$ = 2.4×10$^6$) transport lipids, fat-soluble vitamins, and hormones.

Plasminogen ($M_r$ = 143,000) destroys fibrin in the blood clot, when activated.

The complement components are glycoproteins.

Fibrinogen ($M_r$ = 341,000) is an important factor of the blood clotting system.

The concentration of β-globulins increases at liver cirrhosis, together with that of γ-globulins (β-γ-bridge) (Figure 2.2.5).

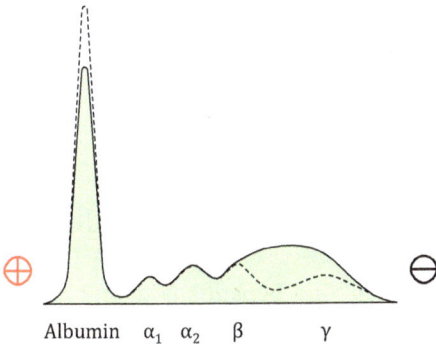

Figure 2.2.5: Beta-gamma bridge in liver cirrhosis.

In more precise agarose gel electrophoresis, β-globulins are separated in two fractions: $β_1$ and $β_2$.

Beta-1-globulins are presented predominantly by transferrin and β-lipoproteins. Increased concentration of transferrin is observed in iron-deficiency anemias, pregnancy, and estrogen therapy. An increased concentration of β-lipoproteins is found in hypercholesterolemia.

Beta-2-globulins are presented predominantly by the complement component 3 (C3) whose concentration increases during the acute phase of inflammation. On the contrary, their concentration decreases in autoimmune diseases since C3 is bound to immune complexes and removed from blood plasma. Fibrinogen also moves with $\beta_2$-globulins, but is absent in the normal serum.

**Intermediate zone of β-γ-globulins**

In the intermediate zone of β-γ-globulins, C reactive protein (an acute phase protein) is present. When its concentration is high, β-γ-fusion (β-γ-bridge) is monitored.

### 2.2.2.5 Gamma-globulins

γ-Globulins (immunoglobulins) represent 10–17% of the total serum protein mass. There are five classes of immunoglobulins (Ig): IgG, IgA, IgM, IgD, and IgE.

IgG ($M_r$ = 160,000) bind specifically to antigens of viral, bacterial, or parasitic origin, also to bacterial toxins, Rh antibodies, insulin, *etc.*

IgA is located in serum as well in secretions. Serum IgA ($M_r$ = 160,000 to 500,000) are the anti-bacterial agglutinins, anti-nuclear, and anti-insulin antibodies, and more. They are the fastest moving of antibodies; therefore, are always in the beginning of gamma zone. When their concentration is increased, IgA cause a β-γ-bridge, too. Their concentration is increased in patients with cirrhosis, respiratory infections, skin disease, or rheumatoid arthritis.

IgM ($M_r$ = 900,000) include antibodies, as ABO isoagglutinines, antibacterial antibodies, Wassermann antibody, anti-thyroglobulin antibodies, *etc.*

The function of IgD ($M_r$ = 150,000) is still unknown.

IgE ($M_r$ = 200,000), known as reagins, play a key role in some allergic diseases, such as hay fever and bronchial asthma.

The concentration of γ-globulins increases in viral hepatitis (Figure 2.2.6) as well as in multiple sclerosis.

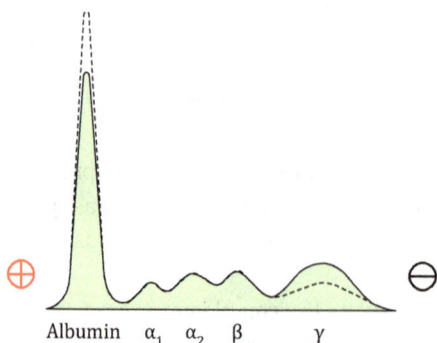

**Figure 2.2.6:** Increased concentration of γ-globulins and decreased concentration of albumin in virus hepatitis.

A spike-like enlargement of the gamma zone is typical for monoclonal gammopathy; a large basis increase indicates polyclonal gammopathy.

The narrow spike in monoclonal gammopathy (Figure 2.2.7) is known as *M spike*. It is malignant or clonal. Myeloma is the most common reason of IgA and IgG spikes. Chronic lymphatic leukemia and lymphosarcoma typically show an increase of IgM-paraprotein concentration. However, there may be up to 8% of healthy adult patients with monoclonal spikes. Waldenström's macroglobulinemia, amyloidosis, and solitary plasmacytomas are also characterized with *M* spikes.

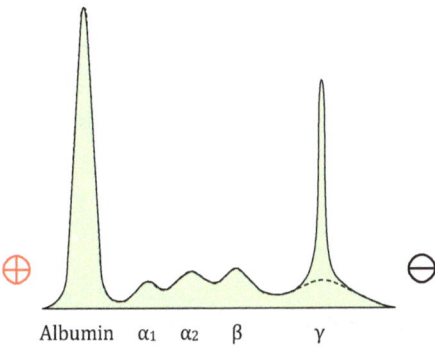

**Figure 2.2.7:** Increased concentration of immunoglobulins in monoclonal gammopathia.

Polyclonal gammopathy is usually a benign condition. It is typical for severe infections, chronic liver disease, rheumatoid arthritis, systemic lupus, and other connective tissue diseases.

The reduction of the gamma-zone is called hypogammaglobulinemia. It is normal for infants, but is a symptom for patients with X-linked agammaglobulinemia. In viral hepatitis, the concentration of γ-globulins is increased, while the concentration of albumin is decreased.

## 2.2.3 Agarose gel electrophoresis of lipoproteins

The lipoproteins are complexes of lipids and proteins. The lipids are represented by fatty acids, triacylglycerols, free and esterified cholesterol, phospholipids, and sphingolipids. They have exogenous origin, i.e. they originate from food, or they have endogenous origin, i.e. they are synthesized in the body.

### 2.2.3.1 High-density lipoproteins

High-density lipoproteins (HDL, $\alpha_1$-lipoproteins) build the fastest lipoprotein fraction. They contain proteins and phospholipids, both in relatively high concentrations. The synthesis of HDL takes place in the mitochondria and microsomes in the liver and small intestine.

### 2.2.3.2 Low-density lipoproteins

Low-density lipoproteins (LDL, $\beta$-lipoproteins) build the largest cholesterol containing fraction. They move behind the pre-$\beta$-lipoproteins. Cholesterol is a widespread steroid in the human body. It is located in the cell membranes. By cholesterol are synthesized bile acids, steroid hormones, and steroid vitamins.

### 2.2.3.3 Intermediate-density lipoproteins

Intermediate-density lipoproteins (IDL) are formed during the degradation of very-low-density lipoproteins (VLDL). Their density is between that of low-density and that of VLDL. IDL are similar to the LDL and like them consist of protein that encircles triacylglycerols and cholesterol esters. The size of IDL is 25–35 nm in diameter. IDL bind to receptors in the plasma membrane of liver cells and with the aid of endocytosis enter the cells where are degraded to form LDL particles.

### 2.2.3.4 Very-low-density lipoproteins

Very-low-density lipoproteins (VLDL, $\alpha_2$-lipoproteins, pre-$\beta$-lipoproteins) carry preferably triacylglycerols of endogenous origin. They move after the $\alpha$-lipoproteins.

### 2.2.3.5 Chylomicrons

The chylomicrons are composed mainly of exogenous triacylglycerols obtained from the food. They are constructed of a lipid droplet, which is surrounded by a thin protein network and phospholipid molecules. The chylomicrons do not move in an electric field.

The triacylglycerols represent glycerol esterified with different fatty acids. The fatty acids are composed usually by 18 straight-chain carbon atoms, containing one or more double bonds and many single bonds. The free fatty acids in blood, which are bound to albumin, serve as a raw material for the synthesis of endogenous triacylglycerols in the liver [10].

### 2.2.3.6 Hyperlipoproteinemias

Hyperlipoproteinemias are diseases that are characterized with increased concentration of lipoproteins. They are subdivided into five types [11].

Hyperlipoproteinemia type I (hyperchylomicronemia) is characterized by high concentration of chylomicrons. In electrophoresis, a strong chylomicron fraction is observed at the application site, LDL and VLDL are weakly visible, HDL remains usually unchanged (Figure 2.2.8).

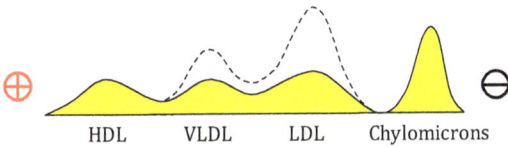

**Figure 2.2.8:** Hyperlipoproteinemia type I.

In hyperlipoproteinemia type II (hyper-LDL-emia), the concentration of LDL (cholesterol) is increased. The HDL band is normal. The hyperlipoproteinemia type II is subdivided into two subtypes: type IIa and type IIb. The pherogram of type IIa shows an increased LDL fraction, whereas the VLDL fraction is usually normal. In type IIb, the concentration of LDL as well as VLDL are increased (Figure 2.2.9).

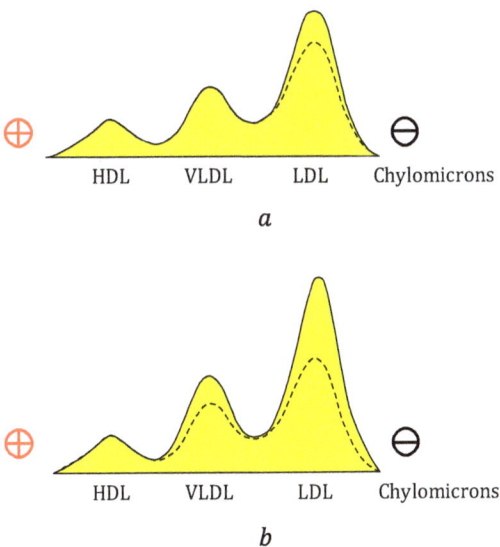

*a*

*b*

**Figure 2.2.9:** Hyperlipoproteinemia type IIa (*a*) and type IIb (*b*).

Hyperlipoproteinemia type III (hyper-LDL-hyper-VLDL-emia) is characterized with a broadband over the LDL and VLDL fractions (increased cholesterol and triacylglycerols), and a normal α-band fraction (Figure 2.2.10).

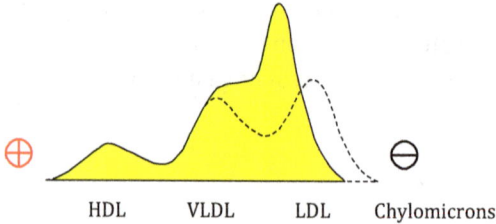

**Figure 2.2.10:** Hyperlipoproteinemia type III.

The hyperlipoproteinemia type IV (hyper-VLDL-emia) is accompanied by an increased VLDL fraction. The pherogram shows normal LDL and HDL bands (Figure 2.2.11).

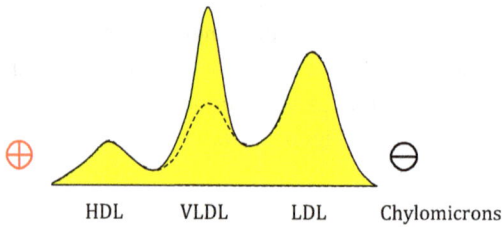

**Figure 2.2.11:** Hyperlipoproteinemia type IV.

In hyperlipoproteinemia type V (hyper-VLDL-emia and hyperchylomicronemia), a simultaneous increase of the concentration of exogenous triacylglycerols (chylomicrons) as well as of endogenous triacylglycerols (VLDL fraction) are seen (Figure 2.2.12).

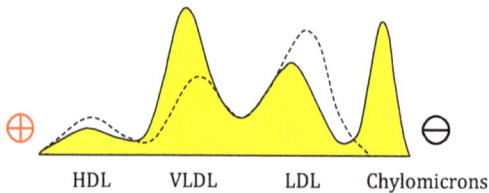

**Figure 2.2.12:** Hyperlipoproteinemia type V.

## 2.2.4 Agarose gel electrophoresis of hemoglobins

Electrophoresis, nearby chromatography, gives best information for hemoglobins (Hb). The hemoglobin electrophoresis is carried out most often in cellulose acetate membranes and agarose gels. Especially, good results are obtained by isoelectric focusing on polyacrylamide gel [12–13].

### 2.2.4.1 Normal hemoglobins

In the embryo, the following hemoglobins can be established: Gower 1 ($\zeta_2\varepsilon_2$), Gower 2 ($\alpha_2\varepsilon_2$), Portland I ($\zeta_2\gamma_2$), and Portland II ($\zeta_2\beta_2$). In the fetus, hemoglobin F ($\alpha_2\gamma_2$) can be found [14]. In adults, three types of hemoglobins exist: hemoglobin A (HbA), hemoglobin (HbA$_2$) and fetal hemoglobin F (HbF).

Hemoglobin A (HbA) is a tetramer with $M_r$ = 67,000 that is composed of four polypeptide chains (globin chains), each of them with $M_r$ = 17,000, and four iron containing prosthetic groups (hemes). The polypeptide chains are designated as $\alpha$- and $\beta$-chains. So, the HbA formula is $\alpha_2\beta_2$. The tertiary structure of every hemoglobin subunit resembles that of myoglobin because they are related genetically.

HbA makes over 95% of the total hemoglobin in a healthy adult; the rest consists of HbA$_2$ and HbF (Figure 2.2.13). HbF dominates during the fetal development, however is replaced later by HbA. The formulas of HbA$_2$ and HbF are $\alpha_2\delta_2$ and $\alpha_2\gamma_2$, respectively. The synthesis of $\delta$ chain of HbA$_2$ begins late in the third trimester. In

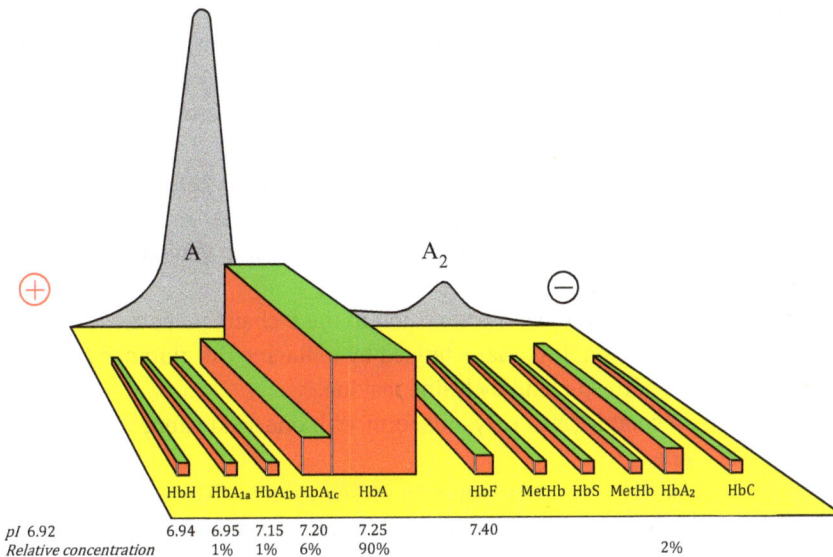

**Figure 2.2.13:** Hemoglobins.

adults, HbF is restricted in a limited population of red cells called F-cells. The concentration of HbF is increased in persons with sickle cell disease and $\beta$-thalassemia.

The hemoglobin concentration is 12.8–17.6 g/dL (on average 15 g/dL) in men, and 11.2–16.0 g/dL (on average 14 g/dL) in women.

HbA is partially glycosylated. The glycosylated hemoglobin, called $HbA_1$, constitutes 8% of the total hemoglobin mass in the blood. It consists of HbA and glucose that is bound non-enzymatically to the valine residue at the N-end of one of the two polypeptide chains.

There are three types of $HbA_1$: $HbA_{1a}$, $HbA_{1b}$, and $HbA_{1c}$. In healthy people, $HbA_{1c}$ constitutes 6% of the total hemoglobin, but its concentration is increased several times in patients with diabetes mellitus [15–16]. The concentration of $HbA_{1c}$ reflects the average glucose concentration in the weeks before testing [17–18], whereas the direct methods determine the glucose concentration just during the testing time. The concentrations of $HbA_{1a}$ and $HbA_{1b}$ are about 1 g/dL each.

### 2.2.4.2 Abnormal and pathological hemoglobins

The globin chains are characterized by their sequence of amino acid residues. If, due to errors in the genetic code, the sequence of the amino acid residues is incorrect, abnormal hemoglobins (hemoglobin variants) occur. If they do not cause disorders, they are classified as abnormal hemoglobins; if they cause illnesses (hemoglobinopathias), they are called pathological hemoglobins [19].

The most widespread pathological hemoglobins are:
- Hemoglobin D-Punjab ($\alpha_2\beta_2^D$).
- Hemoglobin H ($\beta_4$), formed by a tetramer of $\beta$-chains; presents in variants of $\alpha$-thalassemia.
- Hemoglobin Barts ($\gamma_4$), formed by a tetramer of $y$-chains; presents in variants of $\alpha$-thalassemia.
- Hemoglobin S ($\alpha_2\beta_2^S$), a product of a variation in the $\beta$-chain gene; found in people with sickle cell disease; causes sickling of red blood cells.
- Hemoglobin C ($\alpha_2\beta_2^C$), a product of a variation in the $\beta$-chain gene; causes a mild chronic hemolytic anemia.
- Hemoglobin E ($\alpha_2\beta_2^E$), a product of a variation in the $\beta$-chain gene; causes a mild chronic hemolytic anemia. It is characterized by replacing the glutamic acid residue at position 26 of the $\beta$-chain by lysine residue.
- Hemoglobin SC, a complex heterozygous form with one sickle gene and another encoding HbC gene.
- Hemoglobin Hopkins-2 [20], a variant form of hemoglobin that is sometimes established in combination with HbS in the sickle cell disease.

### Sickle cell disease

Sickle cell disease (sickle cell anemia) was first described by Ernest Irons and James Herrick in 1910 [21]. In 1949, Linus Pauling [22] invented the unusual hemoglobin S, and explained the illness with an abnormality in its molecule. The actual molecular change in HbS was described in the late 1950s by Vernon Ingram [23]. It contains, because of a point mutation, a valine residue instead of glutamic acid residue at position 6 of the β-chain.

HBB gene is responsible for sickle cell disease. It is located on the short ($p$) arm of chromosome 11 at position 15.5. People who receive the defective gene from both father and mother (homozygote persons) develop the disease. People who receive one defective and one healthy gene (heterozygote persons) remain healthy.

The homozygote patients with hemoglobin S have a hemoglobin concentration of 6–10 g/dL. Their pherogram shows disturbed synthesis of HbA, compensatory increased synthesis of HbF, and high synthesis of HbS (Figure 2.2.14).

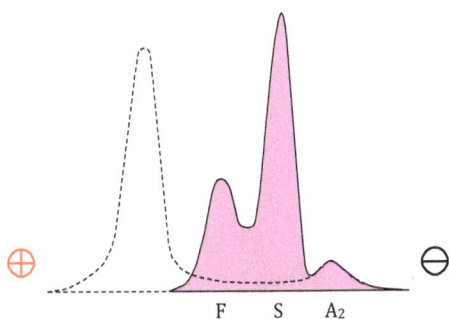

**Figure 2.2.14:** Sickle cell disease.

In 1950, another pathological hemoglobin was identified – hemoglobin C [24]. Its glutamic acid residue at position 6 of the β-chain is exchanged against a lysine residue. This anemia is accompanied by small hematological disturbances. Later, hemoglobin E was identified and afterward hundreds of other hemoglobin variants.

### Thalassemias

Thalassemias are diseases in which the synthesis of globin chains of HbA is disturbed. Instead of this, the body produces the globin chains of $HbA_2$ or HbF. Thalassemias occur more frequently than the hemoglobinopathies. The heterozygous thalassemia is called *thalassemia minor*. It is characterized with a mild anemia and a light increase of the concentrations of $HbA_2$ or HbF. The homozygous thalassemia is called *thalassemia major*. It is a serious disorder that is widespread in the Mediterranean.

*Alpha-thalassemia.* In α-thalassemias, the synthesis of α-chain of HbA is missing or is disturbed. This causes a compensatory excess of β-chains in adults or γ-chains in newborns, which lead to formation of the tetramers $β_4$ (HbH) or $γ_4$ (Hb Barts) in blood. These homotetramers are not useful in the body because they, unlike the heterotetramer, have a very high affinity to oxygen and, as a result, cannot deliver it to tissues. HbH and Hb Barts move in alkaline electrophoresis faster than HbA (Figure 2.2.15).

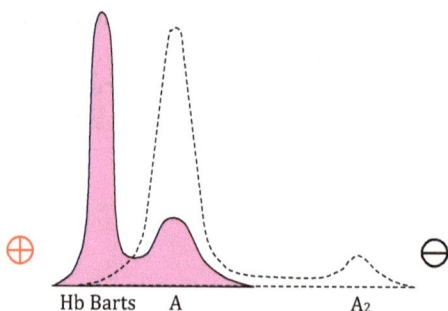

**Figure 2.2.15:** α-Thalassemia.

*Beta-thalassemia.* In β-thalassemias, the β-chains of HbA are not synthesized, which leads to increased production of α-, γ-, or δ-chains.

*β-Thalassemia major*, called Cooley anemia, leads to death in a few years, if not treated. It is accompanied by increased concentrations of $HbA_2$ and HbF (Figure 2.2.16).

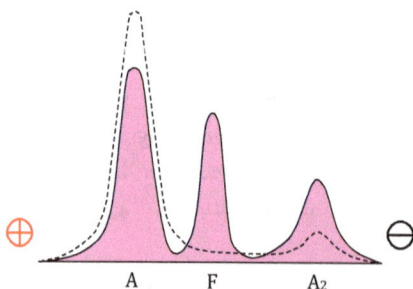

**Figure 2.2.16:** β-Thalassemia.

In addition to β-thalassemia major, β-thalassemia, β-fusion hemoglobinopathia (with Hb Lepore), and various forms of hereditary persistence of hemoglobin F exist. Hb Lepore is a fusion product between a β-chain and a δ-chain. It has the mobility of hemoglobin S. In β-thalassemia, neither β-chains nor δ-chains are synthesized.

*Hemoglobin E disease* appears when a child inherits the HbE gene from both parents. In the first months of life, the fetal hemoglobin disappears and the amount of HbE increases, so that the patients start to have a mild $\beta$-thalassemia. People who are heterozygote for HbE (one normal allele and one abnormal allele) do not show any symptoms (there is no anemia or hemolysis) [25]. Patients who are homozygous for the HbE allele (have two abnormal alleles) suffer from a mild hemolytic anemia and mild enlargement of the spleen.

### 2.2.4.3 Running hemoglobin electrophoresis

Since the pI points of hemoglobins are at pH = 7.0, the hemoglobin electrophoresis should be performed at higher or lower pH. In alkaline buffers, hemoglobins are negatively charged and migrate toward the anode, whereas in acidic buffers they are positively charged and migrate toward the cathode (Figure 2.2.17).

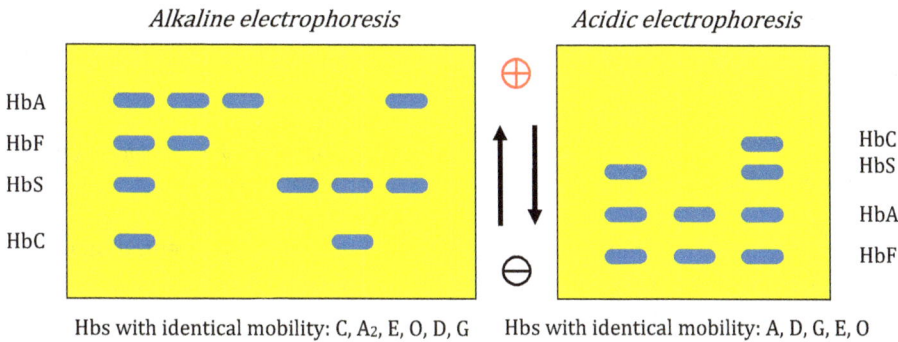

Figure 2.2.17: Hemoglobins obtained by alkaline and acidic electrophoresis.

In the *alkaline agarose gel electrophoresis*, the hemolysates are applied at the cathode. Routinely, it is carried out in TRIS-barbitalate-EDTA buffer with pH = 8.5. The results in the other buffer – TRIS-taurinate-EDTA buffer (also with pH = 8.5) [26] – are even better. In the alkaline hemoglobin electrophoresis, hemoglobins A, S, C, *etc.* can be separated.

The *acidic agarose gel electrophoresis* is used to separate hemoglobins that have equal migration velocities in alkaline electrophoresis. The hemolysates are applied at the anode, and the electrophoresis is carried out in a citrate buffer at pH = 6.1. HbC is separated from HbE, HbS from HbD or HbG, and HbS from HbC. Also, HbF is separated from HbA, which is impossible in the alkaline electrophoresis.

## 2.2.5 Agarose gel electrophoresis of cerebrospinal fluid proteins

The protein concentration in the CSF is about 250 times lower than that in the blood serum. Therefore, CSF should be concentrated prior to electrophoresis. After electrophoresis, the protein bands can be stained with Amido black B10, or Coomassie brilliant blue R-250 or G-250. They can be stained with silver, too.

The CSF represents an ultrafiltrate of the serum plasma. Therefore, the CSF pherogram is similar to the serum pherogram (Figure 2.2.18).

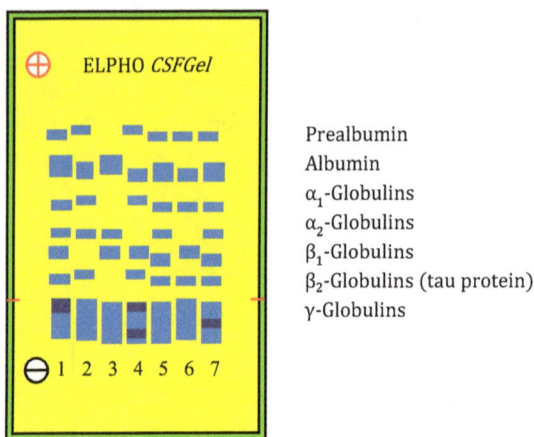

Prealbumin
Albumin
$\alpha_1$-Globulins
$\alpha_2$-Globulins
$\beta_1$-Globulins
$\beta_2$-Globulins (tau protein)
$\gamma$-Globulins

**Figure 2.2.18:** Electrophoretic bands of CSF proteins.

However, there are differences between both pherograms: The concentration of prealbumin in CSF is too high – it represents 4–5% of the total proteins; the concentration of albumin is not as high as in the blood serum; the $\alpha_1$-globulins contain preferably $\alpha_1$-antitrypsin and $\alpha_1$-acidic glycoprotein; the $\alpha_2$-globulins contain preferably 1-1 haptoglobin and ceruloplasmin; the β-band is split up into two subfractions ($\beta_1$ and $\beta_2$), whereby $\beta_1$-subfraction consists mainly of transferrin and hemopexin; and $\beta_2$-subfraction is represented by a specific carbohydrate-arm transferrin called tau-protein ($\tau$-fraction); The $\gamma$-globulins are represented in significantly lower concentration in comparison to the serum $\gamma$-globulins. Often, a post-gamma fraction of low concentration, called gamma C, exists; and the lipoprotein concentration in the CSF is extremely low.

In multiple sclerosis, the agarose gel zone electrophoresis of CSF shows an oligoclonal IgG banding in the gamma aria – up to seven IgG oligoclonal bands [27–28].

## 2.2.6 Electrophoresis of creatine kinase isoenzymes

Creatine kinase (CK, ATP: creatine $N$-phosphotransferase, EC 2.7.3.2) catalyzes the reversible phosphorylation of creatine in the mitochondrial membrane, according to the following reaction:

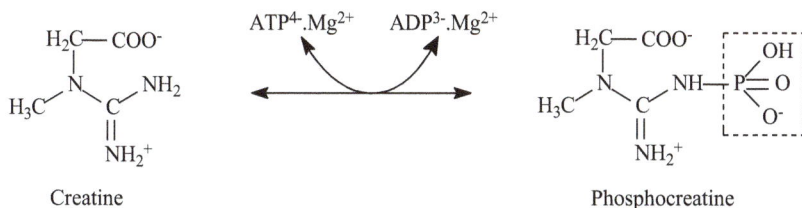

Creatine        Phosphocreatine

CK is composed of two polypeptide chains, B and M, making up of three forms: CK-BB, CK-MB, and CK-MM. They form high-molecular-mass complexes with IgG or IgA.

The isoenzyme CK-BB is contained in the cytosol of brain and nerve cells. It dominates over the other CK isoenzymes in bladder, kidney, prostate, uterus, gastrointestinal tract, lungs, placenta, spleen, liver, pancreas, and thyroid gland [29]. CK-BB is not detectable in normal adult serum. The complex between CK-BB and IgG is a macromolecule with a molecular mass of more than 250,000. It moves on agarose gel electrophoresis between CK-MB and CK-MM [30]. The CK-BB activity in serum is greatly increased in different cancers (of prostate, breast, stomach, lungs, colon, testes, gall bladder, and leukemia) [31].

The isoenzyme CK-MB has its highest activity in the cardiac muscle. It is a specific indicator of the acute myocardial infarction. The elevated serum activity of CK-MB, in addition to the increased $LDH_1/LDH_2$ ratio (see below), the clinical history (previous chest pain), and the abnormal ECG (Q-wave) help to diagnose this severe disease (Figure 2.2.19).

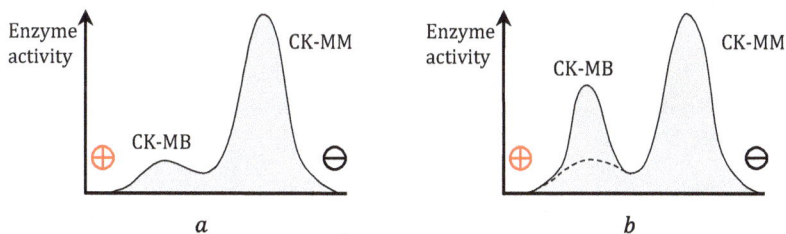

**Figure 2.2.19:** Activity of CK-MB and CK-MM.
(*a*) Activity in a healthy individual; (*b*) activity in a patient with acute myocardial infarction.

In acute myocardial infarction, an interruption or reduction of blood supply in the coronary arteries causes heart muscle necrosis. This damages the plasma membrane of the cardiac cells, which results in a release of CK-MB into the lymphatic system and then into the peripheral blood. The serum concentration of CK-MB increases in 4–8 h after the beginning of infarction, and the high concentration remains up to the 12–18th h [32].

The isoenzyme CK-MM is the most widespread isoenzyme in the skeletal muscle and human serum. It makes more than 50% of the total CK activity.

## 2.2.7 Electrophoresis of lactate dehydrogenase isoenzymes

Lactate **de**hydrogenase (LDH, L-lactate: NAD oxidoreductase, EC 1.1.1.27) is an enzyme, which is located in the cytoplasm. It catalyzes the reduction of pyruvate to lactate according to the following equilibrium, shift to lactate at pH = 7.4:

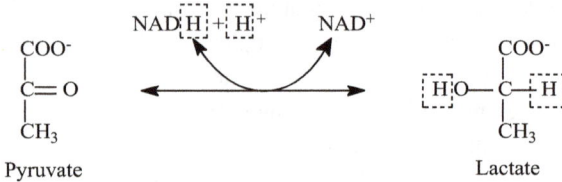

Pyruvate           Lactate

LDH is a tetramer ($M_r$ = 135,000), which consists of two subunits: H (from **heart**) and M (from **muscle**). Five LDH isoenzymes exist that can be separated from each other by electrophoresis: $LDH_1$ ($H_4$), $LDH_2$ ($H_3M$), $LDH_3$ ($H_2M_2$), $LDH_4$ ($HM_3$), and $LDH_5$ ($M_4$). $LDH_1$, the isoenzyme with the highest negative charge, migrates fastest (its mobility is almost equal to that of serum albumin), while $LDH_5$ is the slowest isoenzyme (Figure 2.2.20). $LDH_1$ predominates in heart, kidney cortex, brain, and erythrocytes, while $LDH_5$ predominates in skeletal muscle and liver. $LDH_2$, $LDH_3$, and $LDH_4$ are present in the gall bladder, prostate, and uterus.

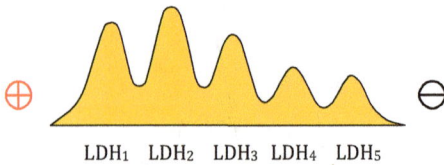

LDH₁  LDH₂  LDH₃ LDH₄  LDH₅

**Figure 2.2.20:** Normal pherogram of LDH isoenzymes.

In serum of a normal adult, the activity of $LDH_2$ is higher than that of the $LDH_1$, and the $LDH_1/LDH_2$ quotient is less than 0.76. In an acute myocardial infarction, $LDH_1$ is released from the myocardium cells into the blood. As a result, the $LDH_1/LDH_2$ quotient becomes in the first 24–48 h higher than 1, which may continue for days or weeks (Figure 2.2.21).

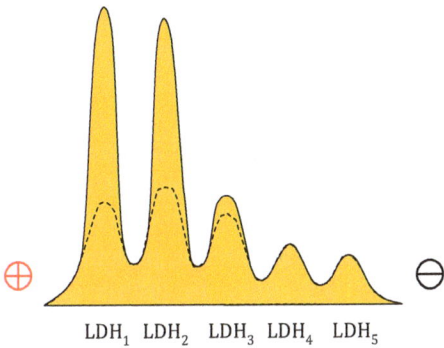

**Figure 2.2.21:** LDH isoenzymes in myocardial infarction.

The activities of $LDH_2$ and $LDH_3$ increase in pulmonary edema; of $LDH_4$ and $LDH_5$ in liver diseases; and of $LDH_5$ in renal diseases. In malignant neoplasms, all LDH isoenzymes show higher activity.

## 2.2.8 Agarose gel electrophoresis of urinary proteins

Normally, the urine contains no proteins, but certain diseases (for example, multiple myeloma) cause a leakage of proteins into the urine. This pathological condition is called *proteinuria*. There are two types of proteinuria: glomerular and tubular (Figure 2.2.22).

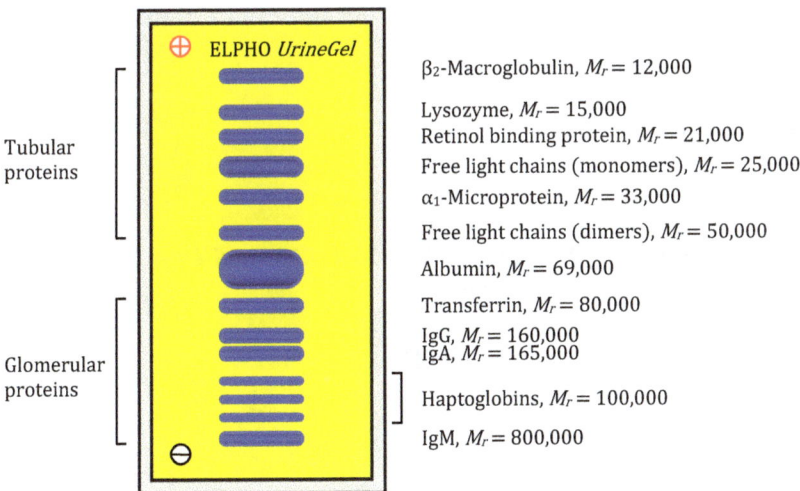

**Figure 2.2.22:** Pherogram of urinary proteins in a patient with mixed proteinuria.

In *glomerular proteinuria*, serum proteins are detected in urine: albumin, $\alpha_1$-acidic glycoprotein, $\alpha_1$-antitrypsin, $\alpha$-lipoproteins, transferrin, haptoglobins, IgA, IgG, and more. They have high-molecular mass, which is greater than the albumin mass. $\beta$-Lipoproteins and IgM are not found [33–34] in the urine.

The *tubular proteinuria* occurs rare, when the tubular absorption fails. In this case, the proteins in urine have molecular masses that are less than the albumin mass, and, therefore, they run before albumin [35–36]: protein hormones [37] (gonadotropins, corticotropin, and erythropoietin), vitamin $B_{12}$-binding protein, blood group substances [38], retinol-binding protein, and some enzymes: lysozyme, amylase, plasmin, trypsin, pepsin, and LDH [39]. Compared to the glomerular proteins, they contain more carbohydrates.

## 2.2.9 Protocols

### Agarose Gel Electrophoresis of Serum Proteins

*The agarose gels for horizontal electrophoresis are located between two polyester films. The support film is connected chemically to the agarose gel, whereas the cover film can be easily removed. The gels have to be stored in a refrigerator at 4 °C.*

**Materials and equipment**
TRIS
NaOH
Formic acid
Taurine
Methylparaben
$NaN_3$
Amido black 10B (Naphthol blue black)
Ethanol
Acetic acid

***TRIS-formate buffer (Gel buffer), pH = 8.8, I = 10 x 0.035 mol/L***

| | |
|---|---|
| TRIS | 2.77 g (0.23 mol/L) |
| NaOH | 1.22 g (0.31 mol/L) |
| HCOOH | 1.39 mL (0.37 mol/L) |
| Taurine | 1.53 g (0.12 mol/L) |
| Methylparaben | 0.30 g (0.02 mol/L) |
| $NaN_3$ | 1.95 g (0.30 mol/L) |
| Deionized water to | 100.00 mL |

*TRIS-taurinate buffer (Electrode buffer), pH = 9.1, I = 10 × 0.06 mol/L*

| | |
|---|---|
| TRIS | 1.84 g (0.15 mol/L) |
| NaOH | 2.37 g (0.59 mol/L) |
| Taurine | 12.26 g (0.98 mol/L) |
| $NaN_3$ | 0.65 g (0.10 mol/L) |
| Deionized water to | 100.00 mL |

*Agarose gel (0.9 g/dL), pH = 8.8, I = 0.035 mol/L*

| | |
|---|---|
| TRIS-formate-buffer, 10× | 10.0 mL |
| Agarose ($-m_r$ = 0.16–0.19) | 0.9 g |
| Sucrose (or sorbitol) | 2.0 g |
| Deionized water to | 100.0 mL |

*Bring the mixture to boil in a microwave oven and cast gels.*

*Electrode buffer, pH = 9.1, I = 0.06 mol/L*

| | |
|---|---|
| TRIS-taurinate buffer, 10× | 10.0 mL |
| Deionized water to | 100.0 mL |

*Staining solution*

| | |
|---|---|
| Amido black 10B | 0.2 g |
| Methanol | 10.0 mL |
| Acetic acid | 10.0 mL |
| Deionized water to | 100.0 mL |

*Destaining solution*

| | |
|---|---|
| Phosphoric acid | 0.05 g |
| Deionized water to | 100.00 mL |

**Procedure**
- Unpack an agarose gel and remove its cover film.
- Blot the gel surface with a filter paper strip where the application template should be laid on.
- Place a polyester application template on the blotted place according to side marks on the support film.
- Press carefully the template with fingers against the gel to remove air bubbles between the template and gel.
- Dilute serum samples 10 times with deionized water, apply 2–5 µL each in the template slots, and let them diffuse in the gel for 5 min.
- Blot the rest of the samples with a filter paper strip and remove the template together with the paper strip.
- Hang the support film with the gel, gel side down, on the bridge of an electrophoresis cell.

- Place the bridge, together with the gel, into the electrode tanks of the electrophoresis cell, so that the gel contact directly with the electrode buffers in the cell (Figure 2.2.23).

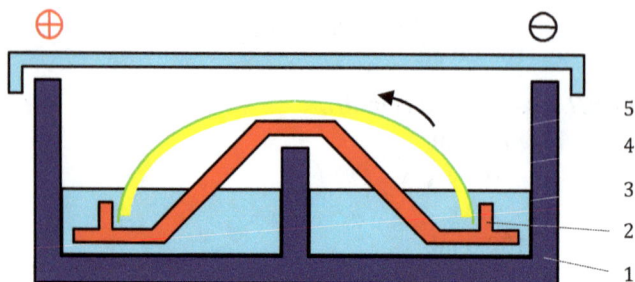

**Figure 2.2.23:** Direct contact between agarose gel and both electrode buffers.
1. Electrophoresis cell; 2. bridge; 3. electrode buffer; 4. agarose gel; 5. support film.

- Cover the electrophoresis cell with its lid, and switch on the power supply.
- Run electrophoresis at constant voltage of 70 V for 20–25 min.
- Fix the gel after electrophoresis in ethanol – acetic acid – water (5:1:4, $V:V:V$) for 5 min.
- Dry the gel with a hair drier or in a drying oven at room temperature.
- Stain the protein bands in the staining solution for 5 min.
- Destain the gel background several times in 0.5 g/dL citric acid.
- Dry the gel for a second time with a hair drier or in a drying oven at room temperature.
- Evaluate the protein bands with a scanner or a densitometer at 600 nm.

**Agarose Gel Electrophoresis of Lipoproteins**

*The agarose gel electrophoresis of lipoprotein is carried out as the agarose gel electrophoresis of serum proteins. However the agarose gel is more concentrated (1.5 g/dL) and contains 5 g/dL bovine serum albumin.*

**Materials and equipment**
TRIS
Formic acid
Agarose ($-m_r = 0.16–0.19$)
**B**ovine **s**erum **a**lbumin (BSA)
Sucrose or sorbitol
Sudan black B
Ethanol

*Agarose gel (1.5 g/dL), pH = 8.8, I = 0.035 mol/L*

| | |
|---|---|
| TRIS-formate buffer, 10× | 10.0 mL |
| Agarose | 1.5 g |
| 5 g/dL BSA | 10.0 mL |
| Sucrose (or sorbitol) | 2.0 g |
| Deionized water to | 100.0 mL |

*Bring the mixture to boil in a microwave oven and cast gels.*

*Electrode buffer, pH = 9.1, I = 0.06 mol/L*

| | |
|---|---|
| TRIS-taurinate buffer, 10× | 10.0 mL |
| Deionized water to | 100.0 mL |

*Staining solution*

| | |
|---|---|
| Sudan black B | 0.2 g |
| Deionized water to *Filter.* | 100.0 mL |

*Destaining solution*

| | |
|---|---|
| Ethanol | 70.0 mL |
| Deionized water to | 100.0 mL |

**Procedure**
- Unpack an agarose gel and remove its cover film.
- Blot the gel surface with a filter paper strip where the application template should be placed on.
- Lay a polyester application template on the blotted place according to the side marks on the support film.
- Press carefully the template with fingers against the gel to remove air bubbles between the template and gel.
- Dilute the serum samples 10 times with deionized water, apply 2–5 µL each into the template slots, and let them diffuse in the gel for 5 min.
- Blot the rest of the samples with a filter paper strip and remove the template together with the paper strip.
- Hang the support film with the gel, gel side down, on the bridge of an electrophoresis cell.
- Fill the electrode tanks of the electrophoresis cell with electrode buffer and place the bridge together with the gel in the cell. As a result, the gel will contact directly the electrode buffers.
- Cover the electrophoresis cell with the lid, and switch on the power supply.
- Run electrophoresis at constant voltage of 70 V for 20–25 min.
- After electrophoresis, fix the gel in ethanol – acetic acid – water (5:1:4, *V:V:V*) for 5 min.
- Dry the gel with a hair drier or in a drying oven at room temperature.
- Mix 5 mL of 1.0 g/dL Sudan black B in ethanol with 45 mL of 60 mL/dL ethanol.

- Stain the lipoproteins in the staining mixture for 10 min. The staining mixture is usable within 24 h.
- Destain the gel three times in 60 mL/dL ethanol for 5 min.
- Rinse the gel in deionized water.
- Dry the gel with a hair dryer or in a drying oven.

**Agarose Gel Electrophoresis of Hemoglobins**

*The agarose gel electrophoresis of hemoglobins is carried out as the agarose gel electrophoresis of serum proteins. Only the agarose gel is more concentrated (1.5 g/dL). The agarose gel electrophoresis of hemoglobins can be run in alkaline or acidic buffers.*

**Alkaline Agarose Gel Electrophoresis of Hemoglobins**

### Materials and equipment
TRIS
Formic acid
Sucrose (or sorbitol)

### 2× *Sample buffer*

| | |
|---|---|
| 1.5 mol/L TRIS, pH = 9.2 | 4.00 mL |
| Glycerol | 8.00 mL |
| Bromophenol blue Na salt | 0.01 g |
| Deionized water | 28.00 mL |

### *Agarose gel (1.5 g/dL), pH = 8.8, I = 0.035 mol/L*

| | |
|---|---|
| TRIS-formate buffer, 10× | 10.0 mL |
| Agarose ($-m_r$ = 0.16–0.19) | 1.5 g |
| Sucrose (or sorbitol) | 2.0 g |
| Deionized water to | 100.0 mL |

*Bring the mixture to boil in a microwave oven and cast gels.*

### *Electrode buffer, pH = 9.1, I = 0.06 mol/L*

| | |
|---|---|
| TRIS-taurinate buffer, 10× | 10.0 mL |
| Deionized water to | 100.0 mL |

**Procedure**

- Wash EDTA containing blood samples in 0.15 mol/L (9.0 g/L) NaCl three times.
- Mix 2 volumes of red cells with 2 volumes of deionized water and 1 volume of $CCl_4$.
- Shake the resulting mixture for 3 min. As a result, the erythrocyte membranes are dissolved (hemolysed).
- Centrifuge after 30 min at 3,000 rev/min.
- Dilute the supernatant in a buffer and KCN, until the Hb concentration reaches $7.75 \times 10^5$ mol/L (10 g/L), and KCN concentration reaches 0.005 mol/L (0.326 g/L).
- Apply on an agarose gel (CA membrane) 10 µL of hemolysate.
- Run electrophoresis as with serum proteins.
- Fix the Hb bands in 0.8 mol/L (130.71 g/L) $CCl_3COOH$ for 15 min.
- Stain in 0.003 mol/L (0.2 g/dL) Bromophenol blue Na salt in ethanol – acetic acid – water (10:1:9, $V:V:V$) for 30 min.
- Destain the background of agarose gel (CA membrane) overnight in ethanol – acetic acid – water (6:1:13, $V:V:V$), or in 0.05 g/dL phosphoric acid.

## Acidic Agarose Gel Electrophoresis of Hemoglobins

**Materials and equipment**
$Na_3$ citrate
Citric acid
D-sorbitol
$NaN_3$

*Solution A (0.2 mol/L $Na_3$ citrate)*

| | |
|---|---|
| $Na_3$ citrate | 5.88 g (0.23 mol/L) |
| $NaN_3$ | 0.07 g (0.01 mol/L) |
| Deionized water to | 100.00 mL |

*Solution B (0.2 mol/L citric acid)*

| | |
|---|---|
| Citric acid | 4.20 g ( mol/L) |
| $NaN_3$ | 0.07 g (0.01 mol/L) |
| Deionized water to | 100.00 mL |

*Sodium citrate buffer, 10×, pH = 6.0, I = 10 × 0.08 mol/L*

| | |
|---|---|
| Solution A | 41.5 mL |
| Solution B | 9.5 mL |
| $NaN_3$ | 0.2 g |
| Deionized water to | 100.0 mL |

*Agarose gel*

| Sodium citrate buffer | 10.0 mL |
|---|---|
| Agarose | 1.0 g |
| D-sorbitol | 10.0 g |
| Deionized water to | 100.0 mL |

*Bring the agarose mixture to boil in a microwave oven and cast gels.*

**Procedure**

*Run electrophoresis as with alkaline hemoglobins. Stain the hemoglobins as the hemo-globins resolved in an alkaline buffer.*

**Agarose Gel Electrophoresis of Cerebrospinal Fluid Proteins**

*The agarose gel electrophoresis of CSF proteins is carried out as the agarose gel electrophoresis of serum proteins.*

**Materials and equipment**

TRIS
Formic acid
Agarose ($-m_r = 0.16–0.19$)
Sucrose or sorbitol
Amido black 10B (Naphthol blue black)
Ethanol
Acetic acid
Centrifuge concentrators

*Agarose gel (0.9 g/dL), pH = 8.8, I = 0.035 mol/L*

| TRIS-formate buffer, 10× | 10.0 mL |
|---|---|
| Agarose | 0.9 g |
| Sucrose (or sorbitol) | 2.0 g |
| Deionized water to | 100.0 mL |

*Bring the mixture to boil in a microwave oven and cast gels.*

*Electrode buffer, pH = 9.1, I = 0.06 mol/L*

| TRIS-taurinate buffer, 10× | 10.0 mL |
|---|---|
| Deionized water to | 100.0 mL |

**Procedure**
- Unpack an agarose gel and remove its cover film.
- Blot the gel surface with a filter paper strip, where samples are to be applied.
- Place an application template on the blotted place according to the side marks on the support film.
- Press gently the template against the gel to remove air bubbles between it and the gel.
- Dilute the cerebrospinal samples with deionized water to a protein concentration of 0.20–0.40 g/L.
- Concentrate 600 µL of a diluted sample in a centrifuge concentrator rotating it in a 10,000 g centrifuge about 10 min, until the sample reaches a volume of 10 µL. So, the concentration of the sample reaches 12 to 24 g/L.
- Apply 2–5 µL of the concentrates each onto the template slots and allow them to diffuse into the gel for 5 min.
- Blot the rest of the samples with a thin filter paper strip and remove the template together with the paper strip.
- Hang the support film with the gel, gel side down, on the bridge of an electrophoresis cell.
- Fill the electrode tanks with electrode buffer and place the bridge with the gel in the cell. So the gel obtains direct contact with the electrode buffers.
- Close the electrophoresis cell with its lid and turn on the power supply.
- Run the electrophoresis at constant voltage of 70 V for 20–25 min.
- After the electrophoresis, fix the gel in a mixture of ethanol – acetic acid – water (5:1:4, $V:V:V$) for 5 min.
- Dry the gel with a hair dryer or in a drying oven at room temperature.
- Stain the bands in 0.1 g/dL Amido black 10B in the fixing mixture for 5 min.
- Destain the gel background several times in 0.5 g/dL citric acid.
- Dry the gel for a second time with a hair drier or in a drying oven at room temperature.

**Agarose Gel Electrophoresis of Urinary Proteins**

**Materials and equipment**
TRIS
Formic acid
Agarose ($-m_r$ = 0.16–0.19)
Sucrose or sorbitol
Amido black 10B (Naphthol blue black)
Ethanol
Acetic acid

**Agarose gel (0.9 g/dL), pH = 8.8, I = 0.035 mol/L**

| | |
|---|---|
| TRIS-formate buffer, 10× | 10.0 mL |
| Agarose ($-m_r$ = 0.16–0.19) | 0.9 g |
| Sucrose (or sorbitol) | 2.0 g |
| Deionized water to | 100.0 mL |

Bring the mixture to boil in a microwave oven and cast gels.

**Electrode buffer, pH = 9.1, I = 0.06 mol/L**

| | |
|---|---|
| TRIS-taurinate buffer, 10× | 10.0 mL |
| Deionized water to | 100.0 mL |

## Procedure

- Unpack an agarose gel and remove its cover film.
- Blot the gel surface with a filter paper strip, where samples should be applied.
- Place an application template on the blotted surface according to the side marks on the support film.
- Press carefully the template with fingers against the gel to remove air bubbles between it and the gel.
- Dialyze and concentrate urine to obtain a protein concentration of 0.2 g/dL.
- Apply 2–5 μL of samples each onto the template slots and let the samples diffuse in the gel for 5 min.
- Blot the rest of the samples with a thin filter paper strip and remove the template together with the paper strip.
- Hang the support film with the gel, gel side down, on the bridge of an electrophoresis cell.
- Fill the electrode tanks of the electrophoresis cell with electrode buffer and place the bridge with the gel in the cell. As a result, the gel obtains a direct contact with the electrode buffers.
- Cover the electrophoresis cell with its lid, and switch on the power supply.
- Run electrophoresis at constant voltage of 70–80 V for 20–25 min.
- After the electrophoresis, fix the gel in ethanol – acetic acid – water (5:1:4, V:V:V) for 5 min.
- Dry the gel with a hair drier or in a drying oven at room temperature.
- Stain the proteins in Amido black 10B solution for 5 min.
- Destain the gel several times in ethanol – acetic acid – water (5:1:4, V:V:V).
- Dry the gel for a second time with a hair drier or in a drying oven at room temperature.

# References

[1]   Felgenhauer K. J Chromatogr, 1979, 173, 299–311.
[2]   Hjerten S. Biochim Biophys Acta, 1961, 53, 514–517.
[3]   Hjerten S. Biochim Biophys Acta, 1962, 62, 445–449.
[4]   Johansson KE, Blomquist J, Hjerten S. J Biol Chem, 1975, 250, 2463–2469.
[5]   Serwer P, Allen JC, Hayes SJ. Electrophoresis, 1983, 4, 232–236.
[6]   Buzas Z, Chrambach A. Electrophoresis, 1982, 3, 121–129.
[7]   Michov BM. Elektrophorese. Theorie Und Praxis. Walter de Gruyter, Berlin – New York, 1996.
[8]   Susann J. The Valley of the Dolls. Gorgi Publ., London, 1966.
[9]   Michov BM. *Protein separation by SDS electrophoresis in a homogeneous gel using a TRIS-formate-taurinate buffer system and homogeneous plates suitable for electrophoresis.* (Proteintrennung durch SDS-Elektrophorese in einem homogenen Gel unter Verwendung eines TRIS-Formiat-Taurinat-Puffersystems und dafür geeignete homogene Elektrophoreseplatten). German Patent 4127546, 1991.
[10]  Havel RJ, Fells JM, Van Duyne M. J Lipid Research, 1962, 3, 297–308.
[11]  Classification of Hyperlipidemias and Hyperlipoproteinemias. Bull World Health Org, 1970, 43, 891–915.
[12]  Jeppson JO. Application Note 307. LKB-Produkter AB, Stockholm, 1977.
[13]  Bunn HF, Gabbay KH, Gallop PM. Science, 1978, 200, 21–27.
[14]  Lanzkron S, Strouse JJ, Wilson R *et al.* Ann Int Med, 2008, 148, 939–955.
[15]  Cohen B, Rubenstein AH. Diabetologia, 1978, 15, 1–8.
[16]  Mortensen HB. J Chromatogr, 1980, 182, 325–333.
[17]  Koenig RJ, Peterson CM, Jones RL, Sandek C, Lehiman PA, Cerami S. New Eng J Med, 1976, 295, 417–420.
[18]  Spicer KM, Allen RC, Bruce MG. Diabetes, 1978, 27, 384–388.
[19]  Huisman THJ. A Syllabus of Human Hemoglobin Variants. Globin Gene Server. Pennsylvania State University, 1996.
[20]  Clegg JB, Charache S. Hemoglobin, 1978, 2, 85–88.
[21]  Herrick JB. Intern Med, 1910, 6, 517–521.
[22]  Pauling L, Itano HA. Science, 1949, 110, 543–548.
[23]  Serjeant GR. Br J Haematol, 2010, 151, 425–429.
[24]  Itano HA, Neel JV. Proc Natl Acad Sci USA, 1950, 36, 613–617.
[25]  Hoffbrand AV, Moss PAH. Hoffbrand's Essential Haematology. 7th ed, Wiley – Blackwell, 2016.
[26]  Michov BM. *Protein separation by SDS electrophoresis in a homogeneous gel using a TRIS-formate-taurinate buffer system and homogeneous plates suitable for electrophoresis.* (Proteintrennung durch SDS-Elektrophorese in einem homogenen Gel unter Verwendung eines TRIS-Formiat-Taurinat-Puffersystems und dafür geeignete homogene Elektrophoreseplatten). German patent, 4127546, 20.08.1991.
[27]  Thompson EJ. Br Med Bull, 1977, 33, 28–33.
[28]  Gearson B, Orr JM. Am J Clin Path, 1980, 73, 87–91.
[29]  Smith AF. Clin Chim Acta, 1972, 39, 351–359.
[30]  Urdal P, Landaas S. Scand J Clin Lab Invest, 1981, 41, 499–505.
[31]  Feld RD, Van Steirteghem AC, Zweig MH *et al.* Clin Chim Acta, 1980, 100, 267–273.
[32]  Roe CR, Limbird LE, Wagner GS, Nerenberg ST. J Lab Clin Med, 1972, 80, 577–589.
[33]  Bruton OC. Pediatrics, 1952, 9, 722–728.

[34]  Cawley LP, Eberhardt L, Schneider D. J Lab Clin Med, 1965, 65, 342–354.

[35]  Axelsson U, Hallen J. Br J Haematol, 1968, 15, 417–420.

[36]  Efremov G, Braend M. Science, 1964, 146, 1679–1680.

[37]  Kohn J. Clin Chim Acta, 1957, 2, 297–303.

[38]  Bartlett RC. Clin Chem, 1963, 9, 317–324.

[39]  Franco JA, Savory J. Am J Clin Path, 1971, 56, 538–542.

# 2.3 Immunoelectrophoresis

2.3.1    Immunodiffusion electrophoresis according to Grabar and Williams —— 110
2.3.2    Rocket immunoelectrophoresis according to Laurell —— 110
2.3.3    Immunofixation and immunoprinting —— 111
2.3.4    Protocols —— 112
         Immunofixation of Serum Proteins —— 112
         Immunofixation of Bence Jones Proteins —— 115
         Immunoprobing with Avidin-biotin Coupling to Secondary Antibody —— 116
         References —— 117

Immunoelectrophoresis as well as affinity electrophoresis are carried out on agarose gels. They pass with binding of proteins to specific ligands. Therefore, both techniques are named *ligand electrophoresis*.

Immunoelectrophoresis presents a combination between agarose gel electrophoresis and immune reactions [1, 2]. It requires immunoglobulins (Ig, antibodies) [3] as ligands. The Ig recognize exogenous antigens (macromolecules in viruses, bacteria, tissues, *etc.*). They are produced by plasma cell lines (clones) of the immune system. The healthy organism possesses over 10,000 different clones. Agarose has been chosen as a gel of choice because it has large pores allowing the proteins to pass. For this technique, 1 g/dL agarose slabs are usually used, which contain buffers with pH = 8.6.

At certain antigen–antibody ratios, net-shaped insoluble immune complexes (immune precipitates) are formed in the gel (Figure 2.3.1). The non-precipitated proteins can be washed out. Immune precipitates are stained with dyes as Amido black 10B or

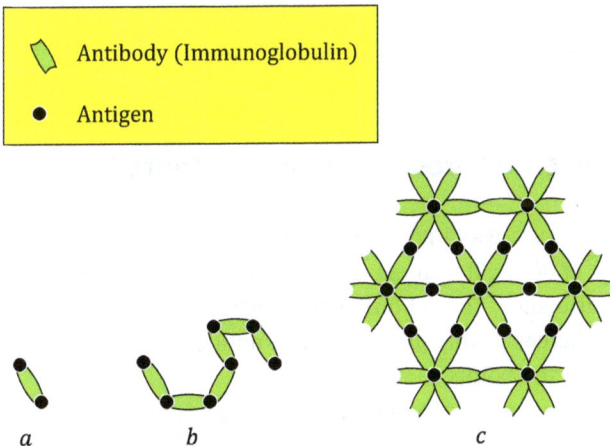

**Figure 2.3.1:** Hypothetical structure of soluble immune complexes (*a* and *b*) and insoluble immune precipitates (*c*). The ratio between the molar concentrations of antibodies and antigens is as follows: (*a*) 1:2; (*b*) 1:1; and (*c*) 3:1.

https://doi.org/10.1515/9783110761641-013

Coomassie brilliant blue R-250. Nowadays, the immunoelectrophoresis is replaced by electroblotting, because it requires fewer antibodies, and has higher sensitivity.

The most significant immunoelectrophoretic methods are immunoelectrophoresis according to Grabar and Williams, rocket-immunoelectrophoresis according to Laurell, immunofixation, and immunoprinting.

## 2.3.1 Immunodiffusion electrophoresis according to Grabar and Williams

Immunodiffusion electrophoresis according to Grabar and Williams [4] is a combination between agarose gel electrophoresis of proteins (antigens) and diffusion of their antibodies: after electrophoresis, antibodies are applied in channels next to the protein bands. The antibodies diffuse against the bands forming with them precipitate arcs (Figure 2.3.2).

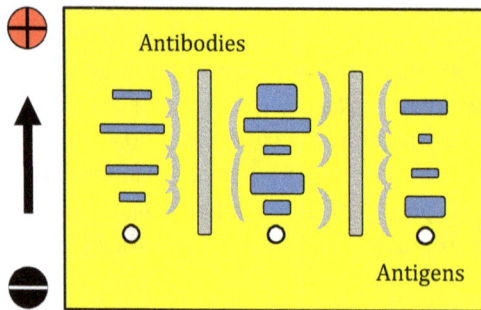

**Figure 2.3.2:** Immunodiffusion electrophoresis according to Grabar–Williams.

## 2.3.2 Rocket immunoelectrophoresis according to Laurell

The rocket immunoelectrophoresis was invented by Laurell [5] and modified by Eriksson *et al.* [6], and Yman *et al.* [7]. It is an electrophoresis of proteins (antigens) on agarose gel containing antibodies. The gel buffer is adjusted to pH = 8.6, so that only the antigens migrate, whereas at this pH value most antibodies cannot move because they are in their isoelectric points.

In the beginning of rocket immunoelectrophoresis, the antigens excess, which results in formation of soluble antigen–antibody complexes. Afterward, immune precipitates are formed, which are rocket-shaped figures whose surface (height) is proportional to the antigen concentrations (Figure 2.3.3). Therefore, the rocket immunoelectrophoresis is a quantitative immunoelectrophoresis, used for quantitation of human serum proteins.

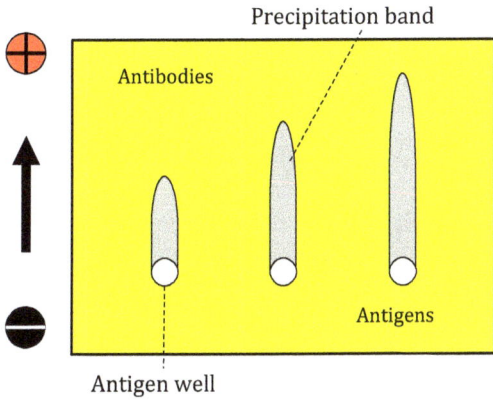

**Figure 2.3.3:** Rocket immunoelectrophoresis according to Laurell.

## 2.3.3 Immunofixation and immunoprinting

The immunofixation was described in 1964 [8, 9]. Later, in modified forms, it was used for identification of different proteins [10–12]. In immunofixation, the proteins first should be separated electrophoretically on agarose gel, and then they should be coated with corresponding antisera. So, immune complexes are built in the agarose gel [13]. The soluble proteins are washed out with physiological solution (Figure 2.3.4).

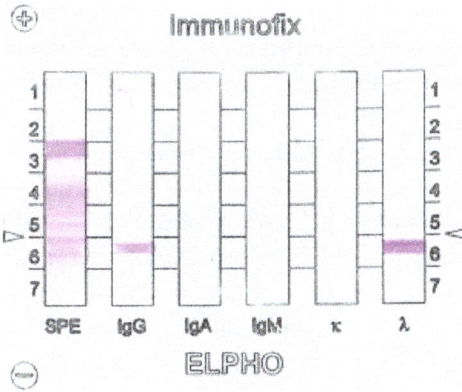

**Figure 2.3.4:** Immunofixation electrophoresis proving IgG paraproteins of λ-type in the blood serum.
SPE – **S**erum protein **e**lectrophoresis

In immunoprinting [14], first an agarose gel electrophoresis of antigens (proteins) is carried out. Then, the gel is overlaid with an antibody-containing agarose gel or cellulose acetate membrane. The resolved antigens diffuse into the antibody containing sheet and form immune precipitates.

## 2.3.4 Protocols

### Immunofixation of Serum Proteins

**Materials and equipment**
TRIS-glycinate or TRIS-taurinate buffer
Agarose gel for immunofixation of serum proteins. It is situated between two films. The lower film is bound to the gel; the upper film can be easily removed from it.
Antisera: Anti-IgG, anti-IgA, anti-IgM, anti-kappa, and anti-lambda
Coomassie violet blue R-200 or Amido black 10B
Pipette

**Staining solution**
Coomassie violet R-200     0.2 g
Deionized water to        100.0 mL

**Destaining solution**
Phosphoric acid            0.05 g
Deionized water to        100.00 mL

**Procedure**

*Preparation and application of sample*
-   Dilute blood sera with saline at least 20 times to obtain a protein concentration of about 0.35 g/dL. Concentrate the cerebrospinal fluid and urine samples to the same protein concentration.
-   Blot carefully the agarose gel surface with a thin filter paper where an application template will be placed.
-   Place an application template on the blotted gel surface according to side markers (Figure 2.3.5) and press it gently against the gel surface to remove air bubbles between the gel and template.
-   Apply 2–3 µL of samples onto the **electro**phoresis (ELP) and next slots, and let the sample diffuse in the gel.
-   After 5 min, blot the remaining sample volumes with filter paper.
-   Remove the application template together with the paper strip from the gel.

*Electrophoresis*
-   Fill the tanks of an ELP cell with electrode buffer.
-   Hang the gel, with its support film up, on the bridge of an ELP cell.

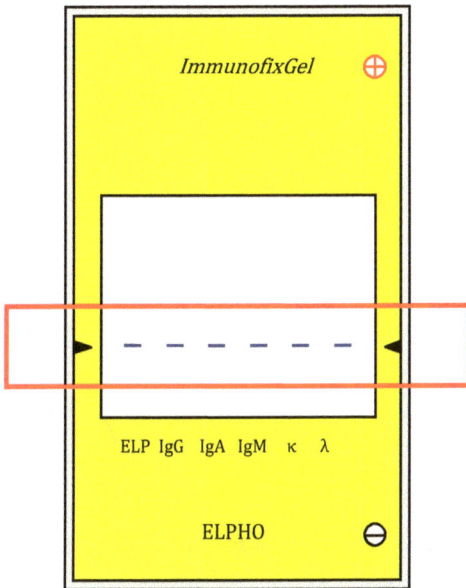

**Figure 2.3.5:** Application of a template for immunofixation onto an agarose gel.

- Insert the bridge with the gel into the tanks of the ELP cell. So, direct contact occurs between the gel and electrode buffers (see *Agarose gel electrophoresis of serum proteins*).
- Cover the ELP cell with its lid, and turn on the power supply.
- Run ELP at constant voltage or electric current (80 V, or 20–25 mA) at room temperature for 20–25 min.

*Immunoprecipitation*
- After the ELP, dry the agarose gel surface with a filter paper.
- Adjust an antiserum template on the gel, according to arrows on the support film, and press it gently against the gel (Figure 2.3.6).
- Apply with a pipette 50 μL of fixing solution (methanol-acetic acid-water, 3:1:6, *V:V:V*) in the ELP bed and 50 μL each of the antisera (G, A, M, κ, and λ) in their beds.
- Incubate the gel with the antisera on a wet filter paper at room temperature for 30 min.
- Remove the antisera template and press the gel dry. The gel dry pressing is carried out in the following way: Place 4–5 filter papers on the gel, a glass plate, and a mass of 1–2 kg for 10 min (Figure 2.3.7).
- Soak the gel in saline (0.9 g/dL NaCl) for 10 min, and press it again.
- Press the gel two more times and afterward dry it with a dryer or in a drying oven.

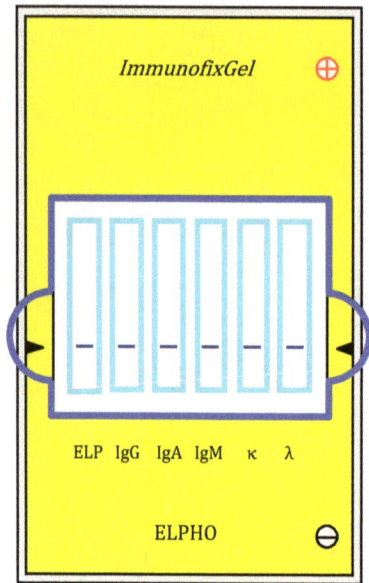

**Figure 2.3.6:** Adjustment of an antiserum template on an agarose gel after electrophoresis.

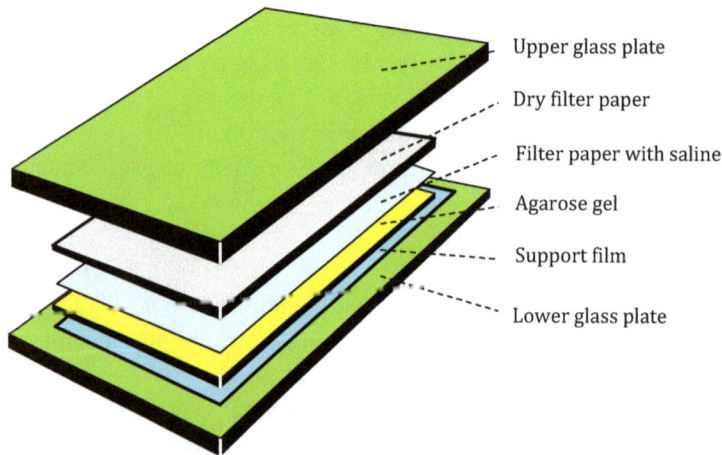

**Figure 2.3.7:** Dry pressing of an agarose gel.

### Staining the protein bands

- Stain the gel in 0.2 g/dL Coomassie violet blue R-200 or 0.2 g/dL Amido black in methanol-acetic acid-water (3:1:6, $V:V:V$) for 10 min.
- Destain the gel two to three times for 2–3 min in 0.5 g/dL citric acid until the background is clear.
- Dry the gel with a dryer or in a drying oven.

*Evaluation*

Compare the precipitation bands in the IFE positions with the ELP reference bands. The lower detection limit for monoclonal IgG, IgA, IgM and κ chains is 0.5 to 1.5 mg/mL, and for λ chains is 1 to 2 mg/mL.

## Immunofixation of Bence Jones Proteins

### Materials and equipment

TRIS-glycinate or TRIS-taurinate buffer

Agarose gel for immunofixation of Bence Jones proteins (light chains of paraproteins). It is located between two films. The lower film is bound to the gel, while the upper film can be easily removed from it.

Coomassie violet blue R-200 or Amido black 10B

### Procedure

*Sample preparing and applying*

- Use fresh urine samples. Most often, due to the low concentration of Bence Jones proteins, urine should be concentrated. A quick method for this purpose is a centrifugation in microcentrifuge concentrators. The final concentration of proteins in urine samples should be 1.5–3.5 g/L.
- Blot with a thin filter paper the gel surface where the samples will be applied.
- Lay down a polyester template onto the blotted gel surface, according to both arrows on the support film (Figure 2.3.5), and press gently the template to remove the air bubbles underneath.
- Apply 2–3 μL of the sample onto the ELP slot as well as onto the slots for GAM, κ, λ, κ free and λ free.
- Wait 5 min until the samples diffuse into the gel.
- Blot the exceeded sample with a filter paper and remove it, together with the template, from the gel.

*Electrophoresis*

- Hang the support film, with the gel down, onto the bridge of an electrophoretic cell.
- Fill the tanks of the electrophoretic cell with electrode buffers and insert the bridge into them. Thus, the gel is directly connected to the electrode buffers (see above).
- Close the electrophoretic cell with its lid and turn on the power supply.
- Run ELP at constant voltage (80 V) or constant electric current (20–25 mA) for 20–25 min at room temperature.

*Immunoprecipitation*
-   After ELP, blot the gel surface with a thin filter paper.
-   Place an antiserum template onto the gel and press it gently (Figure 2.3.6).
-   Apply 50 µL of a polyvalent antiserum into the ELP reference bed and 50 µL each of the appropriate antisera (GAM, κ, κ-free, λ, and λ-free) into the test beds.
-   Incubate the gel with antisera for 30 min on wet filter paper at room temperature.
-   After incubation, remove the antiserum template and press the gel for 10 min by placing a thin filter paper, 4–5 thick filter papers, a glass plate, and a weight of 1–2 kg on it (see above).
-   Soak the gel in saline for 10 min.
-   Press the gel and soak it in saline another two times.
-   Dry the gel with a hairdryer or at room temperature.

*Staining the serum immune complexes*
-   Stain the immune complexes in the gel with Coomassie violet blue R-200 or Amido black blue 10B (see above) for 10 min.
-   Destain the gel background in a destaining solution.
-   Dry the gel with a hairdryer or at room temperature.

*Evaluation*
The results can be evaluated by comparing the precipitation strips in the test beds with the precipitation strips in the reference ELP bed. The lower concentration limit for establishing monoclonal κ-chains is 0.5–1.5 mg/mL, and for α-chains is 1–2 mg/mL.

## Immunoprobing with Avidin-biotin Coupling to Secondary Antibody

### Materials and equipment
TBS (**T**RIS-**b**uffered **s**aline)
TBST – mixture of TBS and **T**ween 20
Avidin
Biotinylated **h**orseradish **p**eroxidase (HRP) or **a**lkaline **p**hosphatase (ALP)
Primary antibody
Biotinylated secondary antibody
Plastic bag
Plastic box

**Procedure**

- Prepare primary antibody in 5 mL TBST (for nitrocellulose or PVDF membranes) or TBS (for nylon membrane).
- Place the membrane with the blotted proteins into a plastic bag containing the primary antibody solution.
- Incubate at room temperature for 30 min.
- Transfer the membrane from the plastic bag into a plastic box.
- Wash the membrane three times for 15 min each in TBST (for nitrocellulose or PVDF membranes) or TBS (for nylon membranes). Add buffer to cover the membrane.
- Transfer the membrane into a plastic bag containing diluted secondary antibody solution in 50 to 100 mL TTBS (for nitrocellulose or PVDF membranes) or TBS (for nylon membranes).
- Incubate at room temperature for 30 min, then wash.
- Prepare avidin-biotin-HRP (or avidin-biotin-ALP) complex.
- Mix two drops of avidin solution and two drops of biotinylated HRP (or ALP) solution into 10 mL TBST (for nitrocellulose or PVDF membranes) or TBS (for nylon membranes).
- Incubate at room temperature for 30 min.
- Add TBST or TBS to 50 mL.
- Transfer the washed membrane to the avidin-biotin-enzyme solution.
- Incubate at room temperature for 30 min.
- Wash for 30 min.
- Develop according to an appropriate visualization protocol.

# References

[1]   Clarke MHG, Freeman T. Clin Sci, 1968, 35, 403–413.
[2]   Axelsen NH, Kroll J, Weeke D. A Manual of Quantitative Immunoelectrophoresis. Methods and Applications. Blackwell Scientific Publ, Oxford – London – Edinburgh – Melbourne, 1973.
[3]   Edelman GM. Science, 1973, 180, 830–840.
[4]   Grabar P, Williams CA. Biochim Biophys Acta, 1953, 10, 193–194.
[5]   Laurell CB. Anal Biochem, 1966, 15, 45–52.
[6]   Eriksson A, Yman IM. Food Safety Quality Assurance Applications Immunoassay Systems. 1992, 65–68.
[7]   Yman IM, Eriksson A, Everitt G, Yman L, Karlsson T. Food Agricultural Immunol, 1994, 6, 167–172.
[8]   Alfonso E. Clin Chim Acta, 1964, 10, 114–122.
[9]   Wilson AT. J Immunol, 1964, 92, 431–434.

[10]  Suzuki K, Harumoto T, Ito S, Matsumoto H. Electrophoresis, 1987, 8, 481–485.
[11]  Pötsch-Schneider L, Klein H. Electrophoresis, 1988, 9, 602–605.
[12]  Bauer K, Molinari E. Immunfixation. Georg Thieme, Stuttgart, 1989.
[13]  Marshall MO. Clin Chim Acta, 1980, 104, 1–9.
[14]  Cotton RGH, Milstein C. J Chromatogr, 1973, 86, 219–221.

# 2.4 Affinity electrophoresis

2.4.1     Theory of affinity electrophoresis —— 119
2.4.2     Lectin affinity electrophoresis —— 119
2.4.2.1   Electrophoresis of alkaline phosphatase isoenzymes on lectin agarose gels —— 120
2.4.3     Saccharide affinity electrophoresis —— 122
2.4.4     Affinity supported molecular matrix electrophoresis —— 123
2.4.5     Phosphate affinity electrophoresis —— 123
2.4.6     Capillary affinity electrophoresis —— 125
2.4.7     Affinity-trap electrophoresis —— 125
2.4.8     Charge shift electrophoresis —— 126
2.4.9     Mobility shift electrophoresis —— 126
2.4.10    Protocols —— 127
          Electrophoresis of ALP Isoenzymes on Lectin Agarose Gels —— 127
          Staining the ALP Isoenzymes —— 128
          References —— 129

The affinity electrophoresis [1, 2] can be used for characterization of polyions, for example, for estimating binding constants in the lectin electrophoresis. It is suitable for studying isoenzymes (isozymes), e.g., of lactate dehydrogenase [3], alcohol dehydrogenase [4], or plasminogen [5], and above all of alkaline phosphatase [6].

## 2.4.1 Theory of affinity electrophoresis

The affinity electrophoresis represents zone electrophoresis in an agarose or polyacrylamide gel, which contains specific ligands, as lectins, enzyme substrates, or other substances. These ligands are not bound to the gel, opposite to the ligands in affinity chromatography. They form complexes with corresponding polyions, which complexes have other electric charges and other mobilities.

A few methods of affinity electrophoresis are known: lectin electrophoresis, saccharide affinity electrophoresis, affinity supported molecular matrix electrophoresis, phosphate affinity electrophoresis, capillary affinity electrophoresis, affinity-trap electrophoresis, charge shift electrophoresis, and mobility shift electrophoresis.

## 2.4.2 Lectin affinity electrophoresis

*Lectins* (phytohemagglutinins) are plant carbohydrate-binding proteins. They show high specificity for foreign glucoconjugates [7–9]. In lectin electrophoresis, wheat germ lectin (**w**heat **g**erm **a**gglutinin, WGA) is used for separating **al**kaline **p**hosphatase (ALP) isoenzymes. ALP (orthophosphate monoester hydrolase, EC 3.1.3.1) is a

https://doi.org/10.1515/9783110761641-014

glycoprotein, which is built of two monomers connected by a $Zn^{2+}$. It catalyzes the hydrolytic splitting of terminal phosphate groups of organic phosphoric esters in alkaline solutions (Figure 2.4.1).

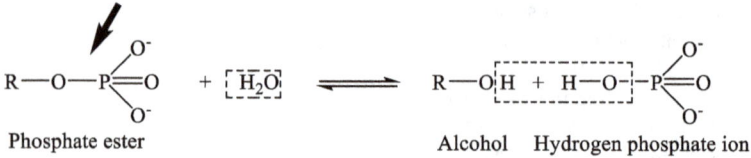

**Figure 2.4.1:** Hydrolysis of a phosphate ester by alkaline phosphatase.

The alkaline phosphatase consists of two polypeptide chains, which are chelated by a $Zn^{2+}$. The polypeptide chains are bound two by two forming a few alkaline phosphatase isoenzymes. They are encoded by three genes: a tissue unspecific gene, a colon gene, and a placental gene.

### 2.4.2.1 Electrophoresis of alkaline phosphatase isoenzymes on lectin agarose gels

Using affinity electrophoresis, following serum ALP isoenzymes are established: liver, bone, small intestine, placental, and macro-ALP isoenzyme (Figure 2.4.2).

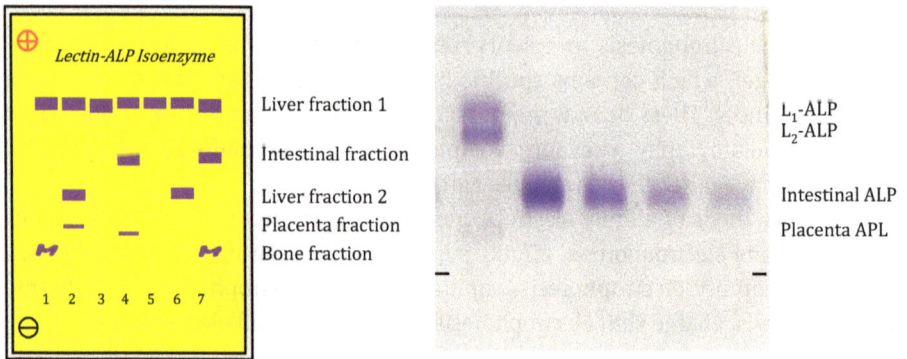

**Figure 2.4.2:** Alkaline phosphatase isoenzymes separated by affinity electrophoresis.

*Liver ALP isoenzyme.* The liver ALP isoenzyme contains *N*-acetylneuraminic (sialic) acid in very low concentration. It is heat stable and is present in two forms: $L_1$ or fast liver ALP; and $L_2$ or slow liver ALP. $L_1$-isoenzyme, together with the small intestine ALP isoenzyme, is found in healthy adult people. $L_2$-isoenzyme is rarely detected. It is located at the cathode side of $L_1$ in a lectin agarose gel. $L_2$ correlates with malignant diseases with or without metastases.

*Bone ALP isoenzyme.* The bone ALP isoenzyme contains residues of *N*-acetylneuraminic (sialic) acid in high concentration. Therefore, it binds immediately to the wheat germ lectin in agarose gels and comes to a standstill in front of the sample application place. The bone ALP isoenzyme is heat labile. Its mass is also twice greater than the mass of the liver ALP isoenzyme in both genders.

*Small intestine ALP isoenzyme.* The small intestine ALP isoenzyme is found in 10% of human population. In the blood serum, up to three bands of it can be detected.

*Placental ALP isoenzyme.* Placental ALP isoenzyme exists in two forms: $P_1$ (90%) and $P_2$ (10%). They appear in blood serum during pregnancy or at cancer of the ovary, pancreas, stomach, and colon, as well as at sarcomas [10, 11].

*Macro-ALP isoenzyme.* The macro ALP isoenzyme is rarely found. It is a complex between alkaline phosphatase and an immunoglobulin. Since its molecular mass is too large, the macro ALP isoenzyme remains at the application place.

As pointed out, the liver and bone ALP isoenzymes differ from each other only by the concentration of their *N*-acetylneuraminic acid. They can be separate electrophoretically from each other in the presence of Triton X-100 and wheat germ lectin. Triton X-100, a nonionic detergent, sets free the ALP isoenzymes from membrane lipoproteins, whereas wheat germ lectin binds specifically to the residues of *N*-acetylneuraminic acid in the bone ALP isoenzyme. As a result, the bone ALP isoenzyme stops in front of the sample application place and forms a tooth-shaped precipitate [12, 13].

After electrophoresis, colorimetric detection of the resolved isoenzyme bands can be carried out. For this purpose, disodium salt of 5-**b**romo-4-**c**hloro-3-**i**ndolyl **p**hosphate ($Na_2$BCIP) is hydrolyzed giving the dimer 5-**b**romo-4-**c**hloro-3-**i**ndol (BCI)$_2$ and disodium hydrogen phosphate [14] (Figure 2.4.3).

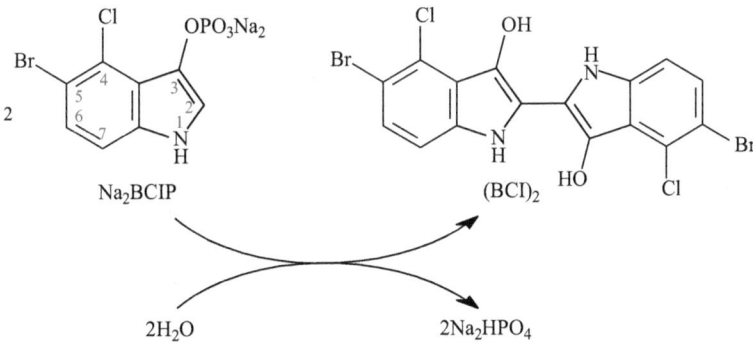

**Figure 2.4.3:** Hydrolysis of $Na_2$BCIP giving a dimer (BCI)$_2$ and disodium hydrogen phosphate.

$Na_2$BCIP is a colorless compound whereas (BCI)$_2$ is a blue dimer. The inorganic phosphate esterifies a serine residue in the neighborhood of the active site of alkaline phosphatase, which transfers afterward the inorganic phosphate onto the aminoalcohol 2-**amino**-2-**methyl**-1-**pro**panol (AMPro). So, the enzyme inhibition is avoided. The colored dimer shows which protein bands contain ALP activity.

The alkaline phosphatase is widely distributed in the human body. Its concentration is age-dependent and is elevated during the active bone growth. The activity of serum ALP is high in all bone disorders accompanied by increased osteoblastic synthesis, for example, in Paget's disease (osteitis deformans), hyperparathyroidism, rickets, osteomalacia, and osteoblastic tumors with metastases. Low serum ALP activity is found in hypothyroidism, hypophosphatasemia, pernicious anemia, and dwarfism.

### 2.4.3 Saccharide affinity electrophoresis

Glycosylation plays an important role in protein modification. To separate glycoproteins, Jackson *et al.* [15] used the boron compound [3-(**methacryloylamino**)-**phenyl**] **b**oronic **a**cid (MPBA) as ligands (probes) (Figure 2.4.4).

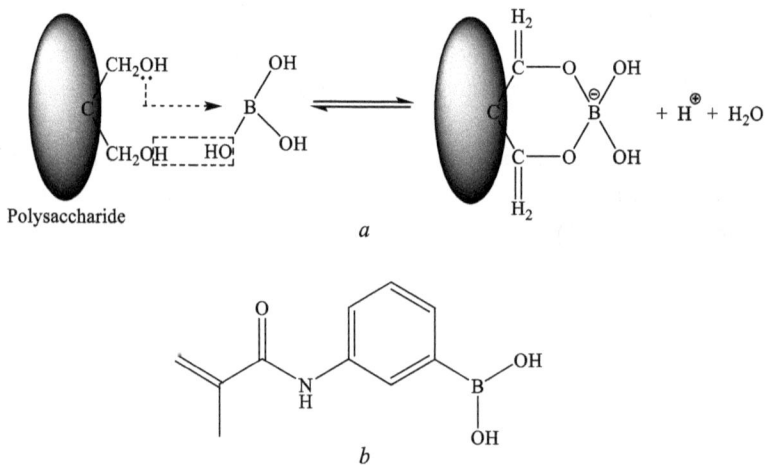

**Figure 2.4.4:** Saccharide affinity electrophoresis.
(*a*) Reversible bonding between a boron compound and a polysaccharide; (*b*) MPBA.

In **b**oronate **a**ffinity **s**accharide **e**lectrophoresis (BASE), MPBA (in concentration of 0.5–1 g/dL) is immobilized by copolymerization in a polyacrylamide gel, where it operates as an affinity probe for fructose and linear polyalcohols of saccharide derivatives containing sialic acids. Phenyl boronic acids also function as saccharide receptors in aqueous solution [16–18] and in polyacrylamide gels when incorporated there [19–21]. They form cyclic boronic esters with various carbohydrates by reversible covalent interactions. As a result, the saccharides migrate slower during gel electrophoresis. Morais *et al.* applied the same technique for serum glycoproteins of diabetic patients [22].

## 2.4.4 Affinity supported molecular matrix electrophoresis

To resolve mucins, Matsuno *et al.* established **s**upported **m**olecular **m**atrix **e**lectrophoresis (SMME) [23–25]. In this method, the fibers of a porous hydrophobic **poly**vinylidene **di**fluoride (PVDF) membrane, used as a support matrix, are coated with a hydrophilic polymer to resolve proteins or lipids. Later, Matsuno and Kameyama [26] combined SMME with affinity electrophoresis, binding to the hydrophobic fibers probes as lectins, glycolipids or antibodies, and hydrophilizing the fibers. Thus, they established the **a**ffinity SMME (ASMME). Migrating through this system, polyions interact with the probes and resolve (Figure 2.4.5).

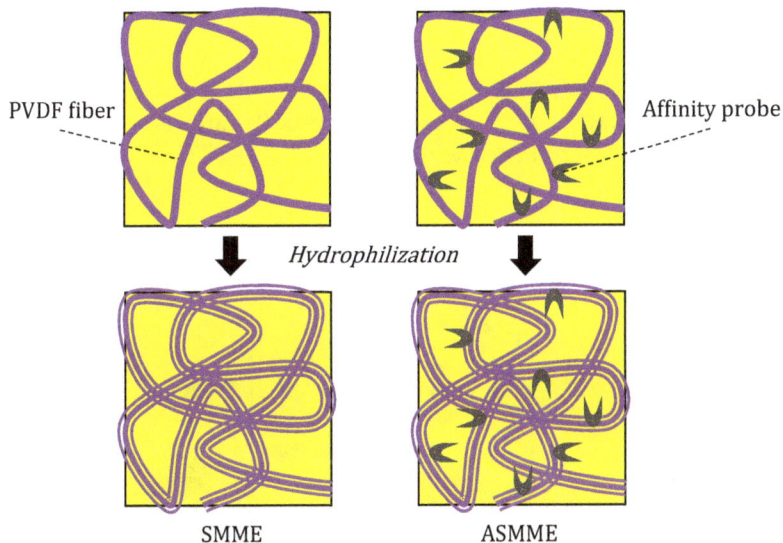

Figure 2.4.5: SMME and affinity SMME.
*Left.* A hydrophobic membrane (PVDF) becomes a separation medium after hydrophilization.
*Right.* Biological components, such as proteins, are adsorbed on PVDF fibers and act as affinity probes after hydrophilization of the PVDF membrane.

## 2.4.5 Phosphate affinity electrophoresis

In **phos**phate affinity electrophoresis (Phos-tag PAGE), Phos-tag [27] binds specifically to the divalent phosphate ions in neutral solutions (Figure 2.4.6). It ($M_r = 595$)

is developed by Hiroshima University for simultaneous analysis of a phosphoprotein isoform and its non-phosphorylated counterpart.

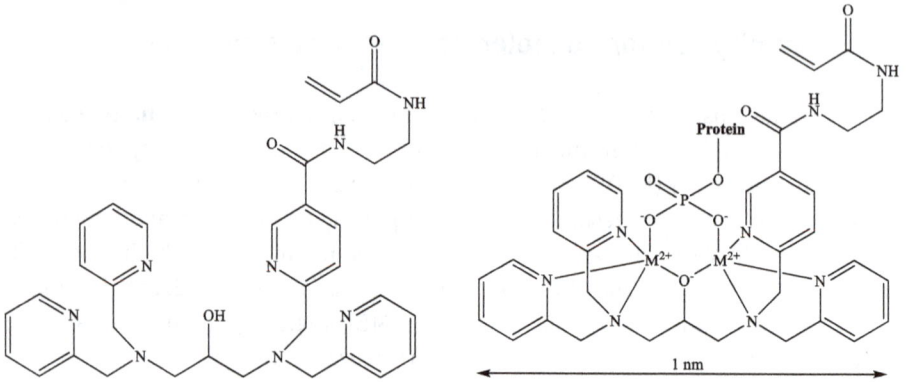

**Figure 2.4.6:** Phos-tag (*left*) and Phos-tag bound to phosphoprotein (*right*).

By applying the Laemmli SDS system [28] in Phos-tag gels, the Phos-tag SDS PAGE is developed [29, 30]. During it, a $M^{2+}$-Phos-tag complex (Figure 2.4.7) is formed, which binds to phosphate groups in gels with pH > 9.

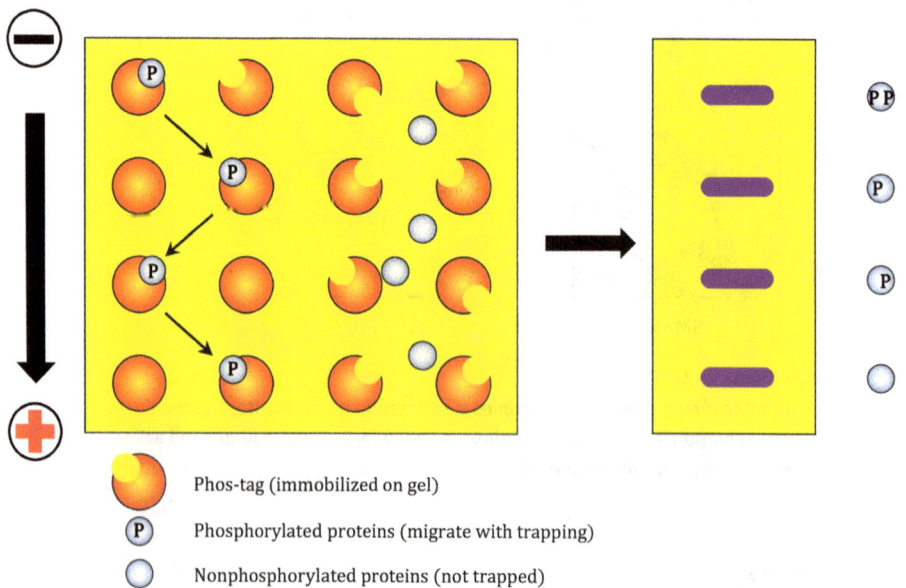

Phos-tag (immobilized on gel)

P Phosphorylated proteins (migrate with trapping)

Nonphosphorylated proteins (not trapped)

**Figure 2.4.7:** Phosphate affinity SDS electrophoresis (Phos-tag SDS PAGE).
The copolymerization of Phos-tag with acrylamide creates a gel that binds phosphorylated proteins and resolves them. $M^{2+} = Zn^{2+}$ or $Mn^{2+}$.

Using $Mn^{2+}$-Phos-tag SDS PAGE, the myosin light chain can be analyzed quantitatively. For better resolution, $Zn^{2+}$ are used in the SDS PAGE system because they form $Zn^{2+}$-Phos-tag complexes. These complexes bind optimally to phosphate groups in neutral solutions.

## 2.4.6 Capillary affinity electrophoresis

In capillary affinity electrophoresis (CAE) samples and probes are mixed, and the resulting complexes are separated by capillary electrophoresis (Figure 2.4.8). The polyions that are bound to their probes migrate; the rest of the polyions does not migrate. A detection system based on scanning laser-induced fluorescence detects the resolved bands.

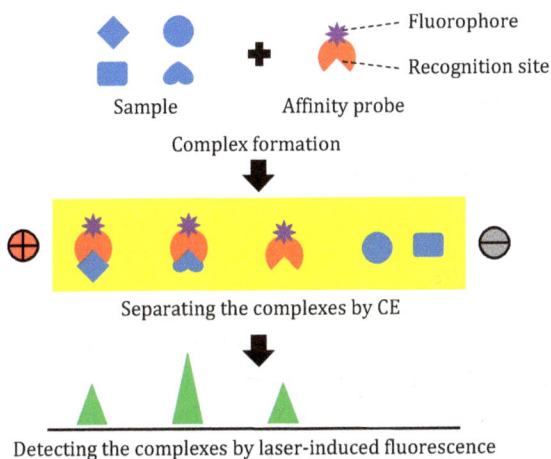

Figure 2.4.8: Principle of capillary affinity electrophoresis.

Shimura and Karger [31] used the fluorophore-labeled fragment antigen-binding (Fab) region of an immunoglobulin as an affinity probe for CAE of human growth hormone. Later, Shimura and Kasai [32] prepared an affinity probe from a recombinant immunoglobulin against human insulin.

## 2.4.7 Affinity-trap electrophoresis

The affinity-trap PAGE (AT-PAGE) was developed by Awada et al. [33]. In this method protein samples are separated by normal PAGE and then transferred onto an affinity-trap polyacrylamide gel with immobilized affinity probes. Proteins that do not have an affinity for the probes pass through the affinity-trap gel; proteins that interact with the probes are trapped and can be stained, or identified by Western blotting or mass spectrometry after in-gel digestion (Figure 2.4.9).

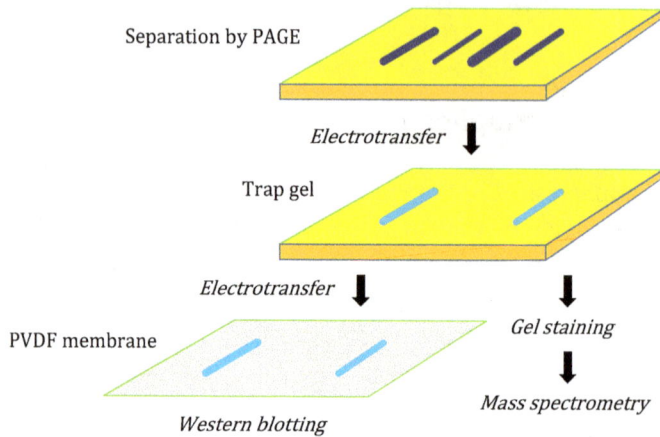

Figure 2.4.9: Principle of affinity-trap electrophoresis.

AT-PAGE is used for analyses of expressional proteins and their modifications.

## 2.4.8 Charge shift electrophoresis

Charge shift electrophoresis (CSE) is a simple and rapid method for distinguishing hydrophilic and amphiphilic proteins. It was introduced by Helenius and Simons [34]. They proved that the electrophoretic mobility of hydrophilic proteins stayed unaffected in detergent solutions; however, the mobility of amphiphilic proteins shifted anodally in Triton X-100-deoxycholate system and cathodally in Triton X-100-cetyltrimethylammonium bromide system.

## 2.4.9 Mobility shift electrophoresis

Mobility shift electrophoresis (MSE, band shift assay) is an affinity electrophoresis technique used to study if a protein or a protein mixture is capable of binding to a given DNA or RNA sequence. With its help, the affinity, abundance, association and dissociation rate constants, binding specificity of DNA-binding proteins, transcription initiation, DNA replication, DNA repair, or RNA processing and maturation can be quantitatively determined. It was described by Garner and Revzin [35], and Fried and Crothers [36].

In the MSE, a protein-nucleate complex is built that is larger than the single protein polyion. Therefore, it moves slower forming another band that is shifted up in the gel (Figure 2.4.10).

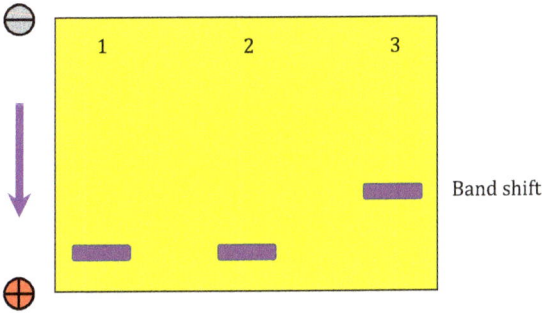

**Figure 2.4.10:** Mobility shift electrophoresis.
Lane 1 contains genetic material. Lane 2 contains a protein as well as a DNA fragment that does not interact with the protein. Lane 3 contains a protein and a DNA fragment that react with each other forming a larger, heavier and slower-moving complex. If not all DNA is bound to the protein, a second band might be seen reflecting the presence of free DNA.

The most common buffers used in the MSE are TRIS-acetate-EDTA and TRIS-borate-EDTA buffers. Inclusion of **b**ovine **s**erum **a**lbumin (BSA) or other carrier proteins in the gel improves the stability of some complexes during electrophoresis.

The MSE is carried out in four steps:

– Preparation of a radioactively labeled DNA probe containing a protein binding site.
– Preparation of a native gel; typical gels have $T = 4$–$5$ g/dL and $C = 0.025$.
– Binding of a protein mixture to a DNA probe.
– Electrophoresis of the protein-DNA complexes in a gel, which afterward is dried and autoradiographed.

To create larger complexes with greater shift, an antibodies against the protein can be added to the protein-nucleate mixture. This method is referred to as a *mobility super-shift electrophoresis*, and is used to identify proteins in protein-nucleic acid complexes.

## 2.4.10 Protocols

### Electrophoresis of ALP Isoenzymes on Lectin Agarose Gels

**Materials and equipment**
Ready-to-use agarose gels

Electrophoresis cell
Power supply

**Procedure**
- Remove the cover film from an ALP isoenzyme agarose gel.
- Blot the gel briefly with a filter paper strip where samples will be applied.
- Lay down an application template on the blotted place adjusting it according to the arrows on the support film.
- Press gently the template with fingers to push out air bubbles between the template and gel.
- Dilute the sera up to an ALP activity of 200–1 000 U/L and mix them with a sample buffer at a volume ratio of 1:1. If cerebrospinal fluid or urine is analyzed, concentrate it up to the same activity and mix it with the sample buffer.
- Apply 5 μL of serum, cerebrospinal fluid or urine in the template slots and let the samples diffuse in the gel for 5 min.
- Blot the rest of the samples with a filter paper strip and remove the template and strip.
- Fill the tanks of an electrophoresis cell with electrode buffer.
- Hang the film-supported gel on the bridge of the electrophoresis cell, the gel side down.
- Place the bridge with the gel into the electrophoresis cell so that the gel ends will contact directly with the electrode buffer in both electrode tanks.
- Run the ALP electrophoresis at constant direct voltage of 80–100 V for 25–30 min.

### Staining the ALP Isoenzymes

### Materials and equipment
To stain ALP isoenzymes, two substrate solutions are needed: solution 1 and solution 2. Substrate solution 1 is an AMPro-Cl buffer (95% 2-**amino**-2-**methyl**-1-**propanol** corrected with HCl to pH = 10.2). Substrate solution 2 contains NBT (**nitro blue tetrazolium** chloride, 4-nitrotetrazolium blue chloride). The solutions are made according to Table 2.4.1:

**Table 2.4.1:** Contents of solutions for ALP isoenzymes.

| Substrate solution 1, 6x | | Substrate solution 2, 24x | |
|---|---|---|---|
| AMPro | 14.23 mL | 98% NBT | 0.83 g |
| 37% HCl | 4.14 mL | Methanol to | 100.00 mL |
| 0.1 g/dL ZnCl$_2$ | 6.82 mL | | |

**Table 2.4.1** (continued)

| Substrate solution 1, 6x | | Substrate solution 2, 24x |
|---|---|---|
| 10 g/dL MgCl$_2$ | 1.02 mL | |
| 10 g/dL Na$_2$BCIP* | 2.57 mL | |
| NaN$_3$ | 0.10 g | |
| Deionized water to | 100.00 mL | |

*Na$_2$BCIP = 5-**b**romo-4-**c**hloro-3-**i**ndolyl **p**hosphate disodium salt.

**Procedure**

- Make 30 mL (4:1:5, $V:V:V$) substrate solution (5.0 mL Substrate solution 1 + 1.25 mL Substrate solution 2 + 23.75 mL deionized water).
- Immerse the gel (gel surface upward) in the substrate solution.
- Incubate at 37 °C for 30 min.
- Inactivate the ALP isoenzymes in methanol-acetic acid-water (4:1:5, $V:V:V$) for 30 min. To make 30 mL solution, mix 12 mL methanol with 3.0 mL acetic acid and 15.0 mL deionized water).
- Wash the gel in tap water for 5 min and rinse it three times in deionized water.
- Dry the gel using a hair dryer, a drying oven, or at room temperature.
- Document the ALP isoenzyme bands using a scanner, photo camera, or densitometer at 580 nm.

# References

[1]   Takeo K, Fugimoto M, Suzuno R, Kuwahara A. Physicochem Biol, 1978, 22, 139–144.
[2]   Takeo K. Electrophoresis, 1984, 5, 187–195.
[3]   Mosbach K (Jakoby WD, Wilchek M (eds.)). Methods in Enzymology. Academic Press, New York, 1974, Vol. 24, Part B, 595–597.
[4]   Andersson L, Jörnvall H, Akeson A, Mosbach K. Biochim Biophys Acta, 1974, 364, 1–8.
[5]   Castellino FJ, Sodetz JM, Sietring GE (Markert CL (ed.)). Isoenzymes. Academic Press, New York, 1975, Vol. 1, 245–258.
[6]   Schreiber WE, Whitta L. Clin Chem, 1986, 32, 1570–1573.
[7]   Van Damme EJM, Peumans WJ, Pusztai A, Bardocz S. Handbook of Plant Lectins: Properties and Biomedical Applications. John Wiley & Sons, 1998, 7–8.
[8]   Rutishauser U, Sachs L. J Cell Biol, Rockefeller University Press, 1975, 65, 247–257.
[9]   Brudner M, Karpel M, Lear C et al. PLoS ONE, 2013, 8, e60838.
[10]  Moss DW. Clin Chem, 1982, 28, 2007–2016.
[11]  Crofton PM. Crit Rev Clin Lab Sci, 1982, 16, 161–194.
[12]  Rosalki SB, Foo AY. Clin Chem, 1984, 30, 1182–1186.
[13]  Schreiber WE, Whitta L. Clin Chem, 1986, 32, 1570–1573.
[14]  McGadey J. Histochemie, 1970, 23, 180–184.

[15] Jackson TR, Springall JS, Rogalle D, Masumoto N, Li HC, D'Hooge F, Perera SP, Jenkins TA, James TD, Fossey JS *et al.* Electrophoresis, 2008, 29, 4185–4191.

[16] James TD, Shinkai S. Top Curr Chem, 2002, 218, 159–200.

[17] Perez-Fuertes Y, Kelly AM, Fossey JS, Powell ME *et al.* Natl Protoc, 2008, 3, 210–214.

[18] Kelly AM, Perez-Fuertes Y, Fossey JS, Yests SL *et al.* Natl Protoc, 2008, 3, 215–219.

[19] Lee MC, Kabilan S, Hussain A, Yang XP *et al.* Anal Chem, 2004, 76, 5748–5755.

[20] Sartain FK, Yang XP, Lowe CR. Anal Chem, 2006, 78, 5664–5670.

[21] Yang XP, Lee MC, Sartain F, Pan XH, Lowe CR. Chem Eur J, 2006, 12, 8491–8497.

[22] Morais MPP, Mackay JD, Bhamra SK, Buchanan JG, James TD, Fossey JS, Van Den Elsen JMH. Proteomics, 2010, 10, 48–58.

[23] Matsuno YK, Saito T, Gotoh M, Narimatsu H, Kameyama A. Anal Chem, 2009, 81, 3816–3823.

[24] Matsuno YK, Dong W, Yokoyama S, Yonezawa S, Saito T, Gotoh M, Narimatsu H, Kameyama A. Electrophoresis, 2011, 32, 1829–1836.

[25] Matsuno YK, Dong W, Yokoyama S, Yonezawa S, Narimatsu H, Kameyama A. J Immunol Methods, 2013, 394, 125–130.

[26] Matsuno YK, Kameyama A. Electrophoresis Lett (Japanese Electrophoresis Society), 2014, 58, 27–29.

[27] Kinoshita E, Takahashi M, Takeda H, Shiro M, Koike T. Dalton Trans, 2004, 21, 1189–1193.

[28] Laemmli UK. Nature, 1970, 227, 680–685.

[29] Shimura R. Progress in Affinophoresis, J Chromatogr, 1990, 510, 251–270.

[30] Kinoshita E, Kinoshita-Kikuta E, Takiyama K, Koike T. Mol Cell Proteomics, 2006, 5, 749–757.

[31] Shimura K, Karger BL. Anal Chem, 1994, 66, 9–15.

[32] Shimura K, Kasai K. Electrophoresis, 2014, 35, 840–845.

[33] Awada C, Sato T, Takao T. Anal Chem, 2010, 82, 755–761.

[34] Helenius A, Simons K. Proc Natl Acad Sci USA, 1977, 74, 529–532.

[35] Garner MM, Revzin A. Nucl Acids Res, 1981, 9, 3047–3060.

[36] Fried M, Crothers DM. Nucl Acids Res, 1981, 9, 6505–6525.

# 2.5 Polyacrylamide gel zone electrophoresis of proteins

2.5.1     Homogeneous gel zone electrophoresis of proteins —— 131
2.5.1.1   Theory of polyacrylamide gel zone electrophoresis of proteins —— 132
2.5.1.2   McLellan buffers —— 133
2.5.1.3   Running electrophoresis —— 134
2.5.2     Gradient gel zone electrophoresis of proteins —— 134
2.5.2.1   Theory of gradient gel zone electrophoresis —— 135
2.5.2.2   Ferguson plots —— 137
2.5.2.3   Determination of Stokes radii and masses of native proteins —— 138
2.5.3     Blue native polyacrylamide gel electrophoresis —— 139
2.5.4     Protocols —— 140
          Horizontal Electrophoresis of Proteins on Gradient Gels —— 140
          Blue Native Polyacrylamide Gel Electrophoresis —— 142
          Preparation of Dialyzed Cell Lysate —— 142
          Casting Gradient BN Gels —— 143
          Running BN Electrophoresis —— 145
          References —— 145

The polyacrylamide gel zone electrophoresis of proteins can be carried out in homogeneous or gradient gels.

## 2.5.1 Homogeneous gel zone electrophoresis of proteins

The **p**oly**a**crylamide **g**el **e**lectrophoresis (PAGE) is used for separating proteins with relative molecular masses $M_r$ from $5 \times 10^3$ to $2 \times 10^6$. The separation depends on the protein mobilities and polyacrylamide gel properties. The *native PAGE* is used for native proteins, for example, enzymes. The *denaturing PAGE* is used to separate proteins, denatured by SDS or other detergents.

The gels for *vertical electrophoresis* usually have direct contact with the tank buffers. The samples are pipetted in gel wells under the upper electrode buffer. The wells are produced by a comb, which was inserted during the polymerization process. The samples should contain 10–20 g/dL glycerol or sucrose to delay their diffusion in the electrode buffer.

The gels for *horizontal electrophoresis* are usually cast on a support film. The samples are applied onto the gel using a silicone template. The horizontal electrophoresis in thin homogeneous or gradient gels, cast on support films, has many advantages over the vertical electrophoresis [1]: they are easily to handle during electrophoresis; permit an usage of electrode strips with buffers; need small sample volumes and adequate cooling; the resolved protein bands are quickly stained; and the electrophoresis can be automated.

https://doi.org/10.1515/9783110761641-015

### 2.5.1.1 Theory of polyacrylamide gel zone electrophoresis of proteins

The polyacrylamide gel zone electrophoresis is running in one buffer. It is contained in the gel, sample, and electrode tanks. The resolution of the electrophoresis depends on the volume and electric charge of the proteins. The electric charge of a protein is a function of the dissociation degree of its chemical groups and depends on the pH value and ionic strength of the buffer.

The ionic strength of a buffer should be relative low to keep heat generation at a minimum. On the other hand, the very low ionic strength lowers electrophoretic resolution. Therefore, the ionic strength of electrophoresis buffers should have middle values in the range of 0.01–0.1 mol/L.

The polyacrylamide gel zone electrophoresis can be carried out in alkaline or acidic buffers. In alkaline buffers, the proteins migrate to the anode, because they are negatively charged; in acidic buffers, they migrate to the cathode, because they are positively charged. The sample is applied on the cathode or anode side of the gel, respectively.

In electrophoresis in alkaline buffers, Bromophenol blue Na salt could be added to the sample; and in electrophoresis in acidic buffers, Fuchsine red could be added to the sample. These dyes overrun the proteins and show their movement in the gel.

The preparation of polyacrylamide gels of different concentrations for polyacrylamide gel zone electrophoresis is shown in Table 2.5.1.

**Table 2.5.1:** Contents of homogeneous polyacrylamide gels of different $T$-concentrations, but same dimensions (120 × 250 × 0.5 mm).

| Solutions | Concentrations | $T = 5$ g/dl | $T = 12.5$ g/dL | $T = 20$ g/dL |
|---|---|---|---|---|
| **Buffer–gel system for acidic proteins** | | | | |
| TRIS-glycinate buffer, 4x (0.8 mol/L, pH = 8.3) or | 0.2 mol/L | 4.0 mL | 4.0 mL | 4.0 mL |
| TRIS-chloride buffer, 4x (1.5 mol/L, pH = 8.8) | 0.375 mol/L | 4.0 mL | 4.0 mL | 4.0 mL |
| Monomeric solution ($T = 50$ g/dL, $C = 0.03$) | $T = 5–20$ g/dL | 1.6 mL | 4.0 mL | 6.4 mL |
| 87% glycerol | 0.2 mol/L | 2.8 mL | 2.8 mL | 2.8 mL |
| 10 g/dL TMEDA | 0.4 mmol/L | 74.4 µL | 74.4 µL | 74.4 µL |
| 10 g/dL APS | 0.2 to 0.15 mmol/L | 64.0 µL | 25.6 µL | 16.0 µL |
| Deionized water to 16.0 mL | – | 7.5 mL | 5.1 mL | 2.7 mL |

**Table 2.5.1** (continued)

| Solutions | Concentrations | $T = 5$ g/dL | $T = 12.5$ g/dL | $T = 20$ g/dL |
|---|---|---|---|---|
| **Buffer–gel system for alkaline proteins** | | | | |
| TRIS-acetate buffer, 4x (0.8 mol/L, pH = 4.7) | 0.2 mol/L | 4.0 mL | 4.0 mL | 4.0 mL |
| Monomeric solution ($T = 50$ g/dL, $C = 0.03$) | $T = 5$–20 g/dL | 1.6 mL | 4.0 mL | 6.4 mL |
| 87% glycerol | 0.2 mol/L | 2.8 mL | 2.8 mL | 2.8 mL |
| 10 g/dL TMEDA | 0.4 mmol/L | 74.4 µL | 74.4 µL | 74.4 µL |
| 10 g/dL APS | 0.2 to 0.15 mmol/L | 64.0 µL | 25.6 µL | 16.0 µL |
| Deionized water to 16.0 mL | – | 7.5 mL | 5.1 mL | 2.7 mL |

$c_{APS}$ was calculated according to the formula $c_{APS} = 0.32/T$, which we propose.

## 2.5.1.2 McLellan buffers

McLellan proposed a set of buffers, which are applicable for electrophoresis of native proteins [2]. These buffers cover a pH range from 3.8 to 10.2, and have relatively low conductivity (Table 2.5.2).

**Table 2.5.2:** Buffers for continuous electrophoresis of native proteins.

| Buffer pH | Basic component, concentration | Concentration for 5x solution | Acidic component, concentration | Concentration for 5x solution |
|---|---|---|---|---|
| 3.8 | β-Alanine, 30 mmol/L | 13.36 g/L | Lactic acid, 20 mmol/L | 7.45 mL/L |
| 4.4 | β-Alanine, 80 mmol/L | 35.64 g/L | Acetic acid, 40 mmol/L | 11.50 mL/L |
| 4.8 | GABA, 80 mmol/L | 41.24 g/L | Acetic acid, 20 mmol/L | 5.75 mL/L |
| 6.1 | Histidine, 30 mmol/L | 23.28 g/L | MES, 30 mmol/L | 29.28 g/L |
| 6.6 | Histidine, 25 mmol/L | 19.40 g/L | MOPS, 30 mmol/L | 31.40 g/L |
| 7.4 | Imidazole, 43 mmol/L | 14.64 g/L | HEPES, 35 mmol/L | 41.71 g/L |
| 8.1 | TRIS, 32 mmol/L | 19.38 g/L | HEPPS, 30 mmol/L | 37.85 g/L |
| 8.7 | TRIS, 50 mmol/L | 30.29 g/L | Boric acid, 25 mmol/L | 7.73 g/L |
| 9.4 | TRIS, 60 mmol/L | 36.34 g/L | CAPS, 40 mmol/L | 44.26 g/L |
| 10.2 | Ammonia, 37 mmol/L | 12.50 mL/L | CAPS, 20 mmol/L | 22.13 g/L |

Other buffers that have been used for native zone electrophoresis are: TRIS-glycinate (pH range 8.3–9.5) [3], TRIS-acetate (pH range 7.2–8.5) [4], and TRIS-borate (pH range 8.3–9.3) [5] buffers. Borate ions $[B(OH)_4^-]$ form complexes with some sugars; therefore, they influence the resolution of glycoproteins.

### 2.5.1.3 Running electrophoresis

The contact between the electrodes and polyacrylamide gel can be realized *via* filter paper bridges impregnated with electrode buffer. They should be parallel to the gel ends, and overlap 1 cm of the gel ends.

An alternative to the paper bridges are the electrode strips. They can be composed of gel (usually polyacrylamide gel) or thick filter paper. The strips contain electrode buffer(s). The electrodes are placed onto the electrode strips (Figure 2.5.1).

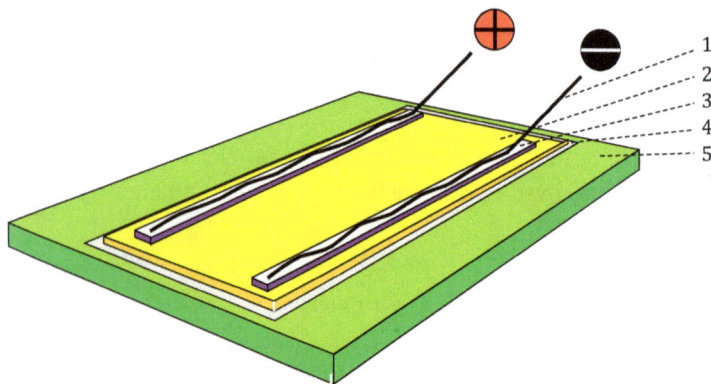

**Figure 2.5.1:** Electrode strips as contacts between gel and electrodes. 1. Electrode; 2. gel; 3. electrode strip; 4. support film; 5. cooling plate.

The electrophoretic conditions depend on buffer composition and gel thickness. The voltage can be in the range 200–500 V and should not be changed during electrophoresis. The cooling temperature is recommended to be 10 °C. If required, the gels can be cooled to 2–4 °C.

## 2.5.2 Gradient gel zone electrophoresis of proteins

The gradient polyacrylamide gels were introduced by Kolin [6] and improved by Margolis and Kenrick [7]. The electrophoresis is usually carried out in same buffers. The $T$-concentration of the polyacrylamide gels increases continuously from one gel end to the other gel end, which results in a continuous decrease of the pore diameters. Therefore, the samples, which are applied on the large porous gel end, first are moving fast, later, however, their velocities decrease gradually because they begin to migrate through the small porous gel.

The gradient gels can be used to identify the molecular masses and radii of native proteins [8], comparing the velocities of tested proteins with the velocities of proteins with known masses and radii.

## 2.5.2.1 Theory of gradient gel zone electrophoresis

Let us consider the movement of a charged particle in a linear concentration gradient. The velocity $v$ of the particle equals the ratio between the distance $l = l_2 - l_1$ and the time $t = t_2 - t_1$. If we express the ratio $v/l$ as $a$, then

$$v = \frac{dl}{dt} = al \qquad (2.5.1)$$

hence

$$\frac{dl}{l} = adt \qquad (2.5.2)$$

The integration of eq. (2.5.1) in the intervals $l_1, l_2$ and $t_1, t_2$ gives the expressions

$$\int_{l_1}^{l_2} \frac{dl}{l} = \int_{t_2}^{t_1} adt \qquad (2.5.3)$$

and

$$\ln \frac{l_2}{l_1} = at \qquad (2.5.4)$$

which can be transformed in

$$e^{at} = \frac{l_2}{l_1} \qquad (2.5.5)$$

It follows from eq. (2.5.1) and (2.5.5) that

$$l_1 = \frac{v_1}{a} \qquad (2.5.6)$$

and

$$l_2 = \frac{v_1}{a} e^{at} \qquad (2.5.7)$$

So, the distance

$$l_2 - l_1 = \frac{v_1}{a}\left(e^{at} - 1\right)$$ (2.5.8)

and the modified linear velocity

$$v_2 - v_1 = v_1\left(e^{at} - 1\right)$$ (2.5.9)

Let us now analyze the movement of the zone $l_1, l_2$ through a linear concentration gradient, if the initial zone width $l_1' - l_1 = d_1$, and the final zone width $l_2' - l_2 = d_2$. Then, it follows from eq. (2.5.5) and eq. (2.5.9) that

$$d_2 = e^{at}\left(l_1' - l_1\right) = d_1 e^{at} = d_1 \frac{v_2}{v_1}$$ (2.5.10)

The polyionic velocity decreases during electrophoresis, i.e. $v_2 < v_1$. Therefore, according to eq. (2.5.10), the final width of the band is narrower than its initial width, i.e. $d_2 < d_1$.

The gradient gel typically has $T = 4–30$ g/dL (0.5 to 4.22 mol/L) (Figure 2.5.2). Margolis and Kenrick used the TRIS-borate-EDTA buffer of pH = 8.3 in the resolving gel as well as in the electrode tank. Other buffers for gradient zone electrophoresis are the TRIS-glycinate, TRIS-barbitalate, and TRIS-taurinate buffers.

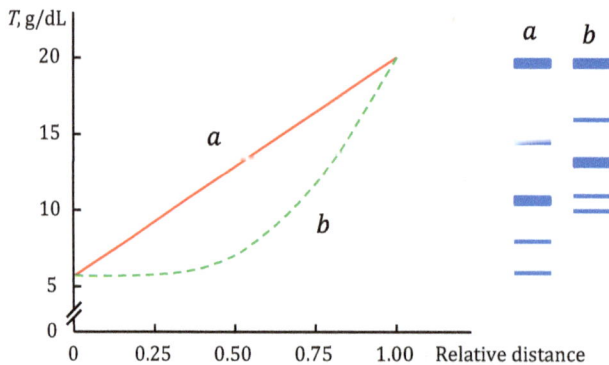

**Figure 2.5.2:** Proteins separated in a linear (*a*) and an exponential (*b*) gradient gels. The polyacrylamide concentration *T* is 5–20 g/dL.

The resolution of the gradient polyacrylamide electrophoresis is high, but the separation takes a long time. Besides, there are difficulties while treating the gel after the electrophoresis – it deforms during drying.

## 2.5.2.2 Ferguson plots

The dependence of the polyionic mobilities on the gel concentration was studied by Kenneth Ferguson who analyzed the movement of the pituitary hormone in starch gels [9]. He described the linear dependence of the mobilities on polyion masses, now referred to as Ferguson plots.

If different polyions are separated under same conditions (in a same buffer at same temperature and in same electric field strength), but at different gel concentrations, different distances $d$ are run, i.e., they have different relative mobilities. The relative mobility $\mu_r$ of a polyion is defined as the ratio between the migration distance $d$ of the polyion (or its mobility $\mu$) in the gel, and the migration distance of the polyion at free electrophoresis $d_0$ (or its absolute mobility $\mu_0$), according to the equation

$$\mu_r = \frac{d}{d_0} = \frac{\mu}{\mu_0} \qquad (2.5.11)$$

The line slope is referred to as the retardation coefficient $K_R$, i.e., the extent, in which the gel matrix affects the polyionic mobility at a certain pH value, ionic strength, and temperature. The absolute mobility $\mu_0$ is related to the electric charge of a protein, while $K_R$ is related to its hydrodynamic properties, which depend on the peptide chain length (its mass).

The relationship between the relative protein mobility and the gel concentration may be expressed by the equation

$$\log \mu_r = \log \mu_{0(r)} - K_R T_r \qquad (2.5.12)$$

where $\mu_{0(r)}$ is the relative absolute mobility of the protein in the absence of any sieving matrix, and $T_r$ is the relative total concentration of the gel, equal to the ratio between $T$, in g/dL, and 1 g/dL. $\mu_r$, $\mu_{0(r)}$, $K_R$, and $T_r$ are dimensionless magnitudes.

According to the location of Ferguson plots, following statements can be made:
- If the lines are parallel, the polyions have identical compositions, but different charges, as the isoenzymes.
- If two non-parallel lines do not intersect, the polyion of the upper straight line is smaller and has a higher net charge than the polyion of the lower straight line.
- If the plots cross, the polyion whose plot crosses the y-axis high up has bigger mass but smaller charge.
- If a lot of plots cross in one point, the protein builds various polymers (Figure 2.5.3).

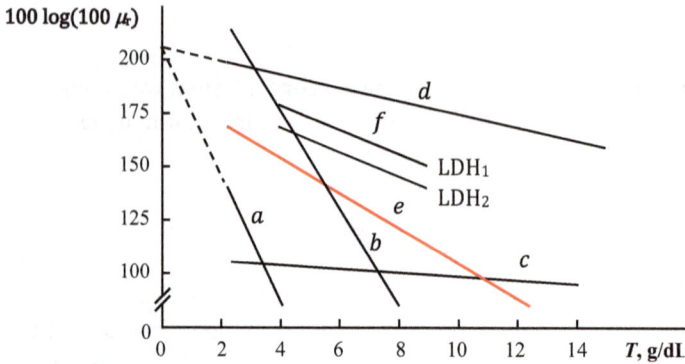

**Figure 2.5.3:** Ferguson plots: relative polyionic mobilities as a function of the polyacrylamide gel concentration $T$. The slopes of the lines are equal to $K_R$. (a) Polyion with a large mass and a low charge; (b) polyion with a large mass and a large charge; (c) polyion with a small mass and a small charge; (d) polyion with a small mass and a large charge; (e) polyion with an intermediate mass and an intermediate charge (typical case); (f) $LDH_1$ and $LDH_2$ – isoenzymes of lactate dehydrogenase, which have same masses but different charges.

### 2.5.2.3 Determination of Stokes radii and masses of native proteins

The Stokes radius $r_s$ of a native protein is linked to its relative maximum migration distance $d_{r(max)}$ in a gel according to the equation

$$\{[\ln(\ln d_{r(max)})]\}^{-1} = \alpha r_s + \beta \qquad (2.5.13)$$

where $\alpha$ and $\beta$ are constants. In order to determine the relative maximum migration distance, linear gradient gels with $T = 3$–$30$ g/dL and constant concentration of the cross-linker (BIS) is necessary. Simultaneously, the Stokes radius of a native protein is linked to its relative molecular mass $M_r$ over the relationship

$$r_s = \varepsilon M_r^{1/3} \qquad (2.5.14)$$

where $\varepsilon$ is a constant. Substituting eq. 2.5.13 in eq. 2.5.14, the equation

$$\ln\{[\ln(\ln d_{r(max)})]\}^{-1} = \gamma(M_r)^{1/3} + \delta \qquad (2.5.15)$$

is obtained where $\gamma$ and $\delta$ are constants. The last equation shows $\gamma$ that the natural logarithm of the reciprocal of $d_{r(max)}$ is linearly correlated with the third root of $M_r$. If the logarithm of $d_{r(max)}$ of calibration proteins is plotted against the logarithm of their relative molecular masses, a straight line is obtained. It allows calculating the relative molecular mass $M_r$ of a protein from the relative maximum migration distance $d_{r(max)}$ (Figure 2.5.4).

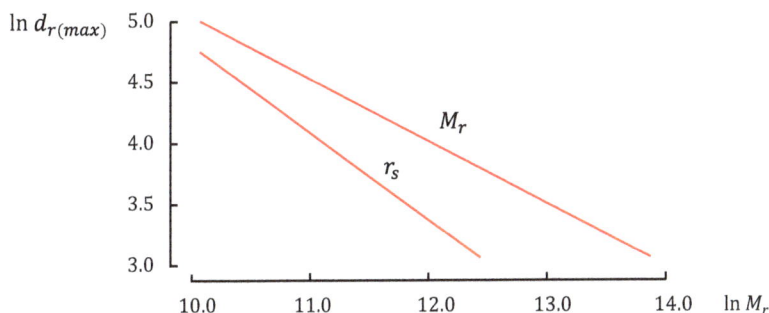

**Figure 2.5.4:** Calibration straight lines used for determination of Stokes radii $r_s$ or relative molecular masses $M_r$ of carbonic anhydrase from mammalian erythrocytes [1].
[1] Wagner H, Blasius E. Praxis der elektrophoretischen Trennmethoden, 1989, 79.

Gradient gels give a better separation and sharpening of protein bands [10] under non-denaturing conditions. Best results are obtained in 3–30 g/dL gels [11] where the relative molecular masses of proteins, in the range of 13,000–950,000, can be determined.

When the mobility and radius of a protein are determined, its electric charge can also be calculated [12].

## 2.5.3 Blue native polyacrylamide gel electrophoresis

Blue native polyacrylamide gel electrophoresis (BN-PAGE) is a technique for precise resolving of proteins in native gels. It uses the anionic blue dye Coomassie brilliant blue G-250. Because of its hydrophobic properties, Coomassie brilliant blue G-250 binds unspecifically to all proteins, inclusively membrane proteins, and covers them with negative charges. The hydrophobic proteins lose their hydrophobicity and convert into water-soluble proteins. At neutral pH values they move through gradient gels toward the anode as blue bands gradually decelerating with the decreasing pore size of the gel [13, 14]. After BN-PAGE, the visualization of proteins can be achieved by additional Coomassie brilliant blue staining, silver staining, or immunoblotting. Then, the protein bands can be analyzed by mass spectrometry [15].

BN-PAGE allows also detection, quantitation, and purification of NADPH-producing enzymes: glucose 6-phosphate dehydrogenase (EC 1.1.1.49), malic enzyme (EC 1.1.1.40), and NADP-dependent isocitrate dehydrogenase (EC 1.1.1.42) [16]. These enzymes, together with phosphogluconate dehydrogenase (EC 1.1.1.44), are involved in the NADPH production. NADPH is a pivot in the lipid synthesis, cellular replication, and functioning of the antioxidative enzymes catalase and glutathione reductase [17, 18, 19].

## 2.5.4 Protocols

### Horizontal Electrophoresis of Proteins on Gradient Gels

**Materials and equipment**
TRIS
HCl
Light and heavy monomeric (acrylamide/BIS) solutions
Sucrose
TMEDA
APS
Casting cassette with glass plates, spacers, and clamps
Gradient maker

*4x TRIS-HCl buffer, pH = 8.8*
TRIS                          182.0 g (1.5 mol/L)
Deionized water      600.0 mL
*Adjust with HCl to pH = 8.8.*
Deionized water to 1,000.0 mL

**Table 2.5.3:** Light monomeric solutions (cited in the text).

| Stock solution | Light monomeric solution (g/dL) | | | | | | | | | |
|---|---|---|---|---|---|---|---|---|---|---|
| | 5 | 6 | 7 | 8 | 9 | 10 | 11 | 12 | 13 | 14 |
| 4x TRIS-HCl pH = 8.8, mL | 3.75 | 3.75 | 3.75 | 3.75 | 3.75 | 3.75 | 3.75 | 3.75 | 3.75 | 3.75 |
| 30 g/dl Acrylamide/ 0.8 g/dL BIS, mL | 2.50 | 3.00 | 3.50 | 4.00 | 4.50 | 5.00 | 5.50 | 6.00 | 6.50 | 7.00 |
| 10 g/dL TMEDA, mL | 0.05 | 0.05 | 0.05 | 0.05 | 0.05 | 0.05 | 0.05 | 0.05 | 0.05 | 0.05 |
| 10 g/dL APS, mL | 0.10 | 0.09 | 0.07 | 0.06 | 0.05 | 0.05 | 0.04 | 0.04 | 0.04 | 0.03 |
| Deionized water,mL, to | 15.00 | 15.00 | 15.00 | 15.00 | 15.00 | 15.00 | 15.00 | 15.00 | 15.00 | 15.00 |

**Procedure**
- Assemble the casting cassette.
- Set up the gradient maker, close all its valves, and place a stir bar in the mixing chamber.
- Place the gradient maker on a magnetic stirrer.

**Table 2.5.4:** Heavy monomeric solutions (cited in the text).

| Stock solution | Heavy monomeric solution (g/dL) | | | | | | | | | | |
|---|---|---|---|---|---|---|---|---|---|---|---|
| | 10 | 11 | 12 | 13 | 14 | 15 | 16 | 17 | 18 | 19 | 20 |
| 4x TRIS-HCl, pH = 8.8,mL | 3.75 | 3.75 | 3.75 | 3.75 | 3.75 | 3.75 | 3.75 | 3.75 | 3.75 | 3.75 | 3.75 |
| 30 g/dL Acrylamide/ 0.8 g/dL BIS, mL | 5.00 | 5.50 | 6.00 | 6.50 | 7.00 | 7.50 | 8.00 | 8.50 | 9.00 | 9.50 | 10.00 |
| Sucrose, g | 2.25 | 2.25 | 2.25 | 2.25 | 2.25 | 2.25 | 2.25 | 2.25 | 2.25 | 2.25 | 2.25 |
| 10 g/dL TMEDA, mL | 0.05 | 0.05 | 0.05 | 0.05 | 0.05 | 0.05 | 0.05 | 0.05 | 0.05 | 0.05 | 0.05 |
| 10 g/dL APS, mL | 0.05 | 0.04 | 0.04 | 0.04 | 0.03 | 0.03 | 0.03 | 0.03 | 0.03 | 0.03 | 0.02 |
| Deionized water, mL, to | 15.00 | 15.00 | 15.00 | 15.00 | 15.00 | 15.00 | 15.00 | 15.00 | 15.00 | 15.00 | 15.00 |

- Connect the outlet tubing of the gradient maker to a micropipette tip and place it onto the casting cassette.
- Pipette the high-concentration (heavy) solution into the mixing chamber and the low-concentration (light) solution into the reservoir chamber, and add APS.
- Open the interconnecting valve and turn on the magnetic stirrer.
- Open the outlet valve and fill the casting cassette from the top. The heavy solution will flow first into the cassette, followed by the light solution.
- Overlay the gradient gel solution with deionized water and allow it to polymerize for 1 h.
- Remove the overplayed water.
- Dissemble the casting cassette and take out the gel with the support film.
- Place the gel on a horizontal separation cell.
- Put polyacrylamide or paper strips, soaked with buffers, onto the gel poles.
- Place a silicone template on the gel in front of the cathode strip.
- Add to samples 10–20 g/dL glycerol and apply them in the slots of the template.
- Run the electrophoresis.
- Stain the protein bands with Coomassie brilliant blue or silver.

**Blue Native Polyacrylamide Gel Electrophoresis**

**Preparation of Dialyzed Cell Lysate**

## Materials and equipment
BISTRIS
6-Aminohexanoic acid (ε-aminocaproic acid)
Sodium chloride
$Na_2EDTA$
Brij 96
Digitonin
Triton X-100
Dialysis membrane for $M_r = 10,000–50,000$
Centrifuge
Microcentrifuge tubes
Magnet stirrer
Parafilm

*Phosphate-buffered saline (PBS) buffer, pH = 7.4*

| | |
|---|---|
| $Na_2HPO_4$ | 1.42 g (10 mmol/L)) |
| $KH_2PO_4$ | 0.24 g (1.8 mmol/L) |
| NaCl | 8.00 g (137 mmol/L) |
| KCl | 0.20 g (2.7 mmol/L) |
| Deionized water to | 1,000.00 mL |

*The buffer should have pH = 7.4, if prepared properly.*

*BN-lysis (BN-dialysis) buffer, pH = 7.0*

| | |
|---|---|
| BISTRIS | 0.42 (20 mmol/L) |
| ε-Aminocaproic acid | 6.56 g (500 mmol/L) |
| $Na_2EDTA$ | 0.07 g (2 mmol/L) |
| NaCl | 0.12 g (20 mmol/L) |
| Glycerol | 10.00 mL |
| Deionized water to | 100.00 mL |

*Adjust to pH = 7.0 with HCl. Store at 4 °C.*

*Detergent*
0.5–2.0 g/dL Digitonin, 0.1–0.5 g/dL Brij 96, 0.1–0.5 mL/dL Triton X-100, 0.1– 0.5 g/dL Dodecylmaltoside
*The detergent should be determined empirically. Store in 5 mL aliquots at –20 °C.*

*Protease and phosphatase inhibitors*

| | |
|---|---|
| PMSF (**p**henyl**m**ethyl**s**ulfonyl **f**luoride) | 0.02 g (1.0 mmol/L) |
| Sodium orthovanadate | 0.01 g (0.5 mmol/L) |
| Deionized water to | 100.00 mL |

## Procedure

- Obtain a cell pellet (10 x $10^6$ cells) at 350 g and 4 °C for 5 min.
- Wash the pellet three times with ice-cold **p**hosphate-**b**uffered **s**aline (PBS) buffer and centrifuge.
- Resuspend the pellet in 250 µL of ice-cold BN-lysis buffer.
- Centrifuge at 13,000 g at 4 °C for 15 min to remove insoluble material.
- Melt a hole in the cap of a 1.5 mL microcentrifuge tube using a heated Pasteur pipette, then place the tube on ice to cool down to 4 °C.
- Transfer the supernatant into the tube through the cap hole.
- Place a dialysis membrane for $M_r$ = 10,000 on top of the opened tube, close the cap, and cut off the excess dialysis membrane.
- Seal the cap with Parafilm.
- Centrifuge the tubes at the lowest speed possible at 4 °C for 10 s.
- Prepare a 100 mL beaker with 10 mL of cold BN-dialysis buffer per 100 µL sample.
- Affix the tube with tape upside-down inside the beaker, and remove air bubbles from the hole beneath the cap using a Pasteur pipette.
- Place the beaker onto the magnet stirrer, switch it on, and leave it run in a cold room overnight.
- Collect the dialyzed cell lysate in a new chilled microcentrifuge tube.


## Casting Gradient BN Gels

## Materials and equipment

BISTRIS
HCl
Acrylamide/BIS solution ($T$ = 40 g/dL, $C$ = 0.03)
TMEDA
APS
Gradient maker
Silicone template


*BN gel buffer, pH = 7.0*

| | |
|---|---|
| BISTRIS | 0.05 mol/L |

*Prepare 80 mL as a 3x stock, adjust pH to 7.0 with HCl, and add deionized water to 100 mL.*

### Stacking gel (T = 4.0 g/dL)

| | |
|---|---|
| 3x BN-gel buffer | 3.00 mL |
| Acrylamide/BIS | 0.90 mL |
| 10 g/dL TMEDA | 0.04 mL |
| 10 g/dL APS | 0.07 mL |
| Deionized water to | 9.00 mL |

### Light separating gel solution (T = 5 g/dL)

| | |
|---|---|
| 3x BN-gel buffer | 5.00 mL |
| Acrylamide/BIS | 1.88 mL |
| 10 g/dL TMEDA | 0.04 mL |
| 10 g/dL APS | 0.10 mL |
| Deionized water to | 15.00 mL |

### Heavy separating gel solution (T = 15 g/dL)

| | |
|---|---|
| 3x BN-gel buffer | 5.00 mL |
| Acrylamide/BIS | 5.63 mL |
| 87% Glycerol | 4.10 mL |
| 10 g/dL TMEDA | 0.04 mL |
| 10 g/dL APS | 0.03 mL |
| Deionized water to | 15.00 mL |

### Procedure

- Close the clamp on the channel between the cylinders of the gradient maker, and the tubing clamp.
- Place a magnetic rod into the mixing cylinder.
- Input a syringe needle between the two glass plates of the casting cassette.
- Add APS to the light and heavy solutions.
- Pour the heavy solution into the mixing cylinder, and the light solution into the reservoir of the gradient maker.
- Switch on the magnetic stirrer.
- Open the clamps and force out the air bubble inside the channel between the cylinders.
- Let the mixed solution flow slowly between the glass plates to fill two-third of the cassette.
- Overlay the gradient solution in the cassette with deionized water.
- Let the solution polymerize at room temperature for 60 min.
- Add APS to the stacking solution.
- Pour the stacking solution with a pipette onto the gradient separating gel.
- Overlay the stacking gel solution with deionized water.

**Running BN Electrophoresis**

**Materials and equipment**
BISTRIS
TRICINE
Ready-made polyacrylamide gel ($T = 40$ g/dL, $C = 0.03$) for BN electrophoresis
Coomassie brilliant blue G-250 (CBB G250)
Silicone template
Gel electrophoresis system

***Electrode buffer, pH = 7.0***
BISTRIS       3.14 g (15 mmol/L)
TRICINE       8.96 g (50 mmol/L)
CBB G250      0.20 g
Deionized water to 1,000.00 mL
    *Store at 4 °C.*

**Procedure**
- Place paper strips, soaked with electrode buffer, onto the anode and cathode ends of the BN gel.
- Place the silicone template onto the stacking gel in front of the cathode strip.
- Load 1–40 µL of the dialyzed lysate and 10–20 µL of a marker mix into the wells of the silicone template.
- Apply 100 V to a minigel or 150 V to a large gel, until the samples have entered the separating gel.
- Increase the voltage to 180 V (minigel) or 400 V (large gel) and run the electrophoresis until the dye front reaches the end of the gel (3–4 h for a minigel, and 18–24 h for a large gel).

Stain, if necessary.

# References

[1]  Görg A, Postel W, Westermeier R, Gianazza E, Righetti PG.. J Biochem Biophys Methods, 1980, 3, 273–284.
[2]  McLellan T.. Anal Biochem, 1982, 126, 94–99.
[3]  Chen B, Griffith A, Catsimpoolas N, Chrambach A, Rodbard D.. Anal Biochem, 1978, 89, 609–615.
[4]  Fairbanks G, Steck TL, Wallach DFH.. Biochemistry, 1971, 10, 2606–2617.
[5]  Margolis, J, Kenrick, KG.. Anal Biochem, 1968, 25, 347–351.
[6]  Kolin A.. Proc Natl Acad Sci USA, 1955, 41, 101–110.
[7]  Margolis J, Kenrick KG.. Anal Biochem, 1968, 25, 347–351.

[8]   Rothe GM, Purkhanbaba M.. Electrophoresis, 1982, 3, 33–42.

[9]   Ferguson KA.. Metabolism, 1964, 13, 985–1002.

[10]  Lambin P, Rochu D. Fine JM. Anal Biochem, 1976, 74, 567–575.

[11]  Lambin P.. Anal Biochem, 1978, 85, 114–125.

[12]  Hedrick JL, Smith AJ.. Arch Biochem Biophys, 1968, 126, 155–164.

[13]  Schägger H, Von Jagow G.. Anal Biochem, 1991, 199, 223–231.

[14]  Schägger H, Cramer WA.. Anal Biochem, 1994, 217, 220–230.

[15]  Camacho-Carvajal MM, Wollscheid B, Aebersold R, Steimle V, Schamel WW.. Mol Cell Proteomics, 2004, 3, 176–182.

[16]  Beriault R, Chenier D, Singh R, Middaugh J, Mailloux R, Appanna V.. Electrophoresis, 2005, 26, 2892–2897.

[17]  Ratledge C.. Biochimie, 2004, 86, 807–815.

[18]  Matsubara S, Takayama T, Iwasaki R, Komatsu N, Matsubara D, Takizawa T, Sato I.. Placenta, 2001, 22, 882–885.

[19]  Lee SM, Koh HJ, Park DC, Song BJ, Huh TL, Park JW.. Free Radie Biol Med, 2002, 32, 1185–1196.

# 2.6 Isotachophoresis of proteins

2.6.1    Theory of isotachophoresis of proteins —— 147
2.6.1.1  Kohlrausch regulating function —— 147
         References —— 150

Isotachophoresis (ITP) is electrophoresis, where all co-ions (ions of same sign charges) migrate with same velocities.

## 2.6.1 Theory of isotachophoresis of proteins

The ITP is based on the theory of Kohlrausch [1]. In 1897, he studied a system of two strong electrolyte solutions, where the solution containing the faster (leading) ion was located in front of the solution containing the slower (trailing) ion. Kohlrausch noted that, if electric current was passed through this system, a moving ionic boundary was formed between the two solutions over the time. He called this phenomenon *persistent function* (now bearing his name – *Kohlrausch regulating function*) and derived electrochemical equations for the ion concentrations at the boundary. The proteins or other polyions arrange themselves between the leading and trailing ions.

Later, Longsworth [2], Svensson [3], Alberty [4], and Dismukes and Alberty [5] obtained equations for moving boundaries composed of weak electrolytes. In 1964, Ornstein [6] stretched the Kohlrausch regulating function onto buffers, and in 1973, Jovin [7] derived equations for moving boundaries between univalent acids and univalent bases. The leading ion is usually smaller, for example, chloride ion; and the trailing ion is usually larger, for example glycinate ion.

### 2.6.1.1 Kohlrausch regulating function

Let us analyze a buffer system consisting of a leading buffer ($a$) and a trailing buffer ($b$). The leading buffer contains the strong electrolyte HA; the trailing buffer contains the weak electrolyte HB. The electrolytes dissociate the ions $A^-$ and $B^-$, respectively; the mobility of ion $A^-$ being higher than the mobility of ion $B^-$, i.e. $|\mu_{A^-}| > |\mu_{B^-}|$. Both buffers contain the same base B that associates a proton to build the counter-ion $HC^+$ (Figure 2.6.1).

https://doi.org/10.1515/9783110761641-016

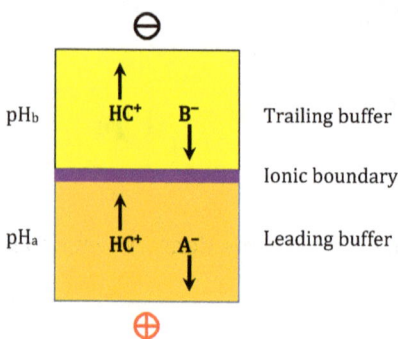

**Figure 2.6.1:** Scheme of buffer system for isotachophoresis.

$A^-$ – leading ion; $B^-$ – trailing ion; $HC^+$ – counter-ion; $pH_a$ and $pH_b$ – pH values of the leading and trailing buffers, respectively

Let us determine when the ions $A^-$ and $B^-$ will move with same effective velocity $v'$, i.e., when

$$v'_{A^-} = v'_{B^-} \tag{2.6.1}$$

It is known that the effective velocity of an ion is equal to the product of its effective mobility $\mu'$ and the electric field strength $E$, i.e.

$$v'_{A^-(B^-)} = \mu'_{A^-(B^-)} E_{a(b)} = \alpha_{HA(HB)} \mu_{A^-(B^-)} E_{a(b)} \tag{2.6.2}$$

where $\alpha_{HA(HB)}$ is the dissociation degree of acid HA (HB) ($\alpha_{HA} = 1$, since HA is a strong acid); $\mu_{A^-(B^-)}$ is the mobility of ion $A^-$ ($B^-$), in $m^2/(s\,V)$; and $E_{a(b)}$ is the electric field strength in the leading (trailing) buffer. The strength of the electric field is given by the following equation

$$E_{a(b)} = \frac{J}{\gamma_{a(b)}} \tag{2.6.3}$$

where $J$ is the density of the total ionic current, in $A/m^2$; and $\gamma_{a(b)}$ is the specific electric conductivity of the leading (trailing) buffer, in $S/m$.

It follows from eqs. (2.6.1)–(2.6.3) that

$$\frac{\alpha_{HB}\mu_{B^-}}{\gamma_b} = \frac{\alpha_{HA}\mu_{A^-}}{\gamma_a} \tag{2.6.4}$$

The specific conductivity

$$\gamma_{a(b)} = F\Sigma c_{A^-(B^-)} z_{A^-(B^-)} \mu_{A^-(B^-)} \tag{2.6.5}$$

where $c_{A^-(B^-)}$ is the concentration of ion $A^-(B^-)$, in $mol/L$; and $z_{A^-(B^-)}$ is the number of electric charges (electrovalence) of ion $A^-$ ($B^-$).

From eqs. (2.6.4) and (2.6.5), the equation

$$\frac{\alpha_{HB}\mu_{B^-}}{\alpha_{HB}C_{HB}z_{B^-} - \mu_{B^-} + \alpha_{Cb}C_{Cb}z_{HC^+}\mu_{HC^+}} = \frac{\alpha_{HA}\mu_{A^-}}{\alpha_{HA}C_{HA}z_{A^-}\mu_{A^-} + \alpha_{Ca}C_{Ca}z_{HC^+}\mu_{HC^+}} \qquad (2.6.6)$$

results, where $c_{Ca(b)}$ is the concentration of base C, in mol/L, in the leading (trailing) buffer.

According to the law of electrical neutrality of a chemical solution,

$$\alpha_{HA(HB)}C_{HA(HB)}z_{A^-(B^-)} + \alpha_{Ca(b)}C_{Ca(b)}z_{HC^+} = 0 \qquad (2.6.7)$$

Hence, eq. (2.6.6) can be transformed into the equation of Ornstein

$$\frac{c_{HB}}{c_{HA}} = \frac{z_{A^-}\mu_{B^-}\left(\mu_{A^-} - \mu_{HC^+}\right)}{z_{B^-}\mu_{A^-}\left(\mu_{B^-} - \mu_{HC^+}\right)} = F_k \qquad (2.6.8)$$

where $F_k$ is the Kohlrausch regulating function.

The equation of Ornstein can be applied not only for anionic but also for cationic buffer systems. If a cationic buffer system is assembled by the leading base A, the trailing base B, and the common acid HC, the Ornstein equation obtains the following form:

$$\frac{c_B}{c_A} = \frac{z_{HA^+}\mu_{HA^+}\left(\mu_{HA^+} - \mu_{C^-}\right)}{z_{HB^+}\mu_{HA^+}\left(\mu_{HB^+} - \mu_{C^-}\right)} = F_k \qquad (2.6.9)$$

At first glance, the Kohlrausch regulating function, i.e. the dependence of the ionic concentration on the ionic mobility, looks incomprehensible. However, it can be explained with the discontinuously distributed electric field strength: its strength in the zone of the trailing buffer is higher than in the zone of the leading buffer. So, a potential gradient is formed in the ionic boundary between the leading and trailing ions.

If a leading ion enters the zone of trailing buffer, it begins to move quicker at the higher voltage and overtakes the ionic boundary. If a trailing ion penetrates into the zone of leading buffer, its velocity decreases due to the lower voltage and the ionic boundary overtakes it. Polyions, which have intermediate effective mobilities, staple in a decreasing turn between the leading and trailing ions and build with them a migration sandwich-like structure (Figure 2.6.2).

To concentrate a polyion, the effective mobility of leading ion $A^-$ has to be greater, and the effective mobility of trailing ion $B^-$ has to be smaller than the effective mobility of the polyion, hence, the following inequation

$$|\alpha_{HA}\mu_{A^-}| > |\alpha_{H_nP}\mu_{P^{n-}}| > |\alpha_{HA}\mu_{B^-}| \qquad (2.6.10)$$

should be respected, where $P^{n-}$ is the polyion to be analyzed. For example, if at 0 °C protein polyion $P^{20-}$ $[\mu_{p20^-} = -5 \times 10^{-9}$ m$^2$/(s V)] has to be concentrated at a front of chloride ion $[\mu_{Cl^-} = -37 \times 10^{-9}$ m$^2$/(s V)] against TRIS-ion $[\mu_{HT^+} = 9 \times 10^{-9}$ m$^2$/(s V)], the Kohlrausch regulating function will be

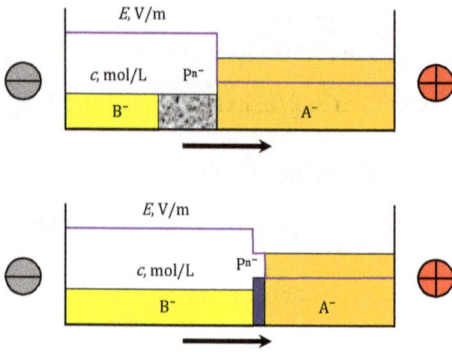

**Figure 2.6.2:** Start of electrophoresis (top) and concentrating of polyions $_{pn-}$ in the moving ionic boundary (bottom). $A^-$ – leading ion; $B^-$ – trailing ion

$$F_k = \frac{c_{H_2O P}}{c_{HCl}} = \frac{(-1)(-5 \times 10^{-9})(-37 \times 10^{-9} + 9 \times 10^{-9})}{(-20)(-37 \times 10^{-9})(-5 \times 10^{-9} + 9 \times 10^{-9})} = 0.022$$

This means that, if the concentration of HCl is 0.06 mol/L, the protein concentration will reach a concentration of

$$0.022 \times 0.06 = 0.0013 \, \text{mol/L}$$

# References

[1] Kohlrausch F. Ann Phys Chem, 1897, 62, 209–239.
[2] Longsworth LG. J Am Chem Soc, 1945, 67, 1109–1119.
[3] Svensson H. Acta Chem Scand, 1948, 2, 841–855.
[4] Alberty RA. J Am Chem Soc, 1950, 72, 2361–2367.
[5] Dismukes FR, Alberty RA. J Am Chem Soc, 1954, 76, 191–197.
[6] Ornstein L. Ann N Y Acad Sci, 1964, 121, 321–349.
[7] Jovin TM. Biochemistry, 1973, 12, 871–879.

# 2.7 Disc-electrophoresis of proteins

2.7.1    Theory of disc-electrophoresis —— 152
2.7.2    Native disc-electrophoresis —— 153
2.7.2.1  Disc-electrophoresis according to Ornstein and Davis —— 153
2.7.2.2  Buffer-gel systems for disc-electrophoresis according to Ornstein and Davis —— 155
2.7.2.3  Disc-electrophoresis according to Allen *et al* —— 156
2.7.2.4  Effect of Hjerten —— 156
2.7.2.5  Buffer-gel systems for disc-electrophoresis according to Allen *et al* —— 156
2.7.2.6  Disc-electrophoresis in one buffer at two pH values according to Michov —— 157
2.7.2.7  Theory of disc-electrophoresis in one buffer at two pH values —— 157
2.7.2.8  Buffer-gel systems for disc-electrophoresis in one buffer at two pH values —— 159
2.7.3    Denatured SDS disc-electrophoresis —— 161
2.7.3.1  Detergents —— 161
         Sodium dodecyl sulfate —— 163
         Sample preparation for SDS polyacrylamide gel electrophoresis —— 165
         Nonreducing sample preparation —— 165
         Reducing sample preparation —— 165
         Reducing sample preparation with alkylation —— 166
2.7.3.2  Practice of SDS disc-electrophoresis —— 167
         SDS disc-electrophoresis in TRIS-chloride-glycinate buffer system according to
         Laemmli —— 167
         SDS disc-electrophoresis in TRIS-acetate-TRICINEate buffer system according to
         Schägger–Jagow —— 168
         SDS disc-electrophoresis in TRIS-formate-taurinate buffer system according to
         Michov —— 169
         SDS disc-electrophoresis in one buffer at two pH values according to Michov —— 170
         SDS disc-electrophoresis in gradient gels —— 171
2.7.3.3  Staining of SDS protein bands —— 171
         Determination of molecular masses of SDS-denatured proteins —— 171
2.7.4    Protocols —— 173
         Disc-electrophoresis in Alkaline Buffer-gel System According to Ornstein–Davis —— 173
         Disc-electrophoresis in One Buffer at Two pH Values According to Michov —— 176
         SDS Disc-electrophoresis According to Laemmli —— 178
         SDS Electrophoresis in TRIS-formate-taurinate Buffer System According to
         Michov —— 180
         Coomassie Brilliant Blue R-250 Staining of Proteins Resolved in SDS Gels —— 183
         References —— 184

**Disc**ontinuous electrophoresis (disc-electrophoresis) is carried out in two or more buffers. The buffers differ in their composition, ionic strength, and pH value, so that the strength of the applied electric field is discontinuously distributed.

The term *discontinuous* buffer system was introduced in 1957 by Poulik [1] after he found that the electrophoretic resolution increases, if the starch gel contains a citrate buffer and the electrode tanks contain a borate buffer. We claim that one buffer can also form a discontinuous buffer system, if the buffer is divided in two or more parts of different ionic strengths and different pH values [2].

https://doi.org/10.1515/9783110761641-017

During disc-electrophoresis, the sample polyions are stacked at first by isotachophoresis, and then are resolved by zone electrophoresis. The isotachophoresis takes place in a large-pore polyacrylamide gel (stacking gel) with a $T$-concentration of 0.4–0.7 mol/L (3–5 g/dL); the zone electrophoresis takes place in a small-pore polyacrylamide gel (resolving gel) with a $T$-concentration of 0.7–4.2 mol/L (5–30 g/dL). A gradient resolving gel can also be used, where the protein bands become sharper [3].

After disc-electrophoresis, the protein bands can be stained or blotted. The bands can be quantified also by a gel imaging device or by visualizing with UV light.

The disc-electrophoresis is used in medicine, molecular biology, genetics, microbiology, biochemistry, and forensics, for separating proteins in blood sera, cerebrospinal fluids, urines, or tissue extracts.

## 2.7.1 Theory of disc-electrophoresis

When the concentration of the leading electrolyte is given, the concentration of the trailing electrolyte can be calculated using the Ornstein equation. To do this, it is necessary to know the mobilities of the leading, trailing, and counter-ions. Thereafter, the pH values and the concentrations of the base and acid in both buffers can be determined.

Let us analyze an acidic buffer system. To determine the pH value of the trailing buffer, the dissociation degree $\alpha_{HB}$ of the trailing acid HB should be calculated. For this purpose, the equation

$$\alpha_{HB} = \frac{\mu'_{B^-}}{\mu_{B^-}} \tag{2.7.1}$$

is used, where $\mu'_{B^-}$ is the effective mobility of the trailing ion B⁻.

The $pH_b$ value of the trailing buffer can be calculated from the Henderson–Hasselbalch equation, according to which

$$pH_b = pK_{c_{HB}} + \log \frac{\alpha_{HB}}{1 - \alpha_{HB}} \tag{2.7.2}$$

The dissociation degree of the base C in the trailing buffer, $\alpha_{Cb}$, can also be determined from the Henderson–Hasselbalch equation

$$\alpha_{C_b} = \left(1 + 10^{pH_b - pK_{c_{HB^+}}}\right)^{-1} \tag{2.7.3}$$

The concentration of the base in the trailing buffer can be calculated from the equation of electrical neutrality of a solution, according to which

$$c_{C_b} = -c_{HB} \frac{\alpha_{HB} z_{B^-}}{\alpha_{C_b} z_{HC^+}} \tag{2.7.4}$$

It follows from the theory of moving ionic boundary and eq. (2.7.4) that

$$\frac{\alpha_{HB}\,c_{HB}}{c_{HA}} = \alpha_{HB}F_k = \frac{[H_3O^+\,{}_b]}{[H_3O^+\,{}_a]} \tag{2.7.5}$$

hence

$$pH_a = pH_b + \log(\alpha_{HB}F_k) \tag{2.7.6}$$

The dissociation degree of the base in the leading buffer, $\alpha_{Ca}$, can also be calculated from the Henderson–Hasselbalch equation

$$\alpha_{Ca} = \left(1 + 10^{pH_a - pK_{C_{HC+a}}}\right)^{-1} \tag{2.7.7}$$

The concentration of the base in the leading buffer can be obtained from the equation of electrical neutrality of a solution

$$c_{Ca} = -c_{HA}\frac{\alpha_{HA}z_{A^-}}{\alpha_{Ca}z_{HC^+}} \tag{2.7.8}$$

The cationic buffer systems, which consist of bases A and B, and a common acid HC, obey similar equations.

The disc-electrophoresis can be carried out in native or denatured (SDS) polyacrylamide gels.

## 2.7.2 Native disc-electrophoresis

Two native disc-electrophoresis methods are known: according to Ornstein–Davis and according to Allen *et al.* We have created a third disc-electrophoresis method, which we called disc-electrophoresis in one buffer at two pH values.

### 2.7.2.1 Disc-electrophoresis according to Ornstein and Davis

The theoretical basis of disc-electrophoresis was developed by Leonard Ornstein [4], and its practical implementation was made by Baruch Davis [5]. It requires three different buffers: stacking, resolving, and electrode buffers.

To explain the disc-electrophoresis according to Ornstein–Davis, we assume that the stacking and resolving buffers are assembled by the strong acid HA and weak base C; and the electrode buffer is composed of the weak acid HB and the same base. Hence, the stacking and resolving buffers contain the leading ion A⁻ and counter-ion HC⁺, and the electrode buffer contains the trailing ion B⁻ and the same counter-ion HC⁺ (Figure 2.7.1).

Let us calculate the concentrations and pH values of a buffer system, which is suitable for disc-electrophoresis of serum proteins at 0 °C. Most serum proteins have

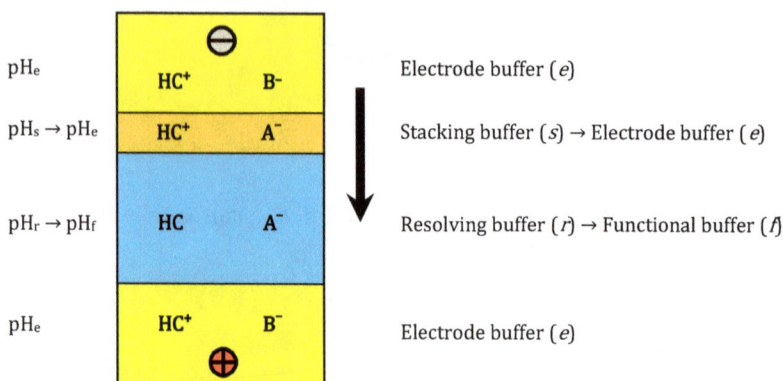

**Figure 2.7.1:** Scheme of the buffer system for disc-electrophoresis according to Ornstein–Davis. A⁻ – leading ion; B⁻ – trailing ion; HC⁺ – counter-ion; pH$_e$, pH$_s$, pH$_r$, and pH$_f$ – pH values of the electrode, stacking, resolving, and functional buffer, respectively. The two electrode buffers can be different or same (as on the figure).

negative total charges at the physiological pH value of the blood, i.e., at pH = 7.40. To charge all of them negatively, it is necessary to dissolve them in a buffer with pH = 9.0–10.0. This buffer should be referred to as functional buffer (f).

The required pH value can be assured by glycine (HG), since the p$K$ value of its protonated amino group is approximately equal to 9.8. At pH = 8.0 and 0 °C, the effective mobilities of serum proteins are between −0.6 × 10⁻⁹ and −7.5 × 10⁻⁹ m²/(s V). This means that the effective mobility of glycinate ion, which serves as a trailing ion, in the electrode buffer should be −0.5 × 10⁻⁹ m²/(s V), and in the functional buffer should be −10 × 10⁻⁹ m²/(s V). However, the effective mobility of the proteins in the resolving gel falls by half; therefore, it is sufficient if the effective mobility of glycinate ion in the resolving gel is −5 × 10⁻⁹ m²/(s V).

The mobility of the glycinate ion $\mu_{G^-}$ at 0 °C is equal to −15 × 10⁻⁹ m²/(s V). So, the dissociation degree of glycine in the electrode buffer should be

$$\alpha_{HGe} = \frac{-0.5 \times 10^{-9}}{-15 \times 10^{-9}} = \frac{1}{30}$$

and the dissociation degree of glycine in the functional buffer should be

$$\alpha_{HGf} = \frac{-5 \times 10^{-9}}{-15 \times 10^{-9}} = \frac{1}{3}$$

From the above-derived equations, it can be calculated that pH$_e$ should be 8.3 and pH$_f$ should be 9.5 (Figure 2.7.2).

TRIS (T) is a suitable buffering base because the p$K$ value of its protonated form (8.1), namely p$K_{HT^+}$, is approximately equal to the calculated pH$_e$ value. Considering

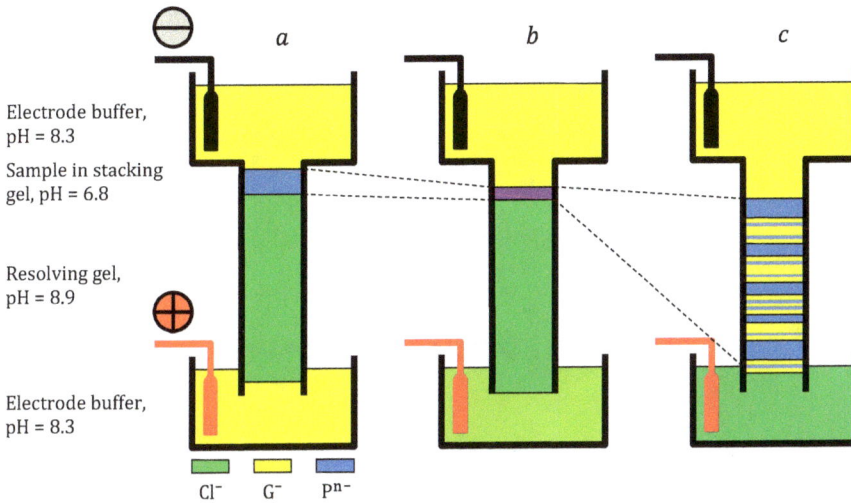

**Figure 2.7.2:** Scheme illustrating disc-electrophoresis of serum proteins. (*a*) Starting the disc-electrophoresis; (*b*) concentrating the proteins in the stacking gel; (*c*) separating the proteins in the resolving gel. $Cl^-$ – chloride ion; $G^-$ – glycinate ion; $P^{n-}$ – protein polyions (polyanions).

that at 0 °C $\mu_{HT^+} = 9 \times 10^{-9}$ m$^2$/(s V), and if leading acid is HCl [$\mu_{Cl^-} = -37 \times 10^{-9}$ m$^2$/(s V) at 0 °C], can be calculated from the above equations that Kohlrausch function $F_k = 0.78$.

It is found *a priori* that good electrophoretic results are obtained when the concentration of the strong acid is 0.05–0.06 mol/L. If $c_{HCl} = 0.06$ mol/L, it follows from the above equations that $c_{HG} = 0.047$ mol/L. It can also be calculated that pH$_s = 6.8$, and pH$_r = 8.9$, $\alpha_{Ce} = 0.37$, $c_{Te} = 0.004$ mol/L, $\alpha_{Cs} = 0.96$, $c_{Ts} = 0.062$ mol/L, $\alpha_{Cr} = 0.14$, and $c_{Tr} = 0.439$ mol/L.

## 2.7.2.2 Buffer-gel systems for disc-electrophoresis according to Ornstein and Davis

Numerous disc-buffer-gel systems have been developed [6] for separation of acidic and basic proteins. The oldest and most widely used of them is the alkaline buffer-gel system of Davis [7]. It is used for separating acidic proteins with relative molecular masses $M_r$ of $10^{4-6}$, stable at pH = 8.9, in 1 mol/L (7 g/dL) polyacrylamide gel. Such proteins are the serum proteins. They are resolved at pH = 9.5. A disc-pherogram obtained with the buffer-gel system of Davis is shown in Figure 2.7.3.

The buffer-gel system of Reisfeld *et al.* [8] is used for separating alkaline proteins with relative molecular masses $M_r$ of approximately $2 \times 10^4$, stable at pH = 4.3, in 2.1 mol/L (15 g/dL) polyacrylamide gel. Such proteins are histones.

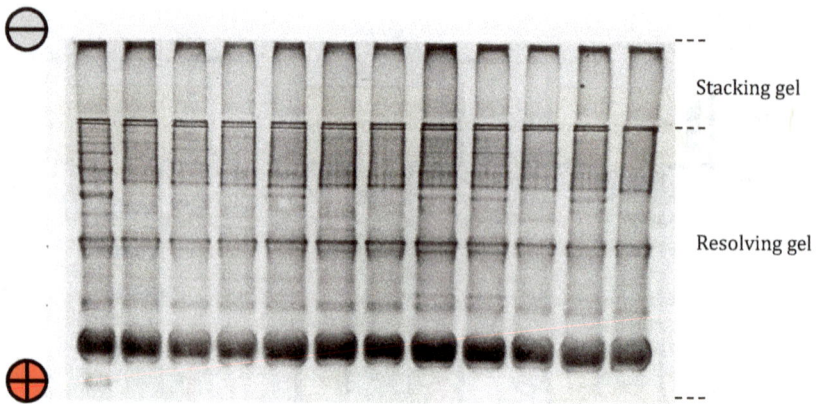

Figure 2.7.3: Disc-pherogram of serum proteins according to Davis.

### 2.7.2.3 Disc-electrophoresis according to Allen *et al.*

During the disc-electrophoresis according to Allen *et al.* [9, 10], the samples are concentrated by the effect of Hjerten and then separated in a step-gradient gel.

### 2.7.2.4 Effect of Hjerten

Hjerten *et al.* [11] proved that biological solutions can be concentrated, if they cross a boundary between a buffer of low ionic strength and a buffer of high ionic strength. Under these conditions, the polyions move quickly through the buffer of low ionic strength (stacking buffer) and diminish their velocity (focus) on the boundary with the buffer of high ionic strength (resolving buffer).

### 2.7.2.5 Buffer-gel systems for disc-electrophoresis according to Allen *et al.*

The buffer system of disc-electrophoresis according to Allen *et al.* also contains leading and trailing ions, but the pH value of the buffer system is everywhere constant. The stacking power of this electrophoresis is lower than this of the disc-electrophoresis according to Ornstein–Davis. However, the gradient resolving gel of a concentration of 0.4–1.7 mol/L (3–12 g/dL) sharpens the bands. In addition, a cover gel is recommended, which prevents the sample to diffuse and reduces the velocity of the trailing ion (Figure 2.7.4).

A few buffer-gel systems for this electrophoresis are created [12]: A sulfate-borate buffer system (pH = 9.0) for resolving plasma and tissue polyions, enzymes, and nucleates; a citrate-borate buffer system (pH = 9.0) for resolving plasma lipoproteins,

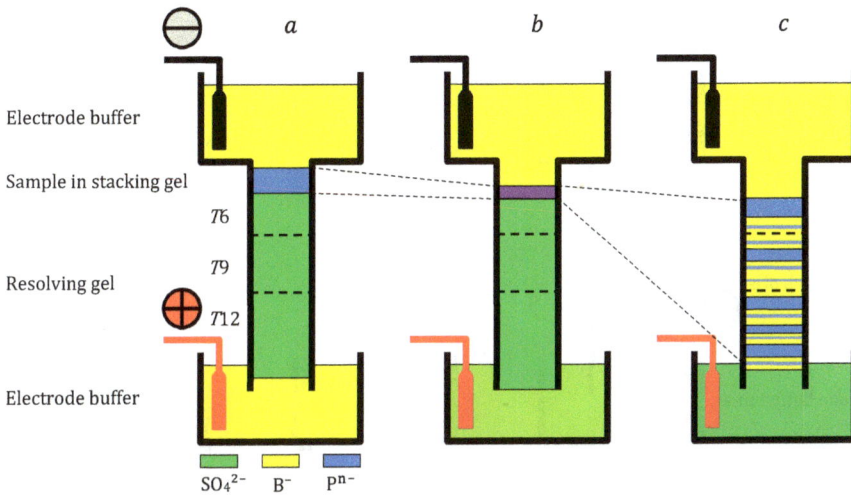

**Figure 2.7.4:** Effects during disc-electrophoresis according to Allen *et al.* (*a*) Starting the electrophoresis; (*b*) protein polyions are stacked by a moving ionic boundary; (*c*) protein polyions are resolved in a step-gradient gel. $SO_4^{2-}$ – sulfate ion; $B^-$ – borate ion; $P^{n-}$ – proteinate polyions.

which show a tendency to aggregate; a chloride-glycinate buffer system (pH = 8.5) for resolving enzymes; and a potassium-β-alanine buffer system (pH = 4.0) for resolving basic proteins.

### 2.7.2.6 Disc-electrophoresis in one buffer at two pH values according to Michov

In the disc-electrophoresis in one buffer at two pH values according to Michov [13], the polyions stack between the same leading and trailing ions and are resolving in a zone electrophoresis.

### 2.7.2.7 Theory of disc-electrophoresis in one buffer at two pH values

The electrophoresis in one buffer at two pH values is similar to the disc-electrophoresis according to Ornstein–Davis. However, its buffer system does not form a moving ionic boundary, but a stationary ionic boundary between stacking and resolving buffers, containing a same ion of the polyion's polarity (Figure 2.7.5).

In the methods of Ornstein–Davis, and Allen *et al.*, as well as in the theoretical works of Jovin [14], and Chrambach *et al.* [15], different leading and trailing ions are used for disc-electrophoresis. We have proved that the leading and trailing ions can

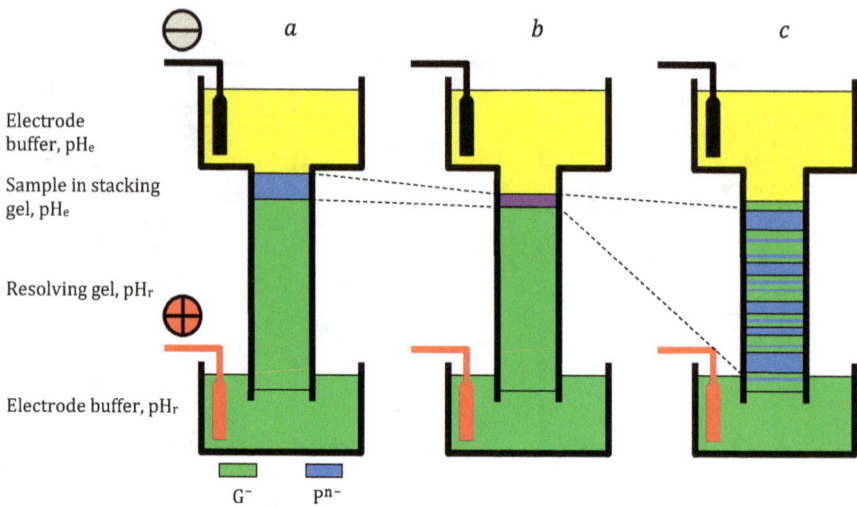

**Figure 2.7.5:** Disc-electrophoresis in one buffer at two pH values. (*a*) Starting the disc-electrophoresis; (*b*) stacking the polyions in a stationary ionic boundary between stacking and resolving gels; (*c*) resolving the polyions in a resolving gel. G⁻, glycinate ion; P$^{n-}$, proteinate polyions.

be the same, e.g., only glycinate or borate ion. So, the polyions are stacking and resolving in only one buffer system consisting of the weak acid HA and the weak base B, but having two pH values.

Let us determine the conditions, under which the same leading and trailing ion A⁻ can form an ionic boundary. To answer this question, we should find when its velocity $v_{A^-}$ in the resolving buffer (*a*) and in the electrode buffer (*b*) is the same, hence when

$$v_{A^-{}_a} = v_{A^-{}_b}$$

(2.7.9)

The velocity of ion i can be expressed by the equation

$$v_i = \mu_i' \, E = \alpha \mu_i E$$

(2.7.10)

where $\mu_i$ and $\mu_i'$ are its mobility and effective mobility, respectively; $E$ is the electric field strength; and $\alpha$ is the ionization degree of the electrolyte that creates ion $i$. If the field strength is represented as a ratio between the density of electric current $J$ and the specific conductivity of the solution $\gamma$, it follows from both equations that

$$\frac{\alpha_{HAa}}{\gamma_a} = \frac{\alpha_{HAb}}{\gamma_b}$$

(2.7.11)

The specific conductivity corresponds to the sum $F\Sigma c_i z_i \mu_i$, where $F$ is the Faraday constant (96,484.55461 C/mol), $c_i$ is the concentration of ion $i$, and $z_i$ is the number of its electric charges (electrovalency). Using this sum and the equation of electrical neutrality of a solution, eq. (2.7.11) can be simplified into

$$c_{HAa} = c_{HAb} = c_{HA} \qquad (2.7.12)$$

The last equation shows that ion $A^-$ moves with the same velocity in all parts of a buffer system, if the concentration of the electrolyte, which builds it, is the same everywhere.

From eqs. (2.7.11) and (2.7.12), the following equation is obtained:

$$\frac{c_{Ba}}{c_{Bb}} = \frac{\alpha_{HAa}\alpha_{Bb}}{\alpha_{HAb}\alpha_{Ba}} \qquad (2.7.13)$$

which describes the stationary ionic boundary.

The stationary ionic boundary can be extended over the whole buffer system [16]. The concentration of the weak acid should be constant; however, the concentration of the weak base should change continuously, which changes the pH value over the buffer system.

To concentrate polyions, $\mu'_{A^-a}$ and $\mu'_{A^-b}$ should be higher and lower, respectively, than the effective mobilities of the polyions. This means that if the protein polyion $P^{n-}$ should be concentrated, then

$$|\mu'_{A^-a}| > |\mu'_{pn^-}| > |\mu'_{A^-b}| \qquad (2.7.14)$$

When these conditions are available, the polyions arrange themselves in the stationary boundary until the equation

$$c_{H_nP} = c_{HA} \frac{\alpha_{HA}z_{A^-}}{\alpha_{H_nP}z_{pn^-}} \qquad (2.7.15)$$

is fulfilled. Here, $c_{H_nP}$ is the concentration of the stacked polyion, $\alpha_{H_nP}$ is its ionization degree, and $z_{pn^-}$ is its electrovalency.

The values of $\alpha_{HAa}$ and $\alpha_{HAb}$ can be calculated from the effective mobilities of ion $A^-$ in the resolving and electrode buffers, i.e., from $\mu'_{A^-a}$ and $\mu'_{A^-b}$, respectively. Then the concentration of the base in the resolving and electrode gels can be determined. The value of $\alpha_{Ba(b)}$ can be computed from the Henderson–Hasselbalch equation.

### 2.7.2.8 Buffer-gel systems for disc-electrophoresis in one buffer at two pH values

Let us derive a buffer system, which is suitable for resolving serum proteins at 25 °C. It is known that at 25 °C and pH $\geq$ 8.0 the effective mobilities of serum protein polyions $\mu'_{pn-}$ are $-15 \times 10^{-9}$ to $-1.2 \times 10^{-9}$ m$^2$/(s V). Since the value of $\mu'_{pn-(max)}$ decreases in the resolving gel, we can assume that the mobility of the same polar ion in the resolving gel $\mu'_{A-a}$ should be $-10 \times 10^{-9}$ m$^2$/(s V), and in the stacking gel $\mu'_{A-b}$ should be $-10^{-9}$ m$^2$/(s V).

These conditions correspond to the TRIS-glycinate buffer. It consists of the weak base TRIS (T) and the weak acid glycine (HG). At 25 °C and an ionic strength of 0.1 mol/L, the absolute mobility of the glycinate ion $\mu_{G-} = -27.87 \times 10^{-9}$ m$^2$/(s V) [17]. Then, it follows from the above equations that $\alpha_{HGa} = 0.3588$, $\alpha_{HGb} = 0.0359$, $c_{HG} = 0.2787$ mol/L, and $I_b = 0.01$ mol/L. Also at 25 °C, p$K_{HG} = 9.78$ and p$K_{HT+} = 8.07$. Hence, p$K_{HGa} = 9.54$, p$K_{HGb} = 9.69$, p$K_{HT+a} = 8.31$, and p$K_{HT+b} = 8.16$, where p$K_{HGa(b)}$ and p$K_{HT+a(b)}$ are the negative decimal logarithms of the concentration ionization constants of glycine and TRIS-ion in the resolving and electrode buffers, respectively. Next we can calculate, also from the above equations, that pH$_a = 9.29$, pH$_b = 8.26$, $\alpha_{Ta} = 0.0952$, and $\alpha_{Tb} = 0.4420$, and that the concentrations of TRIS in the resolving buffer $c_{Ta} = 1.050$ mol/L and in the electrode buffer $c_{Tb} = 0.023$ mol/L.

The APS solution should be freshly prepared and last added. We propose that its concentration (in g/dL) should be calculated according to the formula

$$c_{APS} = \frac{0.32 \ (g/dL)^2}{T} \ [g/dL] \tag{2.7.16}$$

where $T$ is the total monomeric concentration in g/dL.

After disc-electrophoresis in one buffer at two pH values, the protein bands are fixed in a mixture of 1.0 mol/L (16.34 g/dL) trichloroacetic acid and 0.2 mol/L (5.08 g/dL) 5-sulfosalicylic acid dehydrate for 30 min. Then they are stained in filtered 0.001 mol/L (0.17 g/dL) Coomassie brilliant blue R-250, dissolved in the destaining solution, for 30 min. The destaining solution contains methanol, acetic acid, and deionized water in a volume ratio of 3:1:6 (Figure 2.7.6).

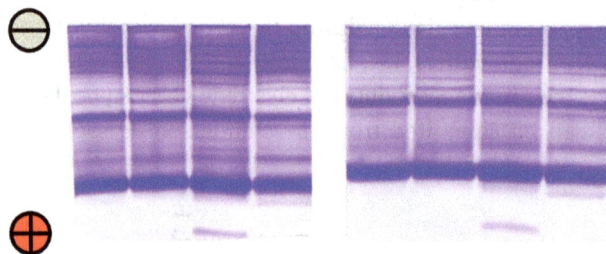

Figure 2.7.6: Serum proteins separated in a TRIS-glycinate buffer expanded stationary boundary of pH = 7.89–9.13, obtained in a 7 g/dL polyacrylamide slab gel. Staining with Coomassie brilliant blue R-250.

## 2.7.3 Denatured SDS disc-electrophoresis

**SDS p**oly**a**crylamide **g**el **e**lectrophoresis (SDS-PAGE) was introduced by Shapiro *et al.* [18]. It is widely used in biochemistry, forensics, genetics, molecular biology, and biotechnology, for separating denatured biological macromolecules, usually proteins or nucleic acids. Whereas the mobility of native polyions depends on their conformation and mass-to-charge ratio, the mobility of denatured polyions depends only on their mass-to-charge ratio.

As the native electrophoresis, SDS electrophoresis can be carried out in one buffer or in a system of two or more buffers. The first method is referred to as SDS zone (continuous) electrophoresis, which can be carried out in homogeneous as well as in gradient gels [19]. The second method is referred to as SDS disc-electrophoresis [20, 21]. SDS disc-electrophoresis does not distinguish from the disc-electrophoresis developed by Ornstein [22] and Davis [23] (see there). The only difference is that it contains SDS in addition. The SDS-protein complexes are concentrated also by a moving ionic boundary in the stacking gel where they arrange themselves according to their decreasing mobilities. Afterward, the SDS-protein complexes migrate through the resolving gel, where they are fractionated depending on their electric charges and masses.

### 2.7.3.1 Detergents

The detergents [24, 25], as all surfactants, lower the surface tension. They are employed in electrophoresis to disrupt protein–lipid and protein–protein bonding. The detergents build micelles in water, whose structure depends on their **c**ritical **m**icelle **c**oncentration (CMC) and aggregation number (Figure 2.7.7).

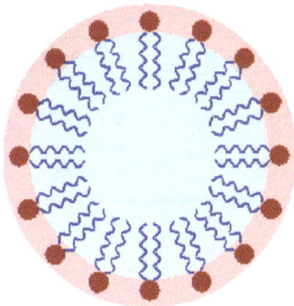

**Figure 2.7.7:** Schematic structure of an oil micelle in water. The detergent hydrophobic tails project into the oil, while the hydrophilic heads remain in contact with water.

CMC is the concentration of a detergent, in which it forms micelles. The lower the CMC value, the greater is the detergent strength. The **a**ggregation **n**umber ($N_a$) is the number of detergent molecules, which form a micelle. The smaller the $N_a$, the less the lipoprotein mass is increased. The CMC and the aggregation number are influenced by the pH value, temperature, and ionic strength.

Depending on their electric charges, the detergents can be divided into three groups: anionic, cationic, and zwitterionic detergents. Additionally, there are detergents that do not have electric charges – nonionic detergents (Table 2.7.1).

**Table 2.7.1:** Detergents used in electrophoresis.

| Detergents | Chemical formulas | Properties at 25 °C |
|---|---|---|
| **Anionic detergents** | | |
| SDS (sodium dodecyl sulfate) | | $M_r = 288.37$ CMC = $8.2 \times 10^{-3}$ mol/L $N_a = 62$ |
| LDS (lithium dodecyl sulfate) | | $M_r = 272.33$ CMC = $8.8 \times 10^{-3}$ mol/L $N_a = 90$ |
| Sodium cholate | | $M_r = 430.57$ CMC = $1.4 \times 10^{-2}$ mol/L $N_a = 3$ |
| **Cationic detergents** | | |
| CTAB (*N*-Cetyl-*N, N, N*-trimethylammonium bromide) | | $M_r = 364.46$ CMC = $0.2 \times 10^{-4}$ mol/L $N_a = 169$ |

**Table 2.7.1** (continued)

| Detergents | Chemical formulas | Properties at 25 °C |
|---|---|---|
| **Zwitterionic detergents** | | |
| CHAPS (3-[(3-cholamidopropyl)-dimethylammonio]-1-propanesulfonate) | 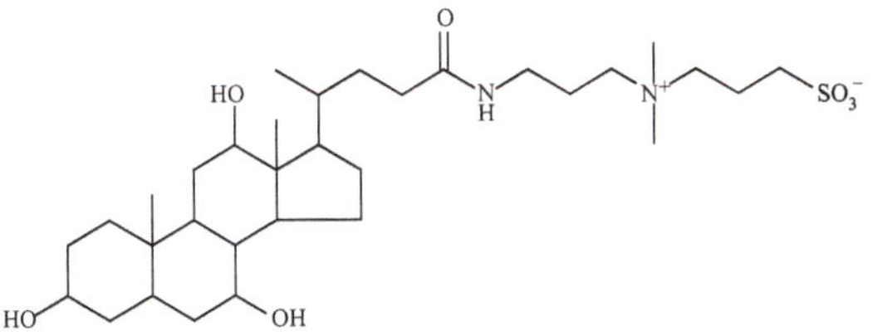 | $M_r = 614.89$ CMC = $4.2 \times 10^{-3}$ mol/L $N_a = 9$–10 |
| **Nonionic detergents** | | |
| Triton X-100 | 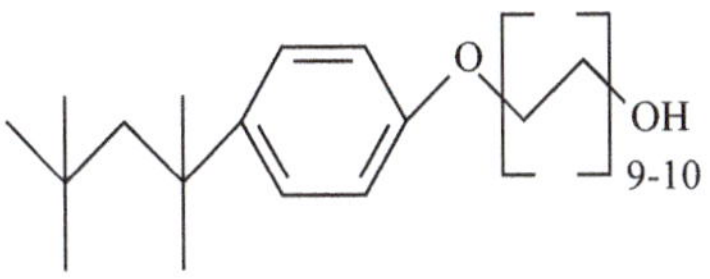 | $M_r = 646.87$ CMC = $3.1 \times 10^{-4}$ mol/L $N_a = 143$ |

$M_r$ – relative molecular mass; CMC – critical micelle concentration; $N_a$ – aggregation number

## Sodium dodecyl sulfate

The most commonly used detergent is sodium **d**odecyl **s**ulfate (SDS, sodium lauryl sulfate, sodium dodecane sulfate) (Figure 2.7.8).

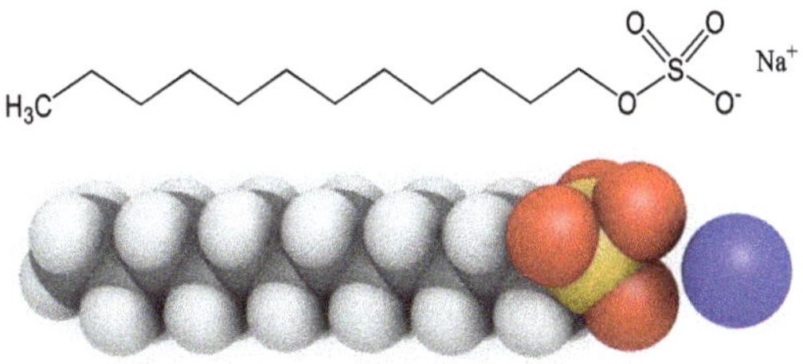

**Figure 2.7.8:** Chemical formula ($NaC_{12}H_{25}SO_4$) and structure of SDS.

SDS is the sodium salt of *dodecyl hydrogen sulfate* – the ester of dodecyl alcohol and sulfuric acid. It consists of a 12-carbon tail bound to a sulfate group that is electrostatically connected to a sodium ion. This structure gives the compound's amphiphilic properties that allow it to form micelles. CMC of SDS in pure water is 8.2 mmol/L at 25 °C [26], its aggregation number is about 62 [27], and its degree of micelle ionization $\alpha$ is around 0.3 (30 g/dL) [28, 29].

SDS ions cover the protein molecules and form with them SDS-protein complexes. Approximately 1.4 g of SDS is connected with 1 g of protein (one SDS ion is bound to two amino acid residues). So the proteins constitute about 42% of the total mass of the SDS-protein complexes [30]. The hydrophobic membrane proteins bind up to 4.5 g of SDS per gram of protein [31]. The glycoproteins depart from this rule since their hydrophilic glycan moiety reduces the hydrophobic interactions between proteins and SDS.

When a protein mixture is heated to 100 °C in the presence of SDS, SDS wraps around the proteins giving them negative electric charges. The hydrophobic tails of SDS ions remain inside the SDS-protein complex, but their heads face outward to water dipoles. The protein polypeptide chains unfold and linearize to become rod-like structures [32, 33] (Figure 2.7.9). With another words, SDS denatures the secondary and tertiary protein structure.

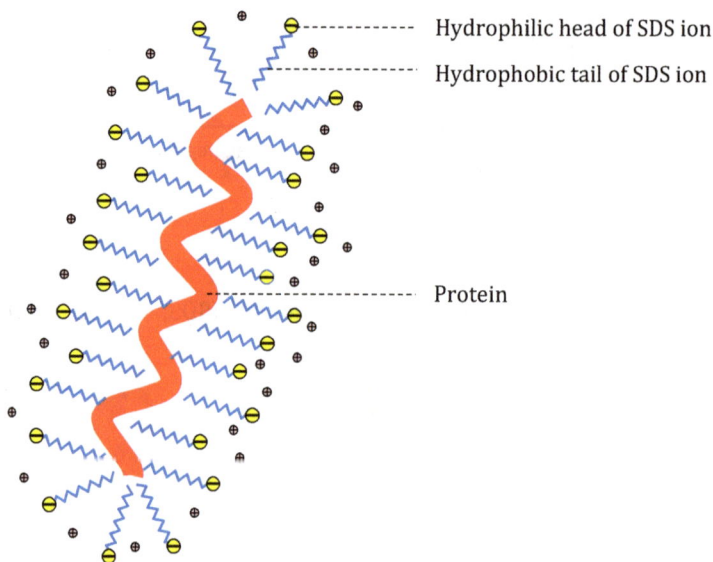

Figure 2.7.9: Structure of SDS-protein complex.

The protein polyions obtain certain numbers of negative electric charges per mass (in C/g) [34, 35], and, therefore, all SDS-protein complexes run to the anode. Since the conformation of all denatured proteins, except histones [36], become similar, the protein velocity through polyacrylamide gel depends only on their masses.

The cationic detergent CTAB has been used as an alternative to SDS for gel electrophoresis of proteins.

## Sample preparation for SDS polyacrylamide gel electrophoresis

If proteins are a part of a solid tissue, the solid tissue should be homogenized, or soni-cated, and then filtered and centrifuged. Afterward, the proteins are mixed with suffi-cient (1–2 g/dL) SDS, and heated at 90–100 °C for 2–5 min [37, 38]. Besides SDS, the samples should contain gel buffer, glycerol, and Bromophenol blue as a trailing dye.

The sample preparation can be nonreducing, reducing, or reducing with alkyl-ation.

## Nonreducing sample preparation

During the nonreducing sample preparation, the disulfide bonds in the proteins re-main intact (Figure 2.7.10). This preparation is carried out in a nonreducing buffer, where the proteins, in a concentration of 0.5–1.0 mg/mL, are incubated at room tem-perature for 1 h. They should not be cooked, because cooking can fragment the pro-teins. Thereafter, the sample should be centrifuged to remove debris.

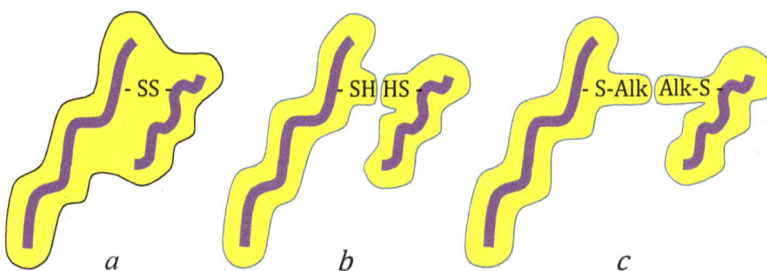

Figure 2.7.10: SDS-treated proteins. The polypeptide chains are presented as thick lines surrounded by SDS ions: (*a*) without reduction; (*b*) with reduction; and (*c*) with reduction and alkylation. –SS–, disulfide bridge; –SH, thiol group; Alk, alkyl group.

## Reducing sample preparation

The reducing sample preparation takes place, if a reducing agent, most often 1,4-**di** **thi**othreitol (DTT), or 2-mercaptoethanol (**beta**-**m**ercapto**e**thanol, BME), or 1,4-**dithi**-o**e**rythritol (DTE), in concentrations of 0.1–1.0 g/dL is added to the sample buffer.

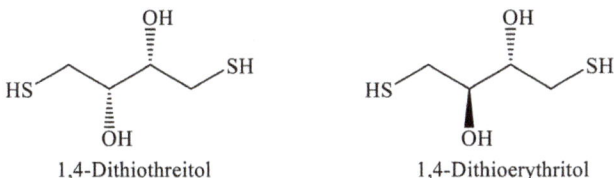

1,4-Dithiothreitol      1,4-Dithioerythritol

The samples should be boiled in a closed vessel containing 1 g/dL SDS and 5 mmol/L DTT for 3–5 min. So the disulfide bridges between the cysteine residues of proteins are cleaved (Figure 2.7.11). As a result, the proteins unfold, stretch their polypeptide chains, and form rod-like structures with SDS ions. The quaternary protein structure (oligomeric structure) breaks too. If proteins should not be heated at 100 °C, the samples are left at room temperature for a few hours or overnight.

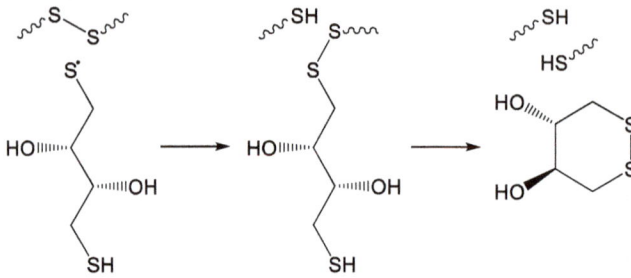

Figure 2.7.11: Reducing a disulfide bond by DTT *via* two sequential thiol-disulfide exchange reactions.

### Reducing sample preparation with alkylation

The SH groups that occur after reducing treatment may be oxidized again by atmospheric oxygen or other oxidizing agents to form disulfide bridges. Therefore, they should be alkylated (to be bound to difficult reacting alkyl groups). The alkylation can be carried out by iodoacetamide, iodoacetic acid, or vinylpyridine [39] (Figure 2.7.12). After alkylation, the proteins increase their molecular masses slightly, because the relative molecular mass $M_r$ of iodoacetamide is 184.96.

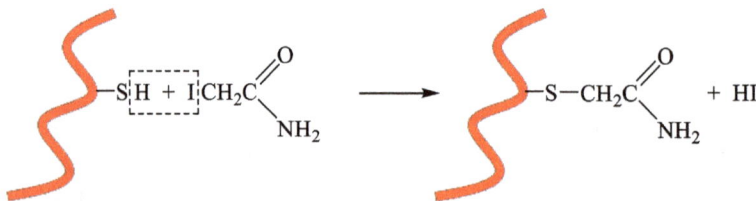

Figure 2.7.12: Alkylation of a thiol group by iodoacetamide. The polypeptide chain is shown as a thick line.

The alkylation is carried out at pH = 8.0 in the following way: 20 g/dL iodoacetamide is added to cool reduced proteins, and after incubation at room temperature for 30 min the sample is centrifuged.

## 2.7.3.2 Practice of SDS disc-electrophoresis

The original SDS electrophoresis was carried out in vertical round gels [40], and later in vertical slab gels [41, 42]. The wells in the vertical slab gels are formed by a comb and are filled with glycerol containing samples under the top (cathode) buffer. The introduction of thin (0.5–1.0 mm) and ultra-thin (0.1–0.5 mm) horizontal polyacrylamide gels [43] on support films [44, 45], e.g., on pretreated polyester films [46], was an important step forward.

The concentration of SDS in the gel should be 0.1 g/dL. The electrode buffer for horizontal electrophoresis can be contained in the tanks of electrophoresis cell or in porous electrode strips. The porous electrode strips can be constructed of gel (polyacrylamide or agarose gel) or paper. The gel and paper strips are 1–2 cm wide and come in direct contact with the electrodes. Since the volume of the strips is smaller than the volume of the electrode tanks, the electrode buffer in the strips should be concentrated.

The most SDS-PAGE separations are carried out under constant electric current. Now three buffer systems are used for SDS disc-electrophoresis in homogeneous gels: TRIS-chloride-glycinate system according to Laemmli, TRIS-acetate-TRICINEate system according to Schägger–Jagow, and TRIS-formate-taurinate buffer system according to Michov. In addition, SDS disc-electrophoresis can be carried out in one buffer at two pH values and in gradient gels.

### SDS disc-electrophoresis in TRIS-chloride-glycinate buffer system according to Laemmli

The SDS disc-electrophoresis was described in 1970 by Laemmli [47] who added SDS to the discontinuous buffer system of Ornstein and Davis. It is characterized by a moving ionic boundary between chloride ions, as leading ions, and glycinate ions, as trailing ions, against TRIS ions as counter-ions, and is running through two gels: stacking and resolving (Figure 2.7.13).

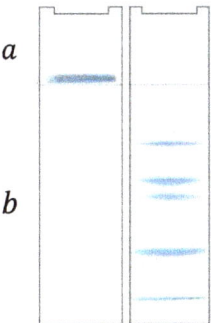

**Figure 2.7.13:** Laemmli electrophoresis: (*a*) stacking gel and (*b*) resolving gel.

The stacking polyacrylamide gel is cast in front of (over) the resolving gel. It is a large pore gel with $T = 4$ g/dL, which contains TRIS-HCl buffer with pH = 6.8. The resolving polyacrylamide gel is a small pore gel with $T = 8$–30 g/dL, which contains TRIS-HCl buffer with pH = 8.8. In both gels, the ratio between acrylamide and BIS concentrations is 30:1. The electrode buffer is a TRIS-glycinate buffer with pH = 8.3.

Protein samples, in concentrations of 0.2–0.6 mg/mL, are dissolved by heating in a sample solution. It contains 50 mmol/L TRIS-HCI, pH = 6.8, 2 mmol/L EDTA, 10 mmol/L DTT, 2 g/dL SDS, 10 mL/dL glycerol, and 0.01 g/dL Bromophenol blue Na salt. The electrophoresis is run at 100–500 V and 20–100 mA. The cooling temperature should be 10–15 °C. The velocity of Bromophenol blue, added to the sample, should be 4–5 cm/h.

Proteins whose $M_r$ is smaller than 12,000 resolve badly in the Laemmli system because they do not leave the band of SDS micelles that are formed behind the leading ions front.


**SDS disc-electrophoresis in TRIS-acetate-TRICINEate buffer system according to Schägger–Jagow**

The SDS disc-electrophoresis in TRIS-acetate-TRICINEate buffer system was empirically developed by Schägger and Jagow [48, 49]. This system uses TRICINEate ion as a trailing ion in place of the glycinate ion in the Laemmli system. It gives better results than the classical TRIS-chloride-glycinate buffer system of Laemmli when low molecular mass proteins ($M_r < 14,000$) are separated. In addition, the gels show a higher storage stability because they contain a TRIS-acetate buffer with pH = 6.4, a pH value at which the polyacrylamide gels hydrolyze very slowly.

The stacking gel is homogeneous ($T = 5$ g/dL, $C = 0.03$), whereas the resolving gel is linearly gradient ($T = 8$–18 g/dL) at the same $C$ value. Both gels contain a TRIS-acetate buffer with pH = 6.4 (0.12 mol/L TRIS, 0.12 mol/L acetic acid, and 1 g/dL SDS). The anode electrode gel ($T = 12$ g/dL, $C = 0.03$) contains a TRIS-acetate buffer with pH = 6.6 (0.45 mol/L TRIS, 0.45 mol/L acetic acid, 0.4 g/dL SDS, and 0.05 g/dL Orange G), and the cathode electrode gel ($T = 12$ g/dL, $C = 0.03$) contains TRIS-TRICINEate buffer with pH = 7.1 (0.08 mol/L TRIS, 0.8 mol/L TRICINE, and 0.6 g/dL SDS).

A disadvantage of the gradient gels according to Schägger and Jagow is that their preparation is more difficult than that of homogeneous gels. In addition, the staining and destaining of gradient gels are irregular and gels roll up during drying.

### SDS disc-electrophoresis in TRIS-formate-taurinate buffer system according to Michov

To eliminate the disadvantages of gradient gels, a new SDS disc-electrophoresis was developed in homogeneous slab gels [50]. This electrophoresis is carried out in a TRIS-formate-taurinate buffer system that has a higher buffer capacity than the prior buffer systems and is suitable for resolving of proteins with relative molecular masses of 6,000–450,000.

The horizontal SDS disc-electrophoresis in TRIS-formate-taurinate buffer system can be carried out using paper electrode strips. The handling of paper electrode strips is easier than the handling of gel electrode strips. Besides, they are nontoxic, while the gel electrode strips contain traces of acrylamide.

The stacking gel, the resolving gel, and the anode strip contain TRIS-formate buffer; the cathode strip contains TRIS-taurinate buffer. Formic acid (HA) forms the leading formate ion (A⁻), taurine (HB) forms the trailing taurinate ion (B⁻), and TRIS forms the counter TRIS-ion (HC⁺). At 25 °C and ionic strength of 0.1 mol/L, the mobilities of the formate, taurinate, and TRIS-ion are as follows: $\mu_{A^-} = -44.22 \times 10^{-9}$, $\mu_{B^-} = -23.53 \times 10^{-9}$, and $\mu_{HC^+} = 17.27 \times 10^{-9}$ m²/(s V) [51].

To assemble a TRIS-formate-taurinate buffer system, we should calculate the function of Kohlrausch [52], the degree of taurine dissociation, the pH value of the trailing buffer ($pH_b$) in the cathode strip, and the pH value of the leading buffer ($pH_a$) in the stacking and resolving gels and in the anode strip. It follows from the above equations that the function of Kohlrausch $F_k = c_{HTa}/c_{HFo} = 0.802$, where $c_{HB}$ and $c_{HA}$ are the molar concentrations of formic acid and taurine. If we assume that $c_{HA} = 0.10$ mol/L, then $c_{HB} = 0.0802$ mol/L.

The SDS-protein complexes move in a $T$13-resolving gel with effective mobilities lower than $-4 \times 10^{-9}$ m²/(s V). Therefore, the effective mobility of taurinate ion $\mu'_{B^-}$ in the stacking gel as well as in the resolving gel should be $-4 \times 10^{-9}$ m²/(s V); the dissociation degree of taurine in the trailing (cathode) buffer should be 0.17; and its ionic strength $I_b$ should be $0.17 \times 0.0802 = 0.0136$ mol/L.

According to the theory of Debye–Hückel [53], at $I_b = 0.0136$ mol/L $pK_{cHC^+b}$ should be equal to 8.17 and $pK_{cHB}$ should be equal to 8.96. Therefore, it follows from the Henderson–Hasselbalch equation that $pH_b = 8.3$ and $pH_a = 7.4$. The stacking gel forms approximately 1/3, and the resolving gel forms approximately 2/3 of the gel volume (Figure 2.7.14).

On the two gel ends, corresponding paper electrode strips are placed, which contain the anode and cathode buffers. On the stacking gel, 10 mm in front of the cathode strip, an application template, best of silicone, should be placed and lightly pressed to create a good contact with the gel. In each template slot, 5–10 µL SDS sample is applied. The samples should contain 20 mL/dL glycerol. Thereafter, the electrodes are placed on the electrode paper strips and the electrophoresis is started. Results obtained by this method are shown in Figure 2.7.15.

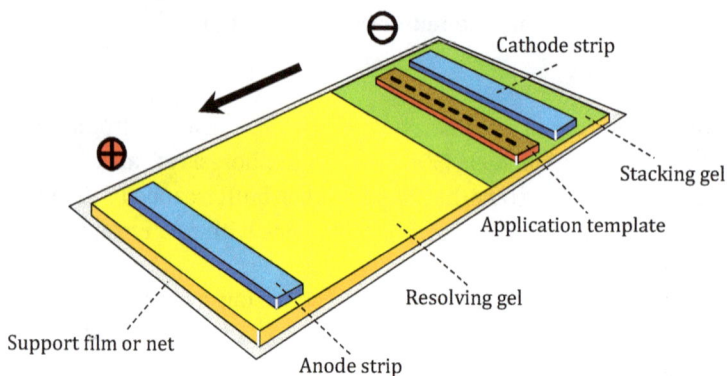

**Figure 2.7.14:** Scheme of a gel for SDS disc-electrophoresis in a TRIS-formate-taurinate buffer system.

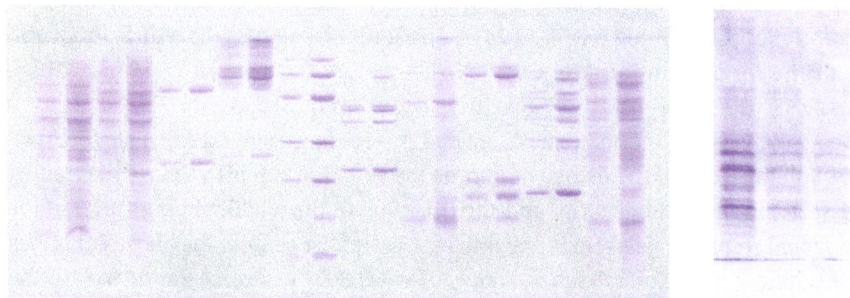

**Figure 2.7.15:** Electrophoretic results in a 13 g/dL homogeneous SDS slab gel according to Michov. *Left*: on a film-supported polyacrylamide gel; *right*: on a net-supported polyacrylamide gel. Applied volumes of 5 μL each. Electrophoresis at constant electric current of 80 mA (100–500 V) and temperature of 10 °C. Running time 90 min. Staining with Coomassie brilliant blue R-250.

**SDS disc-electrophoresis in one buffer at two pH values according to Michov**

In the methods of Laemmli and Schägger–Jagow, different leading and trailing ions are used for SDS disc-electrophoresis: chloride and glycinate ions, and chloride and acetate ions, respectively. On the contrary, in the SDS disc-electrophoresis in one buffer at two pH values after Michov [54], the leading and trailing ions are same, e.g., glycinate or borate ion. So the polyions are stacking and resolving in only one buffer system consisting of a weak acid and a weak base, but having two pH values. So a stationary ionic boundary is formed between a resolving buffer and a stacking buffer, not a moving ionic boundary. The polyions are concentrated at this boundary and then are separated from each other in the resolving gel (see *Native disc-electrophoresis in one buffer at two pH values according to Michov*).

After SDS disc-electrophoresis in one buffer at two pH values the protein bands are fixed in a mixture of 1.0 mol/L (16.34 g/dL) trichloroacetic acid and 0.2 mol/L 5-sulfosalicylic acid (5.08 g/dL 5-sulfosalicylic acid dehydrate) for 30 min. Then they are stained for 30 min in 0.001 mol/L (0.17 g/dL) Coomassie brilliant blue R-250 dissolved in a destaining solution. The destaining solution contains methanol, acetic acid, and water (3:1:6, $V:V:V$).

**SDS disc-electrophoresis in gradient gels**

The pore-gradient gels for SDS disc-PAGE are characterized with continuous decrease of gel pore diameters. As a result, the proteins move in the beginning through the gel with larger pores but stop later in the small pores because they cannot pass through. In this electrophoresis, the discontinuous buffer system of Laemmli is used, too. Gradient gels with $T$ = 4–15 g/dL are suitable for SDS proteins with $M_r$ = 40,000–200,000; and gradient gels with $T$ = 4–20 g/dL are suitable for SDS proteins with $M_r$ = 10,000–200,000.

### 2.7.3.3 Staining of SDS protein bands

After Bromophenol blue has reached the anodic end of the gel, the electrophoresis should be stopped and the protein bands could be stained with Coomassie brilliant blue or silver. A protein band should contain approximately 1 µg of protein to be stained with Coomassie brilliant blue R-250 or 0.1 µg of protein to be stained with silver.

The Coomassie brilliant blue and SDS ions compete for the same binding sites on proteins. Therefore, SDS ions have to be removed from the gel prior to staining. To accelerate this process, the bands can be fixed in methanol–acetic acid. Methanol increases the CMC value (see there) of SDS, so it can be easily washed. However, the low pH values reduce the solubility of SDS, so that the SDS removal proceeds slowly.

Other staining method is the colloidal staining. The colloidal staining has a higher detection sensitivity (approximately 30 ng of protein per band) and the gel background remains clear but the method proceeds overnight [55].

**Determination of molecular masses of SDS-denatured proteins**

SDS–PAGE can be used to estimate the relative molecular masses $M_r$ of denatured proteins [56–58]. This is based on the dependence of protein mobility $\mu$ on the total polyacrylamide gel concentration $T_0$ (undimensional), expressed by the equation

$$\log M_r = a \log T_0 + b \tag{2.7.17}$$

where $a$ and $b$ are undimensional constants. The larger the protein, the slower it migrates in a gel. In practice, the mobilities of investigated proteins are compared to the mobilities of standard proteins (Table 2.7.2) [59, 60].

**Table 2.7.2:** Relative molecular masses ($M_r$) of standard proteins used for determination of the relative molecular masses of SDS-denatured proteins.

| Proteins | $M_r$ | Proteins | $M_r$ |
|---|---|---|---|
| Immunoglobulin M | 950,000 | H chain of IgG | 50,000 |
| 2-Macroglobulin | 380,000 | Ovalbumin | 45,000 |
| Thyroglobulin | 355,000 | Aldolase | 40,000 |
| Ferritin | 220,000 | Alcohol dehydrogenase (liver) | 39,800 |
| Myosin (rabbit muscle) | 205,000 | Lactate dehydrogenase (porcine heart) | 36,000 |
| α-Macroglobulin (reduced) | 190,000 | Pepsin | 34,700 |
| Immunoglobulin A | 160,000 | Carbonic anhydrase | 31,000 |
| RNA polymerase (E. coli) | 160,000 | Carbonic anhydrase (bovine Ery) | 29,000 |
| Immunoglobulin G | 150,000 | Trypsinogen, PMSF treated | 24,000 |
| Ceruloplasmin | 124,000 | Trypsin inhibitor (soybean) | 20,100 |
| β-Galactosidase | 116,300 | β-Lactoglobulin | 18,400 |
| Phosphorylase a (muscle) | 100,000 | Myoglobin (sperm whale) | 16,800 |
| Phosphorylase b (rabbit muscle) | 97,400 | Hemoglobin | 15,500 |
| Lactoperoxidase | 93,000 | Ribonuclease b | 14,700 |
| Plasminogen | 81,000 | Lysozyme (hen egg white) | 14,300 |
| Transferrin | 76,000 | α-Lactalbumin | 14,200 |
| Bovine serum albumin | 66,200 | Cytochrome c (muscle) | 11,700 |
| Catalase (liver) | 57,500 | Insulin (reduced) | 6,600 |
| Glutamate dehydrogenase (liver) | 53,000 | Glucagon | 3,500 |

The relative mobilities $R_f$ of the standard proteins are plotted onto the abscissa axis of a coordinate system, and the logarithms of protein relative molecular masses are plotted on the ordinate axis to build a semilogarithm calibration straight line. The relative mobility is defined as the mobility of a protein divided by the mobility of the ionic front and is calculated as the distance passed by a protein in the resolving gel divided by the distance of the front. When $R_f$ values of unknown proteins are plotted on the calibration straight line, the logarithm of their relative molecular masses can be obtained (Figure 2.7.16).

Useful SDS homogeneous gels for estimating the relative molecular masses of proteins in SDS-PAGE are as follows: $T = 7.5$ g/dL for $M_r = 40–200 \times 10^3$; $T = 10$ g/dL for $M_r = 30–100 \times 10^3$; $T = 12$ g/dL for $M_r = 15–90 \times 10^3$; and $T = 15$ g/dL for $M_r = 10–70 \times 10^3$.

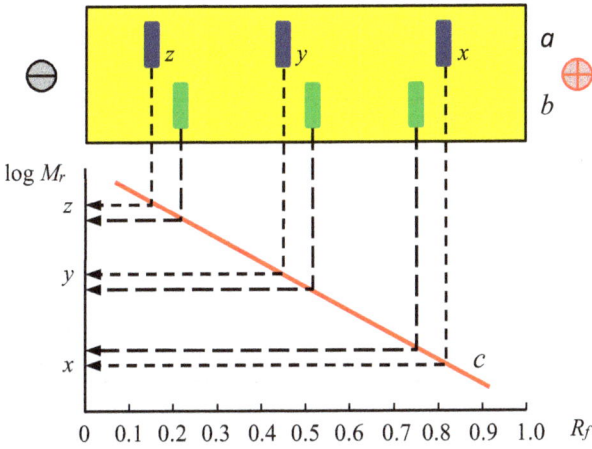

Figure 2.7.16: Determination of relative molecular masses of proteins by SDS electrophoresis using protein standards. (a) Protein bands x, y, and z in the pherogram; (b) bands of protein standards; (c) calibration straight line. $M_r$, relative molecular mass; $R_f$, relative migration distance.

## 2.7.4 Protocols

### Disc-electrophoresis in Alkaline Buffer-gel System According to Ornstein–Davis

**Materials and equipment**
Buffers
Acrylamide and BIS
TMEDA
APS
Trichloroacetic acid (TCA)
5-Sulfosalicylic acid
Coomassie brilliant blue R-250
Methanol
Acetic acid
0.5 mm U-form silicone gasket
Support film for polyacrylamide gel
250 mL side-arm flask for degassing gel solutions
Glass plates
Clamps
Water-jet pump

**Table 2.7.3:** Stacking gel, $T = 4$ g/dL, $C = 0.03$.

| Solutions | Preparation | Ingredients |
|---|---|---|
| TRIS-chloride buffer, 4×<br>  0.25 mol/L TRIS<br>  HCl to pH = 6.8 | Solve 3.03 g TRIS in 80 mL deionized water, titrate with 1 mol/L HCl to pH = 6.8, and fill up with deionized water to 100.0 mL. | 25.00 mL |
| Monomeric solution<br>  $T = 50$ g/dL, $C = 0.03$ | Solve 48.5 g acrylamide and 1.50 g BIS in deionized water and fill up to 100.0 mL. Add 1.0 g Amberlite | 8.00 mL |
| 87% Glycerol | MB-1. | 9.11 mL |
| 10 g/dL TMEDA | | 0.46 mL |
| 10 g/dL APS | | 0.80 mL |
| Deionized water to | | 100.00 mL |

**Table 2.7.4:** Resolving gel, $T = 12.5$ g/dL, $C = 0.03$.

| Solutions | Preparation | Ingredients |
|---|---|---|
| TRIS-chloride buffer, 4×<br>  1.5 mol/L TRIS<br>  HCl to pH = 8.8 | Solve 18.17 g TRIS in 40 mL deionized water, titrate with 1 mol/L HCl to pH = 8.8, and fill up with deionized water to 100.0 mL. | 25.00 mL |
| Monomeric solution<br>  $T = 50$ g/dL, $C = 0.03$ | Solve 48.5 g acrylamide and 1.5 g BIS in deionized water, and fill up to 100.0 mL. Add 1.0 g Amberlite | 25.00 mL |
| 87% glycerol | MB-1. | 9.11 mL |
| 10 g/dL TMEDA | | 0.46 mL |
| 10 g/dL APS | | 0.26 mL |
| Deionized water to | | 100.00 mL |

**Table 2.7.5:** Electrode buffer.

| Solutions | Preparation | Ingredients |
|---|---|---|
| TRIS-glycinate buffer, 10×<br>  0.025 mol/L TRIS<br>  0.192 mol/L glycine<br>  pH = 8.3<br>(The resolving buffer can be used as anode buffer.) | Solve 3.03 g TRIS and 14.41 g glycine in deionized water and fill up to 1,000.0 mL. | 100.00 mL |
| Deionized water to | | 1,000.00 mL |

## Procedure
- Assemble a casting cassette: roll a support film for polyacrylamide gel on a glass plate, place on the film 0.5 mm U-form silicone gasket, put over a second glass plate, and fix the cassette by clamps.
- Make the resolving solution in the side-arm flask and degas it with the water-jet pump.
- Pipette resolving solution with the catalysts into the casting cassette until the solution fills 2/3 of the cassette (Figure 2.7.17).

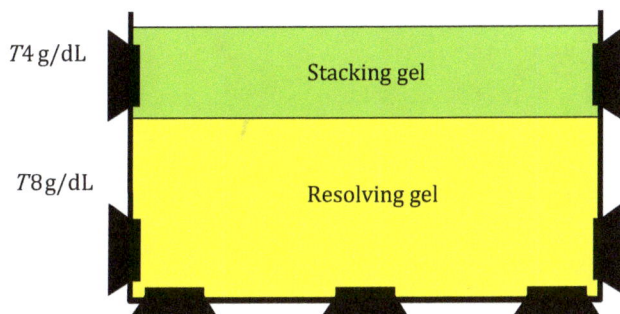

*T*4 g/dL — Stacking gel

*T*8 g/dL — Resolving gel

**Figure 2.7.17:** Casting a homogeneous gel for disc-electrophoresis.

- Pipette 2.0 mL deionized water or resolving buffer onto the resolving solution.
- Let the resolving solution polymerize at a room temperature for 45 min.
- Soak the liquid on the resolving gel using the water-jet pump.
- Pipette the stacking solution onto the resolving gel.
- Let the stacking solution polymerize at room temperature for another 45 min.
- Dissemble the casting cassette and take the polyacrylamide gel.
- Run disc-electrophoresis for 1–2 h.
- Fix the protein bands in a mixture of 1.0 mol/L (16.34 g/dL) trichloroacetic acid and 0.2 mol/L 5-sulfosalicylic acid (5.08 g/dL 5-sulfosalicylic acid dehydrate) for 30 min.
- Stain the bands in a filtered 0.001 mol/L (0.17 g/dL) Coomassie brilliant blue R-250 dissolved in the destaining solution for 30 min.
- Destain the gel background in the destaining solution (methanol–acetic acid–deionized water, 3:1:6, $V:V:V$).
- Dry the gel in air or in an oven.

### Disc-electrophoresis in One Buffer at Two pH Values According to Michov

**Materials and equipment**
Buffers
Acrylamide and BIS
TMEDA
APS
Glycine
Trichloroacetic acid
5-Sulfosalicylic acid
Coomassie brilliant blue R-250
Methanol
Acetic acid
250 mL side-arm flask for degassing gel solutions
Support film for polyacrylamide gel
Glass plates
0.5 mm U-form silicone gasket
Clamps
Water-jet pump

**Table 2.7.6:** Solutions for electrophoresis.

|  | Solution 1, 2×, pH = 9.3 | Solution 2, 2×, pH = 8.3 | Solution 3 |
|---|---|---|---|
| TRIS | 12.72 g | 2.79 g | – |
| Glycine | 2.70 g | 20.92 g | – |
| Acrylamide | – | – | 48.50 g |
| BIS | – | – | 1.50 g |
| Deionized water to | 100.00 mL | 1,000.00 mL | 100.00 mL |

*Solutions 1 and 2 are double concentrated resolving and electrode buffers, respectively. For casting resolving and stacking gels, solution 3 is also needed, where the total monomeric concentration T = 50 g/dL, and the cross-linking degree C = 0.03.*

**Table 2.7.7:** Preparing gels and buffers used for disc-electrophoresis in one buffer at two pH values.

|  | Electrode buffer pH = 8.3 | Stacking gel T = 4 g/dL, pH = 8.3 | Resolving gel T = 7 g/dL, pH = 9.3 | Sample buffer pH = 8.3 |
|---|---|---|---|---|
| Solution 1 | – | – | 50.00 mL | – |
| Solution 2 | 500.00 mL | 50.00 mL | – | 50.00 mL |

**Table 2.7.7** (continued)

| | Electrode buffer pH = 8.3 | Stacking gel $T = 4$ g/dL, pH = 8.3 | Resolving gel $T = 7$ g/dL, pH = 9.3 | Sample buffer pH = 8.3 |
|---|---|---|---|---|
| Solution 3 | – | 8.00 mL | 14.00 mL | – |
| 87% Glycerol | – | 9.11 mL | 9.11 mL | – |
| Bromophenol blue Na salt | – | – | – | 0.001 g |
| 10 g/dL TMEDA | – | 0.40 mL | 0.40 mL | – |
| 10 g/dL APS | – | 0.80 mL | 0.46 mL | – |
| Deionized water to | 1,000.00 mL | 100.00 mL | 100.00 mL | 100.00 mL |

**Procedure**

- Assemble a casting cassette: roll a support film for polyacrylamide gel on a glass plate, place on the film 0.5 mm U-form silicone gasket, put over a second glass plate, and fix the cassette by clamps.
- Make the resolving solution in the side-arm flask and degas it with the water-jet pump.
- Pipette resolving solution with the catalysts into the casting cassette until the solution fills 2/3 of the cassette (see above).
- Pipette 2.0 mL deionized water or resolving buffer onto the resolving solution.
- Let the resolving solution polymerize at a room temperature for 45 min.
- Soak the liquid on the resolving gel using the water-jet pump.
- Pipette the stacking solution onto the resolving gel.
- Let the stacking solution polymerize at room temperature for another 45 min.
- Dissemble the casting cassette and take the polyacrylamide gel.
- Run disc-electrophoresis for 1–2 h.
- Fix the protein bands in a mixture of 1.0 mol/L (16.34 g/dL) trichloroacetic acid and 0.2 mol/L 5-sulfosalicylic acid (5.08 g/dL 5-sulfosalicylic acid dehydrate) for 30 min.
- Stain the bands in a filtered 0.001 mol/L (0.17 g/dL) Coomassie brilliant blue R-250 dissolved in the destaining solution for 30 min.
- Destain the gel background in the destaining solution (methanol–acetic acid–deionized water, 3:1:6, $V:V:V$).
- Dry the gel in air or in an oven.

**SDS Disc-electrophoresis According to Laemmli**

**Materials and equipment**
TRIS
HCl
Acrylamide and BIS
SDS
Glycerol
TMEDA
APS
Glass plates
Support film
0.5 mm U-form spacer
Clamps
Kerosene
Paper strip electrodes
Horizontal electrophoresis unit
Constant current power supply
Coomassie brilliant blue or silver staining

### 4× Stacking gel buffer, pH = 6.8
| | |
|---|---|
| TRIS | 6.05 g (0.5 mol/L) |

*Adjust with 4 mol/L HCl to pH = 6.8*

| | |
|---|---|
| SDS | 0.40 g (0.014 mol/L) |
| Sodium azide | 0.01 g (0.002 mol/L) |
| Deionized water to | 100.00 mL |

*Filter and store at 4 °C up to 2 weeks.*

### 4× Resolving gel buffer, pH = 8.8
| | |
|---|---|
| TRIS | 18.20 g (1.5 mol/L) |

*Adjust with 4 mol/L HCl to pH = 8.8*

| | |
|---|---|
| SDS | 0.40 g (0.014 mol/L) |
| Sodium azide | 0.01 g (0.002 mol/L) |
| Deionized water to | 100.00 mL |

*Filter and store at 4 °C up to 2 weeks.*

### Acrylamide/BIS solution (40 g/dL T, 0.03 C)
| | |
|---|---|
| Acrylamide | 38.8 g |
| BIS | 1.2 g |

Deionized water to 100.0 mL

*Add 1 g Amberlite MB-1, stir for 10 min and filter. The solution can be stored in a refrigerator up to 2 weeks.*

**Table 2.7.8:** Stacking and resolving gels for Laemmli electrophoresis for one casting cassette.

| Stock solution | Stack. gel | Resolving gels | | | | | | | |
|---|---|---|---|---|---|---|---|---|---|
| | 4 | 5 | 7 | 8 | 10 | 12 | 13 | 15 |
| 4× Stacking gel buffer, pH = 6.8, mL | 3.75 | – | – | – | – | – | – | – |
| 4× Resolving gel buffer, pH = 8.8, mL | – | 3.75 | 3.75 | 3.75 | 3.75 | 3.75 | 3.75 | 3.75 |
| Acrylamide/BIS, mL | 1.88 | 1.88 | 2.63 | 3.00 | 3.75 | 4.50 | 4.88 | 5.63 |
| 87% Glycerol | 5.17 | 5.17 | 5.17 | 5.17 | 5.17 | 5.17 | 5.17 | 5.17 |
| 10 g/dL TMEDA, mL | 0.10 | 0.10 | 0.10 | 0.10 | 0.10 | 0.10 | 0.10 | 0.10 |
| 10 g/dL APS, mL | 0.05 | 0.05 | 0.05 | 0.05 | 0.05 | 0.05 | 0.05 | 0.05 |
| Deionized water to, mL | 15.00 | 15.00 | 15.00 | 15.00 | 15.00 | 15.00 | 15.00 | 15.00 |

*Use 5 g/dL gel for SDS-denatured proteins with $M_r$ = 60,000–200,000, 10 g/dL gel for proteins with $M_r$ = 16,000–70,000, and 15 g/dL gel for proteins with $M_r$ = 12,000–45,000.*

### 10× Electrode buffer, pH = 8.3

| | |
|---|---|
| TRIS | 30.28 g (0.25 mol/L) |
| Glycine | 144.00 g (1.92 mol/L) |
| SDS | 10.00 g (0.03 mol/L) |
| Sodium azide | 1.00 g (0.02 mol/L) |
| Deionized water to | 1,000.00 mL |

*Filter and keep at room temperature for up to 2 weeks. Prior to use, mix 100 mL of the electrode buffer with 900 mL of deionized water.*

### Procedure

- Assemble a sandwich of a lower glass plate, a support film for polyacrylamide gel, a 0.5 mm spacer, and an upper glass plate (see above).
- Using a syringe or a pipette pour 10 mL of the resolving gel solution in the casting cassette and overlay with deionized water.
- Allow the solution to polymerize for 30 min.
- Drain off the overlaid water.
- Using a syringe or a pipette pour 5 mL of the stacking gel solution into the casting cassette and overlay with deionized water.
- Allow the solution to polymerize for 60 min at room temperature or at 40 °C in a drying oven.
- Remove the gel from the casting cassette.

- Mix the protein samples with 2× SDS stacking gel buffer (1:1, V:V) and heat in a sealed screw-cap microcentrifuge tubes at 100 °C for 3–5 min. Dissolve molecular mass standards in 1× SDS stacking gel buffer.
- Pipette a few mL of kerosene on the cooling block of the horizontal electrophoresis unit.
- Place the film-supported gel onto the electrophoresis unit.
- Place gel or paper strip electrodes, soaked with 4× electrode buffer onto the gel ends.
- Place a silicon applicator template onto the gel in front of the cathode.
- Apply the samples in the template slots using a micropipette with thin tips.
- Connect the electrophoresis unit to the power supply and run electrophoresis at 10 mA constant electric current until Bromophenol blue enters the resolving gel. Increase then the electric current to 15 mA and run the electrophoresis until Bromophenol blue has reached the anode end of the gel.
- Turn off the power supply.
- Fix the gel in a mixture of ethanol–acetic acid–water (4:1:5, *V:V:V*) for 1 h.
- Stain with either Coomassie brilliant blue or silver nitrate or electroblot the proteins onto a membrane for immunoblotting. If the proteins are radiolabeled, they could be detected by autoradiography.

**SDS Electrophoresis in TRIS-formate-taurinate Buffer System According to Michov**

**Materials and equipment**
TRIS
Formic acid
Taurine
SDS
87% glycerol
TMEDA
APS
Glass plates
0.5 mm U-shaped gasket
Clamps
Horizontal electrophoresis unit
Power supply
BPB (**B**romo**p**henol **b**lue Na salt)

**Table 2.7.9:** Electrode solutions for SDS disc-gel electrophoresis in a TRIS-formate-taurinate buffer system.

|  | 4× Anode buffer<br>pH = 7.8, $I$ = 4 × 0.30 mol/L | 4× Cathode buffer<br>pH = 8.5, $I$ = 4 × 0.06 mol/L |
|---|---|---|
| TRIS | 19.02 g | 8.71 g |
| 99 g/dL formic acid | 4.57 mL | – |
| 99% taurine | – | 12.17 g |
| SDS | 0.24 g | 0.24 g |
| 0.1 g/dL BPB | – | 0.69 g |
| Deionized water to | 100.00 mL | 100.00 mL |

*Nonreducing sample buffer, pH = 7.8*

| 4× Anode buffer, pH = 7.8 | 2.50 mL |
|---|---|
| SDS | 0.10 g |
| 87% Glycerol | 3.00 mL |
| Coomassie brilliant blue G-250 | 0.02 g |
| Deionized water to | 10.00 mL |

*Monomeric solution (T = 50 g/dL, C = 0.03)*

| Acrylamide | 48.5 g |
|---|---|
| BIS | 1.5 g |
| Deionized water to | 100.0 mL |

**Table 2.7.10:** Stacking and resolving gels.

|  | Stacking gel | Resolving gel |  |  |  |  |  |
|---|---|---|---|---|---|---|---|
|  | $T$ = 5 g/dL | $T$ = 9 g/dL | $T$ = 11 g/dL | $T$ = 13 g/dL | $T$ = 15 g/dL | $T$ = 17 g/dL |
| 4× Anode buffer<br>pH = 7.8, mL | 25.00 | 25.00 | 25.00 | 25.00 | 25.00 | 25.00 |
| Monomeric<br>solution, mL | 10.00 | 18.00 | 22.00 | 26.00 | 30.00 | 34.00 |
| 87% Glycerol, mL | 30.00 | 10.00 | 10.00 | 10.00 | 10.00 | 10.00 |
| 10 g/dL TMEDA, mL | 0.45 | 0.45 | 0.45 | 0.45 | 0.45 | 0.45 |
| 10 g/dL APS, mL | 0.36 | 0.20 | 0.16 | 0.14 | 0.12 | 0.11 |
| Deionized water<br>to, mL | 100.00 | 100.00 | 100.00 | 100.00 | 100.00 | 100.00 |

**Table 2.7.11:** Gradient gels.

| | Light solution $T = 6$ g/dL, C = 0.03 | Heavy solution $T = 20$ g/dL, C = 0.03 |
|---|---|---|
| 4× Gel buffer, pH = 7.8 $I$ = 0.30 mol/L, mL | 25.50 | 25.50 |
| Monomer solution, mL | 12.00 | 40.00 |
| 87% Glycerol, mL | – | 30.00 |
| 10 g/dL TMEDA, mL | 0.45 | 0.45 |
| 10 g/dL APS, mL | 0.36 | 0.09 |
| Deionized water to, mL | 100.00 | 100.00 |

*Gradient gels (T = 8–18 g/dL) are cast using a gradient maker. The mixing chamber of the gradient maker is filled with the heavy solution; its reservoir is filled with the light solution. The cassette is filled from below.*

## Procedure

- Place a support film with its hydrophobic side down onto a glass plate, using water as an adhesive agent.
- Roll the support film with a photo roller to remove the air bubbles between the film and glass.
- Place a 0.5 mm U-shaped gasket onto the margins of the support film.
- Smear a second glass plate of same dimensions with a repelling solution to make it hydrophobic.
- Place the second glass plate over and fix all parts with clamps to build a cassette with a volume of 15 mL.
- Pour 10 mL of the resolving gel solution into the casting cassette using a syringe or a pipette and overlay with deionized water.
- Allow the solution to polymerize for 30 min.
- Drain off the overlaid water.
- Pour 5 mL of the stacking gel solution into the cassette using a syringe or a pipette.
- Allow the solution to polymerize at room temperature for 60 min.
- Remove the gel from the casting cassette.
- Dissolve protein samples and mass standards in the sample buffer.
- Heat the mixture in a sealed screw-cap microcentrifuge tube at 100 °C for 3–5 min.
- Apply prior to electrophoresis 0.5–1.0 mL kerosene or silicone oil DC 200 onto the cooling plate of the electrophoretic cell.
- Place the gel with its support film down onto the cooling plate so that the air bubbles between the support film and cooling plate are pushed away.

- Place paper strips with the 4× anode and 4× cathode buffer onto the two opposite ends of the gel.
- Place onto the stacking gel, 10 mm in front to the cathode strip, an application template, best of silicone, and press it gently to create a good contact with the gel.
- Apply 5 µL SDS sample into each slot using a micropipette with thin tips or a microliter syringe.
- Place the electrodes onto the electrode paper strips.
- Connect the electrophoresis cell to the power supply and run electrophoresis at constant electric current density of 0.65 mA/mm$^2$, an electric voltage of 100–500 V, and temperature of 10 °C until Bromophenol blue has reached the anode electrode strip.
- Stain the gel with Coomassie brilliant blue or silver, or electroblot the proteins onto a membrane for immunoblotting. If the proteins are radiolabeled, detect them using autoradiography.

**Coomassie Brilliant Blue R-250 Staining of Proteins Resolved in SDS Gels**

**Materials and equipment**
Fixing solution (5 g/L glutardialdehyde in 30 mL/dL ethanol)
Coomassie brilliant blue R-250
Ethanol
Acetic acid

**Procedure**
- Incubate a SDS polyacrylamide gel with separated proteins in the fixing solution: 10 min for 0.5 mm polyacrylamide gels with $T = 10$ g/dL; and 20 min for 0.5 mm polyacrylamide gels with $T = 16$ g/dL.
- Mix equal volumes of 0.4 g/dL Coomassie brilliant blue R-250 in ethanol–water (6:4, V:V) and 20 mL/dL acetic acid to make the staining solution.
- Stain for twice the time used for fixing.
- Destain the gel several times 30 min each in ethanol–acetic acid–glycerol–water (3:1:1:5, V:V:V:V).
- Dry the polyacrylamide gel under a cellophane membrane (agarose gels without cellophane membrane) at room temperature, under a hair dryer or in a drying oven.

*If polyacrylamide gels are cast on net, the times should be halved.*

## References

[1]    Poulik MD. Nature, 1957, 180, 1477–1479.
[2]    Michov BM. Electrophoresis, 1989, 10, 686–689.
[3]    Hedrick JL, Smith AJ. Arch Biochem Biophys, 1968, 126, 155–164.
[4]    Ornstein L. Ann N Y Acad Sci, 1964, 121, 321–349.
[5]    Davis BJ. Ann N Y Acad Sci, 1964, 121, 404–427.
[6]    Jovin TM. Biochemistry, 1973, 12, 871–879.
[7]    Davis BJ. Ann N Y Acad Sci, 1964, 121, 404–427.
[8]    Reisfeld RA, Lewis VJ, Williams DE. Nature, 1962, 195, 281–283.
[9]    Allen RC, Moore DJ, Dilworth RH. J Histochem Cytochem, 1969, 17, 189–190.
[10]   Allen RC, Budowle B. Gel Electrophoresis of Proteins and Nucleic Acids. Walter de Gruyter, Berlin – New York, 1994.
[11]   Hjerten S, Jerstedt S, Tiselius A. Anal Biochem, 1965, 11, 219–223.
[12]   Maurer HR, Allen RC. Z Klin Chem, 1972, 10, 220–225.
[13]   Michov BM. Electrophoresis, 1989, 10, 686–689.
[14]   Jovin TM. Ann N Y Acad Sci, 1973, 209, 477–496.
[15]   Chrambach A, Jovin TM, Svendsen PJ, Rodbard D. In Catsimpoolas N (ed.). Methods of Protein Separation. Plenum Press, New York, 1979, Vol. 2, 27–144.
[16]   Michov BM. Electrophoresis, 1990, 11, 289–292.
[17]   Michov BM. Electrophoresis, 1985, 6, 471–475.
[18]   Shapiro AL, Vinuela E, Maizel JV. Biochem Biophys Res Commun, 1967, 28, 815–820.
[19]   Servari E, Nyitrai R. Electrophoresis, 1994, 75, 1068–1071.
[20]   King J, Laemmli UK. Mol Biol, 1971, 62, 465–477.
[21]   Maizel JJV. Nature, 1970, 227, 680–685.
[22]   Ornstein L. Ann N Y Acad Sci, 1964, 121, 321–349.
[23]   Davis BJ. Ann N Y Acad Sci, 1964, 121, 404–427.
[24]   Neugebauer JM. In Deutscher MP (ed.). Methods in Enzymology. Acad. Press, San Diego, 1990, Vol. 182, 239–253.
[25]   Hjelmeland LM, Chrambach A. Electrophoresis, 1981, 2, 1–11.
[26]   Rath A, Glibowicka M, Nadeau VG, Chen G, Deber CM. Proc Natl Acad Sci USA, 2009, 106, 1760–1765.
[27]   Turro NJ, Yekta A. J Am Chem Soc, 1978, 100, 5951–5952.
[28]   Bales BL, Messina L, Vidal A, Peric M, Nascimento OR. J Phys Chem B, 1998, 102, 10347–10358.
[29]   Bales BL. J Phys Chem B, 2001, 105, 6798–6804.
[30]   Nielsen TB, Reynolds J. In Hirs CHW, Timasheff SN (eds.). Methods in Enzymology. Acad. Press, New York, 1978, Vol. 48, 3–10.
[31]   Rath A, Glibowicka M, Nadeau VG, Chen G, Deber CM. Proc Natl Acad Sci USA, 2009, 106, 1760–1765.
[32]   Reynolds JA, Tanford C. Proc Natl Acad Sci USA, 1970, 66, 1002–1007.
[33]   Rath A, Glibowicka M, Nadeau VG, Chen G, Deber CM. Proc Natl Acad Sci USA, 2009, 106, 1760–1765.
[34]   Pitt-Rivers R, Impiombato FSA. Biochem J, 1968, 109, 825–830.
[35]   Reynolds JA, Tanford C. Proc Natl Acad Sci USA, 1970, 66, 1002–1007.
[36]   Panyim S, Chalkley R. J Biol Chem, 1971, 246, 7557–7560.
[37]   Weber K, Osborn M. J Biol Chem, 1969, 244, 4406–4412.
[38]   Wyckoff M, Rodbard D, Chrambach A. Anal Biochem, 1977, 78, 459–482.
[39]   Lane LC. Anal Biochem, 1978, 86, 655–664.

[40]  Weber K, Osborn M. J Biol Chem, 1969, 244, 4406–4412.
[41]  Laemmli UK, Favre M. J Mol Biol, 1973, 80, 575–599.
[42]  Cleveland DW, Fischer SG, Kirschner MW, Laemmli UK. J Biol Chem, 1977, 252, 1102–1106.
[43]  Görg A, Postel W, Westermeier R, Gianazza E, Righetti PG. J Biochem Biophys Methods, 1980, 3, 273–284.
[44]  Ansorge W, De Maeyer L. J Chromatogr, 1980, 202, 45–53.
[45]  Olsson I, Axiö-Fredriksson UB, Degerman M, Olsson B. Electrophoresis, 1988, 9, 16–22.
[46]  Pharmacia LKB ExcelGel: Precast Gel and Buffer Strips for Horizontal SDS Electrophoresis. Tryckkontakt, Uppsala, 1989.
[47]  Laemmli UK. Nature, 1970, 227, 680–685.
[48]  Schägger H, Jagow G. Anal Biochem, 1987, 166, 368–379.
[49]  Schägger H. Nat Protoc, 2006, 1, 16–22.
[50]  Michov BM. Protein Separation by SDS Electrophoresis in a Homogeneous Gel Using a TRIS-formate-taurinate Buffer System and Homogeneous Plates Suitable for Electrophoresis. (Proteintrennung Durch SDS-Elektrophorese in Einem Homogenen Gel Unter Verwendung Eines TRIS-Formiat-Taurinat-Puffersystems Und Dafür Geeignete Homogene Elektrophoreseplatten). German Patent 4127546, 1991.
[51]  Michov BM. Electrophoresis, 1985, 6, 471–474.
[52]  Kohlrausch F. Ann Phys, 1897, 62, 209–239.
[53]  Debye P, Hückel E. Phys Z, 1923, 24, 185–206.
[54]  Michov BM. Electrophoresis, 1989, 10, 686–689.
[55]  Neuhoff V, Stamm R, Eibl H. Electrophoresis, 1985, 6, 427–448.
[56]  Andrews AT. Electrophoresis: Theory, Techniques, and Biochemical and Clinical Applications, 2nd edn. Oxford University Press, Oxford, 1986.
[57]  Harnes BD. In Harnes BD, Rickwood D (eds.). Gel Electrophoresis of Proteins: A Practical Approach, 2nd edn. IRL Press, Oxford, 1990, 1.
[58]  Shi Q, Jackowski G. In Harnes BD (ed.). Gel Electrophoresis of Proteins: A Practical Approach, 3rd edn. Oxford University Press, Oxford, 1998, 7.
[59]  Shapiro AL, Vinuela E, Maizel JV. Biochem Biophys Res Commun, 1967, 28, 815–820.
[60]  Weber K, Osborn MJ. Biol Chem, 1969, 244, 4406–4412.

# 2.8 Isoelectric focusing of proteins

2.8.1     Theory of isoelectric focusing —— 188
2.8.2     Isoelectric focusing with carrier ampholytes —— 189
2.8.2.1  Properties of carrier ampholytes —— 189
2.8.2.2  Formation of a pH gradient by carrier ampholytes —— 190
         Separator electrofocusing —— 191
2.8.2.3  IEF with carrier ampholytes on polyacrylamide gels —— 191
         Thin and ultrathin polyacrylamide gels for IEF with carrier ampholytes —— 191
         Rehydratable polyacrylamide gels for IEF with carrier ampholytes —— 191
2.8.2.4  Sample preparation and application —— 192
2.8.2.5  Electrode solutions —— 192
2.8.2.6  Running isoelectric focusing with carrier ampholytes —— 193
2.8.3     Isoelectric focusing in immobilized pH gradients —— 195
2.8.3.1  Properties of immobilines —— 196
2.8.3.2  Casting IPG gels —— 197
         Rehydratable IPG gels —— 198
2.8.3.3  Running isoelectric focusing on IPG gels —— 198
2.8.4     Isoelectric focusing in the clinical laboratory —— 199
2.8.5     Protocols —— 199
         Casting gels for isoelectric focusing with carrier ampholytes —— 199
         Running isoelectric focusing with carrier ampholytes —— 200
         Casting immobiline gels —— 202
         IEF with IPG gel strips —— 202
         References —— 206

Isoelectric focusing (IEF), known also as *electrofocusing,* is an electrophoresis technique for separating proteins in pH gradients. Moving through pH gradients, the proteins reach pH values (their **i**soelectric **p**oints, pI, pH(I)) where they lose their net electric charge and stop. Usually pH gradients are obtained in polyacrylamide gels with low concentration, where any sieving effect is eliminated.

The history of IEF began with Kolin [1]. He obtained unstable pH gradients *via* diffusion of two buffers with different pH values where separated proteins in sharp bands. Later, Svensson (Rilbe) published his theoretical works regarding the carrier (free) ampholytes [2,3]: hundreds of ampholytes with different isoelectric points should form in an electric field a pH gradient. To prove his theory he checked amino acids, synthetic peptides, and hydrolyzed proteins [4] but the problem remained open.

The first successful synthesis of carrier ampholytes (CA) was produced by Vesterberg [5–7], a student of Svensson. He synthesized a mixture of CA from residues of carboxylic acids and polyethylene amines. Later, Grubhofer proposed a method for synthesizing CA from oligoamino and oligosulfonic acid [8] and marketed them as *Servalyts*. The next attempt came from Williams and Söderberg [9, 10] who established the *pharmalyte* CA. Later, CA were synthesized that contained in addition residues of phosphoric and sulfuric acids [11]. LKB offered them as *ampholines*.

https://doi.org/10.1515/9783110761641-018

The next step was the binding of ampholytes to polyacrylamide gel, as a result of which **immobilized pH gradients** (IPG) were created. Nowadays, the IPG gels on plastic strips are preferred, because they can be used also for the second SDS dimension of the two-dimensional electrophoresis or in mass spectrometry.

## 2.8.1 Theory of isoelectric focusing

During IEF, the amphoteric polyions lose their net charges gradually until they stop at their isoelectric points (Figure 2.8.1).

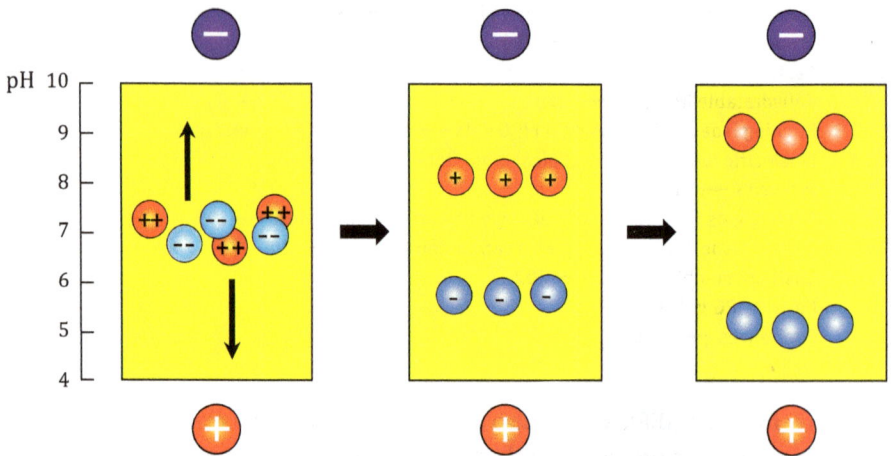

**Figure 2.8.1:** Isoelectric focusing of two proteins with pI values of 9.0 and 5.0.

At pH values above pI, the carboxyl groups are negatively charged ($-COO^-$), but the N-groups ($-NH_2$) carry no charges. As a result, the protein polyions have negative net charges. At pH values below pI, the carboxyl groups ($-COOH$) do not carry any charges; however, the N-groups are positively charged ($-NH_3^+$). As a result, the protein polyions have positive net charges. At the isoelectric points, the net charges of proteins are equal to zero.

During the IEF on the cathode, where electron excess exists, hydroxide ions are formed:

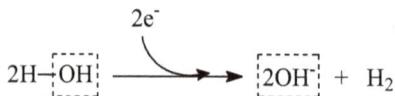

$$2H-OH \xrightarrow{2e^-} 2OH^- + H_2$$

and on the anode, where electron deficiency exists, protons are formed:

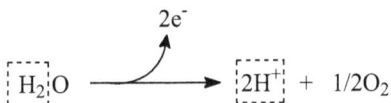

$$H_2O \xrightarrow{\quad 2e^- \quad} 2H^+ + 1/2O_2$$

As a result, the pH value at the anodic end of the gel decreases, and the pH value at the cathodic end of the gel increases. To keep the pH gradient stable in the gel, corresponding electrode solutions are used: an acidic solution at the anode end and an alkaline solution at the cathode end. When an acidic CA meets the anode solution, its basic groups charge positively and will be attracted by the cathode, and *vice versa*.

The direction and the velocity of moving of an amphoteric polyion in a pH gradient depends on its net charge at a given pH value. For example, a protein with a positive net charge migrates toward the cathode through the increasing pH value and loses gradually its net charge through deprotonation. This continues until the numbers of its negative and positive charges become equal and, as a result, its net charge becomes zero. If the polyion moves away from its pI point by diffusion, it obtains again positive or negative charges and moves again to its pI point.

The isoelectric points of most proteins and peptides are located in the pH range of 3–12, but most of them are between pH = 4.0 and 7.0 [12, 13]. Narrow pH gradients can also be obtained. For example, if proteins have pI values from 5.2 to 5.8, they can be focused in a pH gradient of the range 5.0–6.0.

## 2.8.2 Isoelectric focusing with carrier ampholytes

IEF with CA can be carried out in different separation media: polyacrylamide gel [14], agarose gel [15, 16], acetyl cellulose films [17], and granulated gels [18, 19]. They should not have sieving effect.

### 2.8.2.1 Properties of carrier ampholytes

The CA represent a heterogeneous mixture of different low-molecular aliphatic oligoamino-oligocarboxylic acids (zwitterions) with close pI points:

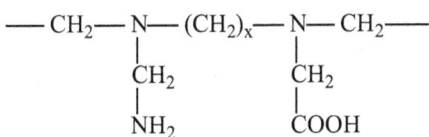

$$-CH_2-N-(CH_2)_x-N-CH_2-$$
$$\quad\quad\quad | \quad\quad\quad\quad\quad | $$
$$\quad\quad\quad CH_2 \quad\quad\quad\quad CH_2$$
$$\quad\quad\quad | \quad\quad\quad\quad\quad | $$
$$\quad\quad\quad NH_2 \quad\quad\quad\quad COOH$$

They arrange themselves in an electric field according to their pI values and build a pH gradient. Since the pK values of CA and proteins to be analyzed are temperature dependent, the IEF should be carried out at a constant temperature, preferably at 10 °C.

### 2.8.2.2 Formation of a pH gradient by carrier ampholytes

At first, the CA are in homogeneous distribution in the gel. After an electric field is applied, a pH gradient occurs in three steps (Figure 2.8.2):

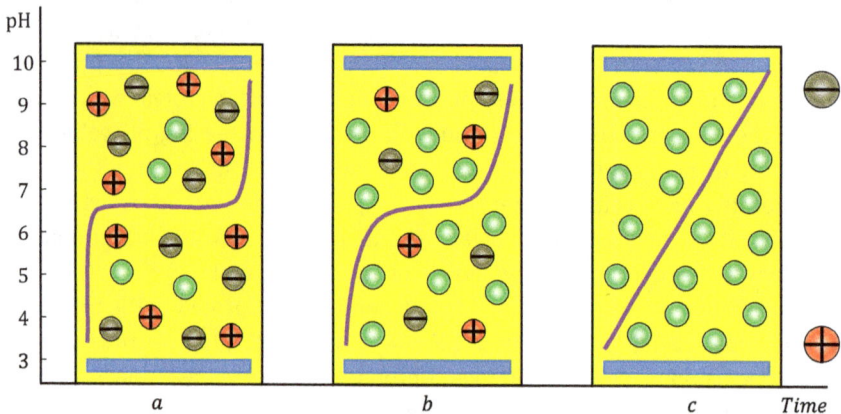

**Figure 2.8.2:** Steps (*a*, *b*, and *c*) of pH gradient building by carrier ampholytes.

- The CA begin to migrate according to their net charges to opposite electrodes. CA with pI lower than the pH value of separation medium carry negative charges; therefore, they migrate to the anode. CA with pI higher than the pH of separation medium have positive charges; therefore, they migrate to the cathode. As a result, two-step pH gradient is created with a pH plateau in the middle.
- The pH plateau disappears and
- Both pH gradients unite to give a continuous linear pH gradient.

Optimal pH gradients are formed by CA with $M_r$ between 300 and 1,000. If $M_r$ is less than 300, the pH gradient is unstable; if they are larger, they cannot be distinguished from proteins to be resolved. The pH gradient depends also on the number of CA molecules. The larger it is, the "smoother" the pH gradient. Ampholyte concentrations of about 2 g/dL give stable pH gradients. Glycerol, sorbitol, or sucrose can be added to the mixture, but they change the pH gradients if their concentrations are higher than 10–20 g/dL.

Diverse pH gradients are used: acidic gradients (pH = 4.0–6.0), very acidic gradients (pH = 2.0–4.0), neutral gradients (pH = 6.0–8.0), alkaline gradients (pH = 8.0–10.0), and very alkaline gradients (pH = 9.0–11.0).

**Separator electrofocusing**

The resolution of a pH gradient can be increased by adding appropriate separators, known as pH gradient modifiers [20,21]. The separators are amino acids or amphoteric electrolytes, which flatten the pH gradient near their isoelectric points. Separators are, e.g., tetraglycine (pI = 5.2), proline (pI = 6.3), threonine (pI = 6.5), β-alanine (pI = 6.9), 5-aminovaleric acid (pI = 7.5), histidine (pI = 7.6), and 6-aminocaproic acid (pI = 8.0). For example, with the help of 0.33 mol/L β-alanine in a pH gradient of 6–8, the glycated $HbA_{1c}$ was separated from the closely adjacent HbA, the major hemoglobin band [22].

IEF is carried out on polyacrylamide gels or rarely on agarose gels.

## 2.8.2.3 IEF with carrier ampholytes on polyacrylamide gels

CA-IEF is usually carried out on polyacrylamide gels: at $T$ = 4–5 g/dL and $C$ = 0.03 for electrofocusing of proteins, and at $T$ = 10 g/dL and $C$ = 0.025 for electrofocusing of oligopeptides. The polyacrylamide gels show no electroosmotic effects. However, they are not suitable for separation of macromolecules with molecular masses more than 500,000. These polyions can be focused only on agarose gels.

**Thin and ultrathin polyacrylamide gels for IEF with carrier ampholytes**

A problem of IEF is the Joule heating, which is produced during electrophoresis. This problem was solved after the gel thickness was reduced, and thin (thickness of 0.5 mm) [23,24] and ultrathin (thickness less than 0.5 mm) [25,26] slab gels were produced on pretreated plastic sheets (support films). Because of their higher surface-to-volume ratio, the thin polyacrylamide or agarose gels dissipate the Joule heating better than the thick ones and, as a result, can be subjected to higher voltage. In addition, the sample volume is reduced, and the times for staining, destaining, and drying of the gel are shortened.

**Rehydratable polyacrylamide gels for IEF with carrier ampholytes**

The polymerization of a monomeric solution in the presence of CA requires higher concentrations of APS. Besides, when a gel was polymerized in the presence of CA, it dissolves partially or completely from the support film in an acidic solution. Nonionic detergents, which are often added to the monomeric solution to increase the solubility of hydrophobic proteins, inhibit the adhesion between the gel and film, too. To improve the adhesion of the gel to the support film, rehydratable gels are used [27, 28].

The rehydratable gels polymerize without addition of CA. Then they are washed with deionized water, are dried, and can be stored indefinitely at −20 °C. Prior to IEF, the rehydratable gels should be swollen (rehydrated) in 2 g/dL CA with 20–30 mL/dL glycerol to their original thickness.

However, the rehydratable gels have some disadvantages: Their production (washing and drying) and their rehydrating take time. Besides, they free water on their surface during electrofocusing.

### 2.8.2.4 Sample preparation and application

The sample proteins should be completely dissolved, and in concentrations at which the pherogram bands will be visible. For Coomassie brilliant blue R-250 staining, the total protein concentration in the sample must be about 0.1 g/dL (1 mg/mL); for silver staining, it must be about 0.01 g/dL (0.1 mg/mL). The application volume of a sample should be between 5 and 10 µL. Highly diluted protein solutions can be concentrated by ultrafiltration.

The hydrophobic proteins, such as membrane proteins, should not be resolved with the help of ionic detergents, as SDS, but with the help of 8–9 mol/L of urea. If the proteins are still unresolved, nonionic detergents (Nonidet NP-40, Triton X-100) at concentrations between 0.5 and 2 mL/dL, or 1–2 g/dL of zwitterionic detergents (CHAPS) come into account [29,30]. Urea and detergents destroy the quaternary structure of the proteins and unfold the polypeptide chains. The disulfide bonds between the polypeptide chains are broken when 1,4-dithiothreitol is added.

The ionic strength (the salt concentration) of the sample should be reduced to a minimum, since when the ionic strength is higher than that in the gel local gradient drifts appear in the gel. As a critical salt concentration in a protein sample, 20 mmol/L was specified [31], but it is better to be decreased to 5 mmol/L. If necessary, the samples can be desalted by dialysis, gel filtration, or ultrafiltration.

Among a lot of methods, the sample application into template slots is the most convenient technique. The template should be removed 10 min after the application. If residual sample is still on the gel, it should be blotted with blotting paper.

### 2.8.2.5 Electrode solutions

The contact between the gel and electrodes is realized *via* electrode solutions whose pH values are higher and lower than the maximum and minimum pI points of the CA. The electrode solutions are dropped onto 5–7 mm wide strips of thick filter paper, placed on a glass plate. The excess of the electrode solutions should be blotted and then the strips are put onto the gel.

The anode strips should contain acidic solution, and the cathode strips should contain basic solution (Table 2.8.1). In addition, the electrode solutions should contain 30–40 mL/dL glycerol, which prevents the drying of electrode strips.

**Table 2.8.1:** Electrode solutions for isoelectric focusing with carrier ampholytes on polyacrylamide and agarose gels.

|  | pH gradient | Anode solution | Cathode solution |
|---|---|---|---|
| Polyacrylamide gel | 3–10 | 0.5 mol/L H$_3$PO$_4$<br>0.25 mol/L acetic acid<br>0.025 mol/L aspartic acid,<br>0.025 mol/L glutamic acid | 0.5 mol/L NaOH<br>0.25 mol/L NaOH)<br>0.025 mol/L lysine<br>0.025 mol/L arginine |
|  | 5–8 | 0.04 mol/L glutamic acid | 0.25 mol/L NaOH |
|  | 3–7 | 0.04 mol/L glutamic acid | 0.2 mol/L histidine |
|  | 6–8 | 0.25 mol/L HEPES | 0.2 mol/L histidine |
| Agarose gel | 3–10 | 0.5 mol/L acetic acid | 0.5 mol/L NaOH |

## 2.8.2.6 Running isoelectric focusing with carrier ampholytes

There is almost no difference between the IEF on polyacrylamide and agarose gels (Table 2.8.2). However, prior to IEF, the surface of agarose gel should be dried with a filter paper to remove the superficial liquid film. In addition, prefocusing of agarose gels should be avoided because the cathode drift here is much stronger than in polyacrylamide gels.

The IEF is run at constant power. In the beginning, the CA wear high net charges and the gel has a high conductivity. When the CA approach their isoelectric points and form a pH gradient, they lose their charges and mobilities. Therefore, the conductivity of the CA lowers, which causes the voltage to increase. Thus, a typical current–voltage relationship is obtained: the electric current falls, the voltage increases, and the electric power remains constant (Figure 2.8.3). In the presence of 8–9 mol/L urea, the temperature should not be below 15 °C to avoid the crystallization of urea.

After IEF, the gels are stained usually with Coomassie brilliant blue R-250, Serva violet 17, or silver (see there). In agarose gels, the silver staining is not enough successful because of the electroosmosis.

**Table 2.8.2:** Electrophoretic conditions for isoelectric focusing on polyacrylamide and agarose gels containing 2 g/dL carrier ampholytes at 10 °C; separation distance of 10 cm.

| pH interval | Time | Power | Voltage | Electric current |
|---|---|---|---|---|
| **Polyacrylamide gels** | | | | |
| *Prefocusing* | | | | |
| pH = 3.0–10.0 | 60 min | So high that the voltage reaches 300 V | Maximum | Maximum |
| Narrow pH gradients | 90 min | So high that the voltage reaches 300 V | Maximum | Maximum |
| (2 pH units) | | | | |
| *Focusing* | | | | |
| pH = 3.0–10.0 | 10 min | As the prefocusing | 200 V | Maximum |
| | 120 min | Increase the power by a factor of 2.5 | Maximum | Maximum |
| Narrow pH gradients | 15 min | As the prefocusing | 200 V | Maximum |
| (2 pH units) | 180 min | Increase the power by a factor of 2.5 | Maximum | Maximum |
| **Agarose gels** | | | | |
| *Focusing* | | | | |
| pH = 3.0–10.0 | 30 min | So high that the voltage reaches 500 V | Maximum | Maximum |
| Narrow pH gradients | 60 min | So high that the voltage reaches 500 V | Maximum | Maximum |
| (2 pH units) | | | | |

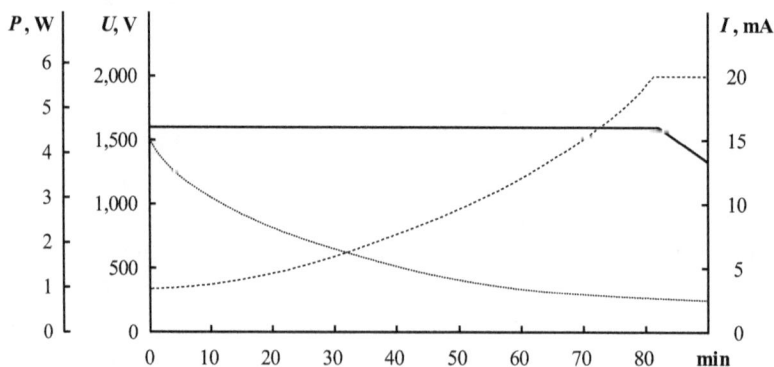

**Figure 2.8.3:** Electric power ($P$), voltage ($U$), and electric current ($I$) during isoelectric focusing in a pH gradient of 3–10 obtained with carrier ampholytes in a polyacrylamide gel with dimensions of 120 × 250 × 0.25 mm, during a run at 10 °C for 90 min. — $P$,----, $U$, ····· $I$.

## 2.8.3 Isoelectric focusing in immobilized pH gradients

Gasparic, Bjellqvist, and Rosengren introduced the IPG [32]. They are acrylamido buffers that are copolymerized by their vinyl groups into polyacrylamide matrix to form IPG gels [33,34] (Figure 2.8.4). First, LKB Produkter AB produced and offered such buffers under the name of *Immobiline*.

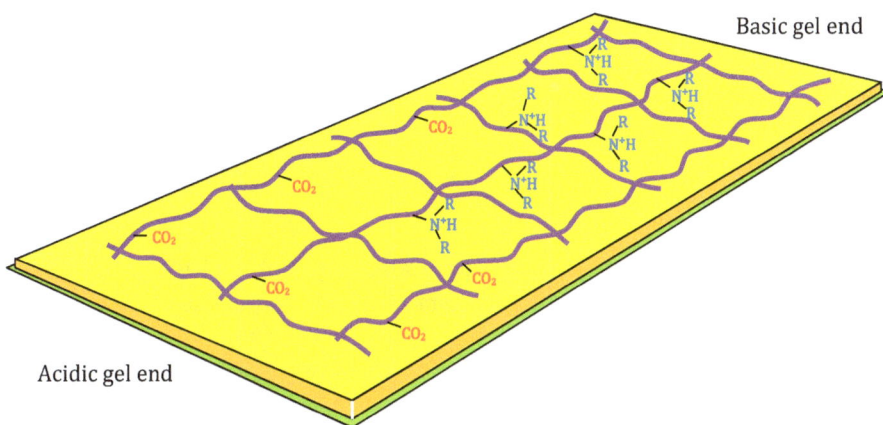

**Figure 2.8.4:** Polyacrylamide gel copolymerized with immobilines.

The distribution of immobilines in polyacrylamide gels takes place during the gel casting. Linear pH gradients are prepared using a gradient maker and two mixtures of immobilines: one relatively basic and one relatively acidic. Ultra-narrow pH gradients (up to 0.01 pH/cm) can also be produced.

The IPGs have several advantages over the pH gradients obtained with CA:

- IPG do not deform by any external influences (electroosmosis, high ionic strength of the samples, or protein overloads).
- The IPGs have a uniform buffering capacity and conductivity along the gel. This cannot be guaranteed with the CA pH gradients because not all of their components are present in same concentration.
- The IPG gels have a low conductivity, as the immobilines cannot move freely. As a result, a little Joule heating is developed during IEF, even at very high voltages.
- In the IPG gels, the cathode drift is excluded, which is available in gels with CA.
- The resolution of IEF in IPG exceeds the resolution of IEF with CA to 10–20 times and allows to resolve proteins that differ from each other by pI values of only 0.001–0.002 pH units [35]. While the resolution of IEF with CA is more than 0.02 pH units/cm, the resolution of IEF in IPG gels may reach 0.001 pH units/cm.

However, the IPG gels also have some disadvantages:
- The preparation of IPG is a complex process.
- The IPG gels can adsorb proteins.
- IEF in IPG gels requires very high voltage.
- The electrophoretic separation takes too long time.

### 2.8.3.1 Properties of immobilines

The immobilines are low-molecular-mass acrylamide derivatives, which carry buffering groups. Their general formula is

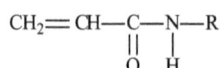

$$CH_2 = CH - C - N - R$$
$$\underset{O}{\overset{\|}{\phantom{C}}} \underset{H}{\overset{|}{\phantom{N}}}$$

where the residue R is either an acidic group (carboxyl group) or a basic group (tertiary amino group). They take part in the copolymerization of acrylamide and BIS by their methylene group.

The immobilines are supplied in concentrations of 0.2 mol/L. They polymerize and hydrolyze spontaneously [36]; therefore, it is recommended [37] that 0.005 mg/mL hydroquinone monomethyl should be added to the acidic immobilines ($pK$ = 1.0, 3.6, 4.4, and 4.6), and the neutral and basic immobilines ($pK$ = 6.2, 7.0, 8.5, and 9.3) to be dissolved in $n$-propanol. The immobiline solutions can be stored for months and years at 4 °C. They should not be frozen.

The structures of the acrylamido buffers used for preparing IPGs are shown in Table 2.8.3.

**Table 2.8.3:** Structures of the acrylamido buffers used for preparing immobilized pH gradients.

| pH | Chemical formula | Name | $M_r$ |
|---|---|---|---|
| **Acidic acrylamido buffers** | | | |
| 1.2 | | 2-Acrylamido-2-methylpropane sulfonic acid | 207 |
| 3.1 | | 2-Acrylamido-glycolic acid | 145 |
| 3.6 | | N-Acryloyl-glycine | 129 |

**Table 2.8.3** (continued)

| pH | Chemical formula | Name | $M_r$ |
|---|---|---|---|
| 4.6 | | 4-Acrylamido-butyric acid | 157 |
| **Neutral and basic acrylamido buffers** | | | |
| 6.2 | | 2-Morpholino propylacrylamide | 184 |
| 7.0 | | 3-Morpholino propylacrylamide | 199 |
| 8.5 | | N,N-Dimethyl aminoethyl acrylamide | 142 |
| 9.3 | | N,N-Dimethyl aminopropyl acrylamide | 156 |
| 10.3 | | N,N-Diethyl aminopropyl acrylamide | 184 |
| >12 | | N,N,N-Triethyl aminoethyl acrylamide | 198 |

## 2.8.3.2 Casting IPG gels

The IPG gels are cast using a gradient maker and immobilines, wherein the heavy (with 25 mL/dL glycerol) solution is acidic and the light (with 5 mL/dL glycerol) solution is basic. The mixing chamber is filled with the heavy solution that builds the lower part of the IPG; the reservoir is filled with the light solution that builds the upper part of the IPG. For IEF of proteins, gels with $T$ = 10 g/dL and $C$ = 0.025 are recommended. In practice, thin (0.5 mm) as well ultrathin (<0.5 mm) IPG gels are used [38].

## Rehydratable IPG gels

After casting, an IPG gel can be used immediately or can be dried. Prior to drying, the gel should be washed 3 times for 20 min each in deionized water and once in 2–3 mL/dL glycerol for 20 min. Then it is dried under a hair dryer and is sealed in a plastic wrap to be stored in the refrigerator at 4–8 °C for several days, or in the freezer at −20 °C for indefinite storing. The dried gel, which is referred to as a *rehydratable gel*, can be later rehydrated (reconstituted) in suitable solutions.

### 2.8.3.3 Running isoelectric focusing on IPG gels

The IPG IEF is carried out in the following way: At 10 °C, the gel end with the low pH value should be connected to the anode, whereas the gel end with the high pH value should be connected to the cathode. If the gel contains urea, the separation temperature should not be below 15 °C, since urea crystallizes at lower temperatures. As electrode solutions for IEF in IPG gels, as well as for hybrid IEF, 0.1 g/dL NaOH or 0.01 mol/L lysine (as a cathode solution), and 0.1 g/dL $H_3PO_4$ or 0.01 mol/L glutamic acid (as an anode solution) are used.

The IEF proceeds in two phases: During the *first phase*, the proteins migrate at a constant voltage into the gel. The electric power and electric current are set at maximum values, but the voltage is set on 120–300 V, so that the field strength should be as low as possible (up to 40 V/cm). If the initial field strength is too high, the proteins may precipitate partially. Since the buffering groups of IPGs are anchored in the gel and do not move freely, the IPG gels have low conductivity and the electric current strength is only 1–2 mA (Figure 2.8.5). This phase continues for 60 min.

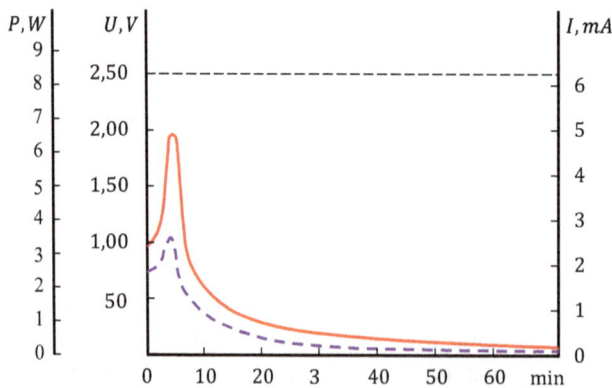

**Figure 2.8.5:** Electric power (*P*), voltage (*U*), and electric current (*I*) during the first 60 min of IEF in an immobilized pH gradient (pH = 4.5–4.8). ─ *P*, –– *U*, ··· *I*.

During the *second phase* of IEF, the voltage should be set on maximum value of 5,000 V (2,000 V in hybrid IEF), the power should be set on maximum value of 10 W; then the electric current reaches usually 15 mA. The electric resistance in the IPG gels is 50–100 times higher than that in gels for IEF with CA. Therefore, the voltage in the IPG IEF should be higher and IEF should continue longer.

In the last years, an IEF microprocessor apparatus was described that is operating at 12,000 V maximum voltage, enabling to run high-speed IEF of proteins within 3 h [39]. In its standard configuration, up to six IPG strips can be run.

The electrofocusing time is reciprocally proportional to the pH units: 3–4 h for broad pH gradients (4–7 pH units), 8 h for narrow pH gradients (2 pH units), and 16 h (overnight) for ultra-narrow pH gradients (1 pH unit or less).

After electrofocusing, the proteins in an IPG gel can be stained with conventional or colloidal Coomassie brilliant blue in phosphoric or sulfuric acid. If higher sensitivity is desired, the IPG gels can be stained with silver.

## 2.8.4 Isoelectric focusing in the clinical laboratory

With the help of IEF on polyacrylamide gels, more than 300 hemoglobin variants [40, 41, 42], lipoproteins [43, 44, 45], phenotypes of $\alpha_1$-antithrypsin [46, 47], urinary proteins [48, 49], globulins [50, 51], salivary proteins [52, 53], catalase [54], and many other clinically important proteins were analyzed. Besides, genetic investigations in forensic medicine [55] were performed. Meat proteins [56, 57], potato varieties [58, 59], and wheat proteins [60] were studied, too.

## 2.8.5 Protocols

### Casting Gels for Isoelectric Focusing with Carrier Ampholytes

**Materials and equipment**
Acrylamide
BIS
TMEDA
APS
40 g/dL Carrier ampholytes
87% Glycerol
0.1 g/dL **Flavin mononucleotide** (FMN, riboflavin-5′-phosphate)
Gel casting cassette
Fluorescent lamp

### *Acrylamide/BIS stock solution (T = 30 g/dL, C = 0.03)*

| | |
|---|---|
| Acrylamide | 29.1 g |
| Bisacrylamide | 0.9 g |
| Deionized water to | 100.0 mL |

### *Monomeric solution*

| | |
|---|---|
| Acrylamide/BIS stock solution | 16.7 mL |
| 40 g/dL Carrier ampholytes | 5.0 mL |
| 87% Glycerol | 10.0 mL |
| 0.1 g/dL FMN | 0.5 mL |
| 10 g/dL TMEDA | 0.4 mL |
| 10 g/dL APS | 0.3 mL |
| Deionized water to | 100.0 mL |

### Procedure

- Pour monomeric solution into the casting cassette.
- Illuminate the monomeric solution in the cassette with the fluorescent lamp for 1 h.
- Open the cassette and illuminate once more the gel face with the fluorescent lamp for 30 min.
- Use the gel immediately or cover it with a polyethylene sheet and store at 4 °C for several days.

## Running Isoelectric Focusing with Carrier Ampholytes

### Materials and equipment

Flat-bed electrophoresis cell
Film-supported gels with carrier ampholytes
Power supply capable of delivering 2,000–3,000 V and 6W
Refrigerated water circulator, if required
1 N NaOH catholyte
1 N $H_3PO_4$ anolyte
Electrode strips
Sample application template

### Procedure

- Pipette 0.5–1.0 mL of kerosene or silicone oil DC 200 onto the cooling plate of an electrophoretic unit.
- Place a gel with its support film down, so that no air bubbles are formed between the film and cooling plate.
- Place the electrode strips with the anode and cathode solutions on both ends of the gel.

- Place an application template onto the gel centrum, and press it gently to provide a good contact.
- Prefocus the gel at 400–500–1,000 V per 10 cm gel for 30–60 min. So a pH gradient will be formed, and the unwanted ions, e.g., APS and TMEDA ions, will migrate toward the electrode strips.

  *Gels with higher glycerol concentration and the rehydrated gels do not require prefocusing.*
- Place an application template onto the gel.
- Apply in the template slots 5–10 µL of desalted samples containing 0.1–0.5 mg/mL proteins (Figure 2.8.6).

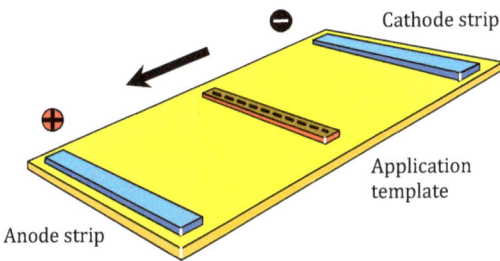

**Figure 2.8.6:** Gel with carrier ampholytes, paper electrode strips, and an application template.

- Increase gradually the field strength from 50–100 to 300–500 V/cm, depending on the pH range, concentration of the carrier ampholytes, gel thickness, glycerol concentration, and cooling temperature. For most proteins, 5,000–20,000 Vh per 10 cm gel is used.
- Interrupt the IEF after the half of electrofocusing time is over.
- Remove the application template and blot the excess solution on the electrode strips with filter paper.
- Continue with the electrofocusing.

**Isoelectric focusing of net-supported gels containing carrier ampholytes**
- Place the net-supported gel onto the cooled to 15 °C cooling plate of the electrophoretic unit.
- Place the electrode strips containing the anode and cathode solutions on both ends of the gel.
- Place an application template onto the middle of the gel and press it gently to provide a good contact.
- Fill the slots of the application template with 5–10 µL of the samples each.
- Start the IEF at 200 V (for gels with separation distance of 10 cm).
- Increase the electric voltage by 400–1,000 V every 30 min until a maximum voltage of 3,000 V is reached.

> *In gels with high glycerol concentration, the voltage can reach 6,000 V.*

– Continue the electric focusing until the desired Vh product.

## Casting Immobiline Gels

### Materials and equipment
Acrylamide
BIS
TMEDA
APS
Immobilines
87% Glycerol
Gel cassette

### Procedure
*IPG gels with linear pH gradients are cast according to the recipes of Gianazza et al. [61, 62] (Table 2.8.4).*

– After polymerization, keep the casting cassette at room temperature for 30 min.
– Remove the IPG gel from the cassette, wash with 2 mL/dL glycerol, and dry at room temperature.
– Cover the gel with a plastic film and store at –20 °C.

## IEF with IPG Gel Strips

### Materials and equipment
Dry IPG strips
Rehydration solution
Rehydration cassette
IEF unit
Power supply

### Sample solubilization buffer

| | |
|---|---|
| Urea | 54.05 g (9.0 mol/L) |
| CHAPS | 2.00 g |
| 40 g/dL pharmalytes 3–10 | 5.00 mL |
| DTT | 0.23 g (15.0 mmol/L) |
| Deionized water to | 100.00 mL |

**Table 2.8.4:** Volumes of the acidic (heavy) and basic (light) immobiline solution, which are used for casting of 15 mL slab gels (120 × 250 × 0.5 mm) with wide pH gradients (2–6 pH units). Both solutions contain monomers in same concentration but the heavy solution contains more glycerol.

|  | Acidic (heavy) solution 0.2 mol/L immobilines with diverse p$K$, in µL | | | | | | pH ranges | Basic (light) solution 0.2 mol/L immobilines with diverse p$K$, in µL | | | | | |
|---|---|---|---|---|---|---|---|---|---|---|---|---|---|
| pH range | 3.6 | 4.6 | 6.2 | 7.0 | 8.5 | 9.3 | | 3.6 | 4.6 | 6.2 | 7.0 | 8.5 | 9.3 |
| 3.5–5.0 | 159 | 119 | 84 | – | – | – | | 113 | 165 | 248 | – | – | – |
| 4.0–6.0 | 303 | 53 | 234 | – | – | – | | 208 | 278 | 147 | – | – | 385 |
| 4.5–6.5 | 221 | 128 | 266 | – | – | – | | – | 304 | 130 | 125 | – | 158 |
| 5.0–7.0 | 37 | 228 | 221 | – | – | – | | – | 253 | 144 | 117 | – | 171 |
| 5.5–7.5 | – | 240 | 189 | 60 | – | – | | 185 | – | 126 | 153 | 151 | – |
| 6.0–8.0 | 232 | – | 172 | 111 | 23 | – | | 153 | – | 93 | 173 | 175 | – |
| 6.5–8.5 | 411 | – | 147 | 99 | 287 | – | | 102 | – | 82 | 148 | 193 | – |
| 7.0–9.0 | 719 | – | – | 145 | 198 | 451 | | 258 | – | – | 124 | 101 | 291 |
| 7.5–9.5 | 356 | – | – | 237 | 121 | 186 | | 110 | – | – | 493 | 74 | 185 |
| 8.0–10.0 | 213 | – | – | 194 | 189 | 50 | | 49 | – | – | 175 | 195 | 154 |
| 4.0–7.0 | 308 | 59 | 240 | – | – | – | | 161 | 394 | 81 | 143 | – | 467 |
| 5.0–8.0 | 374 | 135 | 222 | 71 | 185 | – | | 93 | 66 | 70 | 184 | 185 | – |
| 6.0–9.0 | 415 | – | 214 | 50 | 194 | 43 | | 129 | – | 86 | 239 | 126 | 120 |
| 7.0–10.0 | 289 | – | – | 202 | 187 | – | | 48 | – | – | 173 | 187 | 149 |

(continued)

Table 2.8.4 (continued)

pH ranges

| | Acidic (heavy) solution 0.2 mol/L immobilines with diverse pK, in µL | | | | | | pH range | Basic (light) solution 0.2 mol/L immobilines with diverse pK, in µL | | | | | |
|---|---|---|---|---|---|---|---|---|---|---|---|---|---|
| | 3.6 | 4.6 | 6.2 | 7.0 | 8.5 | 9.3 | | 3.6 | 4.6 | 6.2 | 7.0 | 8.5 | 9.3 |
| | 314 | 135 | 125 | 62 | 91 | – | 4.0–8.0 | – | 295 | 192 | 76 | 178 | 154 |
| | 442 | 310 | 116 | 74 | 424 | 65 | 5.0–9.0 | – | 133 | 140 | 113 | 156 | 122 |
| | 502 | – | 146 | 130 | 139 | 150 | 6.0–10.0 | 53 | – | 178 | 193 | 127 | 174 |
| | 442 | 125 | 124 | 12 | 133 | 118 | 4.0–9.0 | 78 | 226 | 192 | 158 | 38 | 354 |
| | 300 | 247 | 159 | 146 | 121 | 68 | 5.0–10.0 | 11 | 31 | 18 | 224 | 165 | 146 |
| | 588 | – | 243 | 47 | 178 | – | 4.0–10.0 | – | 61 | 27 | 260 | 84 | 190 |

### *Rehydration (reswelling) solution*

| | |
|---|---|
| Urea | 48.05 g (8 mol/L) |
| CHAPS | 0.50 g |
| DTT | 0.23 g (15 mmol/L) |
| 40 g/dL Pharmalyte, pH = 3–10 | 5.00 mL |
| Deionized water to | 100.00 mL |

### Procedure
– Assemble the rehydration cassette with a 0.7 mm thick U-frame (0.5 mm of the plate's U-frame plus two layers of parafilm, 0.1 mm each).
– Clamp the cassette together and pour the rehydration solution.
– Cut dried IPG gels into 3 mm wide strips with the help of a paper cutter. Alternatively, use ready-made IPG strips.
– Introduce the IPG gel strips in the rehydration cassette and rehydrate them to their original thickness (0.5 mm) (Figure 2.8.7) overnight at room temperature.

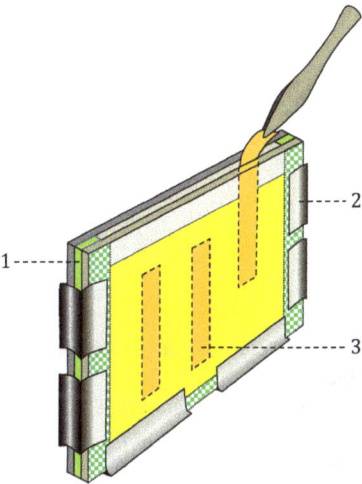

**Figure 2.8.7:** Rehydration of IPG gel strips.
1. U-formed gasket; 2. clamp; 3. gel strip.

– Take the rehydrated gel strips out of the cassette and place them gel up on a sheet of water-saturated filter paper.
– Blot the gel strips gently to remove excess rehydration solution in order to prevent urea crystallization on the gel surface during electrofocusing.
– Pipette 2–3 mL of kerosene onto the flat-bed cooling block of IEF chamber and place the IPG gel strips on it, 2 mm apart from each other. The acidic end of the IPG gel strips should face toward the anode.
– Cut two 3 mm thick filter paper to a length of all IPG gel strips in the tray and use them as electrodes.

- Soak the electrode strips with electrode solutions and remove excessive fluid by blotting with filter paper.
- Place the IEF electrode strips on top of the IPG gel strips at the cathode and anode.
- Set the temperature of the cooling block at 20 °C.
- Dilute samples with the sample buffer and apply 20 μL each. The protein concentration should not exceed 5–10 mg/mL.
- Apply the samples with a pipette.
- Place the lid on the electrofocusing chamber and connect the cables to the power supply.
- Start IEF at low voltage: 150 V for 30 min, then 300 V for 60 min. Raise the voltage at the end to 3,500 V overnight.

# References

[1] Kolin A. In Glick D (ed.). Methods of Biochemical Analysis. Interscience Publishers Inc, New York – London, 1958, Vol. 4, 259–288.
[2] Svensson H. Acta Chem Scand, 1961, 15, 325–341.
[3] Svensson H. Acta Chem Scand, 1962, 16, 456–466.
[4] Svensson H. J Chromatogr, 1966, 25, 266–273.
[5] Vesterberg O. *PhD Dissertation*. Karolinska Institutet, Stockholm. Svenk Kemisk Tidskrift, 1968, 80, 213–225.
[6] Vesterberg O. Acta Chem Scand, 1969, 23, 2653–2666.
[7] Vesterberg O. Ann N Y Acad Sci, 1973, 209, 23–33.
[8] Grubhofer N. *Offenlegungsschrift 2230743*, June 23, 1972.
[9] Williams KW, Söderberg JL. Int Lab, 1979, 9, 45–53.
[10] Söderberg JL. *US Patent 4,333,972*, 1982.
[11] Righetti PG. In Work TS, Burdon RH (eds ). Isoelectric Focusing. Theory, Methodology and Applications. Elsevier Biomedical Press, Amsterdam, 1989.
[12] Wilkins MR, Gooley AA. In Wilkins MR, Williams KL, Appel RD, Hochstrasser DF (eds.). Proteome Research: New Frontiers in Functional Genomics. Springer, Berlin, 1997, 35.
[13] Langen H, Roder D, Juranville JF, Fountoulakis M. Electrophoresis, 1997, 18, 2085.
[14] Leaback DH, Rutter AC. Biochem Biophys Res Commun, 1968, 32, 447–453.
[15] Rosen A, Ek K, Aman P. J Immunol Methods, 1979, 28, 1–11.
[16] Stromski D. Anal Biochem, 1982, 119, 387–391.
[17] Ambler J, Walker B. Clin Chem, 1979, 25, 1320–1322.
[18] Radola BJ. Ann N Y Acad Sci, 1973, 209, 127–143.
[19] Frey MD, Radola BJ. Electrophoresis, 1982, 3, 216–226.
[20] Caspers ML, Posey Y, Brown RK. Anal Biochem, 1977, 79, 166180.
[21] Beccaria L, Chiumello G, Gianazza E, Luppis B, Righetti PG. Am J Hemat, 1978, 4, 367–374.
[22] Jeppson JO, Franzen B, Nilsson VO. Sci Tools, 1978, 25, 69–73.
[23] Vesterberg O. In Arbuthnott JP, Beeley JA (eds.). Isoelectric Focusing. Butterworths, London, 1975, 78–96.
[24] Davies H. In Arbuthnott JP, Beeley JA (eds.). Isoelectric Focusing. Butterworths, London, 1975, 97–113.

[25]   Görg A, Postel W, Westermeier R. Anal Biochem, 1978, 89, 60–70.
[26]   Radola BJ. Electrophoresis, 1980, 1, 43–56.
[27]   Robinson HK. Anal Biochem, 1972, 49, 353–366.
[28]   Sippel TO. Anal Biochem, 1986, 159, 349–357.
[29]   Gianazza E, Astorri C, Righetti PG. J Chromatogr, 1979, 171, 161–169.
[30]   Olsson I, Lääs T. J Chromatogr, 1981, 215, 373–378.
[31]   Kleinert T. Elektrophoretische Methoden in Der Proteinanalytik. Georg Thieme, Stuttgart – New York, 1990.
[32]   Gasparic V, Bjellquist B, Rosengren A. *Swedish patent No.* 14049-1, 1975; *US Patent*, 1978; *Deutsches Patent*, 1981.
[33]   Dossi G, Celentano F, Gianazza E, Righetti PG. J Biochem Biophys Methods, 1983, 7, 123–142.
[34]   Gianazza E, Chillemi F, Duranti M, Righetti PG. J Biochem Biophys Methods, 1983, 8, 339–351.
[35]   Görg A, Postel W, Weser J, Patutschnick W, Cleve H. Am J Human Genet, 1985, 37, 922–930.
[36]   Pietta P, Potaterra E, Fiorino A, Gianazza E, Righetti PG. Electrophoresis, 1985, 6, 162–170.
[37]   Saveby GM, Petterson P, Andrasko J, Ineva-Flygare L, Jahannesson U, Görg A, Postel W, Domscheit A, Mauri PL, Pietta P, Gianazza E, Righetti PG. J Biochem Biophys Methods, 1988, 16, 141.
[38]   Pascali VL, Dobosz M, D´aloja E. Electrophoresis, 1988, 9, 514–519.
[39]   Smejkal GB, Bauer DJ. Am Biotech Lab, 2010, 28, 24–27.
[40]   Jeppsson JO. Application Note 307. LKB-Produkter AB, Stockholm, 1977.
[41]   Basset P, Beuzard Y, Garel MC, Rosa J. Blood, 1978, 51, 971–982.
[42]   Dassett P, Braconnier F, Rosa J. J Chromatogr, 1982, 227, 267–304.
[43]   Utermann G, Hees M, Steinmetz A. Nature, 1977, 269, 604–607.
[44]   Zannis VI, Breslow JL. In Radola BJ (ed.). Electrophoresis ´79. Walter de Gruyter, Berlin, 1980, 437–473.
[45]   Melnik BC, Melnik SF. Anal Biochem, 1988, 171, 320–329.
[46]   Allen RC, Oulla PM, Arnaud P, Baumstark JS. In Radola BJ, Graesslin D (eds.). Electrofocusing and Isotachophoresis. Walter de Gruyter, Berlin – New York, 1977, 255.
[47]   Jeppsson JO. In Radola BJ, Graesslin D (eds.). Electrofocusing and Isotachophoresis. Walter de Gruyter, Berlin, 1977, 273–292.
[48]   Rotbol L. Clin Chim Acta, 1970, 29, 101–105.
[49]   Peisker K. Z Med Lab Diagn, 1985, 26, 28–33.
[50]   Chu JL, Charavi A, Elkon KB. Electrophoresis, 1988, 9, 121–125.
[51]   Mehta PD, Patrick BA, Black J. Electrophoresis, 1988, 9, 126–128.
[52]   Chisholm DM, Beeley JA, Mason DK. Oral Surg, 1973, 35, 620–630.
[53]   Bustos SE, Fung L. In Allen RC, Arnaud P (eds.). Electrophoresis ‘81. Walter de Gruyter, Berlin, 1981, 317–328.
[54]   Ogata M, Satoh Y. Electrophoresis, 1988, 9, 128–131.
[55]   Patzelt O, Geserick G, Schröder H. Electrophoresis, 1988, 9, 393–397.
[56]   King NL. Meat Science, 1984, 11, 59–72.
[57]   Kaiser KP, Krause J. Z Lebensm Unters Forsch, 1985, 180, 181–201.
[58]   Stegemann H, Francksen H, Macko V. Z Naturforsch, 1973, 28, 722–732.
[59]   Kaiser KP, Bruhun LC, Belitz HD. Z Lebensm Unters Forsch, 1974, 154, 339–347.
[60]   Hussein KRF, Stegemann H. Z Acker Pflanzenbau, 1978, 146, 68–78.
[61]   Gianazza E, Celentano F, Dossi G, Bjellqvist B, Righetti PG. Electrophoresis, 1984, 5, 88–97.
[62]   Gianazza E, Astrua-Testori S, Righetti PG. Electrophoresis, 1985, 6, 113–117.

# 2.9 Free-flow electrophoresis of proteins

2.9.1     Theory of free-flow electrophoresis —— 209
2.9.2     Types of free-flow electrophoresis —— 210
2.9.2.1  Free-flow zone electrophoresis —— 211
2.9.2.2  Free-flow isotachophoresis —— 211
2.9.2.3  Free-flow isoelectric focusing —— 212
2.9.3     Device technology of free-flow electrophoresis —— 213
2.9.4     Detection system of free-flow electrophoresis —— 213
2.9.5     Applications of free-flow electrophoresis —— 214
2.9.6     Protocols —— 214
           Free-flow Zone Electrophoresis of Human T and B Lymphocytes —— 214
           References —— 215

Free-flow electrophoresis (FFE) has been developed by Grassmann and Hannig [1–3]. It is a matrix-free electrophoretic technique, which resembles the capillary electrophoresis. Until the 1980s, it was a technology for separation of cells and organelles, but now is used for separation of proteins and other charged particles. Besides the (macro) FFE, miniaturized versions of FFE, called micro-FFE, are also known [4].

The advantage of FFE is the fast separation of proteins, which do not adhere to any matrix structure as agarose or polyacrylamide gel. The separations are highly reproducible and can be carried out under native or denaturing conditions [5]. The electrophoresis continues only several minutes; therefore, most proteins, for example, enzymes, preserve their activity. To succeed this, the pH value during electrophoresis should be kept within physiological limits, and sucrose, glucose, or other compounds should be added to maintain appropriate isosmotic pressure.

## 2.9.1 Theory of free-flow electrophoresis

The sample in FFE is injected in a thin buffer layer (<1 mm) that flows through a separation channel formed between two parallel plates [6, 7]. Perpendicularly to this flow, an electric field is applied in order to deflect charged compounds in different angles according to their electrophoretic mobilities. Therefore, the theory of FFE is based on the fact that two perpendicular forces act on the sample to be analyzed: the electric field on the abscissa axis and the buffer flow on the ordinate axis. As a result, the sample runs the distances $x$ and $y$ after the time $t$. The ratio between these distances is referred to as $tg\theta$, where $\theta$ (deflection angle) is the angle between the resultant line and $y$-axis. The deflection angle increases with the increase in the electric field strength and decreases with the increase in the buffer flow velocity (Figure 2.9.1). The separated compounds are collected continuously as fractions.

https://doi.org/10.1515/9783110761641-019

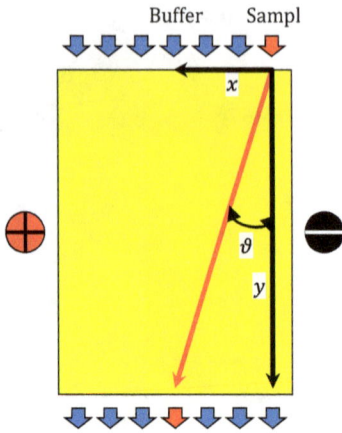

**Figure 2.9.1:** Scheme of free-flow electrophoresis.

The magnitude of $tg\theta$ is equal to the ratio between the distances $x$ and $y$, hence, between the sample velocity $v_s$ and the buffer velocity $v_b$:

$$tg\theta = \frac{x}{y} = \frac{v_s}{v_b} \tag{2.9.1}$$

The electrophoretic velocity of the sample in an electric field could be given by the equation

$$v_s = \mu_s E = \mu_s \frac{J}{\gamma} = \mu_s \frac{I}{q\gamma} \tag{2.9.2}$$

where $\mu_s$ is the sample mobility, $E$ is the electric field strength, $J$ is the density of the electric current, $\gamma$ is the specific conductivity of the buffer, $I$ is the strength of the electric current, and $q$ is the cross section of the electrophoretic channel. From eqs. (2.9.1) and (2.9.2), it follows that

$$tg\theta = \frac{\mu_s I}{q\gamma v_b} \tag{2.9.3}$$

## 2.9.2 Types of free-flow electrophoresis

According to the electrophoretic principles, three types of FFE can be distinguished: free-flow zone electrophoresis (FFZE), free-flow isotachophoresis (FFITP), and free-flow isoelectric focusing (FFIEF).

### 2.9.2.1 Free-flow zone electrophoresis

FFZE is carried out in one (continuous) buffer. The buffer should have sufficient capacity and an appropriate pH value, preferably in neutral range (pH = 7.0–7.4) or in slightly alkaline range (pH = 8.0–9.0). The borate buffer should be carefully used because boric acid forms complexes with polyols [8].

The migration distance $d$ of an analyte moving through the electric field is given by the equation

$$d = \mu_p\, Et \tag{2.9.4}$$

where $\mu_p$ is the electrophoretic mobility of the analyte, $E$ is the electric field strength, and $t$ is the residence time of the analyte in the separation chamber [9].

Several phenomena influence the band broadening ($\sigma_T$): the width of the injected sample $\sigma_{INJ}$, diffusional broadening ($\sigma_D$), hydrodynamic broadening ($\sigma_{HD}$), electrodynamic broadening ($\sigma_{ED}$), Joule heating ($\sigma_{JH}$), and electromigration dispersion ($\sigma_{EMD}$) [10, 11]:

$$\sigma_T^2 = \sigma_{INJ}^2 + \sigma_D^2 + \sigma_{HD}^2 + \sigma_{ED}^2 + \sigma_{JH}^2 + \sigma_{EMD}^2 \tag{2.9.5}$$

The diffusional broadening is related to the residence time $t$ of the analyte in the separation chamber [12]:

$$\sigma_D^2 = 2Dt \tag{2.9.6}$$

where $D$ is the analyte diffusion coefficient [13].

### 2.9.2.2 Free-flow isotachophoresis

In FFITP, discontinuous buffer system is used (Figure 2.9.2). It consists of a leading and trailing buffer. The mobility of the leading ion should be greater, and the mobility of the trailing ion should be smaller than the mobility of the particles to be resolved. During the electrophoresis run, the sample components arrange themselves according to their descending mobilities and form sharp bands whose concentrations depend on the concentration of the leading ion. The counter-ion is usually an ion of a weak base or acid, which has high buffer capacity in the necessary pH range.

To keep the particles to be resolved away from each other, spacer substances are added into the sample. For example, when proteins are to be separated, amino acids are used whose mobility values are between those of the proteins.

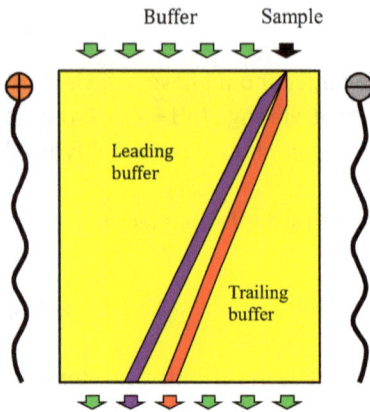

Figure 2.9.2: Scheme of free-flow isotachophoresis of two negatively charged particles.

### 2.9.2.3 Free-flow isoelectric focusing

In FFIEF, the sample is fractionated in a linear pH gradient [14] created by carrier ampholytes (Figure 2.9.3). The proteins to be resolved can be added either as a narrow zone or can be dissolved in the carrier ampholyte solution. When an electric field is applied, a pH gradient is formed, where the proteins are focused at their isoelectric points (pI).

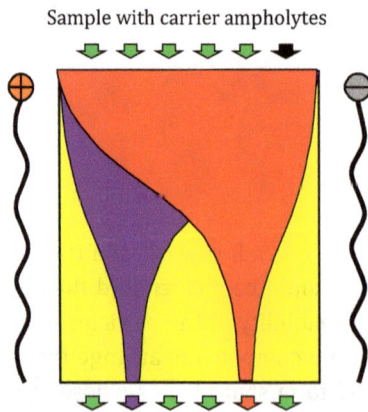

Figure 2.9.3: Free-flow isoelectric focusing in carrier ampholyte pH gradient.

The linear pH gradient is perpendicular to the flow direction. The sample components migrate through the pH gradient, due to the electric field, until they reach their pI. There they loss their electric charges and focus [15].

The equilibrium between the electrophoretic and diffusional transport during IEF is described by the following differential equation, valid under steady-state conditions [16]:

$$\frac{d(c\mu E)}{dx} = \frac{d}{dx}D\frac{dc}{dx}$$ (2.9.7)

where $c$ [in mol/L] is the analyte concentration at position $x$ in the separation chamber, $\mu$ [in m$^2$/(s V)] is the analyte mobility at that point, $E$ [in V/m] is the electric field strength, and $D$ [in m$^2$/s] is the diffusion coefficient.

## 2.9.3 Device technology of free-flow electrophoresis

A FFE device has a front plate and a back plate. The front plate is made of **poly(methylmethacrylate)** (PMMA, Plexiglas). The back plate consists of an aluminum block, covered with a glass mirror, which in turn is covered with a plastic. The front plate contains inlets for samples and buffers, and outlets for fractionation tubes. The distance between the front plate and the back plate is usually 0.1–0.5 mm and can be adjusted by membrane spacers, e.g., of cellulose nitrate.

The separation chamber is divided into three regions: separation region, anode bed, and cathode bed. The electrodes are placed outside the separation region avoiding the disturbance of generated oxygen and hydrogen bubbles. Usually, the device is located on a cooled plate in order to remove the heat generated by the electric current. This principle could be applied for smaller FFE systems that use injection molding and milling [17].

The buffer is fed with the help of a tubing pump and flows vertical in the narrow gap between the two plates [18–20]. The sample is applied in the upper part of the device with the aid of a dossier pump. The electric current is maintained by filter paper strips, which are immersed in lateral electrode tanks. The separated proteins are collected at the lower end of the device. The device is cooled with water.

## 2.9.4 Detection system of free-flow electrophoresis

The detection system of FFE consists of a scanner and computer with printer. The scanner measures the UV absorption of proteins flowing with the buffer. Prior to the electrophoretic separation, the absorbance of the pure buffer is measured to determine a baseline, which should be subtracted from the values obtained. The measurement signals are sent to a computer working with appropriate software.

## 2.9.5 Applications of free-flow electrophoresis

FFE is an excellent method for analyzing human plasma proteins [21]. It can also be used for separation of enzymes [22–23], cell organelles (lysosomes, mitochondria, ribosomes, and more) [24], membranes, chloroplasts, viruses, bacteria, and cells as platelets, lymphocytes, red cells, thymus cells, tumor cells, thyroid cells, and more [25–27]. The FFE is also a good technique for preparative separations of peptides [28–30], proteins [31–33], cellular components [34], and cells [35–37].

Using FFITP, human transferrin was isolated [38], and the purification of ovalbumin, lysozyme [39], and monoclonal antibodies [40] was reported. Using narrow pH gradients in FFIF, antibody isoforms were separated that vary by less than 0.02 pH units [41].

## 2.9.6 Protocols

**Free-flow Zone Electrophoresis of Human T and B Lymphocytes**

**Materials and equipment**
TRIS
Phosphoric acid
NaCl
KCl
T and B lymphocytes
Apparatus for FFE

*Resolving buffer, pH = 7.4*
$NaH_2PO_4\cdot H_2O$          1.38 g (0.01 mol/L)
*Adjust with TRIS to pH = 7.4*
NaCl                          8.77 g (0.15 mol/L)
KCl                           0.22 g (0.003 mol/L)
Deionized water to  1,000.00 mL

*Cathode buffer, pH = 7.4*
$NaH_2PO_4\cdot H_2O$          4.14 g (0.03 mol/L)
*Adjust with TRIS to pH = 7.4.*
NaCl                          26.30 g (0.45 mol/L)
KCl                           0.60 g (0.008 mol/L)
Deionized water to  1,000.00 mL

*Anode buffer, pH = 7.0*
$NaH_2PO_4\cdot H_2O$          2.76 g (0.02 mol/L)
*Adjust with TRIS to pH = 7.0.*
Deionized water to  1,000.00 mL

**Procedure**

- Suspend $2 \times 10^7$ cells of a sample in 1 mL resolving buffer.
- Fill the cathode and anode tanks with cathode and anode buffer, respectively.
- Give 240 V/cm field strength at 20 °C.
- Run FFE for 6 min.

# References

[1]   Grassmann W, Hannig K. Naturwissenschaften, 1950, 37, 397–402.
[2]   Hannig K. Z Anal Chem, 1961, 181, 244–254.
[3]   Barrolier VJ, Watzke E, Gibian HZ. Naturforschung, 1958, 13B, 754.
[4]   Jezierski S, Gitlin L, Nagl S, Belder D. Anal Bioanal Chem, 2011, 401, 2651–2656.
[5]   Patel PD, Weber G. J Biol Phys Chem, 2003, 3, 60–73.
[6]   Wagner H. Nature, 1989, 341, 669–670.
[7]   Roman MC, Brown PH. Anal Chem, 1994, 66, 86A–94A.
[8]   Michov BM. J Appl Biochem, 1982, 4, 436–440.
[9]   Raymond DE, Manz A, Widmer HM. Anal Chem, 1994, 66, 2858–2865.
[10]  Weber G, Bocek P. Electrophoresis, 1998, 79, 1649–1653.
[11]  Gebauer P, Bocek P. Electrophoresis, 2005, 26, 453–462.
[12]  Kohlheyer D, Besselink GAJ, Schlautmann S, Schasfoort RBM. Lab Chip, 2006, 6, 374–380.
[13]  Fonslow BR, Bowser MT. Anal Chem, 2006, 78, 8236–8244.
[14]  Wagner H, Speer W. J Chromatogr, 1978, 157, 259–265.
[15]  Wang Q, Tolley HD, LaFebre DA, Lee ML. Anal Bioanal Chem, 2002, 373, 125–135.
[16]  Righetti PG. Isoelectric Focusing: Theory, Methodology and Applications. Elsevier Biomedical, Amsterdam, 1983.
[17]  Stone VN, Baldock SJ, Croasdell LA, Dillon LA *et al.* J Chromatogr A, 2007, 7755, 199–205.
[18]  Wagner H, Kuhn R, Hofstetter S. In Wagner H, Blasius E (eds.). Praxis Der Elektrophoretischen Trennmethoden. Springer, Heidelberg, 1989, 223–278.
[19]  Hannig K. Hoppe Seyler´s Z Physiol Chem, 1964, 338, 211–227.
[20]  Hannig K. Z Anal Chem, 1961, 181, 244–254.
[21]  Nissum M, Foucher AL. Expert Rev Proteomics, 2008, 5, 571–587.
[22]  Hoffstetter-Kuhn S, Wagner H. Electrophoresis, 1990, 11, 451–456.
[23]  Nath S, Schutte H, Weber G, Hustedt H, Deckwer WD. Electrophoresis, 1990, 11, 937–941.
[24]  Hannig K, Heidrich HG. Free-Flow Electrophoresis. GIT, Darmstadt, 1990, 26–30.
[25]  Grassmann W. Naturwissenschaften, 1951, 38, 200–206.
[26]  Hannig K. Electrophoresis, 1982, 3, 235–243.
[27]  Weber G, Grimm D, Bauer J. Electrophoresis, 2000, 21, 325–328.
[28]  Pruski Z, Kasicka V, Mudra P, Stepanek J, Smekal O, Hlavack J. Electrophoresis, 1990, 11, 932–986.
[29]  Kasicka V, Prusik Z. Am Lab, 1994, 26, 22–28.
[30]  Kasicka V, Prusik Z, Sazelova R, Jiracek J, Barth T. J Chromatogr A, 1998, 796, 211–220.
[31]  Knisley KA, Rodkey LS. Electrophoresis, 1990, 11, 927–931.
[32]  Clifton MJ, Jouve N, De Balmann H, Sanchez V. Electrophoresis, 1990, 11, 913–919.
[33]  Poggel M, Melin T. Electrophoresis, 2001, 22, 1008–1015.
[34]  Kessler R, Manz HJ. Electrophoresis, 1990, 11, 979–980.
[35]  Zeiller K, Loser R, Pascher G, Hannig K. Hoppe-Syler's Z Physiol Chem, 1975, 356, 1225–1244.

[36]   Rodkey LS. Appl Theor Electrophor, 1990, 1, 243–247.
[37]   Bauer J. J Chromatogr B, 1999, 722, 55–69.
[38]   Caslavska J, Thormann W. Electrophoresis, 1994, 15, 1176–1185.
[39]   Caslavska J, Gebauer P, Thormann W. J Chromatogr, 1991, 585, 145–152.
[40]   Schmitz GB, Böttcher A, Kahl H, Brüning T. J Chromatogr, 1987, 431, 327–342.
[41]   Patel PD, Weber G. J Biol Phys Chem, 2003, 3, 60–73.

# 2.10 Capillary electrophoresis of proteins

2.10.1    Theory of capillary electrophoresis of proteins —— 217
2.10.2    Instrumentation —— 218
2.10.2.1  Coating the capillaries —— 219
2.10.2.2  Sieving matrix in capillary electrophoresis —— 220
2.10.3    Practice of capillary electrophoresis —— 222
2.10.3.1  Injection —— 222
2.10.3.2  Separation —— 222
2.10.3.3  Detection —— 223
2.10.4    Types of capillary electrophoresis —— 224
2.10.5    Applications of capillary electrophoresis —— 224
2.10.5.1  Serum protein analysis —— 224
2.10.5.2  Hemoglobins —— 225
2.10.5.3  Isoenzymes —— 226
2.10.5.4  Immune complexes —— 226
2.10.5.5  Single cell analysis —— 227
2.10.6    Protocols —— 227
          Capillary IEF of Proteins —— 227
          References —— 228

Capillary electrophoresis (CE) [1, 2] can be carried out in buffers or on gels containing native or denaturing buffers [3]. The interest in CE is based on its high resolution and speed. It is possible to separate reproducibly with the help of CE attomole quantities ($10^{-18}$ to $10^{-21}$ mol) of proteins, nucleic acids, carbohydrates, and other compounds in minutes up to an hour [4, 5].

## 2.10.1 Theory of capillary electrophoresis of proteins

The theory of CE does not differ from the general electrophoretic theory [6, 7]. The analytes migrate in an electric field $E$ with the velocity

$$v_{ep} = \mu_{ep}E = = \frac{QE}{6\pi r\eta} \tag{2.10.1}$$

where $\mu_{ep}$ is the electrophoretic mobility, $E$ is the field strength, $Q$ is the net charge of the analyte, $r$ is its Stokes radius, and $\eta$ is the dynamic viscosity of the solvent. Other variables such as shape and hydrophobicity have also been shown to affect the velocity [8].

Simultaneously in a fused silica capillary, silanol (–Si–OH) groups ($pK \approx 3$) attached to the internal wall of the capillary are ionized giving negatively charged silanolate groups (–Si–O⁻). The positive ions of the buffer are attracted to the negative silanolate groups. Therefore, an electric double layer with fixed and diffuse positive counter-ions is built (Figure 2.10.1). When an electric field is applied in a capillary,

https://doi.org/10.1515/9783110761641-020

the diffuse positive counter-ions (cations) migrate to the negatively charged cathode through the electric field. Since they are hydrated, the water of the buffer migrates too, which causes **electroosmotic flow** (EOF). Therefore, in CE two opposite processes are acting: the electrophoretic velocity of polyions and EOF of the buffer [9].

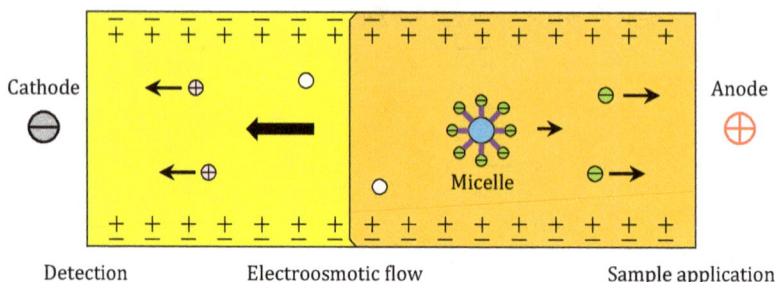

**Figure 2.10.1:** Depiction of the interior of a fused silica capillary filled with a buffer during electrophoresis.

The velocity of the EOF $v_0$ can be expressed as

$$v_{eo} = \mu_{eo} E \tag{2.10.2}$$

where $\mu_0$ is the electroosmotic mobility. It is defined as

$$\mu_{eo} = \frac{\varepsilon \zeta}{\eta} \tag{2.10.3}$$

where $\varepsilon$ is the permittivity of the buffer, $\zeta$ is the zeta potential of the capillary wall, and $\eta$ is the dynamic viscosity. Hence, the velocity $v$ of an analyte in CE is

$$v = v_{ep} + v_{eo} = \left(\mu_{ep} + \mu_{eo}\right) E = \left(\frac{Q}{6\pi r} + \varepsilon \zeta\right) \frac{E}{\eta} \tag{2.10.4}$$

Without electroosmosis, the negative ions should migrate to the anode and never reach the detector at the cathode end of the capillary. The detector is reached first by cations, then by neutral molecules, and finally by anions.

## 2.10.2 Instrumentation

A CE instrument consists of a source vial, sample vial, capillary, destination vial, detector, high-voltage power supply, and recording device. The source vial, capillary, and destination vial are filled with a buffer. The capillary is flexible and made usually of fused silica. It is from 7 to 100 cm long (usually 50 cm), and has about 375 μm outer diameter and usually 50 μm inner diameter. Teflon capillaries are also available.

The outside of the capillaries is coated with a thin layer of polyimide to get flexible. The inside of the capillaries is coated with polyacrylamide, methylcellulose, or other compounds. Thus, the adsorption of polyions on the capillary surface is avoided (Figure 2.10.2).

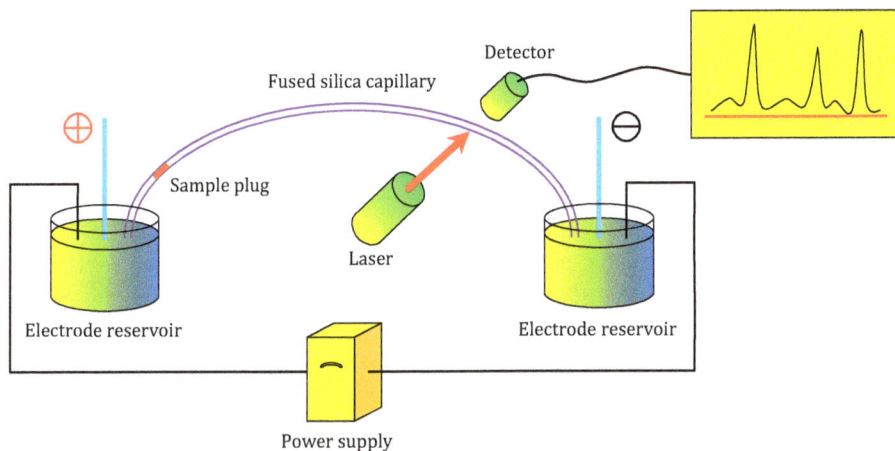

**Figure 2.10.2:** Capillary electrophoresis instrumentation.

The sample is introduced into the capillary by pressure, siphoning, or electrokinetically. Then the capillary is returned to the source vial. After electric voltage is applied, the sample components separate from each other and are detected at the outlet end of the capillary. The results are presented as peaks in a pherogram.

## 2.10.2.1 Coating the capillaries

During CE, noxious interactions appear between the proteins to be resolved and the capillary. For example, basic proteins adsorb onto the negatively charged internal wall of the fused silica capillary, which leads to impaired efficiency and poor resolution [10, 11]. In addition to the direct interaction with the capillary wall, the proteins are moved by EOF, which also hinders the resolution of CE. The electroosmotic flow can be slowed using ionic additives, such as $Mg^{2+}$ and hexamethonium, which bind to the negatively charged silanolates and, as a result, lower the electric charge of the capillary

Hexamethonium

Another strategy to slow EOF is to coat the capillary wall with polymers. There are two types of coating: permanent and dynamic [12].

*Permanent capillary coating* is a complex process. Prior to it, the capillary surface should be cleaned and activated by etching, leaching, dehydration, and silylation [13]. The etching can be done with sodium hydroxide; the leaching can be carried out with hydrochloric acid; and to improve the silylation reaction, all water has to be removed from the surface at 160 °C overnight.

*Dynamic capillary coating* is the simplest approach to diminish EOF. During this coating, polymers are adsorbed on capillaries *via* physical bonds. The dynamic capillary coating can be performed using neutral or cationic polymers. Surfactants can also be used.

Neutral polymers [14] quench the protein binding to the silica wall in 50–60%; only the neutral hydrophobic polymer poly(*N,N*-**dim**ethylacrylamide) [poly(DMA)] shows higher inhibition (ca. 85%). To the group of cationic polymers belong **polye**thylenimine (PEI) [15–17], **poly**(*N*-**hydroxyethylacrylamide**) (PHEA) [18, 19], chitosan [20, 21], **poly(dim**ethylsiloxane) (PDMS) [22, 23], *N*-**me**thyl**poly**vinyl**py**ridinium quaternary ion (PVPy-Me) [24], and more.

Surfactants that can be used as coating agents could be divided into two groups: nonionic and zwitterionic surfactants (Table 2.10.1).

Besides the above-shown coating polymers, fluorinated polymers are also used as coating substances, for example, **poly(t**etrafluoroethene) (PTFE, or Teflon) and **fluoroc**arbon (FC). They do not swell in the presence of organic solvents and are optically transparent to lower wavelengths. Teflon is the most widely used polymer for capillary coating but its mechanical softness and absorptivity in the low-UV spectral region can be problematic [25, 26].

### 2.10.2.2 Sieving matrix in capillary electrophoresis

A sieving matrix that was proposed for CE is the polymer **poly(2-ethyl-2-ox**azoline) (PEOX). It has a relative molecular mass $M_r$ from 50,000 to 500,000 and can be used in concentrations of 6–12 g/dL for SDS CE [27]. PEOX has good hydrolytic stability because the amide is highly substituted.

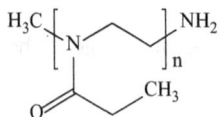

Poly(2-ethyl-2-oxazoline)

**Table 2.10.1:** Chemical formulas of nonionic and zwitterionic surfactants.

| Detergent | Formula | CMC (mmol/L) | $M_r$ |
|---|---|---|---|
| *Nonionic surfactants* | | | |
| Triton X-100 (polyethylene glycol tert-octylphenyl ether) | | 0.02-0.09 | 625 |
| Brij 35 (polyoxyethylene lauryl ether) | $(C_2H_4O)_{23}H$ | 0.05-0.1 | 1,225 |
| Tween 20 (polyoxyethelene sorbitanmonolaurate) | $HO(C_2H_4O)_w$  $(OC_2H_4)_xOH$  $w+x+y=21$  $CH(OC_2H_4)_yOH$  $CH_2O(C_2H_4O)_zC_2H_4OCOCH_2(CH_2)CH_3$ | 0.06 | 1,228 |
| *Zwitterionic surfactants* | | | |
| Lauryl sulfobetaine SB-12 (N-dodecyl-N,N-dimethyl-3-ammonio-1-propanesulfonate) | | 2-4 | 335.6 |
| Palmityl sulfobetaine SB-16 (N-hexadecyl-N,N-dimethyl-3-ammonio-1-propanesulfonate | | 0.01-0.06 | 391.6 |
| CHAPS (3-[(3-cholamidopropyl) dimethylammonio]-1-propanesulfonate) | | 6-10 | 614.9 |

A homolog of PEOX is **poly(***N***,***N***-dimethylpropionamide)** (PDMPA).

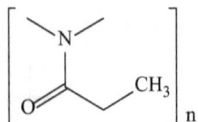

Poly(N,N-dimethylpropionamide)

Hydroxypropyl cellulose was also used as a sieving matrix [28]. Another polymer that was used for separating single cells is **poly(ethylene oxide)** (PEO) [29]. It has $M_r = 100,000$ and low viscosity, which allows a hydrodynamic injection of proteins or cells into a coated capillary.

## 2.10.3 Practice of capillary electrophoresis

CE is carried out in three steps: injection of a sample into the capillary, separation, and detection of protein bands.

### 2.10.3.1 Injection

The injection of a sample is carried out by electromigration, gravity, or pressure [30]. To inject a sample into a capillary by electromigration, a sample applier is needed. When the sample is applied at the anode, no positive ions are allowed to be present in the sample. If this is undesirable, the injection can be carried out by gravity. Then the sample vial and the corresponding capillary end should be raised to a higher level than the other capillary end, whereby the sample is sucked by the siphon effect.

### 2.10.3.2 Separation

The migration of proteins is due to a high-voltage power supply which creates electric field between the source and destination vials. All ions, positive or negative, are pulled through the capillary in the same direction by EOF and are detected near its outlet end. The output of the detector is connected with a computer or an integrator. The Joule heating is removed by a blower.

## 2.10.3.3 Detection

There are diverse techniques for detecting the separated bands: **u**ltraviolet (UV) detection, laser-**i**nduced **f**luorescence (LIF) detection, **m**ass **s**pectrometry (MS) detection, **c**hemi**l**uminescence (CL) detection, and **w**hole-**c**olumn **i**maging **d**etection (WCID).

The UV detection is the most common detection. It is based on the UV absorbance and is used for analyzing the hemoglobin variants. Frequently employed wavelengths are 210, 280, and 415 nm. Zhu *et al.* [31–33] have found that the maximal absorbance of the heme group is at 415 nm wavelength.

The LIF is a method, in which an atom or a molecule is excited to a higher energy level by absorption of laser light followed by spontaneous emission of light [34–36]. Hb exhibits native fluorescence that relies on the fluorescence of aromatic amino acid residues [37].

The MS provides extremely high sensitivity of CE for very small sample concentrations. It includes ionization of chemical species and sorting the ions based on their mass-to-charge ratio. As a result, the compound masses within a sample are measured. MS analysis is used to measure Hb variants and amino acid sequences [38].

The CL detection is another method widely used in combination with chromatography, spectrometry, and immunoassay [39]. It is characterized by excellent sensitivity and selectivity, allowing high resolution and precise quantification [40].

The WCID has been successfully employed for the electrofocusing analysis of Hb. Here a short capillary (a few cm long) is used as a separation channel, which is connected with two pieces of a capillary through two dialysis hollow fiber junctions. The dialysis hollow fiber junctions contact the electrolyte reservoirs and provide electric conduction and passage of small ions (such as protons and hydroxide ions), but confine large molecules (such as proteins) inside the capillary. When high voltage is applied, the large molecule in the separation channel will be focused at their pI values. Then the focused bands are imaged with a **c**harge **c**oupled **d**evice (CCD) camera (Figure 2.10.3).

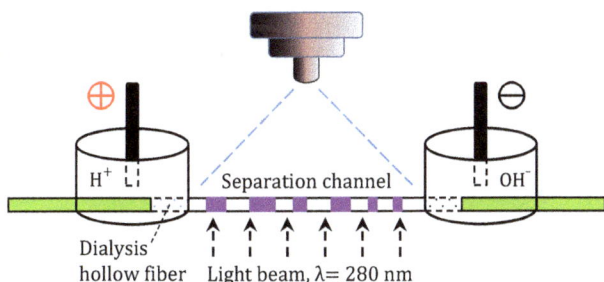

**Figure 2.10.3:** Principle of whole-column imaging detection.

## 2.10.4 Types of capillary electrophoresis

According to the main electrophoretic methods, three CE methods are distinguished: capillary **z**one **e**lectrophoresis (CZE), **c**apillary **i**sotachophoresis (CITP), and capillary **i**soelectric **f**ocusing (CIEF).

CZE was carried out in capillaries with an inner diameter of 40 μm [41]. It was run in phosphate buffer [42]. The phosphate buffer can cover a broad range of pH value (2.55–11.43) due to the three dissociation constants of phosphoric acid and its ions (p$K_{H_3PO_4}$ = 2.0, p$K_{H_2PO_4^-}$ = 7.2, and p$K_{HPO_4^{2-}}$ = 11.0).

CITP [43–44] needs two buffers, which contain a leading ion (e.g., formate ion) and a trailing ion (e.g., taurinate ion). It is carried out in capillaries with inner diameters of 0.8–0.2 mm.

In CIEF, proteins with different isoelectric points are separated in a pH gradient [45]. It is important that EOF is equal to zero. This can be achieved by a modification of the capillary surface or at a suitable pH value of the buffer [46]. If the capillaries are not coated, the separation medium viscosity has to be increased by adding glycerol [47] or methylcellulose [48].

## 2.10.5 Applications of capillary electrophoresis

CE is widely used in the clinical and forensic medicine [49, 50]. An instrument for clinical **CE** (cCE) should have the capability for rinsing, washing, thermostating, and easy replacement of the capillaries as well as of accurate applying of nanoliter volumes of a sample. In addition, the cCE design should include dilution of the sample and automated buffer replenishment.

### 2.10.5.1 Serum protein analysis

Chen *et al.* [51] showed first that the CE is an alternative to agarose electrophoresis. Figure 2.10.4 shows the comparative profiles obtained with scanning densitometry of an agarose gel and direct analysis of a pooled serum sample by CE.

*Prealbumin (transthyretin).* Prealbumin transports thyroxine and triiodothyronine. Its concentration is an important indicator of nutritional status, inflammation, malignancy, liver cirrhosis, or Hodgkin's disease. It occurs in a relatively low concentration in "normal" serum (200–360 mg/L) but can be detected by CE at 214 or 200 nm wavelength.

*Lipoproteins.* The serum lipoproteins transport lipids in the blood. They have spherical molecules composed of a hydrophobic core and a polar shell. The lipoproteins are subdivided into five density classes: **h**igh-**d**ensity lipoproteins (HDL), **l**ow-**d**ensity lipoproteins (LDL), **v**ery-**l**ow-**d**ensity lipoproteins (VLDL), intermediate-

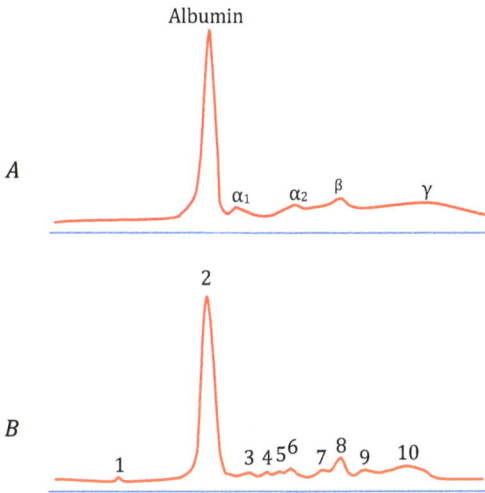

**Figure 2.10.4:** Serum protein analysis by agarose gel electrophoresis and capillary electrophoresis. (*A*) Agarose zone electrophoresis of normal serum proteins giving five fractions. (*B*) Capillary zone electrophoresis of the same proteins giving more fractions in TRIS-borate buffer, pH = 8.3. Detection at 214 nm. 1. Prealbumin; 2. albumin; 3. $\alpha_1$-acidic glycoprotein; 4. $\alpha_1$-antitrypsin; 5. β-lipoprotein; 6. haptoglobin; 7. $\alpha_2$-macroglobulin; 8. transferrin; 9. complement C3; 10. γ-globulin.

density lipoproteins (IDL), and chylomicrons (see there). LDL are the major cholesterol depot in the blood. They carry cholesterol to the tissues [52]. HDL carry cholesterol in the reverse direction – from the tissues to the liver for excretion [53].

The concentrations of LDL and HDL are highly important in the clinical diagnostics. However, the adsorption of lipoproteins onto the fused silica capillary wall makes their separation difficult [54, 55]. Attempts have been made to diminish the adsorption by dynamic or permanent coating of capillaries [56, 57]. The dynamic coating is made with methylglucamine or SDS or polymers such as **p**oly(**e**thyleneox-ide) (PEO) or hydroxypropyl methylcellulose [58, 59].

*Immunoglobulin subtypes.* Klein and Jolliff [60] have shown that immunoglobulin subtypes (IgG, IgA, IgM, heavy chains, and light chains κ and λ) can be identified by CE. The serum samples should be incubated with a solid phase to which specific antibodies are bound. During the incubation, the immunoglobulin subtypes bind specifically to their antibodies.

## 2.10.5.2 Hemoglobins

**H**emoglobins (Hb) ($M_r$ = 64,500) are the major proteins in **r**ed **b**lood **c**ells (RBC). CE has important significance in their analysis [61–63] (Figure 2.10.5) because of its high speed, low sample consumption, and high resolution.

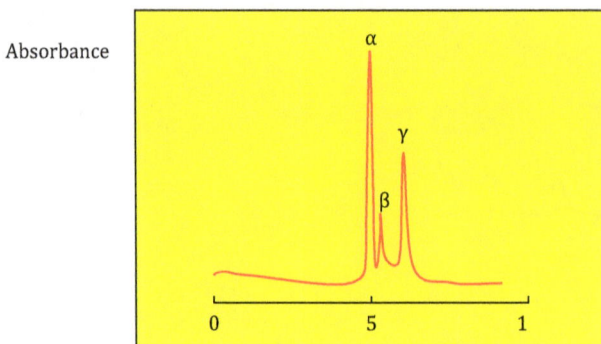

**Figure 2.10.5:** Separation of fetal α-, β-, and γ-globin chains.

The CZE methods for Hb separation use strongly acidic buffers (pH = 2.0–2.5) [64–66] or strongly alkaline buffers (pH = 11.8) [67]. Under these circumstances, the globin chains dissociate from the heme during electrophoresis [68, 69]. So the Hb variants $A_2$, F, and $A_{1c}$ were studied [70]. Zhu *et al.* [71] demonstrated that using capillary isoelectric focusing α-thalassemias can be identified. The computer-assisted CIEF is also an excellent method for the clinical assessment of hemoglobinopathies because of its automatization, high resolution, and possibility of simultaneous quantification of Hb variants [72–74].

### 2.10.5.3 Isoenzymes

CE can also be used for isoenzyme assays [75], for example, of alkaline phosphatase, creatine kinase, A and B forms of O–N-acetylglucosaminidase, P and S types of amylase, proteolytic isoenzymes, γ-glutamyl transpeptidase, kallikrein, renin, cathepsin B, and 5′-nucleotidase.

### 2.10.5.4 Immune complexes

CE can be used for separation of immune complexes from unbound antibody and antigen [76]. Ultratrace detection is also possible when using fluorescence-tagged monoclonal antibodies and laser-induced fluorescence (LIF) detection [77]. As a result, the detection of cancer biomarkers in nanomolar or even femtomolar concentration was done. In addition, it was demonstrated that linear **polyacrylamide** (LPA)-coated capillaries are suitable for CE-MS analysis of IgG subunits [78]: the heavy and light chains of IgG molecules were separated on LPA-coated capillaries by more than 3 min, with peak width of about 45 s.

## 2.10.5.5 Single cell analysis

Jorgenson *et al.* [79] first demonstrated the potential of CE for analyzing cells differing in age, differentiation, subtype, *etc.* Olefirowicz and Ewing [80, 81] furthered these studies by using special capillaries for studying the cytoplasm of dopaminergic neurons.

Other applications of CE were the separations of human chorionic gonadotropin and interferon-$\beta_1$ proteoforms. For human chorionic gonadotropin (a highly glycosylated protein), the identification of over 20 glycoforms has been reported using polyvinyl alcohol-coated capillary [82, 83]. For interferon-$\beta_1$ proteoforms, instead of regular PEI-coated capillary, a cross-linked polyethylenimine coating has been used on the quantitative CE-MS analysis [84].

## 2.10.6 Protocols

### Capillary IEF of Proteins

#### Materials and equipment
IEF markers
Ampholyte mixture, pH = 3.0–10.0
Phosphoric acid
Sodium hydroxide
50 μm i.d. coated (for unknown pI) or uncoated (for known pI) fused silica capillary
CE instrument
Power supply delivering voltage of 2,000–3,000 V and power of 6 W

***Sodium borate buffer, 5 mmol/L, pH = 8.0***
Sodium borate anhydrate      1.01 g (5 mmol/L)
*Adjust pH to 8.0 using HCl*
Deionized water to      1,000.00 mL

***Sodium borate buffer, 50 mmol/L, pH = 8.0***
Sodium borate anhydrate      10.06 g (50 mmol/L)
*Adjust pH to 8.0 using HCl*
Deionized water to      1,000.00 mL

***Sodium borate buffer, 500 mmol/L, pH = 8.0***
Sodium borate anhydrate      10.06 g (500 mmol/L)
*Adjust pH to 8.0 using HCl*
Deionized water to      100.00 mL

***10 mmol/L phosphoric acid***
85% $H_3PO_4$      2.3 mL
Deionized water to      100.0 mL

**20 mmol/L sodium hydroxide**

| | |
|---|---|
| NaOH | 4.0 g |
| Deionized water to | 100.0 mL |

**Procedure**

- Dilute a sample with 50 mmol/L sodium borate buffer to give a final concentration of 1 mg/mL.
- Fill a coated column with 500 mmol/L sodium borate buffer.
- Inject the sample.
- Separate the proteins at 10 kV and 25 °C for 30 min.
- Add ampholyte mixture to 0.5 mL protein sample to give a final concentration of 2.5 g/dL ampholytes. If desired, add IEF markers to a final concentration of 0.1 mg/mL to calibrate the column.
- Fill the capillary by pressurizing the reservoir.
- Place 10 mmol/L phosphoric acid in the anode reservoir, and 20 mmol/L NaOH in the cathode reservoir.
- Focus the sample at constant 10 kV for 5 min.
- Wash the column after each run with 10 mmol/L phosphoric acid for 1 min.
- Store the column in deionized water at room temperature.

# References

[1]   Landers JP. Clin Chem, 1995, 41, 495–509.
[2]   Engelhardt H, Beck W, Kohr J, Schmitt T. Angew Chemie, 1993, 105, 659–804.
[3]   Landers JP, Oda RP, Spelsberg TC, Nolan JA, Ulfelder KJ. BioTechniques, 1993, 14, 98–109.
[4]   Hjerten S. Chromatogr Rev, 1967, 9, 122.
[5]   Jorgenson JW, Lukacs KD. Science, 1983, 222, 266–272.
[6]   Vespalec R, Bocek P. Electrophoresis, 1994, 15, 755–762.
[7]   Hilser VJ, Freire E. Anal Biochem, 1995, 224, 465.
[8]   Oda RP, Madden BJ, Morris JC, Spelsberg TC, Landers JP. J Chromatogr, 1994, 680, 341–351.
[9]   Knox H, Grant JH. Chromatogr, 1987, 24, 135.
[10]  Rodriguez I, Li SFY. Anal Chim Acta, 1999, 383, 1–26.
[11]  Dolnik V, Hutterer KM. Electrophoresis, 2001, 22, 4163–4178.
[12]  Doherty EAS, Meagher RJ. Electrophoresis, 2003, 24, 34–54.
[13]  Cifuentes A, Canalejas P, Ortega A, Diez-Masa JC. J Chromatogr A, 1998, 823, 561–571.
[14]  Verzola B, Gelfi C, Righetti PG. J Chromatogr A, 2000, 874, 293–303.
[15]  Horvath J, Dolnik V. Electrophoresis, 2001, 22, 644–655.
[16]  Erim FB, Cifuentes A, Poppe H, Kraak JC. J Chromatogr A, 1995, 708, 356–361.
[17]  Thakur D et al. Anal Chem, 2009, 81, 8900–8907.
[18]  Chiari M, Nesi M, Sandoval JE, Pesek JJ. J Chromatogr A, 1995, 717, 1–13.
[19]  Albarghouthi MN, Stein TM, Barron AE. Electrophoresis, 2003, 24, 1166–1175.
[20]  Huang XJ, Wang QQ, Huang BL. Talanta, 2006, 69, 463–468.
[21]  Yao YJ, Li SFY. J Chromatogr A, 1994, 663, 97–104.
[22]  Jorgenson JW, Lukacs KD. Science, 1983, 222, 266–272.

[23]  Badal MY, Wong M, Chiem N, Salimi-Moosavi H, Harrison DJ. J Chromatogr A, 2002, 947, 277–286.

[24]  Sebastiano R, Mendieta ME, Contiello N, Citterio A, Righetti PG. Electrophoresis, 2009, 30, 2313–2320.

[25]  Lukacs KD, Jorgenson JW. HRC CC, 1985, 8, 407–411.

[26]  De Mello A. Lab Chip, 2002, 2, 31N–36N.

[27]  Bernard R, Loge G. Electrophoresis, 2009, 30, 4059–4062.

[28]  Hu S, Zhang Z, Cook LM, Carpenter EJ, Dovichi NJ. J Chromatogr A, 2000, 894, 291–296.

[29]  Hu S, Jiang J, Cook LM, Richards DP, Hortick L, Wong B, Dovichi NJ. Electrophoresis, 2002, 23, 3136–3142.

[30]  Eby MJ. Biotechnology, 1989, 7, 903.

[31]  Bolger CA, Zhu M, Rodriguez R, Wehr T. J Liq Chromat, 1991, 14, 895–906.

[32]  Zhu M, Rodriguez R, Wehr T, Siebert C. J Chromatogr, 1992, 608, 225–237.

[33]  Zhu M, Wehr T, Levi V, Rodriguez R et al. J Chromatogr, 1993, 652, 119–129.

[34]  Tango WJ. J Chem Phys, 1968, 49, 4264.

[35]  Kinsey JL. Annu Rev Phys Chem, 1977, 28, 349–372.

[36]  Zare RN. Annu Rev Anal Chem, 2012, 5, 1–14.

[37]  Wong KS, Yeung ES. Mikrochim Acta, 1995, 120, 321–327.

[38]  Li MX, Liu L, Wu JT, Lubman DM. Anal Chem, 1997, 69, 2451–2456.

[39]  Garcia-Campaha AM, Baeyens WRG. Chemiluminescence in Analytical Chemistry. Marcel Dekker, New York, 2001.

[40]  Liu YM, Cheng JK. J Chromatogr A, 2002, 959, 1–13.

[41]  Jorgenson JW. Science, 1983, 222, 266–272.

[42]  Gebauer P, Bocek P. Electrophoresis, 2000, 21, 2809–2813.

[43]  Burgi DS. Anal Chem, 1993, 65, 3726–3729.

[44]  Mazereeuw M, Tjaden UR, Van Der Greet J. J Chromatogr A, 1994, 677, 151–157.

[45]  Silvertand LHH, Sastre Torano J, Van Bennekom WP, De Jong GJ. J Chromatog A, 2008, 1204, 157–170.

[46]  Hjerten S. J Chrom, 1985, 347, 191.

[47]  Busnel JM, Varenne A, Descroix S, Peltre G et al. Electrophoresis, 2005, 26, 3369–3379.

[48]  Kilar F, Vegvari A, Mod A. J Chromatogr A, 1998, 813, 349–360.

[49]  Thormann W, Molteni S, Caslavska J, Schmutz A. Electrophoresis, 1994, 15, 3.

[50]  Monnig CA, Kennedy RT. Anal Chem, 1994, 66, 280.

[51]  Chen FTA, Liu CM, Hsieh YZ, Sternberg JC. Clin Chem, 1991, 37, 14–19.

[52]  Brown MS, Goldstein JL. New Eng J Med, 1981, 305, 515–517.

[53]  Wang MH, Briggs MR. Chem Rev, 2004, 104, 119–137.

[54]  Ruiz-Jimenez J, Kuldvee R, Chen J, Oorni K, Kovanen P, Riekkola ML. Electrophoresis, 2007, 28, 779–788.

[55]  Kuldvee R, Wiedmer SK, Oorni K, Riekkola ML. Anal Chem, 2005, 77, 3401–3405.

[56]  Schmitz G, Möllers C, Richter V. Electrophoresis, 1997, 18, 1807–1813.

[57]  Weiller BH, Ceriotti L, Shibata T, Rein D, Roberts MA, Lichtenberg J, German JB et al. Anal Chem, 2002, 74, 1702–1711.

[58]  Ping G, Zhu B, Jabasini M, Xu F, Oka H, Sugihara H, Baba Y. Anal Chem, 2005, 77, 7282–7287.

[59]  Wang H, Wang HM, Jin QH, Cong H, Zhuang GS, Zhao JL, Sun CL et al. Electrophoresis, 2008, 29, 1932–1941.

[60]  Jones WR. In Landers JP (ed.). Handbook of Capillary Electrophoresis. Boca Raton, FL, CRC Press, 1993, 209–232.

[61]  Clarke GM, Higgins TN. Clin Chem, 2000, 46, 1284–1290.

[62]  Hedlund B. In Fairbanks VF (ed.). Hemoglobinopathies and Thalassemias. Marcel Decker, New York, 1980, 14–17.
[63]  Doelman CJA, Weykamp CW. Ned Tijdchr Klin Chem, 2000, 25, 229–232.
[64]  Ferranti R, Malorni A, Pucci R, Fanali S, Nardi A, Ossicini L. Anal Biochem, 1991, 194, 1–8.
[65]  Zhu M, Rodriguez R, Wehr T, Siebert C. J Chromatogr, 1992, 608, 225–237.
[66]  Cotton F, Lin C, Fontaine B, Gulbis B, Janssens J, Vertongen F. Clin Chem, 1999, 45, 237–243.
[67]  Ong CN, Liau LS, Ong HY. J Chromatogr, 1992, 576, 346–350.
[68]  Shihabi ZK, Hinsdale ME. Electrophoresis, 2005, 26, 581–585.
[69]  Wang J, Zhou S, Huang W, Liu Y, Cheng C, Lu X, Cheng J. Electrophoresis, 2006, 27, 3108–3124.
[70]  Jenkins M, Ratnaike S. Clin Chem Lab Med, 2003, 41, 747–754.
[71]  Zhu M, Wehr T, Levi V, Rodriguez R, Shiffer K, Zhu AC. J Chromatogr, 1993, 652, 119–129.
[72]  Hempe JM, Craver RD. Clin Chem, 1994, 40, 2288–2295.
[73]  Craver RD, Abermanis JG, Warrier RP, Ode DL, Hempe JM. Am J Clin Pathol, 1997, 107, 88–91.
[74]  Hempe JM, Granger JG, Craver RD. Electrophoresis, 1997, 18, 1785–1795.
[75]  Landers JP, Schuchard M, Sismelich T, Spelsberg TC. J Chromatogr, 1992, 603, 247–257.
[76]  Nielsen RG, Rickard EC, Santa PF, Sharknas DA, Sittampalam GS. J Chromatogr, 1991, 539, 177–185.
[77]  Shimura K, Karger BL. Anal Chem, 1994, 66, 9–15.
[78]  Han M, Rock BM, Pearson JT, Rock DA. J Chromatogr B, 2016, 1011, 24–32.
[79]  Kennedy RT, Oates MD, Cooper BR, Nickerson B, Jorgenson JW. Science, 1989, 246, 57–63.
[80]  Olefirowicz TM, Ewing AG. J Neurosci Methods, 1990, 34, 11–15.
[81]  Olefirowicz TM, Ewing AG. Chimia, 1991, 45, 106–108.
[82]  Thakur D et al. Anal Chem, 2009, 81, 8900–8907.
[83]  Belder D, Deege A, Husmann H, Kohler F, Ludwig M. Electrophoresis, 2001, 22, 3813–3818.
[84]  Bush DR, Zang L, Belov AM, Ivanov AR, Karger BL. Anal Chem, 2016, 88, 1138–1146.

# 2.11 Two-dimensional electrophoresis

2.11.1   Theory of 2D electrophoresis —— 231
2.11.2   Isoelectric focusing in the first dimension —— 232
2.11.2.1 Sample preparation —— 232
2.11.2.2 ISO-DALT and IPG-DALT —— 233
2.11.3   SDS disc-electrophoresis in the second dimension —— 234
2.11.4   Detection and evaluation of proteins in 2D pherograms —— 236
2.11.4.1 Autoradiography and fluorography —— 236
2.11.4.2 Two-dimensional gel image analysis —— 236
2.11.5   Protocols —— 237
         Two-dimensional Gel Electrophoresis Using the O'Farrell System —— 237
         Sample preparation —— 237
         First-dimensional gels (isoelectric focusing) —— 237
         Second-dimensional gels (SDS electrophoresis) —— 238
         References —— 239

The **two-d**imensional (2D) gel electrophoresis (2D PAGE) is a combination of two high-resolution electrophoretic methods: IEF and SDS-PAGE. It was introduced in 1975 by O'Farrell [1] and Klose [2].

## 2.11.1 Theory of 2D electrophoresis

In the first direction (dimension) of the 2D electrophoresis, the positively charged proteins will run toward the negative end of the gel, and the negatively charged proteins will run to the positive end of the gel. Afterward, the proteins are resolved in the second direction (dimension) by denaturing SDS discontinuous electrophoresis (disc-electrophoresis) according to their masses: Prior to the SDS disc-electrophoresis, the gel strip with the resolved proteins should be treated with SDS. This procedure unfolds (denatures) the proteins into long straight polyions and binds a number of SDS ions to them proportional to their masses. Since the SDS ions are negatively charged, all proteins will become approximately the same mass-to-charge ratio (Figure 2.11.1).

The IEF of the 2D electrophoresis can be done in pH gradient gels with carrier ampholytes or in IPG gels [3]. The 2D electrophoresis with carrier ampholytes is characterized by a shorter focusing run than the 2D electrophoresis in IPG gels; however, the 2D electrophoresis in IPG gels has higher reproducibility. The IEF of the 2D electrophoresis can also be carried out in hybrid IPG gels, which contain additional carrier ampholytes [4].

As a result of 2D electrophoresis, the protein polyions form a 2D pherogram of spots that looks like a geographical map. The position of each protein spot can be coordinated in the rectangular (Cartesian) system (Figure 2.11.2).

https://doi.org/10.1515/9783110761641-021

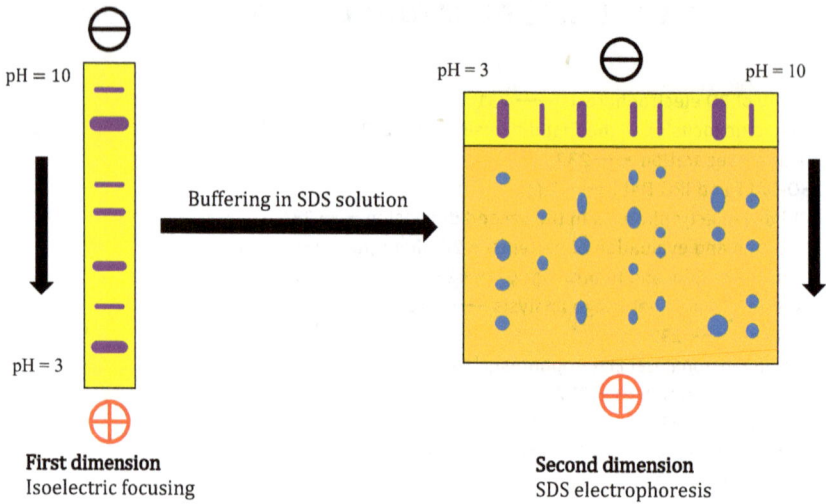

**First dimension**
Isoelectric focusing

**Second dimension**
SDS electrophoresis

**Figure 2.11.1:** Principle of two-dimensional electrophoresis.

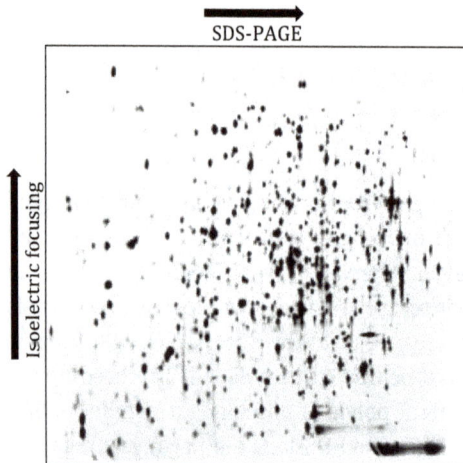

**Figure 2.11.2:** Protein spots obtained by 2D electrophoresis.

## 2.11.2 Isoelectric focusing in the first dimension

Prior to IEF in the first dimension, the protein sample should be appropriately prepared.

### 2.11.2.1 Sample preparation

Tissue or blood samples for a 2D electrophoresis need to be processed and solubilized before applying onto an IEF gel. Solubilization requires lysis reagents such as neutral detergents (e.g., CHAPS). The protein concentration in the sample should be about 10 mg/mL. In addition, 8.0–9.0 mol/L urea, 1.0–2.0 mL/dL Nonidet NP-40, 0.5 g/dL dithiothreitol, and 0.8–1.0 g/dL carrier ampholytes (if the first-dimension gel contains carrier ampholytes) should be added to the sample. The sample buffer should have low ionic strength.

### 2.11.2.2 ISO-DALT and IPG-DALT

If the IEF is performed by carrier ampholytes, the 2D electrophoresis is referred to as ISO-DALT system [5]. **Iso** is for the pI separation by IEF using carrier ampholytes, and **Dalt** is for the mass resolving by SDS disc-electrophoresis (Dalton is the old unit for mass, which has found no place in SI system). If the IEF is performed in an IPG gel, the 2D electrophoresis is referred to as IPG-DALT system.

The gels for IEF with carrier ampholytes have $T$ concentration of 4–5 g/dL and a degree of cross-linking $C = 0.03$, and contain 8–9 mol/L urea and 1–2 g/dL carrier ampholytes. Their ends should be covered with paper or gel strip electrodes; the anode strip could contain 10 mmol/L glutamic acid or 0.8 mol/L phosphoric acid (54.3 mL of 85 g/dL phosphoric acid to 1,000.0 mL deionized water); and the cathode strip could contain 10 mmol/L lysine, or 0.8 mol/L ethylenediamine (53.7 mL ethylenediamine to 1,000.0 mL deionized water), and 10 mL/dL glycerol. Then a silicone template could be laid down onto the gel, and 10 μL of samples each could be applied into its slots.

For IPG-Dalt system, immobilized pH gradient of pH = 4.0–10.0, 4.0–7.0, or 7.0–10.0 is used. They can be prepared as follows (Table 2.11.1)

The IPG gel is prepared using a gradient maker at room temperature. Thereafter, the gel is removed from the casting cassette, and washed 3 times for 20 min each in deionized water, 1 time in 2 mL/dL glycerol, and finally dried at room temperature using a fan. At the end, the gel is put in a plastic wrap and stored at –20 °C.

After rehydration at room temperature for 3–24 h, the swollen IPG gel strips are put between two filter papers to remove the excess rehydration solution so that urea should not crystallize on their surface. Then the strips are placed, the acidic ends facing the anode, onto the coated with kerosene cooling block of a horizontal chamber. Next, the IPG gel strips are covered with paper or gel electrode strips (Figure 2.11.3). At the end, the samples are applied, using a silicone template, onto the gel strips.

The IEF is carried out at 10–15 °C, wherein a prefocusing is needed only in IEF with carrier ampholytes; prefocusing for IEF on IPG gels is unnecessary. After IEF, the gel with carrier ampholytes or the IPG gel strips can be used immediately for the second dimension of the 2D electrophoresis, or can be stored in a freezer at –80 °C or in liquid nitrogen.

**Table 2.11.1:** Preparation of solutions for 250 × 120 × 0.5 mm IPG gels ($T$ = 4 g/dL, $C$ = 0.03) with different pH gradients.

| | IPG gradient | | | | | |
| --- | --- | --- | --- | --- | --- | --- |
| | pH = 4.0–10.0 | | pH = 4.0–7.0 | | pH = 7.0–10.0 | |
| | Light solution | Heavy solution | Light solution | Heavy solution | Light solution | Heavy solution |
| Immobilines p$K$ = 3.6, µL | – | 588.0 | 161.0 | 308.0 | 48.0 | 289.0 |
| p$K$ = 4.6, µL | 61.0 | – | 394.0 | 59.0 | – | – |
| p$K$ = 6.2, µL | 27.0 | 243.0 | 81.0 | 240.0 | – | – |
| p$K$ = 7.0, µL | 260.0 | 47.0 | 143.0 | – | 173.0 | 202.0 |
| p$K$ = 8.5, µL | 84.0 | 178.0 | – | – | 187.0 | 187.0 |
| p$K$ = 9.3, µL | 190.0 | – | 467.0 | – | 149.0 | – |
| Monomeric solution, mL ($T$ = 50 g/dL, $C$ = 0.03) | 0.64 | 0.64 | 0.64 | 0.64 | 0.64 | 0.64 |
| 87% Glycerol, mL | 0.3 | 2.0 | 0.3 | 2.0 | 0.3 | 2.0 |
| 10 g/dL TMEDA, µL | 37.0 | 37.0 | 37.0 | 37.0 | 37.0 | 37.0 |
| 10 g/dL APS, µL | 37.0 | 37.0 | 37.0 | 37.0 | 37.0 | 37.0 |
| Deionized water to, mL | 8.0 | 8.0 | 8.0 | 8.0 | 8.0 | 8.0 |

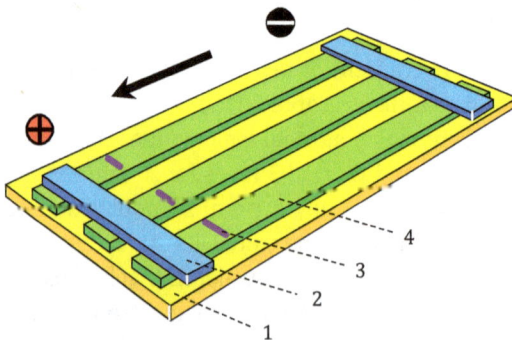

**Figure 2.11.3:** IEF on IPG gel strips.
1. Separation chamber; 2. electrode strip; 3. sample; 4. IPG gel strip.

## 2.11.3 SDS disc-electrophoresis in the second dimension

After IEF in the first dimension, SDS electrophoresis should be run in the second dimension. So the focused proteins are separated according to their molecular masses.

Prior to the SDS electrophoresis, the gel strip with the focused proteins should be equilibrated in test tubes containing SDS solution. IEF gel strips, which contain carrier ampholytes, should be equilibrated for 2 min, and IPG gel strips should be equilibrated twice for 15 min each. After the second equilibration of IPG gel strips, 0.26 mol/L 2-iodoacetamide is added to the equilibration solution to alkylate the protein SH groups and avoid streaking in the gel during silver staining. Thereafter, the equilibrated IEF or IPG gel strips are sandwiched for 1 min between two filter papers to suck the excess of the equilibration solution and is placed on the stacking SDS gel.

The SDS electrophoresis can be run in the TRIS-chloride-glycinate system of Ornstein [6], Davis [7], and Laemmli [8]; in the TRIS-chloride-TRICINEate system of Schäger [9]; or in the TRIS-formate-taurinate system of Michov [10]. The electrophoresis is fulfilled at 10–15 °C. After 60 min at 30–80 mA, 200 V, and 6–16 W (for a standard gel of 250 × 120 × 0.5 mm), the SDS electrophoresis is continued at 30–80 mA, 800 V, and 15–40 W, depending on the buffer, for further 1.5–5 h, until the Bromophenol blue front reaches the anodic strip (Figure 2.11.4).

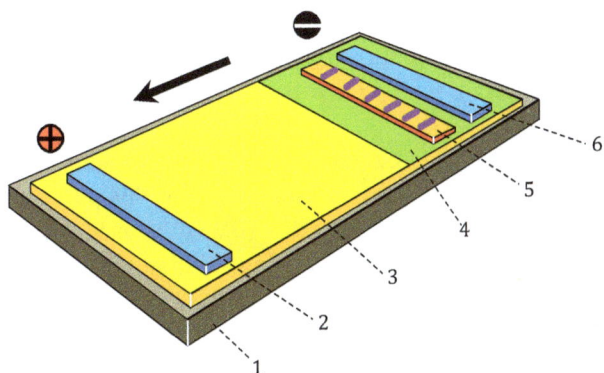

**Figure 2.11.4:** SDS electrophoresis after placing an equilibrated gel strip on an SDS stacking gel. 1. Separation chamber; 2. anode strip; 3. separating gel; 4. stacking gel; 5. gel strip with protein bands separated by IEF; 6. cathode strip.

For the second dimension of 2D electrophoresis, an SDS gradient or homogeneous gel [11] can be used. In both cases, the gel consists of a stacking part and a resolving part. The SDS stacking gel has usually $T = 5$ g/dL and $C = 0.02–0.04$, and contains 0.125 mol/L TRIS-chloride buffer with pH = 6.8. The SDS resolving gel contains usually $T = 10–15$ g/dL and $C = 0.02–0.04$, and contains 0.375 mol/L TRIS-chloride buffer of pH = 8.8.

After the electrophoresis, SDS is removed from the gel by incubating in a solution of Triton X-100.

## 2.11.4 Detection and evaluation of proteins in 2D pherograms

The 2D electrophoresis has the highest resolution available today. It is capable of re-solving over 3,000–4,000 protein spots in a single gel and allows a separation of two protein polyions that differ one from another by only one charged amino acid residue. Therefore, the 2D electrophoresis is used for studying proteins of microorganisms [12, 13], plants [14, 15], or milk [16, 17], and is of great importance for the clinical diagnostics [18, 19].

### 2.11.4.1 Autoradiography and fluorography

The autoradiography and fluorography are the most sensitive detection methods for 2D pherograms. The *autoradiography* is referred to as emission of radioactively labeled proteins onto an X-ray film. For this purpose, proteins should be labeled with different isotopes: with $^3$H-amino acids and with $^{14}$C-amino acids or $^{35}$S-methionine.

In *fluorography*, a visible light, produced by interaction between a radioactive radiation and a scintillation substance, blackens an X-ray film. For this purpose, 15 g/dL 2,5-diphenyloxazole (PPO) in **dim**ethyl **s**ulf**o**xide (DMSO) is used. To do this, the gel after the 2D electrophoresis is stored in a fixing solution for 1 h, in a destaining solution for 10 min, in DMSO overnight, in 15 g/dL PPO in DMSO for 3.5 h, and in 5 g/dL glycine for 1 h. Then it is dried, as 10 wet sheets of filter paper, a wet cellophane membrane, the gel, a second wet cellophane membrane, and po-rous polyethylene sheets are placed on a perforated metal plate and covered with a rubber flap. All layers should be compressed under vacuum produced by a water-jet pump. The drying is carried out overnight at 50 °C. Then an X-ray film is placed on the dried gel and the film exposure takes place at –70 °C for 2–3 days. Finally, the spots on the film are obtained as follows: 6 min developing, 10 min fixing, and 10–20 min water cleaning.

### 2.11.4.2 Two-dimensional gel image analysis

Two-dimensional electrophoresis is used to obtain practical proteomics information. Modern 2D gel analysis software can rapidly analyze spots in the gel. With appropri-ate computer program, the entire analysis process from background correction to spot matching results takes minutes. Robots are also used for isolation of protein spots from 2D gels.

## 2.11.5 Protocols

### Two-dimensional Gel Electrophoresis Using the O'Farrell System

### Sample preparation

#### Materials and equipment
Buffer
Tissue sample
Homogenizer with a pestle
Centrifuge

#### Procedure
- Place the tissue sample in the homogenizer.
- Add 1.5–2.0 mL buffer per 100 mg tissue and homogenize using strokes with a pestle.
- Transfer an aliquot to a 200 µL centrifuge tube and centrifuge at 200,000 g for 1 h.
- Load the supernatant onto the first-dimensional gel.

### First-dimensional gels (isoelectric focusing)

#### Materials and equipment
Acrylamide
Bisacrylamide
TMEDA
APS
Urea
Ampholytes, pH = 4.0 to 8.0
Nonidet P-40
NaOH
$H_3PO_4$
Casting cassette
Electrofocusing unit
Power supply

#### Monomeric solution, T = 30.8 g/dL, C = 0.03
Acrylamide 30.0 g
Bisacrylamide 0.8 g
Deionized water to 100.0 mL

**Procedure**

- Mix 8.25 g urea, 2.0 mL of the monomeric solution, 0.75 mL ampholytes (pH = 4.0–8.0), and 6.0 mL deionized water in a small vacuum flask.
- Place the flask in a warm water bath on a magnetic stirrer and stir until the urea is dissolved.
- Deaerate the solution by applying vacuum for 2–3 min.
- Add 0.3 mL Nonidet P-40, swirl until dissolved, and pass through a filter.
- Add TMEDA and APS.
- Pipette the mixture into a casting cassette.
- Allow the monomeric solution to polymerize for 1 h.
- Fill one reservoir of the electrofocusing unit with 0.085 g/dL phosphoric acid and the other reservoir with 0.02 mol/L NaOH.
- Prefocus the gel at 200 V for 1 h.
- Apply 10–30 µL protein samples onto the gel.
- Switch on the power supply.
- Focus at constant voltage for 16 h.

**Second-dimensional gels (SDS electrophoresis)**

**Materials and equipment**

First-dimensional gel
Acrylamide
Bisacrylamide
SDS
TMEDA
APS
Equilibration buffer
Electrode buffer
Casting cassette
Horizontal electrophoresis apparatus
Power supply

**Procedure**

- Prepare gel mixing buffer, monomeric solution, SDS, TMEDA, APS, and deionized water (Table 2.11.2). Deaerate using vacuum for 5 min.
- Assemble glass plates with 1.0 mm spacers to produce a casting cassette.
- Pour the monomeric solution in the casting cassette to 5 mm below the top and overlay with deionized water.
- Allow the solution to polymerize for 1.5 h.
- Using a spatula, slide off a strip from the first-dimensional gel and equilibrate it in the equilibration buffer.
- Place the equilibrated gel strip on the SDS gel.

**Table 2.11.2:** Solutions for second-dimensional gels.

| Stock solutions | Final monomeric concentration, g/dL | | | |
|---|---|---|---|---|
| | 7.5 | 10 | 12.5 | 15 |
| Gel buffer | 12.50 mL | 12.50 mL | 12.50 mL | 12.50 mL |
| 30 g/dL acrylamide/0.8 g/dL bisacrylamide | 12.50 mL | 16.70 mL | 20.80 mL | 25.10 mL |
| 10 g/dL SDS | 0.50 mL | 0.50 mL | 0.50 mL | 0.50 mL |
| 10 g/dL TMEDA | 0.26 mL | 0.26 mL | 0.26 mL | 0.26 mL |
| 10 g/dL APS | 0.25 mL | 0.23 mL | 0.21 mL | 0.19 mL |
| Deionized water to | 50.00 mL | 50.00 mL | 50.00 mL | 50.00 mL |

- Mount the gel in an electrophoresis unit.
- Fill the tanks with electrode solutions.
- Start cooling at 10–20 °C.
- Switch on the power supply.
- Electrophorese at 15–20 mA/gel until the tracking dye reaches the opposite end of the gel.
- Stain the gel and process it for immunoblotting or autoradiography.

# References

[1]   O'Farrell PH. J Biol Chem, 1975, 250, 4007–4021.
[2]   Klose J. Humangenetik, 1975, 26, 231–243.
[3]   Bjellqvist B, Ek K, Righetti PG, Gianazza E, Görg A, Westermeier R, Postel W. J Biochem Biophys Methods, 1982, 6, 317–339.
[4]   Altland K, Rossmann U. Electrophoresis, 1985, 6, 314–325.
[5]   Anderson NG, Anderson NL. Anal Biochem, 1978, 85, 331–354.
[6]   Ornstein L. Ann N Y Acad Sci, 1964, 121, 321–349.
[7]   Davis BJ. Ann N Y Acad Sci, 1964, 121, 404–427.
[8]   Laemmli UK. Nature, 1970, 227, 680–685.
[9]   Schägger H. Nat Protoc, 2006, 1, 16–22.
[10]  Michov BM. Protein Separation by SDS Electrophoresis in a Homogeneous Gel Using a TRIS-formate-taurinate Buffer System and Homogeneous Plates Suitable for Electrophoresis. (Proteintrennung Durch SDS-Elektrophorese in Einem Homogenen Gel Unter Verwendung Eines TRIS-Formiat-Taurinat-Puffersystems Und Dafür Geeignete Homogene Elektrophoreseplatten). German Patent 4127546, 1991.
[11]  Michov BM. GIT Fachz Lab, 1992, 36, 746–749.
[12]  Pollard JW. Celis J, Brava R (eds.). Two-dimensional Gel Electrophoresis of Proteins. Academic Press, San Diego, 1983, 363–395.
[13]  Lossius I, Siastad K, Haarr L, Kleppe K. J Gen Microbiol, 1964, 130, 3153.

[14]  Remy R, Ambard-Bretteville F. Colas Des Francs C, Electrophoresis, 1987, 8, 528–532.
[15]  Holloway PJ, Arundel PH. Anal Biochem, 1988, 172, 8–15.
[16]  Marshall T, Williams KM. Electrophoresis, 1988, 9, 143–147.
[17]  Anderson NG, Powers MT, Tollaksen SL. Clin Chem, 1982, 28, 1045–1055.
[18]  Marshall T. Anal Biochem, 1984, 139, 506–509.
[19]  Wirth PJ, Yuspa SH, Thorgeirsson SS, Hennings H. Cancer Res, 1987, 47, 2831–2838.

# 2.12 Preparative electrophoresis of proteins

2.12.1   Preparative disc-electrophoresis —— 241
2.12.1.1 Elution of proteins during electrophoresis —— 241
2.12.1.2 Elution of proteins after electrophoresis —— 242
         Elution by diffusion —— 243
         Elution by gel dissolving —— 243
         Electroelution —— 243
2.12.2   Preparative isoelectric focusing —— 245
2.12.2.1 Preparative IEF with carrier ampholytes in granulated gels —— 245
2.12.2.2 Preparative IEF in immobilized pH gradients —— 247
2.12.2.3 Recycling isoelectric focusing —— 248
2.12.3   QPNC-PAGE —— 248
2.12.4   Protocols —— 249
         QPNC-PAGE —— 249
         References —— 250

Preparative electrophoresis is a valuable method for isolation of diverse proteins, for example, ATP-dependent enzymes and respiratory complexes. Three types of preparative electrophoresis are used: disc-electrophoresis, isoelectric focusing (IEF), and **q**uantitative **p**reparative **n**ative **c**ontinuous **p**oly**a**crylamide **g**el **e**lectrophoresis (QPNC-PAGE).

## 2.12.1 Preparative disc-electrophoresis

The separated proteins can be eluted from the gel either during or after the electrophoresis.

### 2.12.1.1 Elution of proteins during electrophoresis

Elution of proteins from a gel during electrophoresis can be carried out, when the proteins are removed by a buffer flowing to a collector. For this purpose, a cassette with a narrow channel in its lower part can be used [1]. The channel is connected with a peristaltic pump. The pump pushes the buffer through the narrow channel and transports with it the proteins to the collector (Figure 2.12.1).

Another construction for protein elution during electrophoresis uses also a gel-casting cassette [2]: The cassette is divided by three vertical spacers (two outer spacers and one inner spacer), which form two compartments: a large compartment for the preparative gel and a small compartment for the detection gel (Figure 2.12.2). On both sides of the preparative polyacrylamide gel, polyethylene tubings are attached reaching its bottom. The detection gel is used to determine the protein

https://doi.org/10.1515/9783110761641-022

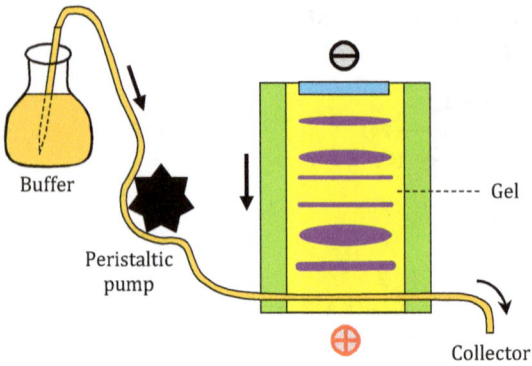

**Figure 2.12.1:** Scheme of a simple device for protein elution during electrophoresis.

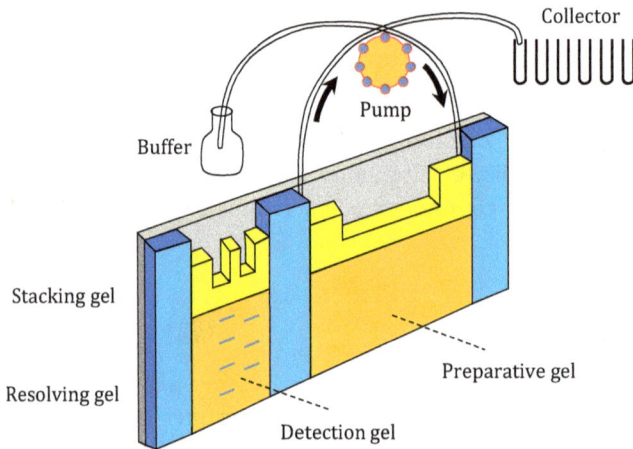

**Figure 2.12.2:** Cassette for elution of proteins during electrophoresis.
The cassette has two sections: a large section for the preparative gel and a small section for the detection gel. The lower end of the preparative gel is wrapped in a dialysis membrane. With the help of a two-channel peristaltic pump, the protein fractions are transported from the dialysis membrane to a collector.

positions. When disc-electrophoresis is carried out, both gels consist of a stacking gel ($T$ = 4 g/dL, $C$ = 0.03) and a resolving gel ($T$ = 10 g/dL, $C$ = 0.03). The protein yield can reach 89–99%.

### 2.12.1.2 Elution of proteins after electrophoresis

Prior to elution, the protein bands after electrophoresis should be located in the gel. Then the bands could be excised and the proteins eluted. If the proteins are colored (e.g., hemoglobins), it is easy to find them. If they are not colored, they should be stained or their location could be defined using UV absorbance [3, 4], fluorescence imaging [5–7], or phosphorescence at low temperatures [8]. An elegant way to identify enzymes in a gel is to use in-gel assays for enzyme activities [9–11].

The elution of proteins after electrophoresis can be done by diffusion, gel dissolving, or electroelution.

## Elution by diffusion

The elution by diffusion is the simplest method to free proteins from a gel [12]. For this purpose, buffers without detergent should be used [13]. The monomers, TMEDA, and APS, which diffuse together with the proteins, can be removed using ion exchange chromatography [14, 15]. However, the elution by diffusion is time-consuming and incomplete because proteins still remain in the gel – the yield is low [16].

## Elution by gel dissolving

Elution by gel dissolving (depolymerizing) is a complicated method leading to chemical modification of proteins. To obtain proteins from gels with the standard cross-linker BIS, harsh conditions are needed. For example, if a polyacrylamide gel is dissolved by incubation in 30 mL/dL hydrogen peroxide ($H_2O_2$) at 50 °C [17], irreversible damages of all proteins take place. Another alternative is to use a special cross-linker, for example, $N,N'$ diallyltartardiamide. This gel should be solubilized in 2 g/dL periodic acid ($H_5IO_6$), which is suited for elution of high-molecular-mass proteins [18]. Unfortunately, the usage of periodic acid is not adequate for most proteins [19–21].

## Electroelution

The electroelution of proteins from a preparative gel is carried out in electric field driving the proteins to migrate in direction opposite to the electrophoresis direction into a dialysis bag [22–24]. This technique is characterized by a high yield. So insulin, myoglobin, and bovine serum albumin were eluted from the gel up to 95%.

A few types of electroelution are known: vertical-type elution, horizontal-type elution, bridge-type elution, discontinuous conductivity gradient elution, and steady-state stacking elution [25] (Figure 2.12.3). Finally, the protein eluate can be concentrated and lyophilized.

An apparatus for *vertical-type elution* (Figure 2.12.3a) consists of a tube with buffer, sieve, a dialysis bag, and a tank with buffer [26]. The elution is carried out in the following steps: the dialysis bag is attached on the sieved tube, which is filled with buffer,

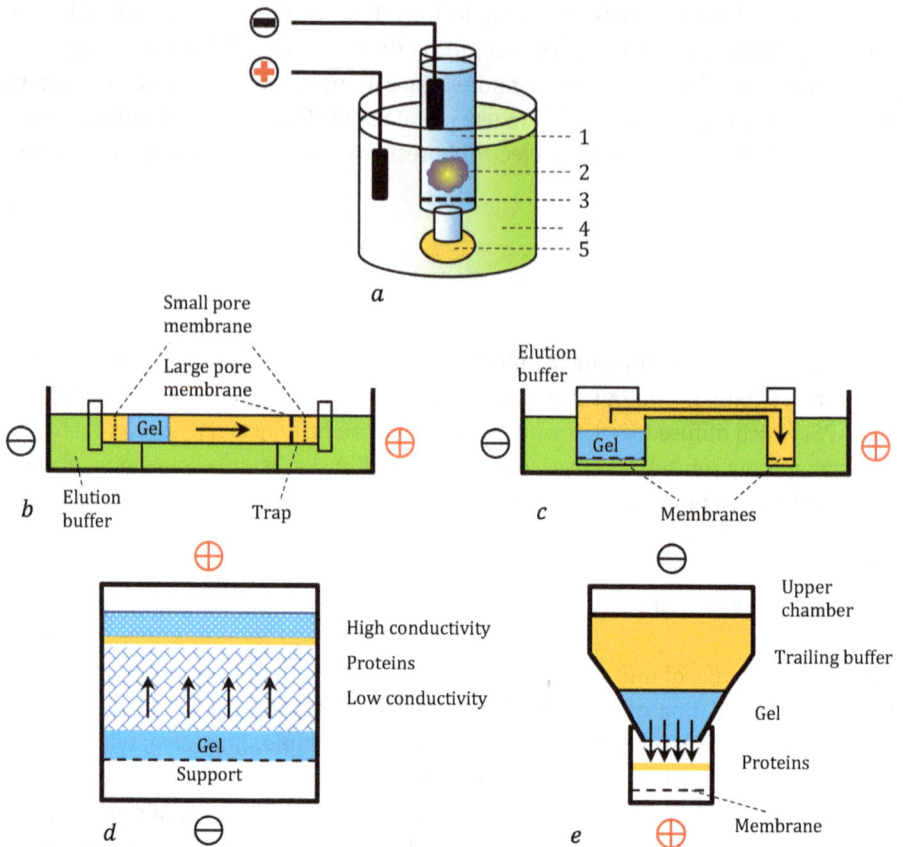

**Figure 2.12.3:** Different types of electroelution of proteins. The proteins are collected on a small-pore membrane that supports the gel.
(*a*) Vertical-type elution: 1. tube with buffer; 2. gel; 3. sieve; 4. tank with buffer; 5. dialysis bag.
(*b*) Horizontal-type elution: small-pore and large-pore membranes are used to collect the eluted proteins. (*c*) Bridge-type elution: The electric field drives proteins out from the bigger chamber through a bridge into the smaller chamber. (*d*) Discontinuous conductivity gradient elution: The trapping process is based on the migration of proteins in a high conductivity layer. (*e*) Steady-state stacking elution: isotachophoresis is expanded to elute proteins in a funnel-shaped device.

and placed in a tank with the same buffer. The polyacrylamide gel is cut into small pieces and put into the tube. Thereafter, electric current is applied, which drives the proteins to penetrate through the sieve into the dialysis bag. The dialysis membrane can be omitted [27] using a layer of hydroxylapatite, which captures the protein. The bound proteins can then be eluted by phosphate buffer with other pH values.

For *horizontal-type elution* (Figure 2.12.3b), a flat-bed electrophoresis tank and a special column are used [28, 29]. Both ends of the column are sealed with semi-permeable membranes to form an elution chamber that is divided by a large-pore

membrane into two compartments. The gel pieces are placed in the large compartment. After applying an electric field, proteins migrate into the small compartment
and are captured there by a small-pore membrane.

The construction for *bridge-type elution* (Figure 2.12.3c) consists of two separate
horizontal chambers, which are connected by a bridge [30, 31]. Both chambers are
sealed by dialysis membranes. The bigger chamber contains the gel material with
the proteins, and the smaller chamber collects the eluate. The proteins migrate
electrophoretically out of the bigger chamber, pass the bridge, and are captured in
the smaller chamber.

The *discontinuous conductivity gradient elution* (Figure 2.12.3d) does not use a
dialysis membrane but a combination of two solutions. A low-conductivity glycerol
layer is surrounding the gel. A second high-conductivity glycerol layer is set upon
the first one [32–34]. Proteins in the low-conductivity zone migrate rapidly out of the
gel. When they reach the upper zone, their migration slows down because of the
highly concentrated solution. However, the high concentrated layer can cause a salt-
ing-out effect, which diminishes the yield.

The *steady-state stacking elution* (Figure 2.12.3e) is based on the isotachophoresis, which can be expanded [35] to separate proteins in a funnel-shaped device [36,
37]. Standard electrophoresis equipment can also be applied for steady-state stacking elution, if gel pieces are embedded in a new stacking gel overlaid with a glycerol layer [38]. As a result, proteins leave the gel pieces to be concentrated into the
glycerol solution.

## 2.12.2 Preparative isoelectric focusing

Preparative IEF can be carried out with carrier ampholytes or in immobilized pH
gradients.

### 2.12.2.1 Preparative IEF with carrier ampholytes in granulated gels

The preparative IEF using carrier ampholytes is accomplished in horizontal granular
gels [39, 40] made of the dextran products Sephadex G-75, G-200, or best Ultrodex
[41]. It proceeds in the following steps: producing a granulated gel, introducing a
sample into the granulated gel, IEF, and elution of proteins.

Prior to preparing the gel, Sephadex G-75 and G-200 should be washed with deionized water. When the difference between the conductivities of water before and
after washing has reached a minimum, the washing process can be stopped. If Ultrodex is used, washing is omitted.

The preparation of a horizontal granulated layer involves three phases: 1. mixing the washed gel with carrier ampholytes; 2. inserting electrode strips into the

mold and casting the gel with carrier ampholytes; and 3. drying partially the gel and carrier ampholytes (Figure 2.12.4).

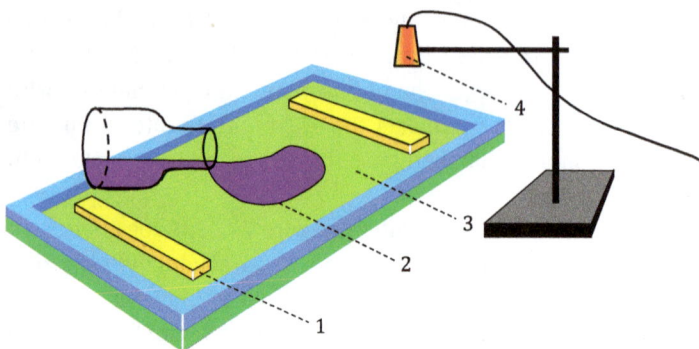

**Figure 2.12.4:** Producing a horizontal layer of granulated gel for preparative IEF with carrier ampholytes.
The granulated gel is mixed with carrier ampholytes. Then electrode strips (1) are inserted, the mixture (2) is cast into a mold (3), and the gel layer is partially dried using a hair dryer (4).

The sample can be uniformly distributed in the gel layer, or introduced as a zone in the gel layer. In the first case, the sample is added when the gel is mixed with the carrier ampholytes. In the second case, the gel layer, containing ampholines, is cast into a rectangular frame, and a part of it is taken off using a spatula forming a well. The gel part is transferred into a Becher glass where the mixture is mixed with the sample. Then the mixture is poured back into the well (Figure 2.12.5).

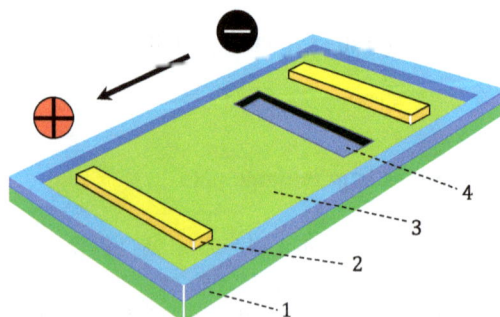

**Figure 2.12.5:** Application of sample as a zone in a granulated gel layer.
1. Glass plate with a silicone frame; 2. electrode strip; 3. granulated gel; 4. well in the gel layer.

Before starting the preparative IEF, moist filter paper strips should be placed on both electrodes. The anode filter paper strip is impregnated with 0.5 mol/L $H_3PO_4$, and the cathode filter paper strip is impregnated with 0.5 mol/L NaOH. Thereafter, the mold with the granulated gel layer is placed on the cooling plate of an electro-phoresis unit. Afterward, the separation unit is covered with the safety lid and the electrophoresis unit is connected to the power supply.

Based on the newly established electrophoretic method called **divergent flow IEF (DF IEF)**, a DF IEF instrument was proposed, which operates without carrier ampholytes [42]. In DF IEF, the proteins are separated, desalted, and concentrated in one step, and the proteins have been concentrated up to 16.8-fold.

After preparative electrofocusing, the visualization of protein bands in the mold could be done placing a filter paper onto the granulated gel, staining the pro-tein blot, then placing it under the mold, and finally taking out the gel segment containing the desired protein. Then the protein can be extracted from the gel seg-ment and analyzed.

Finally, the proteins should be freed from the accompanying carrier ampho-lytes. For this purpose, different methods are used to separate the low mass carrier ampholytes ($M_r \approx 800$) from the high mass proteins: dialysis, gel filtration, and ul-trafiltration. At the end, the proteins could be lyophilized and frozen.

## 2.12.2.2 Preparative IEF in immobilized pH gradients

The preparative IEF can be carried out also in immobilized pH gradients [43–45]. An interesting technique of it is the channel focusing (Figure 2.12.6). It proceeds in the following steps: casting a preparative IPG gel, analytical IEF in the margins of the IPG gel, digging channels, running the preparative IEF, and elution of proteins.

*Casting a preparative IPG gel.* An IPG gel for preparative IEF is cast as an IPG gel for analytical IEF. Then two short slots for analytical electrophoresis of proteins of interest are made in the gel margins, and a long slot is made in the medium of the gel for the application of sample.

*Analytical IEF in the margins of the IPG gel.* Margin strips containing the appli-cation slots are cut from the dried IPG gel and rehydrated. Then a small portion of the sample underwent an analytical electrofocusing, and the obtained protein bands are stained to determine the protein position.

*Digging channels.* The remaining gel is rehydrated as the margin strips and is placed on a glass plate next to the strips with the stained protein bands. Using a spatula, a channel is cut into the gel against the desired protein. The channel is filled with a swollen granular gel, which should collect the protein at its migration. If necessary, more channels can be cut for different proteins.

*Running preparative IEF.* The preparative IEF in an IPG gel is running under the same conditions as the analytical IEF in IPG gels.

*Elution of proteins.* After the preparative electrofocusing in an IPG gel, the con-tent of the channel is taken up with a spatula and is transferred to an elution col-umn. The proteins are eluted, frozen, or lyophilized. The gel is cleaned with a

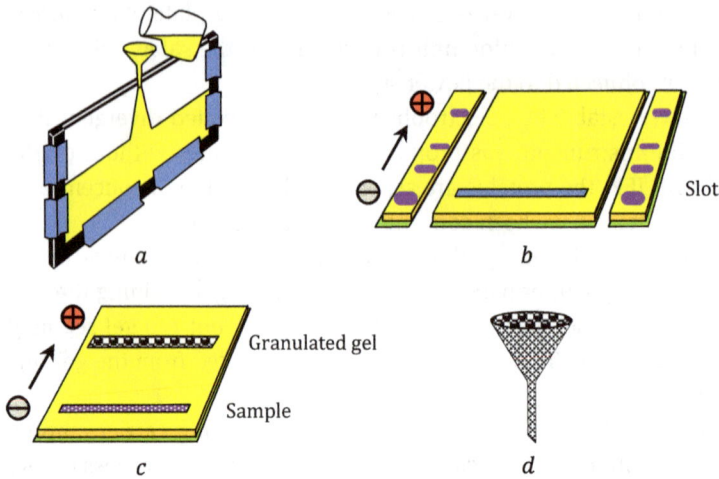

**Figure 2.12.6:** Channel focusing.
(*a*) Casting an IPG gel; (*b*) cutting margins of the IPG gel and running an analytical IEF to determine the position of proteins of interest; (*c*) making a well (channel), filling it with the granulated gel, and carrying out a preparative focusing; (*d*) extracting and eluting proteins from the granulated gel.

buffer from the rest of proteins, washed, and dried. It can be used for next channel IPG electrofocusing.

### 2.12.2.3 Recycling isoelectric focusing

The recycling **IEF** (RIEF) according to Bier *et al.* [46, 47] is based on the principle of continuous recirculation of a separation medium in a special apparatus. It consists of an IEF unit, a multichannel peristaltic pump, pH monitor, and a thermostat. The pH value, the conductivity, the temperature, and the optical density of the protein fractions are measured and controlled with the aid of a computer.

The IEF unit is filled with carrier ampholytes and the sample, which are then recycled through the pump, the pH monitor, and the thermostat. When a pH gradient between the columns is formed, the proteins migrate through the membranes until they reach their isoelectric points. The Joule heating, which is produced during the RIEF, is removed by the thermostat.

## 2.12.3 QPNC-PAGE

QPNC-PAGE (**q**uantitative **p**reparative **n**ative **c**ontinuous **p**olyacrylamide **g**el **e**lectrophoresis) is a variant of polyacrylamide gel zone electrophoresis. It is used to isolate metalloproteins in biological samples and to resolve properly and improperly folded proteins or protein isoforms [48]. During the electrophoresis, the metalloproteins are not dissociated into apoproteins and metal cofactors.

QPNC-PAGE is based on the **t**ime of **pol**ymerization of **a**crylamide (tpolAA). The time of polymerization of a gel may affect the peak elution times of separated metalloproteins in a pherogram due to the compression of the gel and its pores on proteins. In order to ensure maximum reproducibility in the gel pore size and to obtain a fully polymerized large-pore gel for a PAGE, the polyacrylamide gel is polymerized for 69 h at room temperature (tpolAA = 69 h). The heat generated by the polymerization process is dissipated constantly. As a result, the prepared gel is homogeneous, stable, hydrophilic, electrically neutral, free of monomers or radicals, and does not bind proteins [49].

QPNC-PAGE is used for isolating **c**opper **c**haperone for **s**uperoxide dismutase (CCS), **s**uper**o**xide **d**ismutase (SOD), prions, transport proteins, amyloids, and metalloenzymes, which are present in brain blood or other samples in Alzheimer's disease or amyotrophic lateral sclerosis [50]. CCS or SOD molecules control the concentrations of essential metal ions (e.g., $Cu^+$, $Cu^{2+}$, $Zn^{2+}$, $Fe^{2+}$, $Fe^{3+}$, $Ni^{2+}$, $Mo^{2+}$, $Pd^{2+}$, $Co^{2+}$, $Mn^{2+}$, $Pt^{2+}$, $Cr^{3+}$, and $Cd^{2+}$) in organisms and thus balance oxidative and reductive processes in the cytoplasm [51].

## 2.12.4 Protocols

### QPNC-PAGE

**Materials and equipment**
TRIS
HCl
Acrylamide
BIS
TMEDA
**A**mmonium **p**eroxydi**s**ulfate (APS)
Sodium azide
Buffer recirculation pump

*Stock solutions*
200 mmol/L TRIS-HCl, 10 mmol/L $NaN_3$, pH = 10.0
200 mmol/L TRIS-HCl, 10 mmol/L $NaN_3$, pH = 8.0
Acrylamide/BIS, $T$ = 40 g/dL, $C$ = 0.027
10 g/dL APS. Prepare freshly.

*Electrophoresis buffer*
20 mmol/L TRIS-HCl, 1 mmol/L $NaN_3$, pH = 10.0. Keep at 4 °C.

*Monomeric solution, T = 4 g/dL, C = 0.027*
Acrylamide/BIS 10.0 mL
10 g/dL TMEDA 0.5 mL
10 g/dL APS 0.5 mL
    *Add prior to gel casting.*
Electrophoresis buffer to 100.0 mL

*Elution buffer*
20 mmol/L TRIS-HCl, 1 mmol/L NaN$_3$, pH = 8.0. Keep at 4 °C.

**Procedure**
- Pipette the monomeric solution to a level of 40 mm in a graduated glass column with an inner diameter of 28 mm. Finally, add APS. The time of polymerization is 69 h at room temperature.
- Mix 3.0 mL of a sample with 0.3 mL glycerol.
- Apply the mixture under the upper electrophoresis buffer and run the electrophoresis at 4 °C.
- Elute the separated proteins continuously in a special elution chamber and transport them to a fraction collector.

# References

[1]   Kyriakopoulos A, Kalcklösch M, Weiß-Nowak C, Behne D. Electrophoresis, 1993, 14, 108–111.
[2]   Lim YP, Callanan H, Lin SH, Thompson NL, Hixson DC. Anal Biochem, 1993, 214, 156–164.
[3]   Bartolini P, Arkaten R, Ribela MT. Anal Biochem, 1989, 176, 400–445.
[4]   Yamamoto H, Nakatani M, Shinya K, Kim BH, Kakuno T. Anal Biochem, 1990, 191, 58–64.
[5]   Leibowitz MJ, Wang RW. Anal Biochem, 1984, 137, 161–163.
[6]   Roegener J, Lutter R, Reinhardt R, Bluggel M *et al.* Anal Chem, 2003, 75, 157–159.
[7]   Riaplov E, Li Q, Seeger S. Protein Pept Lett, 2007, 14, 712–715.
[8]   Mardian JK, Isenberg I. Anal Biochem, 1978, 91, 1–12.
[9]   Gabriel O, Gersten DM. Anal Biochem, 1992, 203, 1–21.
[10]  Zerbetto E, Vergani L, Dabbeni-Sala F. Electrophoresis, 1997, 78, 2059–2064.
[11]  Manchenko GP. Handbook of Detection of Enzymes on Electrophoretic Gels, 2nd edn. CRC Press, Boca Raton, FL, 2003.
[12]  Lewis UJ, Clark MO. Anal Biochem, 1963, 6, 303–315.
[13]  Anderson JM, Waldron JC, Thorne SW. Plant Sci Lett, 1980, 77, 149–157.
[14]  Brooks KR, Sander EG. Anal Biochem, 1980, 107, 182–186.
[15]  Brysk MM, Barlow E, Bell T, Rajaraman S, Stach RW. Prep Biochem, 1988, 18, 217–225.
[16]  Shoji M, Kato M, Hashizume S. J Chromatogr A, 1995, 698, 145–162.
[17]  Young RW, Fulhorst HW. Anal Biochem, 1965, 77, 389–391.
[18]  Hahn EC, Hahn PS. J Virol Methods, 1987, 75, 41–52.
[19]  Anker HS. FEBS Lett, 1970, 7, 293.
[20]  Baumann G, Chrambach A. Anal Biochem, 1976, 70, 32–38.
[21]  Späth PJ, Koblet H. Anal Biochem, 1979, 93, 275–285.

[22] Abramovitz AS, Randolph V, Mehra A, Christakos S. Prep Biochem, 1984, 74, 205–221.
[23] Ahmadi B. Anal Biochem, 1979, 97, 229–231.
[24] Paszkiewicz-Gadek A, Gindzienski A, Porowska H. Anal Biochem, 1995, 226, 263–267.
[25] Seelert H, Krause F. Electrophoresis, 2008, 29, 2617–2636.
[26] McDonald C, Fawell S, Pappin D, Higgins S. Trends Genet, 1986, 2, 35.
[27] Ziola BR, Scraba DG. Anal Biochem, 1976, 72, 366–371.
[28] Jacobs E, Clad A. Anal Biochem, 1986, 754, 583–589.
[29] Tuszynski GR, Damsky CH, Führer JR, Warren L. Anal Biochem, 1977, 83, 119–129.
[30] Allington WB, Cordry AL, McCullough GA, Mitchell DE, Nelson JW. Anal Biochem, 1978, 85, 188–196.
[31] Hunkapiller MW, Lujan E, Ostrander F, Hood LE. Methods Enzymol, 1983, 91, 227–236.
[32] Otto M, Snejderkova M. Anal Biochem, 1981, 777, 111–114.
[33] Stralfors R, Beifrage R. Anal Biochem, 1983, 128, 7–10.
[34] Knoetzel J, Braumann T, Grimme LH. J Photochem Photobiol B Biol, 1988, 7, 475–491.
[35] Baumann G, Chrambach A. Proc Natl Acad Sci USA, 1976, 73, 732–736.
[36] Wachslicht H, Chrambach A. Anal Biochem, 1978, 84, 533–538.
[37] Nguyen NY, DiFonzo J, Chrambach A. Anal Biochem, 1980, 706, 78–91.
[38] Mendel-Hartvig I. Anal Biochem, 1982, 727, 215–217.
[39] Radola BJ. Biochim Biophys Acta, 1975, 386, 181–195.
[40] Frey MD, Radola BJ. Electrophoresis, 1982, 3, 216–226.
[41] Radola BJ. Methods Enzymol, 1984, 104, 256–275.
[42] Stastna M, Slais K. Electrophoresis, 2010, 31, 433–439.
[43] Ek K, Bjellquist B, Righetti PG. Biochim Biophys Methods, 1983, 8, 135–155.
[44] Gelfi C, Righetti PG. J Biochim Biophys Methods, 1983, 8, 157–172.
[45] Righetti PG. Electrophoresis, 2006, 27, 923–938.
[46] Bier M, Egen NB, Allgyer TT, Twitty GE, Mosher RA. Gross E, Meienhofer J (eds.). Peptides. Structure and Biological Function. Pierce Chemical Co., Rockford, Illinois, 1979, 79–89.
[47] Bier M. Electrophoresis, 1998, 79, 1057–1063.
[48] Seelert H, Krause F. Electrophoresis, 2008, 29, 2617–2636.
[49] Garfin DE. Expert Rev Proteomics, 2009, 6, 239–241.
[50] Fitri N, Kastenholz B, Buchari B, Amran MB, Warganegara FM. Anal Letters, 2008, 41, 1773–1784.
[51] Robinson NJ, Winge DR. Annu Rev Biochem, 2010, 79, 537–562.

# 2.13 Microchip electrophoresis of proteins

2.13.1    Microchip materials —— 254
2.13.1.1  PDMS —— 254
2.13.1.2  PMMA —— 255
2.13.1.3  PC —— 255
2.13.2    Microchip fabrication —— 256
2.13.2.1  Fabrication of channel plate —— 257
2.13.2.2  Fabrication of cover plate —— 258
2.13.2.3  Wall coating —— 258
          Dynamic wall coating —— 258
          Permanent wall coating —— 259
2.13.2.4  Bonding the plates —— 259
2.13.3    Zone electrophoresis on microchip —— 261
2.13.3.1  Free-flow electrophoresis on microchip —— 261
2.13.3.2  Affinity- and immunoelectrophoresis on microchip —— 262
2.13.4    Isotachophoresis on microchip —— 262
2.13.4.1  Disc-electrophoresis on microchip —— 262
2.13.5    Isoelectric focusing on microchip —— 262
2.13.6    Two-dimensional electrophoresis on microchip —— 263
2.13.7    Protein separation technique on microchip —— 263
2.13.7.1  Concentrating the protein samples prior to microchip electrophoresis —— 263
2.13.7.2  Running protein electrophoresis —— 264
2.13.7.3  Detecting the proteins —— 264
          Staining the proteins —— 264
          Fluorescence detection —— 264
          Chemiluminescence detection —— 265
          Mass spectrometry detection —— 265
2.13.8    Microchips in clinical diagnostics —— 265
          References —— 266

The **microfluidic** chip (microchip) **electrophoresis** (ME) resembles a miniaturized capillary electrophoresis in a planar minidevice. Since the pioneering works of Harrison *et al.* [1, 2], the microchips are of considerable interest, owing to their portability, small sample consumption, and high speed of electrophoresis [3–5]. They are used in clinical and forensic diagnostics [6], proteomics [7], pharmaceutical analysis, environmental monitoring, *etc.*

A microchip contains narrow channels with diameters of about 50 μm, volumes of 10–50 μL, and size of 1–10 cm. They are filled with buffers, are arranged in the form of a cross, and are connected with four reservoirs (Figure 2.13.1). The short channel is dedicated to sample injection, whereas the long one represents the separating channel [8].

The injection of the sample into the microchip is achieved by the electrokinetic mode: First, the sample is pipetted into the sample load reservoir; then voltage is applied between the sample load reservoir and sample waste reservoir. When the

https://doi.org/10.1515/9783110761641-023

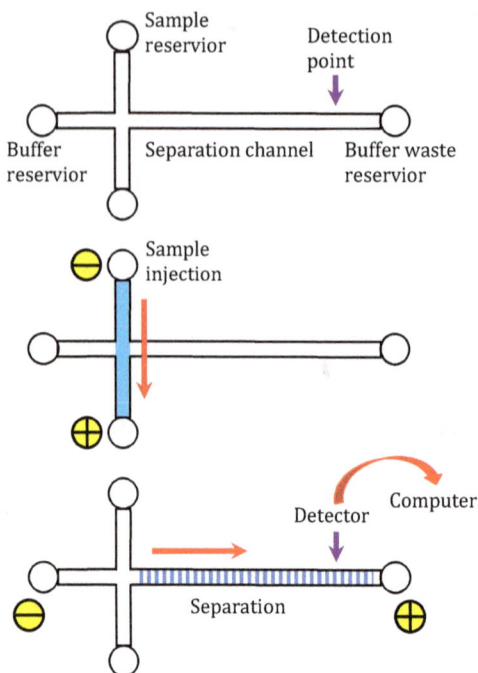

**Figure 2.13.1:** Events in microchip zone electrophoresis.

electrophoresis is run, the analytes resolve as a result of their mobilities and the electroosmotic flow, driven by the surface charges of the channel.

## 2.13.1 Microchip materials

The first microchips were fabricated from glass [9–11] or quartz [12–14] using photolithographic techniques [15]. However, their application was limited because of high cost and complicated procedures [16]. Nowadays, the microchips are made of rigid polymers [17–19]. They can be fabricated using *in situ* polymerization [20], laser ablation [21], imprinting [22], injection molding [23, 24], hot embossing [25–27], *etc.* The most popular polymers for preparing microchips are PDMS (**polydim**ethylsilox-ane), PMMA (**poly(m**ethyl **m**ethacrylate)), and PC (**p**oly**c**arbonates).

### 2.13.1.1 PDMS

PDMS belongs to the group of polymeric organosilicon compounds that are referred to as silicones [28, 29]. PDMS has the chemical formula $CH_3[Si(CH_3)_2O]_nSi(CH_3)_3$, where $n$ is the number of repeating monomer units [30].

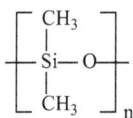

$$\left[\begin{array}{c} CH_3 \\ | \\ -Si-O- \\ | \\ CH_3 \end{array}\right]_n$$

PDMS monomer

PDMS has many advantages over other polymer materials: it is nontoxic, inert, optically transparent, flexible, and nonflammable [31–33]. However, PDMS has also some disadvantages. The most prominent among them is its hydrophobicity [34, 35]. This hinders the introduction of aqueous solutions into the microchannels. To decrease the hydrophobicity of PDMS, oxygen plasma [36], or UV/ozone [37] treatment can be used. During oxygen plasma treatment, surface silanol groups (–SiOH) are added to the siloxane backbone, which makes the polymer hydrophilic and transparent down to 280 nm [38].

## 2.13.1.2 PMMA

PMMA, also known as acrylic glass as well as Plexiglas, Crylux, Acrylite, Lucite, and Perspex, was developed in 1928 in different laboratories by William Chalmers, Otto Röhm, Walter Bauer, and more. It is a transparent thermoplastic often used as an alternative to glass [39–41]. Chemically, PMMA is a synthetic polymer of methyl methacrylate.

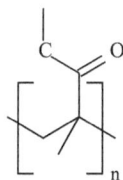

PMMA monomer

PMMA is an ideal polymer for the fabrication of microfluidic chips because of low price, optical transparency, and good mechanical properties [42, 43]. Its surface can be modified by aminolysis, reduction, and photoactivation (Figure 2.13.2) [44].

## 2.13.1.3 PC

PC are a group of thermoplastic polymers containing carbonate groups [45–47]. They are strong, tough, robust, molded, and thermoformed materials. PC are pliable and resistant to chemicals. The precursor monomer of PC is **bisp**henol **A** (BPA).

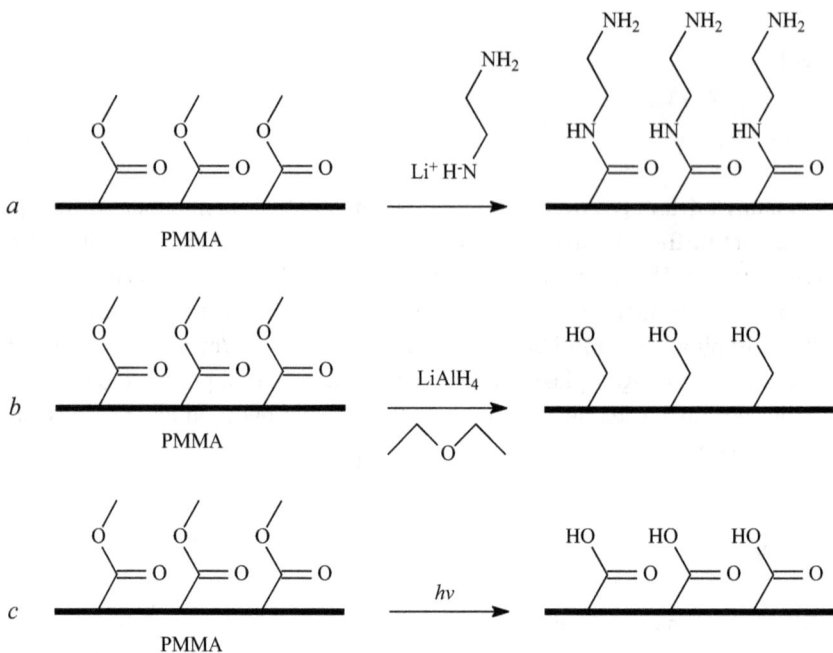

Figure 2.13.2: Modification of PMMA surface using aminolysis (*a*), reduction (*b*), and photoactivation (*c*).

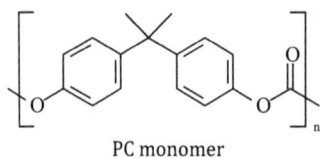

PC monomer

Other polymers used for preparation of microchips are **poly**ethylene **t**erephthalate **g**lycol (PETG), **poly**ethylene **t**erephthalate (PET) [48, 49], **poly**styrene (PS) [50–52], Mylar [53], **poly**imide (PI) [54, 55], **c**yclic **o**lefin **c**opolymer (COC), and more.

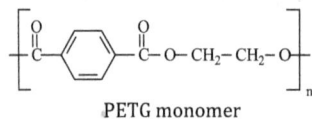

PETG monomer

## 2.13.2 Microchip fabrication

The microchip fabrication includes fabrication of channel and cover plates, wall coating, and bonding the plates.

## 2.13.2.1 Fabrication of channel plate

The channel plate can be fabricated by lithography, hot embossing, room-temperature imprinting, injection molding, laser ablation, *in situ* polymerization, solvent etching, *etc.*

The technology of soft lithography is used for fabrication of PDMS microchips [56]: A computer-aided design (CAD) program is employed for the design of microchips. The CAD-generated patterns are printed on a transparency that is used as a photomask in UV photolithography to generate a master. The resulting relief structure serves as a master for fabricating PDMS molds into which a liquid PDMS prepolymer is poured.

The *in situ* polymerization of methyl methacrylate takes place in molds with the aids of UV light [57] and heat [58]. To define the dimension of the PMMA channel plates, a rigid rectangle-shaped frame with a rectangular cavity is sandwiched between glass plate and a silicon or stainless-steel template to form a mold. Prepolymerized methyl methacrylate molding solution is injected into the rigid mold. The solution in the mold is exposed to UV light (365 nm lamp) within 4–6 h or heated in a water bath within 11–12 h. Finally, the formed channel is removed from the mold by sonicating in water bath (Figure 2.13.3).

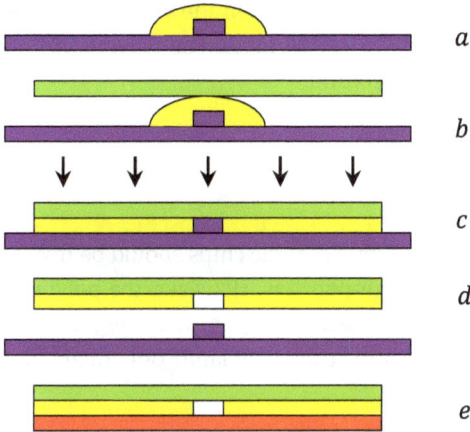

**Figure 2.13.3:** Scheme of microfabrication process.
(*a*) Casting methyl methacrylate prepolymer molding solution on a PDMS template; (*b*) sandwiching between PMMA plate and PDMS template; (*c*) exposing the sandwich mold to UV light; (*d*) demolding the PMMA channel plate; (*e*) covering and thermal sealing of PMMA channel plate to form a complete microchip.

Chen *et al.* [59] have developed a simple method for fabricating fiber-based PMMA microchip: Methyl methacrylate molding solution containing UV initiator was sandwiched between a PMMA cover plate and a PMMA base plate bearing glycerol-permeated fiberglass bundles and exposed to UV light. When the glycerol in the fiberglass bundles was flushed away with water, the obtained porous fiberglass-packed channels could be employed to perform electrophoresis separation. So the fiber-based microchips were fabricated without the expensive lithography-based techniques.

### 2.13.2.2 Fabrication of cover plate

The cover plate of PMMA microdevices can be fabricated by injection molding [60, 61]: The PMMA pellets are melted and injected under pressure into a mold cavity containing an inserted master. The masters are employed to define the structure of the microchips. Finally, the cavity is cooled to allow the channel plate to demold from the mold.

Laser ablation was also employed to fabricate PMMA microdevices using software [62, 63]. In this process, a beam of high-energy laser breaks bonds in polymer molecules. The PMMA plate is mounted on a table that moves in the $X$ and $Y$ directions while the focused laser beam scans.

Rapid fabrication of PMMA microfluidic chips has been made based on the *in situ* surface polymerization of methyl methacrylate [64]. Methyl methacrylate containing 2,2′-azo-bis-isobutyronitrile was allowed to prepolymerize in a water bath to form a viscous solution that was mixed with methyl methacrylate containing a redox initiation couple of benzoyl peroxide and *N,N*-dimethyl-aniline. Then the viscous solution was sandwiched between a silicon template and a PMMA plate. The polymerization was complete within 50 min under ambient temperature.

### 2.13.2.3 Wall coating

For separation of proteins, the channel surface of microfluidic chips should be hydrophilic to limit the protein adsorption during electrophoresis. However, the polymers used for fabrication of microchips are hydrophobic. Therefore, they ought to be coated with hydrophilic substances. The wall coating can be dynamic or permanent.

#### Dynamic wall coating

In dynamic wall coating, hydrophilic substances in the buffer coat the wall of the microchip channels during electrophoresis. Diverse methods for dynamic wall coating are known: for glass microchips, PMMA microchips, and more.

For glass microchips, PDMA is used simultaneously as a wall coating material and a sieving matrix. Its presence reduces the electroosmosis to $0.5 \times 10^{-9}$ m$^2$/(s V) [65]. Dynamic coatings of the channel surface of PMMA microchips were made by dissolving positively or negatively charged surfactants, or hydrophilic neutral polymers in the electrophoresis buffers [66, 67]. A dynamic coating process using 2 g/dL **hydroxyethyl** cellulose (HEC) for the surface modification of PMMA microfluidic chips was developed [68]. Mohamadi *et al.* [69] have described a dynamic coating method using methylcellulose and the nonionic detergent Polysorbate 20, which suppresses protein adsorption onto PMMA channel wall.

**Permanent wall coating**

Permanent wall coating happens when hydrophilic substances coat the wall of the microchip channels. Diverse methods for permanent wall coating are used for glass microchips, PMMA microchips, and other microchips.

The classical wall coating of glass microchips proposed by Hjerten was applied with some modifications [70–72]. After flushing channels with NaOH, they are filled with γ-methacryloxypropyltrimethoxysilane in diluted acetic acid and acetonitrile for 1 h. Then an aqueous solution of acrylamide with ammonium persulfate and TMEDA is pumped into the channel and polymerized at room temperature. Finally, the channels are flushed with water and dried by vacuum.

One of the earliest methods for modification of the PMMA surface was proposed by Wichterle [73]. His principle consists in reesterification of PMMA with polyfunctional hydroxyl compounds, such as ethylene glycol, glycerol, mannitol, and saccharose. First, the PMMA surface was covered with ethylene glycol or another polyalcohol; then it was treated with hot sulfuric acid to perform reesterification and replace methanol rests in PMMA with ethylene glycol rests. After neutralization with sodium hydrogen carbonate, a hydrophilic transparent layer is formed on the surface of PMMA. Later, the method was modified [74]: The polymer was hydrolyzed first with hot sulfuric acid containing sodium or potassium hydrogen sulfate and then the hydrolyzed PMMA was esterified with glycerol.

A covalent hydrophilization of channels was introduced by Soper *et al.* [75, 76] who imparted amine or octadecyl groups to the surface of PMMA microchannel *via* aminolysis and reactions with *n*-octadecylisocyanate. Using atom-transfer radical polymerization, **p**oly**e**thylene **g**lycol was grafted onto the surface of PMMA channels, which also reduced the electroosmotic flow and the adsorption of proteins on the PMMA surface [77]. Kitagawa *et al.* [78] have developed a one-step covalent immobilization of **p**oly(**e**thyleneimine) (PEI) onto PMMA substrates to achieve an efficient separation of basic proteins in microchip electrophoresis.

## 2.13.2.4 Bonding the plates

The microchannels have to be closed without changing their physical parameters, or altering their dimension. A variety of bonding techniques have been developed: thermal, solvent, polymerization, microwave bonding, and room-temperature imprinting.

In the *thermal bonding* the channel and cover plates are assembled and heated to 105 °C in a convection oven and pressed together using a bonding device [79]. However, the pressure and temperature may cause microchannel deformation. In an attempt to overcome these limitations, hot-press bonding conducted in a vacuum [80] or in a boiling water bath [81] was developed.

*Solvent bonding* has also been reported for PMMA microchips [82–84]. Klank *et al.* [85] employed a plasma-enhanced ethanol bonding for such purpose. Lin *et al.* [86] sealed PMMA microfluidic chips utilizing a solvent composed of ethanol and 1,2-dichloroethane at room temperature.

The *polymerization bonding* is another method for bonding PMMA microchips [87]: Methyl methacrylate containing initiators was allowed to prepolymerize in an 85 °C water bath for 8 min to produce a bonding solution. Prior to bonding, the cover plate was coated with a thin layer of the bonding solution and was bonded to the channel plate at 95 °C for 20 min. Finally, the bonding device was put in a 95 °C convection oven when the monomer methyl methacrylate was polymerized to realize the bonding.

*Microwave bonding* of PMMA microchips at low temperature was proposed by Lei *et al.* [88]. Later, Yussuf *et al.* [89] have reported bonding of PMMA microfluidic chips using microwave energy and conductive polyaniline.

Xu *et al.* [90] have fabricated PMMA microchips by *room-temperature imprinting*: The PMMA plate was placed on a silicon template, and the assembly was inserted between two polished aluminum plates and was hydraulic pressed at room temperature. After the pressure was released, the open channels on the plastic plate were sealed with a PDMS film. Later, Woolley *et al.* [91] developed a method for rapid prototyping of PMMA microfluidic chips using SU-8 photoresist: SU-8 is an epoxy-based negative photoresist. It was employed as a template for solvent imprinting on a glass slide while pressed into a solvent-wetted surface.

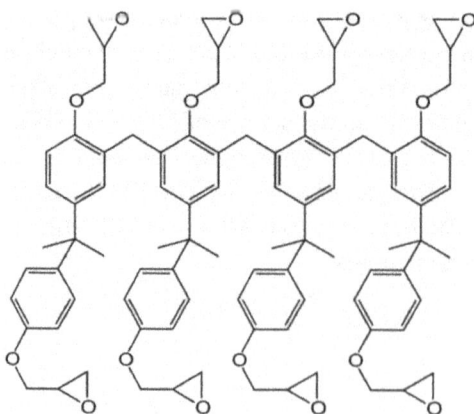

SU-8

## 2.13.3 Zone electrophoresis on microchip

The principle of the microchip zone electrophoresis is same as the principle of conventional electrophoresis: the separations of proteins are based on differences in their electrophoretic mobility, which generally correlate to their charge-to-mass ratio.

### 2.13.3.1 Free-flow electrophoresis on microchip

The first device for free-flow zone electrophoresis on microchips was developed by Raymond *et al.* [92, 93]. Using this device, the authors separated three rhodamine-B isothiocyanate-labeled amino acids: lysine, glutamine, and glutamic acid, and a mixture of human serum albumin, bradykinin, and ribonuclease A.

Later, Fonslow and Bowser [94] presented a free-flow electrophoresis (FFE) microchip fabricated from two glass plates. This device contained closed electrode side channels, where gas bubbles were flushed out by a pressure-driven flow passing the integrated gold electrodes. The channels were separated by connecting side channel arrays.

Next, a device for FFE on glass microchip using laser-printing toner as a structural material was proposed [95]; The separation channel is 8 μm deep and has an internal volume of 1.42 μL. The Joule heating dissipation was found to be very efficient up to an electric current density of 8.83 mA/mm$^2$ that corresponds to power dissipation per unit volume of running electrolyte of 172 mW/μL. The electrophoresis was run at a maximum voltage and electric current of 500 V and 100 mA, respectively. The pumping of the running buffer and the sample solutions through the separation channel was performed by connecting the outlet reservoir of the microchip to the air entrance of a small air compressor working as a vacuum pump (Figure 2.13.4).

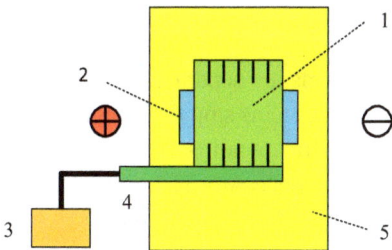

**Figure 2.13.4:** Diagram of free-flow electrophoresis on microchip.
1. Separation channel; 2. side reservoirs; 3. air compressor; 4. outlet reservoir; 5. glass plate.

### 2.13.3.2 Affinity- and immunoelectrophoresis on microchip

Affinity- and immunoelectrophoresis on microchips take place in affinity, most often in immune interaction [96]. Immunoelectrophoresis on microchips has allowed the fast quantitation of inflammatory cytokines in the cerebrospinal fluid of patients presenting a head trauma [97]. In this procedure, the reservoir was silanized to exhibit the amino terminal group and so to obtain an affinity area between the Fab fragments and fluorescently tagged analytes. Then a low-pH buffer was added to dissociate the antibody–antigen complexes and the electrophoresis was run.

## 2.13.4 Isotachophoresis on microchip

During isotachophoresis (ITP) the samples are concentrated between a leading ion and a trailing ion. The leading ion is chosen to have a higher mobility than any ions (of the same polarity) in the sample. The trailing ion is chosen to have a lower mobility than any ions (of the same polarity) in the sample. After applying an electric field, equilibrium is reached, in which all ions have the same velocity. The concentration of ions in the sample arises and is a consequence of the concentration of leading ion.

### 2.13.4.1 Disc-electrophoresis on microchip

Lin *et al.* [98] have carried out disc-electrophoresis on microchips integrating ITP preconcentration with zone gel electrophoresis. Each microchip was designed with a sample injection channel to allow stacking the sample into a narrow zone. Thus, they separated carbonic anhydrase, ovalbumin, BSA, and conalbumin.

## 2.13.5 Isoelectric focusing on microchip

In isoelectric focusing on microchip, the charged proteins are separated according to their pI in a pH gradient generated by ampholytes. The proteins stop their migration in the separation channel once they reach the pH zone corresponding to their pI. Next, mobilization of the focused sample zones [99, 100] is necessary to perform an in-channel detection of proteins. To avoid mobilization step, whole-column imaging (WCI) method may be employed. WCI detection is carried out using an epifluorescence microscope with a xenon lamp [101], a mercury lamp [102, 103], or LED [104–106].

## 2.13.6 Two-dimensional electrophoresis on microchip

Li *et al.* [107] combined IEF with SDS electrophoresis on a plastic planar microchip with dimensions of $2 \times 3$ cm. Thus, they carried out two-dimensional (2D) protein separation in less than 10 min. Later, a miniaturized instrument capable of performing 2D electrophoresis was described [108]. It consists of a compartment for the first-dimensional IEF polyacrylamide gel, which is connected to a second-dimensional SDS polyacrylamide gel. The focused samples are automatically transferred from the IEF gel to the SDS polyacrylamide gel by electromigration. Separated protein spots can be excised from stained gels, digested with trypsin, and identified by mass spectrometry.

Shadpour and Soper [109] carried out 2D electrophoretic separation of proteins using an embossed PMMA microchip. The separation in the first dimension was based on SDS electrophoresis, and the second dimension was carried out using a fast **micellar electrokinetic chromatography** (MEKC), which sorted the proteins depending on the differences in their interaction with SDS. Also, microfluidic 2D separation of proteins was described, which combined temperature gradient focusing with SDS polyacrylamide gel electrophoresis [110].

## 2.13.7 Protein separation technique on microchip

The protein electrophoresis on microchips, similar to the common gel electrophoresis, was carried out in three steps: concentrating the samples, running electrophoresis, and detecting separated proteins.

### 2.13.7.1 Concentrating the protein samples prior to microchip electrophoresis

The samples for microchip electrophoresis can be concentrated by filtering [111]: A 355-nm laser beam was used to excite a photoinitiator in a monomer/solvent solution, leading to polymerization and phase separation of a thin membrane that was covalently bonded to the acrylate-functionalized silica channel surface. The irradiated region was defined by shaping the adjustable slit with cylindrical and spherical optics. The resulting membranes were 50 μm thick. Upon application of voltage, linear electrophoretic concentration of charged proteins is achieved at the membrane surface because buffer ions can pass through the membrane while proteins larger than the molecular mass cutoff of the membrane are retained.

Wang *et al.* [112] have developed a microfluidic sample concentration device based on the electrokinetic trapping mechanism enabled by nanofluidic filters. Flat nanofluidic channels filled with buffer solution are used as an ion-selective membrane to generate an ion-depletion region for electrokinetic trapping. The field in the nanofluidic channel is used to generate the ion-depletion region and to extend the charge layer that traps the biomolecules.

### 2.13.7.2 Running protein electrophoresis

Various injection procedures are proposed, such as pinched or gating modes [113]. They provide alternative routes to prevent from sample zone dispersion by tuning the electric potential value in the reservoirs. The double T intersection design is often preferred, because it improves the control over the injected sample volume.

### 2.13.7.3 Detecting the proteins

The detection of resolved proteins can be done by staining, fluorescence, chemiluminescence (CL), and mass spectrometry.

#### Staining the proteins

The binding between dyes and proteins can be covalent or non-covalent.

*Covalent binding of proteins* can be realized between protein molecules and amine or other groups (carboxylic or thiol groups) of the dyes. Gottschlich *et al.* [114] have developed a microchip where naphthalene-2,3-dicarboxaldehyde was used for labeling of the insulin chains A and B, produced after disulfide reducing.

*Non-covalent binding of proteins* is performed with fluorogenic reagents. The labeling reagent should have low fluorescence in the unbound state and a high fluorescence enhancement when bound to protein. Bousse *et al.* [115] have developed a glass microchip for non-covalent fluorescent labeling method. Denatured SDS–protein complexes bound the fluorescent dye when the separation begins. The SYPRO Orange and Agilent dye exhibited a fluorescent enhancement upon binding to SDS micelles or SDS–protein complexes. At the end of the separation channel, in front of the detection point, an intersection was used to dilute SDS below its critical micelle concentration.

#### Fluorescence detection

The fluorescence detection is the most widely used method for protein detection in microchip electrophoresis due to its high sensitivity. The laser-induced fluorescence (LIF) detection is the important method for protein detection after electrophoresis on microchips [116, 117]. According to this, the laser is focused on very small protein volumes to obtain high irradiation. Lamp-based fluorescence detection methods, following LIF detection, are used for optical detection of separated proteins. They use microscope-based detector setups using xenon or mercury lamps. For this purpose, epifluorescence microscopes were combined with **photomultiplier tube** (PMT) detection [118].

Vieillard *et al.* [119] have reported a method combining microfluidic chip with integrated optics. A channel with optical waveguides was prepared on a soda-lime glass plate, on which the microfluidic system was fabricated. Using this technique, the separation and detection of β-lactoglobulin A and carbonic anhydrase II needed less than 1 min.

### Chemiluminescence detection

Another detection method used in microfluidic electrophoresis is the **chemiluminescence** (CL) detection. Ren *et al.* [120] have described a new strategy to afford on-line CL detection of heme proteins on microchips. The detection principle is based on the catalytic effect of heme proteins on the CL reaction of luminol–$H_2O_2$ enhanced by 4-iodophenol. Cytochrome *c*, myoglobin, and horse radish peroxidase were separated within 10 min on a glass microchip.

### Mass spectrometry detection

The possibility to integrate a microfluidic chip separation with **mass** spectroscopy (MS) detection has been studied by many researchers [121, 122]. A PMMA microfluidic-based gel protein recovery system, in which fluid transports protein from a PAGE gel piece to a collection reservoir *via* a microfluidic channel, has been developed [123]. The sample proteins were mobilized out of the gel into a microfluidic channel by electric field. Afterward, the proteins were detected using **matrix-assisted laser desorption/ionization** MS (MALDI MS) over gel loads of 0.1–10 pg.

## 2.13.8 Microchips in clinical diagnostics

A few cases of microfluidic analysis of proteins from blood, urine, cerebrospinal fluid, saliva, tears, and more were described. For example, four human serum proteins (IgG, transferrin, $\alpha_1$-antitrypsin, and albumin) were resolved on microchip, in less than 60 s [124].

Giordano *et al.* [125] have developed a method for dynamic labeling of proteins with NanoOrange in microchip SDS zone electrophoresis of serum albumin. Later, the group of Chan [126] has proposed a method for quantifying urinary albumin, a marker of microalbuminuria, which is a risk factor of cardiovascular diseases.

MALDI-based techniques were successfully used with the goal of identifying proteins in connection with HER-2/neu-positive, aggressive-type breast cancer [127]. Tumor cells were collected by laser-captured microdissection (LCM) from HER-2/neu-positive and -negative tumors.

The protein microarray system could simultaneously determine the concentration of two viral antigens (HBsAg and HBeAg) and seven antiviral protein antibodies (HBsAb, HBcAb, HBeAb, HCVAb, HDVAb, HEVAb, and HGVAb) of human hepatitis viruses in human sera within 20 min. The results were confirmed by ELISA [128].

## References

[1] Harrison DJ, Manz A, Fan Z, Ludi H, Widmer HM. Anal Chem, 1992, 64, 1926–1932.
[2] Harrison DJ, Flury K, Seiler K, Fan Z et al. Science, 1993, 261, 895–897.
[3] Woolley AT, Mathies RA. Proc Natl Acad Sci USA, 1994, 91, 11348–11352.
[4] Wang J, Chen G, Chatrathi MR, Musameh M. Anal Chem, 2004, 76, 298–302.
[5] Xu GX, Wang J, Chen V, Zhang LY et al. Lab Chip, 2006, 6, 145–148.
[6] Verpoorte E. Electrophoresis, 2002, 23, 677–712.
[7] Lion N, Rohner TC, Dayon L, Arnaud IL et al. Electrophoresis, 2003, 24, 3533–3562.
[8] Guber AE, Heckele M, Herrmann D, Muslija A et al. Chem Eng J, 2004, 101, 447–453.
[9] Flarrison DJ, Manz A, Fan Z, Lüdi H, Widmer HM. Anal Chem, 1992, 64, 1926–1932.
[10] Shinohara E, Tajima N, Suzuki H, Funazaki J. Anal Sci Suppl, 2001, 17, i441–i443.
[11] Fonslow BR, Bowser MT. Anal Chem, 2005, 77, 5706–5710.
[12] Jacobson SC, Moore AW, Ramsey JM. Anal Chem, 1995, 67, 2059–2063.
[13] He B, Tait N, Regnier F. Anal Chem, 1998, 70, 3790–3797.
[14] Tokeshi M, Minagawa T, Kitamori T. Anal Chem, 2000, 72, 1711–1714.
[15] Dolnik V, Liu SR, Jovanovich S. Electrophoresis, 2000, 21, 41–54.
[16] Becker H, Gartner C. Electrophoresis, 2000, 21, 12–26.
[17] Osiri JK, Shadpour H, Park S, Snowden BC, Chen ZY, Soper SA. Electrophoresis, 2008, 29, 4984–4992.
[18] Fuentes HV, Woolley AT. Anal Chem, 2008, 80, 333–339.
[19] Marchiarullo DJ, Lim JY, Vaksman Z, Ferrance JP, Putcha L, Landers JP. J Chromatogr A, 2008, 1200, 198–203.
[20] Chen ZF, Gao YH, Su RG, Li CW, Lin JM. Electrophoresis, 2003, 24, 3246–3252.
[21] Roberts MA, Rossier JS, Bercier P, Girault HH. Anal Chem, 1997, 69, 2035–2042.
[22] Martynova L, Locascio LE, Gaitan M, Kramer GW et al. Anal Chem, 1997, 69, 4783–4789.
[23] McCormick RM, Nelson RJ, AlonsoAmigo MG, Benvegnu J, Hooper HH. Anal Chem, 1997, 69, 2626–2630.
[24] Martynova L, Locascio LE, Gaitan M, Kramer GW et al. Anal Chem, 1997, 69, 4783–4789.
[25] McCormick RM, Nelson RJ, Alonso-Amigo MG, Benvegnu DJ, Hooper HH. Anal Chem, 1997, 69, 2626–2630.
[26] Verpoorte E. Electrophoresis, 2002, 23, 677–712.
[27] Becker H, Locascio LE. Talanta, 2002, 56, 267–287.
[28] Raymond D, Manz A, Widmer HM. Anal Chem, 1994, 66, 2858–2865.
[29] Raymond D, Manz A, Widmer HM. Anal Chem, 1996, 68, 2515–2522.
[30] Mark JE, Allcock HR, West R. Inorganic Polymers. Prentice Hall, Englewood, NJ, 1992.
[31] Effenhauser CS, Bruin GJM, Paulus A, Ehrat M. Anal Chem, 1997, 69, 3451–3457.

[32] Albrecht JW, El-Ali J, Jensen KF. Anal Chem, 2007, 79, 9364–9371.
[33] Ng JM, Gitlin I, Stroock AD, Whitesides GM. Electrophoresis, 2002, 23, 3461–3473.
[34] Duffy DC, McDonald JC, Schueller OJA, Whitesides GM. Anal Chem, 1998, 70, 4974–4984.
[35] McDonald JC, Duffy DC, Anderson JR, Chiu DT, Wu H, Schueller OJA, Whitesides GM. Electrophoresis, 2000, 21, 27–40.
[36] Duffy DC, McDonald JC, Schueller OJA, Whitesides GM. Anal Chem, 1998, 70, 4974–4984.
[37] Xiao D, Le TV, Wirth MJ. Anal Chem, 2004, 76, 2055–2061.
[38] Makamba H, Kim JH, Lim K, Park N, Hahn JH. Electrophoresis, 2003, 24, 3607–3619.
[39] Chen YH, Chen SH. Electrophoresis, 2000, 27, 165–170.
[40] Martynova L, Locascio LE, Gaitan M, Kramer GW et al. Anal Chem, 1997, 69, 4783–4789.
[41] Liu Y, Ganser D, Schneider A, Liu R et al. Anal Chem, 2001, 73, 4196–4201.
[42] Soper SA, Ford SM, Qi S, McCarley RL et al. Anal Chem, 2000, 72, 643A–651A.
[43] Grass B, Neyer A, Jöhnck M, Siepe D et al. Sens Actuators B, 2001, 72, 249–258.
[44] Liu J, Lee ML. Electrophoresis, 2006, 27, 3533–3546.
[45] Henry AC, Tutt TJ, Galloway M, Davidson YY et al. Anal Chem, 2000, 72, 5331–5337.
[46] Liu Y, Ganser D, Schneider A, Liu R et al. Anal Chem, 2001, 73, 4196–4201.
[47] Vreeland WN, Locascio LE. Anal Chem, 2003, 75, 6906–6911.
[48] Malmstadt N, Yager R, Hoffman AS, Stayton PS. Anal Chem, 2003, 75, 2943–2949.
[49] Munce NR, Li J, Herman PR, Lüge L. Anal Chem, 2004, 76, 4983–4989.
[50] Barker SLR, Ross D, Tarlov MJ, Gaitan M, Locascio LE. Anal Chem, 2000, 72, 5925–5929.
[51] Stachowiak TB, Rohr T, Hilder EF, Peterson DS et al. Electrophoresis, 2003, 24, 3689–3693.
[52] Yang Y, Li C, Lee KH, Craighead HG. Electrophoresis, 2005, 26, 3622–3630.
[53] Macounova K, Cabrera CR, Holl MR, Yager P. Anal Chem, 2000, 72, 3745–3751.
[54] Yin H, Killeen K, Brennen R, Sobek D et al. Anal Chem, 2004, 77, 527–533.
[55] Gobry V, Van Oostrum J, Martinelli M, Rohner TC et al. Proteomics, 2002, 2, 405–412.
[56] Mcdonald JC, Whitesides GM. Acc Chem Res, 2002, 35, 491–199.
[57] Muck A, Wang J, Jacobs M, Chen G et al. Anal Chem, 2004, 76, 2290–2297.
[58] Chen ZF, Gao YH, Su RG et al. Electrophoresis, 2003, 24, 3246–3252.
[59] Chen Z, Zhang LY, Chen G. Electrophoresis, 2007, 28, 2466–2473.
[60] McCormick RM, Nelson RJ, AlonsoAmigo MG, Benvegnu J, Hooper HH. Anal Chem, 1997, 69, 2626–2630.
[61] Piotter V, Hanemann T, Ruprecht R, Hausselt J. Microsys Technol, 1997, 3, 129–138.
[62] Chenga JY, Wei CW, Hsua KH, Young TH. Sens Actuators B, 2004, 99, 186–196.
[63] Yuan DJ, Das SJ. Appl Phys, 2007, 101, 24901.
[64] Chen J, Lin YH, Chen G. Electrophoresis, 2007, 28, 2897–2903.
[65] Bousse L, Mouradian S, Minalla A, Yee H, Williams K, Dubrow R. Anal Chem, 2001, 73, 1207–1212.
[66] Dang F, Zhang L, Hagiwara H, Mishina Y, Baba Y. Electrophoresis, 2003, 24, 714–721.
[67] Zhang Y, Ping GC, Zhu BM, Kaji N et al. Electrophoresis, 2007, 28, 414–421.
[68] Du XG, Fang ZL. Electrophoresis, 2005, 26, 4625–4631.
[69] Mohamadi MR, Mahmoudian L, Kaji N, Tokeshi M, Baba Y. Electrophoresis, 2007, 28, 830–836.
[70] Schmalzing D, Adourian A, Koutny L, Ziaugra L, Matsudaira R, Ehrlich D. Anal Chem, 1998, 70, 2303–2310.
[71] Salas-Solano O, Schmalzing D, Koutny L, Buonocore S, Adourian A, Matsudaira R, Ehrlich D. Anal Chem, 2000, 72, 3129–3137.
[72] Liu Y, Ganser D, Schneider A, Liu R, Grodzinski P, Kroutchinina N. Anal Chem, 2001, 73, 4196–4201.
[73] Wichterle O. *US Patent 3,895,169*, 1975.

[74]  Suie J, Krcova Z. *US Patent 5,080,683*, 1992.
[75]  Soper SA, Henry AC, Vaidya B, Galloway M *et al*. Anal Chim Acta, 2002, 470, 87–99.
[76]  Llopis SL, Osiri J, Soper SA. Electrophoresis, 2007, 28, 984–993.
[77]  Liu JK, Pan T, Woolley AT, Lee ML. Anal Chem, 2004, 76, 6948–6955.
[78]  Kitagawa F, Kubota K, Sueyoshi K, Otsuka K. Sci Technol Adv Mat, 2006, 7, 558–565.
[79]  Galloway M, Stryjewski W, Henry A, Ford SM *et al*. Anal Chem, 2002, 74, 2407–2415.
[80]  Chen ZF, Gao YH, Lin JM, Su RG, Xie Y. J Chromatogr A, 2004, 1038, 239–245.
[81]  Kelly RT, Woolley AT. Anal Chem, 2003, 75, 1941–1945.
[82]  Wang J, Pumera M, Chatrathi MR, Escarpa A *et al*. Electrophoresis, 2002, 23, 596–601.
[83]  Kelly RT, Pan T, Woolley AT. Anal Chem, 2005, 77, 3536–3541.
[84]  Peeni BA, Lee ML, Hawkins AR, Woolley AT. Electrophoresis, 2006, 27, 4888–4895.
[85]  Klank H, Kutter JP, Geschke O. Lab Chip, 2002, 2, 242–246.
[86]  Lin CH, Chao CH, Lan CW. Sens Actuators B, 2007, 121, 698–705.
[87]  Chen G, Li JH, Qu S, Chen D, Yang PY. J Chromatogr A, 2005, 1094, 138–147.
[88]  Lei KF, Ahsan S, Budraa NW, Li J, Mai JD. Sens Actuators A, 2004, 114, 340–346.
[89]  Yussuf AA, Sbarski L, Hayes JR, Solomon M, Tran N. J Micromech Microeng, 2005, 15, 1692–1699.
[90]  Xu JD, Locascio L, Gaitan M, Lee CS. Anal Chem, 2000, 72, 1930–1933.
[91]  Sun XH, Peeni BA, Yang WC, Becerril HA, Woolley AT. J Chromatogr A, 2007, 1162, 162–166.
[92]  Raymond DE, Manz A, Widmer HM. Anal Chem, 1994, 66, 2858–2865.
[93]  Raymond DE, Manz A, Widmer HM. Anal Chem, 1996, 68, 2515–2522.
[94]  Fonslow BR, Bowser MT. Anal Chem, 2005, 77, 5706–5710.
[95]  De Jesus DP, Blanes L, Do Lago CL. Electrophoresis, 2006, 27, 4935–4942.
[96]  Heegaard NHH, Kenedy RT. J Chromatogr A, 2002, 768, 93–103.
[97]  Phillips TM. Electrophoresis, 2004, 25, 1652–1659.
[98]  Huang H, Xu F, Dai ZR, Lin BC. Electrophoresis, 2005, 26, 2254–2260.
[99]  Tan W, Fan ZH, Qiu CX, Ricco AJ, Gibbons I. Electrophoresis, 2002, 23, 3638–3645.
[100] Yao B, Yang H, Luo G, Wang L *et al*. Anal Chem, 2006, 78, 5845–5850.
[101] Mao Q, Pawliszyn J. Analyst, 1999, 124, 637–641.
[102] Cui H, Horiuchi K, Dutta R, Ivory CF. Anal Chem, 2005, 77, 1303–1309.
[103] Cui H, Horiuchi K, Dutta P, Ivory CF. Anal Chem, 2005, 77, 7878–7886.
[104] Wu XZ, Sze NSK, Pawliszyn J. Electrophoresis, 2001, 22, 3968–3971.
[105] Yao B, Luo G, Feng X, Wang W *et al*. Lab Chip, 2004, 4, 603–607.
[106] Yao B, Luo G, Wang L, Gao Y *et al*. Lab Chip, 2005, 5, 1041–1047.
[107] Li Y, Buch JS, Rosenberger F, DeVoe DL, Lee CS. Anal Chem, 2004, 76, 742–748.
[108] Demianova Z, Shimmo M, Poysa E, Franssila S, Baumann M. Electrophoresis, 2007, 28, 422–428.
[109] Shadpour H, Soper SA. Anal Chem, 2006, 78, 3519–3527.
[110] Shameli SM, Ren CL. Anal Chem, 2015, 87, 3593–3597.
[111] Song S, Singh AK, Kirby BJ. Anal Chem, 2004, 76, 4589–4592.
[112] Wang YC, Stevens AL, Han J. Anal Chem, 2005, 77, 4293–4299.
[113] Heeren F, Verpoorte E, Manz A, Thormann W. Anal Chem, 1996, 68, 2044–2053.
[114] Gottschlich N, Culbertson CT, McKnight TE, Jacobson SC, Ramsey JM. J Chromatogr B, 2000, 745, 243–249.
[115] Bousse L, Mouradian S, Minalla A, Yee H *et al*. Anal Chem, 2001, 73, 1207–1212.
[116] Das C, Fan ZH. Electrophoresis, 2006, 27, 3619–3626.
[117] Liu X, Liang AY, Shen Z, Zhang Y *et al*. Electrophoresis, 2006, 27, 3125–3128.
[118] Chirica G, Lachmann J, Chan J. Anal Chem, 2006, 78, 5362–5368.
[119] Vieillard J, Mazurczyk R, Morin C, Hannes B *et al*. J Chromatogr B, 2007, 845, 218–225.

[120] Huang XY, Ren JC. Electrophoresis, 2005, 26, 3595–3601.

[121] Sung WC, Makamba H, Chen SH. Electrophoresis, 2005, 26, 1783–1791.

[122] Freire SLS, Wheeler AR. Lab Chip, 2006, 6, 1415–1423.

[123] Razunguzwa TT, Biddle A, Anderson H, Zhan D, Powell M. Electrophoresis, 2009, 30, 4020–4028.

[124] Colyer CL, Mangru SD, Harrison DJ. J Chromatogr A, 1997, 781, 271–276.

[125] Giordano B, Jin L, Couch AJ, Ferrance JP, Landers JP. Anal Chem, 2004, 76, 4705–4714.

[126] Chan OTM, Herold DA. Clin Chem, 2006, 52, 2141–2146.

[127] Zhang D, Tai LK, Wong LL, Chiu LL *et al.* Mol Cell Proteomics, 2005, 4, 1686–1696.

[128] Xu R, Gan X, Fang Y, Zheng S, Dong Q. Anal Biochem, 2007, 362, 69–75.

# 2.14 Blotting of proteins

2.14.1    Theory of protein blotting —— 271
2.14.2    Blot membranes —— 272
2.14.3    Transfer of proteins —— 272
2.14.3.1  Electrotransfer of proteins —— 272
          Tank blotting of proteins —— 273
          Semidry blotting of proteins —— 273
2.14.3.2  Capillary transfer of proteins —— 274
2.14.4    Blocking —— 275
2.14.5    Detection —— 276
2.14.5.1  Detection by dyes —— 276
2.14.5.2  Detection by probes —— 276
2.14.5.3  Immunoblotting —— 277
2.14.6    Making the blot membranes transparent —— 277
2.14.7    Blotting techniques —— 278
2.14.7.1  Western blotting —— 278
2.14.7.2  Far-Western blotting —— 279
2.14.7.3  Southwestern blotting —— 279
2.14.7.4  Northwestern blotting —— 279
2.14.7.5  Eastern blotting —— 280
2.14.8    Protocols —— 280
          Western Blotting —— 280
          Semidry Blotting —— 282
          Capillary Blotting —— 283
          India Ink Staining of Proteins on Membrane —— 284
          Detecting Proteins by Immunoblotting —— 285
          Immunoprobing with Avidin–biotin Coupling to Secondary Antibody —— 286
          References —— 287

Electrophoretically separated proteins can be visualized directly in the gel or indirectly after blotting onto a membrane. The proteins on a blot membrane are immobilized and are accessible to reagents or high-molecular-mass ligands such as lectins, antibodies, antigens, DNA, or RNA.

## 2.14.1 Theory of protein blotting

Although blotting can be carried out in an electric field, it is not an electrophoresis method because it does not lead to separation of charged particles.

Blotting is carried out in three steps: transfer of proteins onto a blot membrane, blocking the free binding sites on the blot membrane, and detection of the blotted proteins (Figure 2.14.1). The unoccupied binding sites on the blot membrane have to be blocked with substances, which do not participate in the detection reaction. The blotted proteins are detected with high-molecular-mass ligands (probes), or with unspecific dyes.

https://doi.org/10.1515/9783110761641-024

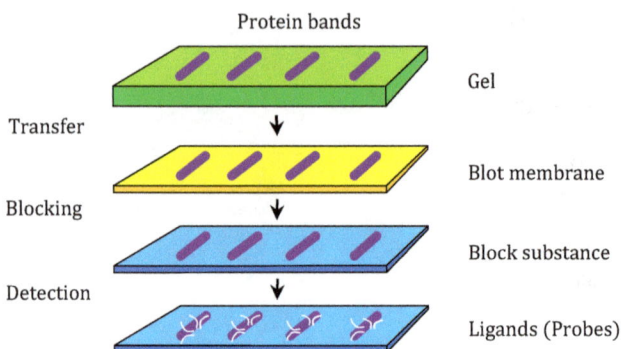

**Figure 2.14.1:** Steps of blotting.

## 2.14.2 Blot membranes

For blotting, special blot membranes are used: **nitro**cellulose (NC), **p**oly**v**inylidene **di**fluoride (PVDF) membrane, nylon, **d**iazo**b**enzyloxy**m**ethyl (DBM) cellulose, **d**iaz-**o**phenylthioether (DPT) cellulose, or activated glass fiber paper:

- The NC membrane is the most commonly used blot membrane [1, 2]. Typically, its pores are 0.2 or 0.45 µm wide. The smaller the pores, the greater the binding capacity of the blot membrane.
- PVDF membrane possesses high binding capacity and is mechanically stable [3, 4]. It allows easier stripping than the NC membrane.
- Nylon has good mechanical stability [5]. It binds the proteins electrostatically by hydrophobic interactions. This causes a strong background when staining the proteins.
- Diazobenzyloxymethyl paper [6] and diazophenylthioether paper [7] bind proteins electrostatically and covalently. Today, they are replaced by nylon.

## 2.14.3 Transfer of proteins

The most widespread transfer of proteins is the electrotransfer. Another transfer process is the capillary transfer. It is used also for transferring of nucleic acids (see there).

### 2.14.3.1 Electrotransfer of proteins

The electrotransfer (electroblotting) was patented in 1987 by William Littlehales [8]. It uses electric current to pull proteins from the gel onto a PVDF or an NC membrane.

The electrotransfer is the fastest way of blotting and can be carried out with poly-acrylamide or agarose gels.

The film-supported gels are problematic for electrotransfer, since the film should be removed after electrophoresis [9]. Meanwhile, there are net-supported gels [10, 11], which are suitable for electroblotting.

The electroblotting has two variants: tank blotting and semidry blotting.

## Tank blotting of proteins

To carry out tank blotting [12], the gel with the electrophoretically separated proteins and the blot membrane are placed vertically between filter papers and set down in a tank with buffer. Thereafter, the gel is blotted between two electrodes (Figure 2.14.2). The transfer continues usually overnight.

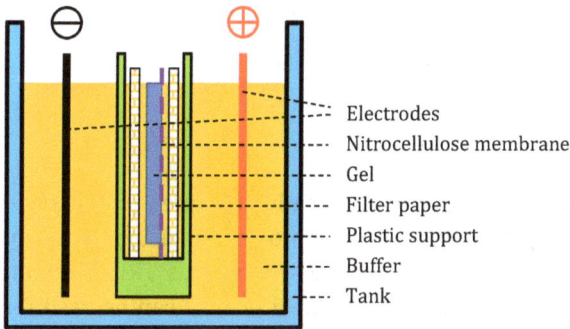

- - - Electrodes
- - - Nitrocellulose membrane
- - - Gel
- - - Filter paper
- - - Plastic support
- - - Buffer
- - - Tank

**Figure 2.14.2:** Tank blotting.
The blot sandwich is constructed of a buffer-saturated sheet of filter papers, gel, membrane, and another buffer-saturated sheet of filter papers. It is placed in a tank containing transfer buffer. The transfer is carried out at a voltage of 100 V for 1–2 h or at 15 V overnight.

## Semidry blotting of proteins

The semidry blotting is very popular [13, 14]. The gel and blot membrane are placed between filter papers soaked with transfer buffer and the resulting blot sandwich is inserted between the horizontal graphite plates of a blotter. Blotters with glassy carbon plates are available too, which are more solid than the graphite plates.

The semidry blotting can be carried out, similarly to electrophoresis, in a continuous buffer or in a disc-buffer system.

In the *continuous semidry buffer*, a same buffer is used on both sides of the blot membrane. Under these conditions, the blotted bands on the membrane are not enough sharp.

In the *disc-semidry blotting*, between both plates of the blotter, following matters are placed: a lower thick filter (anode) paper soaked with anode buffer, a blot membrane, a net gel (with the gel side down), a cellophane membrane, and an upper thick filter paper soaked with the cathode buffer. Then the upper (cathode) plate of the blotter is weighted with 1 kg mass. Finally, the power supply is turned on to start the blotting (Figure 2.14.3).

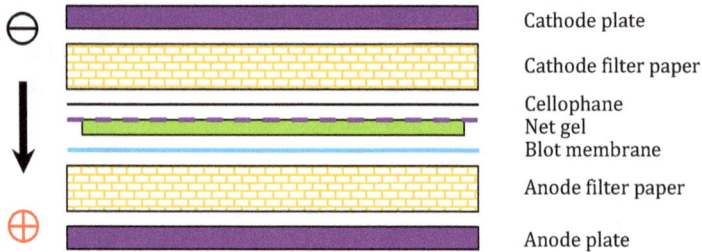

Figure 2.14.3: Disc-semidry blotting.

During the disc-semidry blotting, the electrophoretically separated proteins are overtaken by a moving ionic boundary. Therefore, the protein bands are seen very sharp on the blot membrane. We recommend the discontinuous TRIS-formate-taurinate buffer system for semidry blotting [15] (Table 2.14.1). It can be used for native proteins (after disc-electrophoresis or isoelectric focusing) or denatured proteins (after SDS electrophoresis).

### 2.14.3.2 Capillary transfer of proteins

The capillary transfer was introduced by Edwin Southern (Southern blotting) in 1975 [16], when he transferred DNA fragments onto an NC membrane. In this procedure [17], a blot membrane is placed onto the gel with the resolved bands, and the blot membrane is covered with a stack of dry papers. The paper stack acts with capillary force onto the gel and sucks its buffer together with the resolved bands (Figure 2.14.4). The transfer needs a long time, usually overnight.

After the protein transfer is complete, the blot membrane should be blocked and the proteins detected.

**Table 2.14.1:** TRIS-formate-taurinate buffer system for disc-semidry blotting.

| Anode buffer: TRIS-formate buffer, pH = 7.8, I = 0.06 mol/L | | |
|---|---|---|
| TRIS | 9.36 g | 0.077 mol/L |
| 99 g/dL formic acid | 2.29 mL | 0.018 mol/L |
| SDS | 0.58 g | 2.0 mmol/L) |
| Methanol | 200.00 mL | 5.0 mol/L |
| Sodium azide | 0.10 g | 1.5 mmol/L |
| Deionized water to | 1,000.00 mL | – |
| **Cathode buffer: TRIS-taurinate buffer, pH = 8.5, I = 0.01 mol/L** | | |
| TRIS | 4.20 g | 0.035 mol/L |
| Taurine | 6.02 g | 0.048 mol/L |
| SDS | 0.58 g/L | 2.0 mmol/L) |
| Methanol | 200.00 mL | 5.0 mol/L |
| Sodium azide | 0.10 g | 0.0015 mol/L |
| Deionized water to | 1,000.00 mL | – |

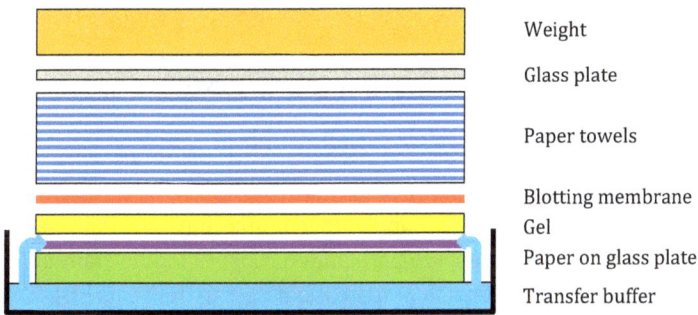

Weight

Glass plate

Paper towels

Blotting membrane
Gel
Paper on glass plate

Transfer buffer

**Figure 2.14.4:** Capillary transfer of proteins.

## 2.14.4 Blocking

To cover free binding sites on the blot membrane, suitable macromolecular substances (blocking reagents) are used. They should not react with the transferred proteins or with the detection reagents. Usually 3–5 g/dL **b**ovine **s**erum **a**lbumin (BSA) is used as a blocking reagent. Other blocking reagents are 5 g/dL skimmed milk powder, 3 g/dL fish gelatin, 0.05–0.1 g/dL Tween 20 or Triton X-100, 1.0 g/dL casein, and 10 g/dL fetal calf serum. For PVDF membranes, liquid fish gelatin is preferred [18].

## 2.14.5 Detection

After transfer and blocking, the blotted proteins can be detected by dyes or high-molecular-mass ligands (probes).

### 2.14.5.1 Detection by dyes

Nonspecific staining of proteins on blot membranes can be carried out with the anionic dyes such as Ponceau S, Amido black 10B, Coomassie brilliant blue R-250, India ink, Fast green FCF, with chelates, and metals (see there).

The staining with Ponceau S [19–21] is easy to use. The staining with Amido black 10B (Naphthol blue black) is most prevalent [22–24]. The staining with Coomassie brilliant blue R-250 [25–27] has a limited use for the blot membranes, since it binds to most membranes and cannot be completely freed from them.

Simple to use is the India ink staining of proteins blotted on NC membrane [28, 29]. It is more sensitive than the staining with Coomassie brilliant blue R-250, Amido black 10B, and Fast green, and its sensibility is almost as high as the silver staining method. The proteins on membrane appear as black bands on a gray background.

Fast green FCF is used in a concentration of 0.1 g/dL dissolved in 1 mL/dL acetic acid [30, 31].

Chelates can also serve as dyes for blotted proteins. Such an ability has, e.g., the purple-colored chelate ferrozine/ferro ion [32].

The metal (colloidal gold and silver) staining of membranes has the highest sensitivity but long incubation and troublesome preparation. Copper iodide staining is also sensitive but needs complicated steps to prepare the necessary reagents [33].

### 2.14.5.2 Detection by probes

Probes are antibodies, antigens, lectins, nucleic acids, and more. A few methods of detection by probes are used: colorimetric, chemiluminescent, radioactive, and fluorescent detection.

During the *colorimetric detection*, the blot membrane is probed for proteins of interest with antibodies, linked to enzymes (e.g., peroxidase). The enzyme drives a colorimetric reaction converting an appropriate substrate into a colored product. Next, the blot is washed away from the soluble dye and the protein concentration is evaluated through densitometry. In enzyme blotting, native enzymes are blotted and detected using specific coupled color reactions [34, 35].

The *chemiluminescence detection* needs the incubation of Western blot with a substrate that will luminesce when exposed to a reporter on a secondary antibody. The light is detected by **charge-coupled d**evice (CCD) cameras, which capture a digital image of the Western blot or photographic film. The image is analyzed by densitometry, which evaluates the stained protein. Nowadays, software allows even molecular mass analysis (Figure 2.14.5).

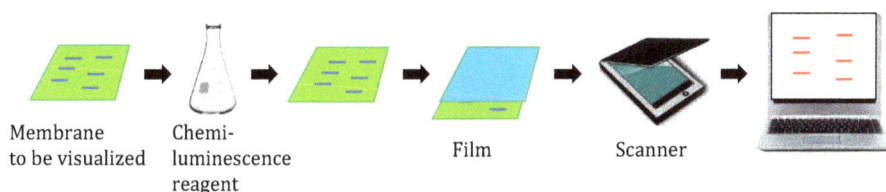

Membrane to be visualized    Chemi-luminescence reagent       Film       Scanner

**Figure 2.14.5:** Chemiluminescence detection.

The *radioactive detection* does not require enzyme substrates but labeled proteins. After placing of an X-ray film onto the Western blot, dark regions are created, which correspond to the proteins of interest. The importance of this method is declining due to its hazardous radiation; therefore, it is rarely used now.

For *fluorescent detection*, a fluorescently labeled probe is excited by light and the appeared emission is detected by a photosensor such as a CCD camera equipped with appropriate filters. This technique is considered to be one of the best methods for quantification, but is less sensitive than chemiluminescence method [36].

### 2.14.5.3 Immunoblotting

In the immunoblotting [37, 38], the antigens (proteins) are transferred to the blot membrane. There they form precipitation bands with specific immunoglobulins. The precipitation bands can be visualized, if the proteins are labeled, for example, with $^{125}$I [39].

## 2.14.6 Making the blot membranes transparent

After staining, a NC blot membrane can be made transparent without changing the color intensity of the stained bands [40]. For this purpose, the blot membrane should be impregnated with a monomeric solution, which has the same refractive index as NC. Thereafter, the monomer is polymerized using a photoinitiator and UV irradiation.

The transparency is achieved by the following procedure:

- A few drops of the monomeric solution (usually of 2 g/dL benzoin methyl ether in **trimethylolpropane trimethacrylate**, TMPTMA) with a photoinitiator are pipetted on the blot membrane and a second PVC film is placed on.

- The air bubbles between the blot membrane and the PVC films are removed with a photo roller;
- Both sides of the resulting sandwich are irradiated for 15 s each with UV light.

## 2.14.7 Blotting techniques

Different blotting techniques for proteins are known: **W**estern **b**lotting (WB), **F**ar**w**estern **b**lotting (FWB), **S**outhwestern **b**lotting (SWB), **N**orthwestern **b**lotting (NWB), and **E**astern **b**lotting (EB).

### 2.14.7.1 Western blotting

WB (protein immunoblot) is a technique for transferring native or denatured proteins onto a NC or PVDF membrane separated by gel electrophoresis, where they bind to specific antibodies (Figure 2.14.6). It originates from the laboratory of George Stark at Stanford University [41] and the laboratory of Harry Towbin at the Friedrich Miescher Institute [42]. The name of Western blotting was given by Neal Burnette [43] as a nod to Southern blotting, developed earlier by Edwin Southern [44].

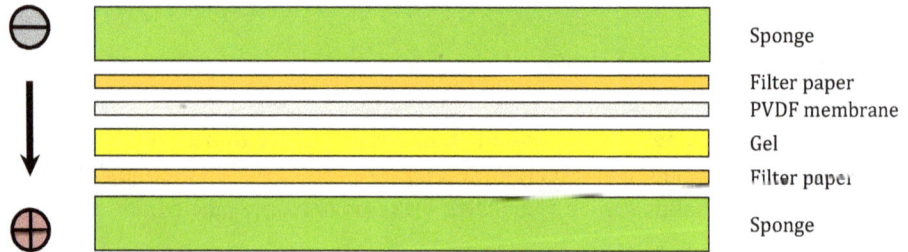

| | Sponge |
| | Filter paper |
| | PVDF membrane |
| | Gel |
| | Filter paper |
| | Sponge |

Figure 2.14.6: Scheme of Western blotting.

The HIV test contains a WB procedure to detect anti-HIV antibody in human serum samples. Proteins from HIV-infected cells are electrophoretically separated and blotted onto a blot membrane. Then, the serum to be tested is incubated with primary antibodies; the free antibodies are washed away. Finally, a secondary anti-human antibody linked to an enzyme signal is added.

The WB is also used as a definitive test for **b**ovine **s**pongiform **e**ncephalopathy (BSE, commonly referred to as *mad cow disease*). It can be used also for diagnosing some forms of Lyme disease (Lyme borreliosis), hepatitis B infection, and HSV-2 infection caused by herpes type 2.

## 2.14.7.2 Far-Western blotting

FWB was derived from the standard WB to detect protein–protein interactions *in vitro*. After proteins are transferred on the membrane, it is blocked and probed, usually with purified bait proteins. The bait proteins form complexes with prey proteins [45]. Then, an antibody probe is used to detect the presence of these complexes.

Unlike most methods using cell lysates (e.g., co-**immunop**recipitation, co-IP) or living cells (e.g., **f**luorescent **r**esonance **e**nergy **t**ransfer FRET), FWB determines whether two proteins bind to each other directly.

## 2.14.7.3 Southwestern blotting

SWB is based on Southern blotting, which is used for DNA detection, and on WB, which is used for protein detection. It was first described by B. Bowen and colleagues in 1980 [46, 47]. The SWB is carried out on NC membranes. With its aid, DNA-binding proteins are identified and characterized by their ability to bind to specific oligonucleotide probes.

Proteins are separated on a polyacrylamide gel containing SDS, renatured by removing SDS in the presence of urea, and blotted onto an NC membrane by diffusion. The genomic DNA region of interest is digested by restriction enzymes. In their activity, they produce fragments of different sizes, which are end-labeled and allowed to bind to the separated proteins. The specifically bound DNA is eluted from the protein–DNA complexes and analyzed by polyacrylamide gel electrophoresis.

## 2.14.7.4 Northwestern blotting

NWB is a hybrid of Western and Northern blotting. It is used to detect interactions between proteins that are immobilized on an NC membrane and RNA for studying the gene expression.

After separating by electrophoresis on agarose or polyacrylamide gel, RNA-binding proteins are transferred onto an NC membrane [48]. The blot is soaked in a blocking solution, such as TRIS-HCl (pH = 7.5), containing Mg acetate, DDT, Triton X-100, nonfat milk, or BSA [49]. Then a specific competitor RNA is applied, and the blot is incubated at room temperature [50]. During this time, the competitor RNA binds to the RNA-binding proteins on the membrane. After the incubation, the blot is washed in order to remove the unbound RNA. Common wash solutions are **p**hosphate-**b**uffered **s**aline (PBS) buffer or a Tween 20 solution [51]. Finally, the blot is developed by X-ray or similar autoradiography methods [52].

### 2.14.7.5 Eastern blotting

EB is a biochemical technique used to analyze protein posttranslational modifications leading to synthesis of lipoproteins, phosphoproteins, and glycoproteins: phosphorylation, acetylation, acylation, alkylation, nitroalkylation, arginylation, biotinylation, formylation, geranylgeranylation, glutamylation, glycosylation, hydroxylation, isoprenylation, lipoylation, methylation, selenation, succinylation, sulfation, transglutamination, ubiquitination, and more [53, 54]. It is similar to lectin blotting (i.e., detection of carbohydrate epitopes on proteins) [55].

## 2.14.8 Protocols

### Western Blotting

**Materials and equipment**
NC or PVDF membrane
Filter paper
Semidry transfer unit
Power supply

***10× TBST (TRIS-buffered saline containing Tween 20), pH = 7.4***

| | |
|---|---|
| NaCl | 80.0 g (1.37 mol/L) |
| KCl | 2.0 g (0.03 mol/L) |
| TRIS | 30.0 g (0.25 mol/L) |
| Tween | 10.0 mL (0.02 mol/L) |

*Dissolve in 800 mL deionized water. Adjust pH to 7.4 with HCl. Add deionized water to 1 L.*

***TBST buffer, pH = 8.0***

| | |
|---|---|
| TRIS | 0.30 g (0.02 mol/L) |

*Adjust pH to 8.0 with HCl.*

| | |
|---|---|
| Sodium chloride | 0.82 g (0.14 mol/L) |
| Potassium chloride | 0.02 g (0.003 mol/L) |
| Tween-20 | 0.05 mL (0.1 mmol/L) |
| Deionized water to | 100.00 mL |

***Cell lysis buffer***

| | |
|---|---|
| TRIS-HCl (pH = 7.5) | 0.24 g (20 mmol/L) |
| NaCl | 8.77 g (150 mmol/L) |
| NP-40 | 1.00 mL |

| EDTA | 0.58 g (2.0 mmol/L) |
|---|---|
| Leupeptin | 100.00 µg |
| Aprotinin | 100.00 µg |
| $Na_3PO_4$ | 0.02 g (1 mmol/L) |
| PMSF | 0.02 g (1 mmol/L) |
| NaF | 0.02 g (5 mmol/L) |
| $Na_4P_2O_7$ | 0.08 g (3 mmol/L) |
| Deionized water to | 100.00 mL |

### Transfer buffer

| TRIS | 3.0 g (0.025 mol/L) |
|---|---|
| Glycine | 14.4 g (1.19 mol/L) |
| Methanol | 200.0 mL |
| Deionized water to | 1,000.0 mL |

### Blocking buffer, and primary and secondary antibody dilution buffer
TBST with 5% nonfat dry milk

### Procedure
- Place cells in a microcentrifuge tube and centrifuge to collect cell pellet.
- Lyse the cell pellet with 100 µL of cell lysis buffer on ice for 30 min (for $10^6$ cells use 100 µL of cell lysis buffer).
- Centrifuge at 14,000 rpm (16,000 g) for 10 min at 4 °C.
- Transfer the supernatant to a new tube and discard the pellet. Remove 20 µL of supernatant and mix with 20 µL of 2× sample buffer.
- Boil for 5 min. Cool at room temperature for 5 min. Microcentrifuge for 5 min.
- Load up to 40 µL of sample to each well of a 1.5 mm thick gel.
- After electrophoresis transfer proteins onto an NC or PVDF membrane according to the manufacturer instructions.
- Place the blot membrane in blocking buffer consisting of 5% nonfat dry milk/ TBST.
- Incubate the blot membrane with agitation at room temperature for 1 h, or at 4 °C overnight.
- Dilute the primary antibody to the recommended concentration in 5% nonfat dry milk/TBST. Place the membrane in the primary antibody solution and incubate at room temperature for 2 h, or at 4 °C overnight with agitation.
- Wash three times for 5 min each with wash buffer (TBST, pH = 7.4).
- Incubate the membrane at room temperature for 30 min with horseradish peroxidase (HRP)-conjugated secondary antibody, diluted to 1:1,000–1:5,000 in 5% nonfat dry milk/TBST.

- Wash 4 times each with TBST for 10 min and once with PBS for 2 min.
- Incubate membrane (protein side up) with 10 mL of enhanced chemilumines-cence substrate for 1–2 min. The final volume required is 0.125 mL/cm$^2$.
- Drain off the excess detection reagent and wrap up the blots.
- Place the wrapped blots (protein side up) in an X-ray film cassette and expose to X-ray film from 5 s to 60 min.

**Semidry Blotting**

**Materials and equipment**

Transfer buffer
Transfer membrane: 0.45 μm NC, PVDF, and neutral or positively charged nylon
Methanol
Acetic acid
Coomassie brilliant blue G-250
Filter paper
Semidry transfer unit
Power supply

**Procedure**

- Soak a stack of chromatography papers with electrode buffer and place half on the lower electrode of a semidry blotter (the cathode).
- Wet a PVDF membrane with methanol and incubate it in electrode buffer until the membrane is submerged in the buffer.
  *If NCs are used, do not use methanol because it dissolves NC.*
- Place a gel with separated proteins on top of the chromatography papers and cover it with the PVDF membrane.
- Put on top the remaining half of the chromatography paper stack soaked with electrode buffer.
- Place the anode.
- Place on the anode a 5 kg weight.
- Set the voltage to 15 V, limit the electric current to 0.4 mA per cm$^2$ of gel area, and electroblot at room temperature for 16–24 h.
- Stain the wet PVDF membranes in a mixture of 25 mL/dL methanol, 10 mL/dL acetic acid, and 0.02 g/dL Coomassie brilliant blue G-250 for 5 min.
- Destain twice with 25 mL/dL methanol and 10 mL/dL acetic acid for 10 min.
- Rinse in deionized water and let the PVDF membrane dry.

## Capillary Blotting

**Materials and equipment**
NC paper (pore size 0.20–0.45 µm)
Transfer buffer
Transfer membrane: 0.45 µm NC, PVDF, and neutral or positively charged nylon
Methanol
Filter paper
Glass plates
Weight of 500–1,000 g

*Blotting buffer (pH = 8.3)*

| | |
|---|---|
| TRIS | 2.42 g (0.02 mol/L) |
| Glycine | 11.26 g (0.15 mol/L) |
| Methanol | 200.00 mL |
| Deionized water to | 1,000.00 mL |

Store at 4 °C.

*Staining solution*

| | |
|---|---|
| Amido black 10B | 0.1 g |
| Methanol | 40.0 mL |
| Acetic acid | 10.0 mL |
| Deionized water to | 100.0 mL |

*Destaining solution*

| | |
|---|---|
| Methanol | 40.0 mL |
| Acetic acid | 10.0 mL |
| Deionized water to | 100.0 mL |

**Procedure**
- Place two trays with blotting buffer on the table.
- Place onto the trays a 25 × 20 cm glass plate and six layers of chromatographic 3 mm filter paper on it.
- Dip the two ends of the paper in the blotting buffer and allow the papers to wet completely.
- Place the gel with proteins on the wetted filter paper.
- Wet an NC membrane of the same size as the gel that is to be blotted in the blotting buffer.
- Place the gel onto the wetted NC membrane, avoiding air bubbles trapped in between.
- Lay six layers of chromatographic 3 mm filter paper on the gel and place over another glass plate.

- Place a weight of approximately 500 g over the setup and leave the set for 1–2 days.
- Free the NC membrane and immerse it in the Amido black solution for 10 min with shaking.
- Destain in the destaining solution.
- Dry the blot between sheets of filter paper.

### India Ink Staining of Proteins on Membrane

**Materials and equipment**
TBST
India ink
Acetic acid
Tween 20
Rocking platform

*TBST buffer, pH = 8.0*

| | |
|---|---|
| TRIS | 0.30 g (0.02 mol/L) |
| *Adjust pH to 8.0 with HCl.* | |
| Sodium chloride | 0.82 g (0.14 mol/L) |
| Potassium chloride | 0.02 g (0.003 mol/L) |
| Tween | 200.05 mL |
| Deionized water to | 100.00 mL |

*Staining solution*

| | |
|---|---|
| India ink | 0.10 mL |
| Acetic acid | 1.00 mL |
| TBST solution to | 100.00 mL |
| *Stir for 60 min and filter.* | |

**Procedure**
- Soak the blot membrane in deionized water after proteins have been transferred. *PVDF membranes have to be prewetted in methanol.*
- Incubate the membrane in TBST buffer in a glass tray on a rocking platform for 60 min.
- Stain in the staining solution for 2 h or overnight.
- Destain the membrane several times in TBST buffer.
- Wash the membrane 2 × 2 min with deionized water.
- Dry the membrane for long-term storage.

## Detecting Proteins by Immunoblotting

### Materials and equipment
TBST
Methanol
SDS
Recombinant protein or unlabeled translated protein for probe
5 g/dL nonfat instant dry milk in TBST
Primary antibody specific for protein probe
**Al**kaline **p**hosphatase (ALP)-conjugated secondary antibody against Ig of species, from which specific antibody was obtained.
0.10 mol/L EDTA, pH = 8.0
Developing solution
PVDF membrane
Paper sheets
Semidry transfer equipment

### *Semidry transfer buffer*

| | |
|---|---|
| TRIS | 0.58 g (48 mmol/L) |
| Glycine | 0.29 g (39 mmol/L) |
| Methanol | 20.00 mL |
| SDS | 0.10 g |
| Deionized water to | 100.00 mL |

*Adjust volume to 1 L with deionized water. Store at room temperature.*

### Procedure
- Transfer the proteins onto a blot membrane and block the free places.
- Incubate the blot in 5 mL/dL nonfat milk in TBST at room temperature for 1 h, on an orbital shaker.
- Wash three times in TBST for 10 min by agitating on an orbital shaker.
- Dilute the ALP-conjugated secondary antibody in 5 mL/dL milk in TBST.
- Incubate the blot for 1 h.
- Wash the blot in TBST 6 times for 5 min each, with agitation.
- Rinse the blot in ALP buffer.
- Incubate the blot in developing solution for 1–15 min.
- Rinse the blot with 100 mL deionized water.
- Wash the blot with 0.10 mol/L EDTA, pH = 8.0 for 5 min, to stop the development reaction.
- Rinse with deionized water, dry, and photograph.

**Immunoprobing with Avidin–biotin Coupling to Secondary Antibody**

**Materials and equipment**
NC or PVDF membrane
Blocking buffer
TBS (TRIS-buffered saline)
TBST
Avidin
Biotinylated HRP or ALP
Primary antibody
Biotinylated secondary antibody
Heat sealable plastic bags

**Procedure**
– Place the membrane with the transferred proteins in a heat-sealable plastic bag with blocking buffer.
– Seal the bag and incubate on a shaker or rocking platform. For NC or PVDF membranes, incubate at room temperature for 30–60 min. For nylon membrane, incubate at 37 °C for 2 h.
– Prepare primary antibody in 5 mL TBST (for NC or PVDF membrane) or TBS (for nylon membrane). Dilutions generally range from 1/100 to 1/100,000.
– Open the bag, remove the blocking buffer, and replace with primary antibody solution.
– Incubate at room temperature for 30 min with gentle rocking.
– Remove the membrane from the plastic bag and place it in a plastic box.
– Wash the membrane three times in TBST (for NC or PVDF membrane) or TBS (for nylon membrane) for 15 min.
– Dilute two drops of the biotinylated secondary antibody in 50–100 mL of TBST (for NC or PVDF membrane) or TBS (for nylon membrane).
– Transfer the membrane into a plastic bag containing secondary antibody solution.
– Incubate at room temperature with slow rocking for 30 min and wash.
– While the membrane is being incubated with the secondary antibody, prepare avidin–biotin–HRP or –ALP complex. Dilute two drops of avidin solution and two drops of biotinylated HRP or ALP in 10 mL TBST (for NC or PVDF membrane) or TBS (for nylon membrane).
– Incubate at room temperature for 30 min, then dilute with TBST or TBS to 50 mL.
– Wash and transfer the membrane into the avidin–biotin–enzyme solution.
– Incubate at room temperature for 30 min with slow rocking.
– Wash 2–3 times for 30 min each.
– Develop according to the appropriate visualization protocol.

# References

[1]   Hansen JS, Heegaard PM, Jensen SP, Norgaard-Pedersen B, Bog-Hansen TC. Electrophoresis, 1988, 9, 273–278.

[2]   Erlich HA, Levinson JR, Cohen SN, McDevitt HO. J Biol Chem, 1979, 254, 12240–12247.

[3]   Pluskal MB, Przekop MB, Kavonian MR, Vecoli C, Hicks DA. BioTechniques, 1986, 4, 272–282.

[4]   Reigh JA, Klein DC. Appl Theor Electrophoresis, 1968, 1, 59.

[5]   Gershoni JM, Palade GE. Anal Biochem, 1982, 124, 396–405.

[6]   Burnette WN. Anal Biochem, 1981, 112, 195–203.

[7]   Reiser J, Wardale J. Eur J Biochem, 1981, 114, 569–575.

[8]   Littlehales WJ Electroblotting Technique for Transferring Specimens from a Polyacrylamide Electrophoresis or like Gel onto a Membrane. Patent US 4840714A, 1987.

[9]   Jägersten C, Edström A, Olsson B, Jacobson G. Electrophoresis, 1988, 9, 662–665.

[10]  Nishizawa H, Murakami A, Hayashi N, Iida M, Abe YI. Electrophoresis, 1985, 6, 349–350.

[11]  Kinzkofer-Peresch A, Patestos NP, Fauth M, Kegel F, Zok R, Radola BJ. Electrophoresis, 1988, 9, 497–511.

[12]  Bittner M, Kupferer P, Morris CF. Anal Biochem, 1980, 102, 459–471.

[13]  Kyhse-Andersen J. J Biochem Biophys Methods, 1984, 10, 203–209.

[14]  Tovey ER, Baldo BA. Electrophoresis, 1987, 8, 384–387.

[15]  Michov BM Protein Separation by SDS Electrophoresis in a Homogeneous Gel Using a TRIS-formate-taurinate Buffer System and Homogeneous Plates Suitable for Electrophoresis. (Proteintrennung Durch SDS-Elektrophorese in Einem Homogenen Gel Unter Verwendung Eines TRIS-Formiat-Taurinat-Puffersystems Und Dafür Geeignete Homogene Elektrophoreseplatten.) German Patent 4127546, 1991.

[16]  Southern EM. J Mol Biol, 1975, 98, 503–517.

[17]  Olsson BG, Weström BR, Karlsson BW. Electrophoresis, 1987, 8, 377–464.

[18]  Lee SL, Stevens J, Wang WW, Lanzillo JJ. Biotechniques, 1994, 17, 60–62.

[19]  Salinovich O, Montelaro RC. Anal Biochem, 1986, 156, 341–347.

[20]  Aebersold R, Leavitt J, Saavedra R, Hood L, Kent S. Proc Natl Acad Sci USA, 1987, 84, 6970–6974.

[21]  Montelaro R. Electrophoresis, 1987, 8, 432–438.

[22]  Towbin H *et al.* Proc Natl Acad Sci USA, 1979, 76, 4350–4354.

[23]  Gershoni JM, Palade GE. Anal Biochem, 1983, 131, 1–15.

[24]  Soutar AK, Wade DR, Creighton TE (eds.). Protein Function: A Practical Approach. Wiley–Liss, New York, 1989, 55.

[25]  Fazekas De St. Groth S, Webster RG, Datyner A. Biochim Biophys Acta, 1963, 71, 377–391.

[26]  Meyer TS, Lambert BL. Biochim Biophys Acta, 1965, 107, 144–145.

[27]  Burnette WN. Anal Biochem, 1981, 112, 195–203.

[28]  Hancock K, Tsang VC. Anal Biochem, 1983, 133, 157–162.

[29]  Hughes JH, Mack K, Hamparian VV. Anal Biochem, 1988, 173, 18–25.

[30]  Elliott DG, Conway CM, Applegate LJ. Dis Aquat Organ, 2009, 84, 139–150.

[31]  Luo S, Wehr NB, Levine RL. Anal Biochem, 2006, 350, 233–238.

[32]  Patton WF, Lam L, Su Q, Lui M, Erdjument-Bromage H, Tempst P. Anal Biochem, 1994, 220, 324–335.

[33]  Root DD, Reisler E. Anal Biochem, 1989, 181, 250–253.

[34]  McLellan T, Ramshaw JA. Biochem Genetics, 1981, 19, 647–654.

[35]  Sock J, Rohringer R. Anal Biochem, 1988, 171, 310–319.

[36]  Mathews ST, Plaisance EP, Kim T. Met Mol Biol, 2009, 536, 499–513.

[37]  Karcher D, Lowenthal A, Thormar H, Noppe M. J Immun Methods, 1981, 43, 175.

[38]  Holmquist L. Electrophoresis, 1988, 9, 511–513.

[39]  Renart J, Reiser J, Stark GR. Proc Natl Acad Sci USA, 1979, 76, 3116–3120.

[40]  Pharmacia LKB Development Technique, File No. 230, PhastSystem, 1989.

[41]  Renart J, Reiser J, Stark GR. Proc Natl Acad Sci USA, 1979, 76, 3116–3120.

[42]  Towbin H, Staehelin T, Gordon J. Proc Natl Acad Sci USA, 1979, 76, 4350–4354.

[43]  Burnette WN. Anal Biochem, 1981, 112, 195–203.

[44]  Southern EM. J Mol Biol, 1975, 98, 503–517.

[45]  Wu Y, Li Q, Chen XZ. Nat Protoc, 2007, 2, 3278–3284.

[46]  Bowen B, Steinberg J, Laemmli UK, Weintraub H. Nucl Acids Res, 1980, 8, 1–20.

[47]  Philippe J. Harwood AJ (ed.). Methods in Molecular Biology. Humana Press Inc., Totowa, 1994, Vol. 31, 349–361.

[48]  Verena B, Walz A, Bickel M. Nucl Acids Res, 1997, 25, 2417–2423.

[49]  Liao HJ, Kobyashi R, Mathews MB. Proc Natl Acad Sci USA, 1998, 95, 8514–8519.

[50]  Franke C, Grafe D, Bartsch H, Bachmann M. Met Mol Biol, 2009, 536, 441–449.

[51]  Schumacher J, Lee K, Edelhoff S, Braun R. J Cell Biol, 1995, 129, 1023–1032.

[52]  Stohlman SA, Baric RS, Nelson GN, Soe LH, Welter LM, Deans RJ. J Virol, 1988, 62, 4288–4295.

[53]  Mann M, Jensen ON. Nat Biotechnol, 2003, 21, 255–261.

[54]  Walsh CT, Garneau-Tsodikova S, Gatto GJ Jr. Angewandte Chemie, 2005, 44, 7342–7372.

[55]  Freeze HH. Preparation and Analysis of Glycoconjugates. Current Protocols in Molecular Biology. 1993, Chapter 17, 17.7.1–17.7.8.

# 2.15 Evaluation of protein pherograms

2.15.1    Qualitative evaluation of protein pherograms —— 289
2.15.1.1  Fixing of proteins —— 290
2.15.1.2  Staining of proteins —— 290
          Anionic dye staining of proteins —— 291
          Ponceau S staining of proteins —— 291
          Amido black staining of proteins —— 291
          Coomassie brilliant blue staining of proteins —— 292
          Conventional Coomassie brilliant blue staining of proteins —— 292
          Colloidal Coomassie brilliant blue staining of proteins —— 293
          India ink staining of proteins —— 293
          Counter-ionic dye staining of proteins —— 293
          Metal staining of proteins —— 294
          Silver staining of proteins —— 294
          Chelate dye staining of proteins —— 295
          Staining of glycoproteins —— 295
          Staining of lipoproteins —— 296
          Staining of enzymes —— 296
          Autoradiography of proteins —— 297
          Fluorography of proteins —— 297
          SYPRO Ruby staining of proteins —— 298
          Destaining of gel background —— 298
          Drying of gels —— 299
2.15.2    Quantitative evaluation of a protein pherogram —— 299
2.15.2.1  Densitometry —— 299
          Densitometers —— 301
2.15.2.2  Scanning —— 302
2.15.3    Protocols —— 302
          Coomassie Brilliant Blue Staining of Proteins in Polyacrylamide Gels —— 302
          Colloidal Coomassie Staining —— 303
          Rapid Silver Staining —— 304
          SYPRO Ruby Staining of Proteins —— 305
          Staining of Lipoproteins with Sudan Black B —— 305
          References —— 306

## 2.15.1 Qualitative evaluation of protein pherograms

Most proteins are colorless in nature. Therefore, they are usually stained after electrophoresis. The nonspecific staining involves fixing, staining, destaining of the background, and drying of the gel.

https://doi.org/10.1515/9783110761641-025

### 2.15.1.1 Fixing of proteins

The fixing causes denaturation and precipitation of proteins. For this purpose, methanol, ethanol, acetic acid, **trichloroa**cetic acid (TCA), 5-sulfosalicylic acid, or glutardialdehyde are used.

*Mixture of alcohol, acetic acid, and water.* The alcohol, methanol or ethanol, dehydrates the proteins, while acetic acid diminishes the pH value of the solution. Thus, the hydration shell of proteins is damaged, and most of them are put at or near their pI points. As a result, the protein solubility decreases and the proteins precipitate. Usually, two mixtures are used: methanol–acetic acid–water (3:1:6, *V:V:V*) or ethanol–acetic acid–water (4:1:5, *V:V:V*).

*Trichloroacetic or (and) 5-sulfosalicylic acid.* TCA in a concentration of 10–20 g/dL settles all proteins having relative molecular mass Mr above 5,000. 5-Sulfosalicylic acid in a concentration of 2–5 g/dL precipitates also most proteins, except serum $\alpha_1$-acidic glycoprotein. Better results are obtained when using a mixture of 10 g/dL TCA and 5 g/dL 5-sulfosalicylic acid.

*Glutardialdehyde.* Glutardialdehyde, in a concentration of 0.5–1.0 mL/dL, cross-links small protein molecules ($M_r = 1,000–10,000$). Its aldehyde groups react with the amino groups of adjacent proteins and form an insoluble network (Figure 2.15.1). This reaction takes place at pH = 6.0–7.0.

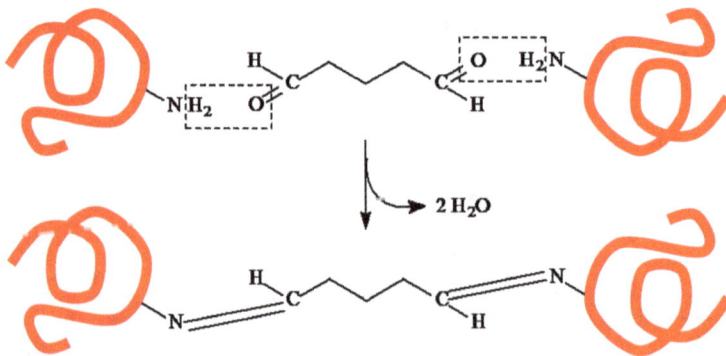

**Figure 2.15.1:** Glutardialdehyde as a cross-linker of proteins.

### 2.15.1.2 Staining of proteins

The proteins in a gel can be found (visualized) using anionic dye staining, counterionic dye staining, metal staining, autoradiography, fluorography, immunoprobing, or other methods.

## Anionic dye staining of proteins

To the anionic dye staining methods belong the staining methods with Ponceau S, Amido black 10B, Coomassie brilliant blue (CBB) R-250, Fast green FCF, India ink, and more.

## Ponceau S staining of proteins

The staining with Ponceau S, $M_r = 760.58$, is used usually after electrophoresis on cellulose acetate membranes [1–3].

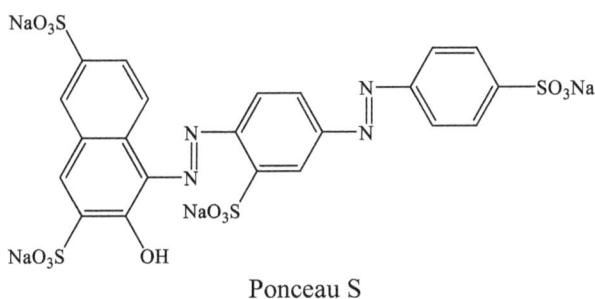

Ponceau S

The fixing and staining are carried out in 0.3 g/dL Ponceau S in 3 g/dL TCA; the destaining in 0.5 g/dL TCA; and drying at room temperature.

## Amido black staining of proteins

The staining with Amido black 10B (Naphthol blue black), $M_r = 616.50$, is prevalent for agarose and polyacrylamide gels [4, 5], rarely for nitrocellulose blot membranes [6].

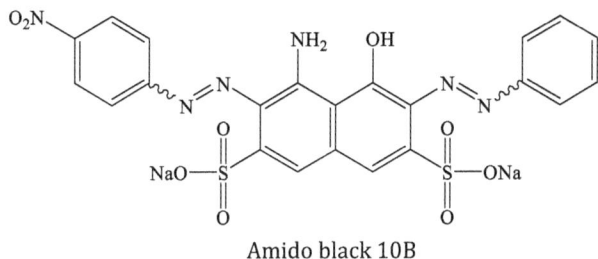

Amido black 10B

Prior to staining, the agarose gel should be dried after fixation. Thus, the proteins will be stained faster, and the gel absorbs less dye.

**Coomassie brilliant blue staining of proteins**

Originally, the Coomassie dyes were used for staining silk and wool. They were introduced to color electrophoretically separated proteins in the 1960s – first CBB R-250 [7–9], then CBB G-250 [10], and finally Coomassie violet R-200 [11–13] (Table 2.15.1).

**Table 2.15.1:** Coomassie dyes used in electrophoresis.

| Dyes | Chemical formulas | Properties |
|---|---|---|
| Coomassie brilliant blue R-250 (Acid blue 83), sodium salt | | $M_r = 825.97$<br>$\lambda = 560$ nm<br>Soluble in methanol and ethanol<br>Sensitivity: 10–50 ng/mm$^2$ |
| Coomassie brilliant blue G-250 (Acid blue 90), sodium salt | | $M_r = 854.03$<br>$\lambda = 580$ nm<br>Soluble in hot water, methanol, and ethanol<br>Sensitivity: 15–50 ng/mm$^2$ |
| Coomassie violet R-200 (Acid violet 17), sodium salt | | $M_r = 761.92$<br>Soluble in water, methanol, and ethanol<br>Sensitivity: 50–100 ng/mm$^2$ |

CBB R-250 is the most popular protein stain. In acidic solutions, it binds to the amino groups of proteins, which are contained in lysine, arginine, and histidine residues. The high intensity of CBB R-250 can be explained with the secondary binding between the dye molecules. The proteins are detected as blue bands on a clear background.

There are two staining methods with Coomassie dyes: conventional and colloidal staining.

**Conventional Coomassie brilliant blue staining of proteins**

The conventional Coomassie staining methods are relatively fast and less complicated. Their sensitivity reaches 100 ng protein/band.

First, Fazekas de St. Groth *et al.* described a procedure for staining proteins on a cellulose acetate sheet using CBB R-250 [14]. They suggested that the negatively charged sulfonate groups of CBB R-250 bind electrostatically to the positively charged protonated amino groups of proteins and that the positively charged quaternary ammonium groups of CBB R-250 bind to the negatively charged protein carboxyl groups. In these processes, hydrophobic interactions and hydrogen bonding take part, too. Later, Meyer and Lambert [15] and Chrambach *et al.* [16] employed CBB R-250 for staining of proteins in polyacrylamide gels.

Usually, CBB R-250 is dissolved in two solutions: a mixture of methanol–acetic acid–water, or a solution of TCA. In the first case, the gel background remains bluish; in the second case, it remains clear. Later, methanol, a compound with toxic properties, was replaced by ethanol [17].

### Colloidal Coomassie brilliant blue staining of proteins

Chrambach *et al.* [18] and Diezel *et al.* [19] replaced CBB R-250 by colloidal CBB G-250 dissolved in 12.5 g/dL TCA, to reduce the gel background. Neuhoff *et al.* [20, 21] employed the same staining principle including ammonium sulfate and phosphoric acid in the protocol. Their modification alleviated the problem of background staining and enabled the detection of very weak spots. Neuhoff dye ("blue silver") recipe contains 2 g/dL $H_3PO_4$, 10 g/dL $(NH_4)_2SO_4$, 20 mL/dL methanol, and 0.1 g/dL CBB G-250. The sensitivity of this method is more than 30 ng protein/band (0.7 ng protein/$mm^2$). However, the CBB G-250 staining needs a long time [22].

### India ink staining of proteins

The staining with carbon (India ink) [23, 24] is more sensitive than the Coomassie staining and comes closely to the sensitivity of silver staining. It is used mostly for staining of blot membranes (see there).

### Counter-ionic dye staining of proteins

The counter-ionic dye staining method employs two oppositely charged dyes (a negatively charged acidic dye and a positively charged basic dye), which form an ion-pair complex [25]. The ion-pair complex forms a colloidal staining solution. During the counter-ionic dye staining, the equilibrium exists between the ion-pair complexes and the free forms of dyes.

A counter-ionic dye staining method is the EV-ZC staining [26]. It uses the basic dye **e**thyl **v**iolet (EV), and the acidic dye **z**in**c**on (ZC). The detection limit of this method is 4–8 ng of protein in about 60 min.

### Metal staining of proteins

To the metal staining of proteins belong the silver staining and chelate dye staining.

### Silver staining of proteins

The vast application of silver staining to detect proteins after electrophoresis was established by Merril and colleagues who used formaldehyde to reduce silver ions into silver atoms under alkaline conditions [27, 28]. It is some 100 times more sensitive than Coomassie staining at detection limit of 1–0.1 ng protein per $mm^2$ of band area (50 ng protein in a band). The main disadvantage of silver staining is the poor correlation between staining intensity and protein concentration, because this method is not an end-point staining method as the less sensitive Coomassie or SYPRO Ruby staining [29]. Besides, there are proteins that react little or do not react with the silver ions [30] because staining depends on the complexation of $Ag^+$ with glutamate, aspartate, and cysteine residues [31].

The mechanisms of silver staining are similar to those of a photographic process [32]: The gel matrix with the separated proteins is saturated with $Ag^+$, which bind preferentially to basic amino acid residues of proteins. When $Ag^+$ are reduced to elementary Ag, the proteins obtain a typical brownish-gray-black color [33,34].

The silver staining comprises the following steps: fixing, sensitizing, developing, and stopping. Finally, the gel is dried.

The *fixing* step does not differ from that at Coomassie staining. It denatures the proteins to precipitate. For this purpose, weak acids, e.g., acetic acid and alcohol, are usually used.

During the *sensitizing* step (impregnating), the silver ions are bound to the precipitated proteins to form silver nuclei.

The *developing* step (visualizing) requires a reducing agent, such as formaldehyde, which reduces the silver ions to metallic silver (Figure 2.15.2). The reaction takes place near the silver nuclei much faster than in the gel, since the silver nuclei catalyze the reduction and transform the protein bands into dark brown to black bands on a slightly yellowish to colorless background. The developing step takes place at higher pH values, for example, in a sodium carbonate solution.

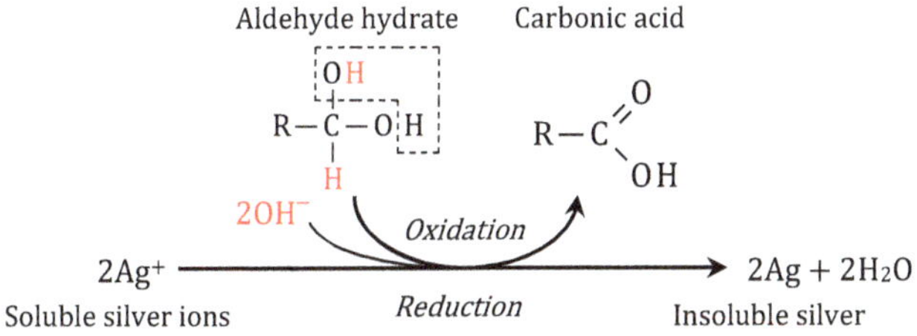

**Figure 2.15.2:** Mechanism of silver staining.

After the developing step, the reduction should be *stopped*. Otherwise, all silver ions will be reduced to metallic silver in the gel, which will obtain a black color. In principle, the stopping step is caused by a strong decrease in the pH value.

The *drying* step of the colored pherogram is similar to that after the Coomassie staining. It is carried out usually under a hair dryer. If the polyacrylamide concentration is high, the wet gel should be covered, free of air bubbles, with water-swollen cellophane.

## Chelate dye staining of proteins

Metal chelates are organometallic complexes that bind to proteins [35]. They appear at acidic pH and can be used to detect proteins on nitrocellulose, polyvinylidene difluoride, and nylon membranes, as well as in polyacrylamide gels. An example for the chelate dye staining of proteins is the zinc-imidazole staining [36]. Similarly, a zinc-imidazole staining procedure was proposed for **p**oly(**e**thylene **g**lycol) (PEG)-linked proteins separated in SDS polyacrylamide gels [37–39]. After electrophoresis, the gels were rinsed with deionized water and incubated in an imidazole-SDS solution. Then, they were soaked in zinc sulfate solution. The PEGylated protein bands remained transparent.

## Staining of glycoproteins

The detection of polysaccharides in glycoproteins, glycolipids, mucins, and glycogen is carried out with the **p**eriodic **a**cid–**S**chiff (PAS) test [40]. The Schiff reagent is a reaction product of fuchsine and sodium bisulfite. Periodic acid oxidizes the vicinal diols in the sugars, usually breaking up the bond between two adjacent carbon atoms not involved in the glycosidic linkage, and creating a pair of aldehydes at the two free tips of each broken monosaccharide ring. The aldehydes react with the Schiff reagent to give a purple-magenta color.

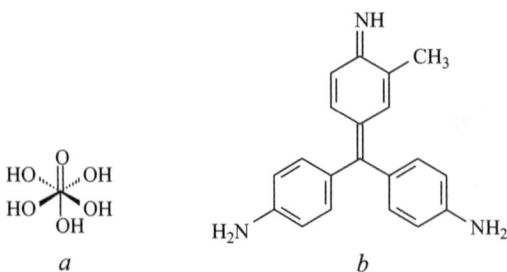

Periodic acid (*a*) and fuchsine (*b*).

A new variant of PAS test is the staining test of Hart *et al.* [41]. It uses the dye Pro-Q Emerald 488, which reacts with periodic acid-oxidized carbohydrate groups, generating a bright green-fluorescent signal on glycoproteins. This dye permits detection of less than 5–18 ng of glycoprotein per band, depending on the nature and degree of protein glycosylation, making it 8–16-fold more sensitive than the standard colorimetric PAS test. The green-fluorescent signal from Pro-Q Emerald 488 dye may be visualized using charge-coupled device/xenon arc lamp-based imaging systems or 470–488 nm laser-based gel scanners.

### Staining of lipoproteins

The staining of lipoproteins in a gel is carried out with Sudan Black B. First, the gel is dried at 60 °C, and then it is stained in freshly prepared staining solution (Sudan black and sodium chloride in ethanol). If a precipitate is formed, ethanol should be added until the precipitate dissolves. Finally, the gel is destained in a solution of NaCl in ethanol. After washing in deionized water, the gel is dried again at 60 °C.

### Staining of enzymes

Enzymes may be detected directly in the gel or indirectly by color reactions. In the direct detection, the gel with the electrophoretically separated enzymes is placed in a solution with specific substrates coupled to a diazo dye. In the indirect detection, the gel is covered with an additional carrier, for example, chromatography paper, cellulose acetate film, or agarose gel, which are impregnated with specific substrates.

Specific reactions are used for the detection of different enzymes: alkaline phosphatase [42, 43], alcohol dehydrogenase [44, 45], amylase [46, 47], cellulase [48], deoxyribonuclease [49], glucosidase [50], glycosyl transferase [51, 52], catalase [53, 54], creatine kinase [55], lactate dehydrogenase [56], malate dehydrogenase [57], peroxidase [58, 59], phosphoglucomutase [60], peptide hydrolases [61, 62], acid phosphatase [63, 64], trypsin [65], and more.

## Autoradiography of proteins

*Autoradiography* is the most sensitive detection method for proteins. In this method, X-ray film is used to visualize radioactive molecules that have been electrophoresed on agarose or polyacrylamide gels, or hybridized (e.g., on immunoblots). A photon or a $\beta$-particle from a radioactive molecule reduces silver bromide crystals on a film emulsion forming grains of silver (Table 2.15.2).

**Table 2.15.2:** Isotopes and their sensitivities.

| Isotope | Sensitivity |
|---|---|
| $^{32}P$ | 50 |
| $^{125}I$ | 100 |
| $^{14}C$ | 400 |
| $^{35}S$ | 400 |
| $^{3}H$ | 8,000 |

The autoradiography can be used directly or indirectly.

For *direct autoradiography*, the proteins are marked with the isotopes $^{14}C$, $^{35}S$, $^{32}P$, or $^{125}I$ [66]. After electrophoresis, the resolving gel is brought into contact with an X-ray film for 24 h. The blackening on the film is made by the direct irradiation of the radioactively labeled proteins in the gel.

In *indirect autoradiography*, the radioactive radiation is amplified with the aid of a calcium tungstate screen. Therefore, it is several times more sensitive than the direct autoradiography: After electrophoresis, a "sandwich" consisting of the separation gel, an X-ray film, and an intensifying screen is incubated at $-70$ °C. The radiation that penetrates the film hits the screen and excites emission of light quanta. They enhance the blackening on the film.

## Fluorography of proteins

Fluorescent staining is more sensitive than Coomassie staining, and often as sensitive as silver staining. Fluorescence occurs when a molecule absorbs light of a certain wavelength and releases this energy by emitting photons. Most commonly utilized fluorogenic reagents are **o**-**ph**thaldialdehyde (OPA) and **n**aphthalene **d**icarboxaldehyde (NDA) [67]. OPA is used in the presence of a thiol, e.g., **β**-**m**ercaptoethanol (BME) [68], 3-mercapto-1-propanol [69], **d**ithiothreitol (DTT) [70], ethanethiol [71], or *N*-**a**cetyl-L-**c**ysteine (NAC) [72].

In *fluorography*, a gel with the electrophoretically separated proteins, which have been previously labeled with $^3$H, $^{14}$C, or $^{35}$S, is impregnated with scintillating substances such as 2,5-diphenyloxazole [73] or sodium salicylate [74]. The gel is dried and is let to act on X-ray film at −70 °C. The interaction between the radioactive radiation and the scintillating substance produces visible light that blackens the film. It is also possible to treat the sample, prior to electrophoresis, with a fluorescent marker [75, 76], which makes the bands visible under UV light.

**SYPRO Ruby staining of proteins**

SYPRO Ruby staining was first introduced in 2000 by Berggren *et al.* [77]. Its sensitivity is better than other protein stainings and as sensitive as the best silver staining procedures [78]. SYPRO Ruby stain is a fluorescent ruthenium-based metal chelate dye, which interacts with basic amino acid residues (of lysine, arginine, or histidine). It is less sensitive if the proteins are native. Therefore, it is better if the native gels are soaked in 0.05 g/dL SDS for 30 min and then stained with SYPRO Ruby stain.

SYPRO Ruby stain is compatible with **m**atrix-**a**ssisted **l**aser **d**esorption/**i**onization (MALDI) mass spectroscopy. MALDI mass spectrometry requires little sample preparation and can be automated using robotic liquid-handling systems. MALDI mass spectroscopy is utilized for biomarkers, for example, prostate cancer biomarker [79].

**Destaining of gel background**

Generally, the destaining of gel background is carried out in the solution where the stain was dissolved.

Cellulose acetate membranes, stained with Ponceau S, are destained in 0.5 g/dL TCA.

Agarose gels, stained with Amido black 10B, are destained in 10 mL/dL acetic acid.

Agarose gels, stained for glycoproteins with PAS reagent, are destained in 7 mL/dL acetic acid.

Agarose gels, stained for lipoproteins with Sudan black B, are destained in a mixture of ethanol and 2 g/dL NaCl (55:45, *V:V*).

Polyacrylamide gels, stained with CBB R-250, are destained in a mixture of methanol–acetic acid–water (3:1:6, *V:V:V*) or of ethanol–acetic acid–water (4:1:5, *V:V:V*).

Polyacrylamide gels, stained with colloidal CBB G-250, are destained in 0.1 mol/L TRIS-hydrogen phosphate buffer with pH = 6.5.

Polyacrylamide gels, stained with silver, are destained in deionized water.

Polyacrylamide gels, stained with SYPRO Ruby stain, are destained by incubating the gel in 0.1 mL/dL Tween 20 or in 7.5 mL/dL acetic acid overnight.

**Drying of gels**

Gel drying can be carried out in air, under a hair fan, or in a drying room. If the polyacrylamide concentration is higher than 10 g/dL, prior to drying, the wet gel should be covered with a cellophane membrane. The gradient polyacrylamide gels are placed on the support film down on a glass plate and are covered free of air bubbles with a water-swollen cellophane membrane, wherein the protruding margins of the cellophane membrane should be folded onto the underside of the glass plate.

## 2.15.2 Quantitative evaluation of a protein pherogram

The presence of a stained band could be detected with the naked eye. However, it is difficult or impossible to determine the differences between the stain intensities (concentrations) of the bands. This has to be done by densitometry of pherograms or scanning them.

### 2.15.2.1 Densitometry

Densitometry means quantitative measurement of the absorbance (**o**ptical **d**ensity, OD) of a material when light passes through it [80]. The maximum and minimum densities that can be measured are called $D_{max}$ and $D_{min}$, respectively.

Densitometry is carried out on dry gels or gel photos. As a result, a densitogram is obtained. The densitogram is a curve that reflects the concentrations of bands obtained by cellulose acetate, agarose gel, or polyacrylamide gel electrophoresis (Figure 2.15.3).

When a light beam (radiant flux) of a certain wavelength and intensity $I_0$, in cd, irradiates a substance, a part of it is absorbed, resulting in a reduction of its intensity into $I$, in cd. According to the Beer–Lambert law [81, 82], the natural logarithm of the ratio between the irradiated and transmitted light is referred to as absorbance $A$ (OD):

$$A = \ln \frac{I_0}{I} \qquad\qquad (2.15.1)$$

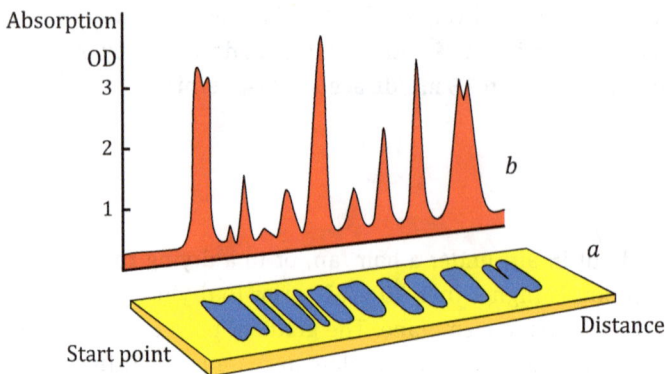

**Figure 2.15.3:** Densitometry: (*a*) pherogram and (*b*) densitogram.

In densitometry, white light or laser light is used. The white light should be adjusted, using a suitable filter, because the dyes have different absorption maxima (Figure 2.15.4). On the contrary, the laser beam has a fixed wavelength, an extremely narrow width, and only one spectral line. For example, the wavelength of the helium–neon laser is 632.5 nm.

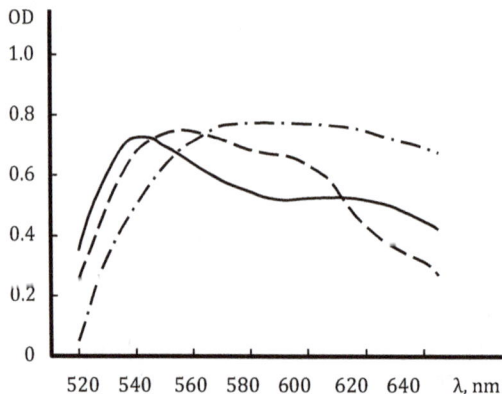

**Figure 2.15.4:** Absorption spectra of different dyes. — Coomassie brilliant blue R-250; — — Coomassie brilliant blue G-250; — ·— Amido black 10B (Naphthol blue black B).

The baseline of a densitogram defines its background. Two methods are used to define the baseline (Figure 2.15.5): In the first method, the baseline is drawn as a horizontal line through the lowest point of a densitogram (horizontal baseline). This method is used for pherograms with homogeneous background. In the second method, the baseline connects each band minimum (manual baseline). This method is used for unevenly colored background. The first method is simpler, but the manual baseline brings precise results, as it excludes the influence of the background.

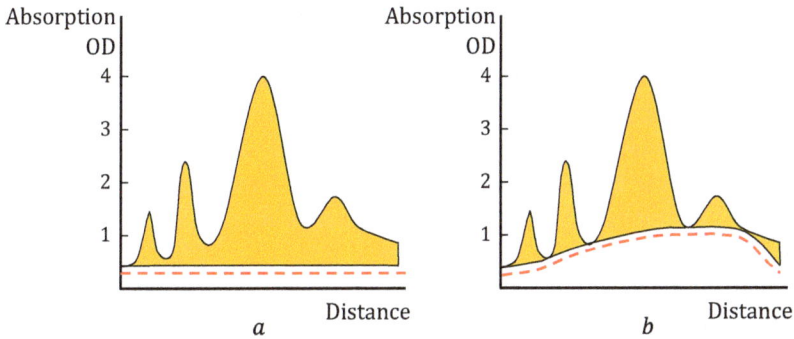

**Figure 2.15.5:** Different baselines. (*a*) Horizontal baseline; (*b*) manual baseline. – – Baseline.

Evaluating a densitogram, absolute or relative protein concentrations can be determined. In the first case, a calibration curve is required; in the second case, a relationship between the total band surface (total concentration) and a band surface is needed. The relative concentration is often more important than the absolute concentration. In the regions of low protein concentrations, a linear relationship exists between the concentration and the corresponding band area.

In the one-dimensional (1D) densitometry, the bands of a pherogram are scanned only in one direction. In the two-dimensional (2D) densitometry, the spots of a pherogram are scanned in two directions; then the information is translated into digital signals and processed in a computer [83, 84]. In the 1D densitometry, a mountain-shaped curve is obtained on one baseline; in the 2D densitometry, a mountain-shaped curve is obtained on a base surface. Therefore, the 2D densitogram looks like a geographic map.

In the 1D densitometry, the evaluation of protein concentrations uses the proportionality between the protein concentration and the band surface; in the 2D densitometry, the protein concentration is calculated from the proportionality between the protein concentration and the band volume. Therefore, appropriate computer software is needed for the 2D densitometry, which takes into account the base surface and the spot volumes [85, 86].

## Densitometers

The densitometers work usually in a transmission mode, since gels, blot membranes, or autoradiograms are usually transparent. It this case, it determines the OD of a sample placed between a light source and a photoelectric cell [87]. A high-resolution densitometer can scan and analyze bands using a computer and an appropriate software. The molecular masses and isoelectric points of proteins can also be determined.

The resolution of a densitometer depends above all on the width of light beam.

The *width of light beam* has a crucial importance for the resolution of a densitometer when bands lie close together. The wider the light beam, the more the measured absorption differs from the band reality on the gel. The light beam, however, should not be below a certain minimum width, as this should reduce its intensity. Therefore, a compromise has to be taken in the choice of beam width. A white light beam should not be thinner than 100 μm, and a laser beam should not be thinner than 50 μm.

On the market, *microdensitometers* are also available. They measure OD in microscopic domains [88, 89]. An alternative version to microdensitometers is the beam expanded laser microdensitometer [90, 91]. It has increased data collection speed, increased depth of focus, and superior signal-to-noise ratios [92].

### 2.15.2.2 Scanning

Using a standard scanner and appropriate software, the resolved bands on a gel can be quantified. The image produced by a scanner should be an accurate representation of both positions and intensities of the bands.

A few software programs for scanning pherograms are offered. One of them is the software program UN-SCAN-IT (Silk Scientific, Inc., Orem, USA). This program allows automatically analyzing gel electrophoresis images turning a scanner into a densitometer. It can also quantify western blots, agarose gels, PCR gels, *etc*. UN-SCAN-IT can save the data in ASCII and clipboard format, calibrate the image intensity to normalize results, export gel data to other spreadsheet, data analysis, and graphics programs, and digitize the graphs.

## 2.15.3 Protocols

### Coomassie Brilliant Blue Staining of Proteins in Polyacrylamide Gels

#### Materials and equipment
TCA
CBB R-250
Acetic acid
Methanol or ethanol

#### *Fixing solution*
| | |
|---|---|
| Ethanol | 50.0 mL |
| Acetic acid | 10.0 mL |
| Deionized water to | 100.0 mL |

### Solution A

| | |
|---|---|
| CBB R-250 | 0.2 g |
| Methanol or ethanol | 60.0 mL |
| Deionized water to | 100.0 mL |

### Solution B

| | |
|---|---|
| Acetic acid | 20.0 mL |
| Deionized water to | 100.0 mL |

### Procedure

- Fix a polyacrylamide gel with separated protein bands in the fixing solution for 20 min. Gels that contain nonionic detergents (Nonidet NP-40, Triton X-100) are to be fixed in 30 mL/dL isopropanol.
- Rinse the gel with tap water.
- Stain the protein bands with a mixture of solution A and solution B (1:1, $V:V$) for 20 min.
- Destain the gel several times in 10 mL/dL acetic acid or methanol–acetic acid–water (3:1:6, $V:V:V$). The last destaining solution should contain 5 mL/dL glycerol. The fixation, staining, and destaining can be accelerated, if the temperature is increased to 60 °C.
- Place the polyacrylamide gel onto a glass plate, with its support film down.
- Cover the gel with a cellophane membrane swollen in water.
- Remove the air bubbles between the gel and cellophane membrane using a roller. Place the edges of the cellophane membrane around the glass plate.
- Dry at room temperature.

## Colloidal Coomassie Staining

### Materials and equipment

Glutardialdehyde
CBB G-250
Ammonium sulfate [$(NH_4)_2SO_4$]
Phosphoric acid ($H_3PO_4$)
Methanol or ethanol

### Purification of CBB G-250

Solve 4 g of CBB G-250 in 250 mL of 10 mL/dL acetic acid and heat up to 60 °C. After cooling, filter or centrifuge the solution and use the sediment.

### Preparing the staining solution
Add 10.0 g of ammonium sulfate to 98 mL of 2 g/dL $H_3PO_4$ until it dissolves completely. Mix with 2 mL of 5 g/dL CBB G-250. Do not filter.

### Procedure
- Fix a polyacrylamide gel with separated protein bands in 10 g/dL TCA for 1 h.
- Stain the proteins in the staining solution overnight.
- Rinse the gel in 0.1 mol/L TRIS-hydrogen phosphate buffer with pH = 6.5 for 2 min and in 25 mL/dL methanol for 1 min.
- Stabilize the protein–dye complexes in 20 g/dL ammonium sulfate.

### Rapid Silver Staining
*This protocol is rapid but is not enough sensitive in detecting very small protein bands.*

### Materials and equipment
Formaldehyde fixing solution
$AgNO_3$
Sodium thiosulfate ($Na_2S_2O_3$)
Citric acid
Dialysis membrane
Glass plates

### Procedure
- Place a polyacrylamide gel with separated proteins in a plastic container, add formaldehyde fixing solution, and agitate on a shaker at room temperature for 10 min.
- Wash the gel twice with deionized water for 5 min.
- Soak the gel in 0.2 g/L $Na_2S_2O_3$ for 1 min.
- Wash the gel twice in deionized water for 20 s.
- Soak the gel in 0.1 g/dL $AgNO_3$ for 10 min.
- Wash the gel with deionized water.
- Wash the gel with fresh 0.2 g/dL sodium thiosulfate developing solution and agitate until the band intensities are adequate.
- Add 2.3 mol/L citric acid and agitate for 10 min.
- Wash the gel in deionized water, agitating for 10 min.
- Sandwich the gel between two pieces of wet dialysis membrane on a glass plate and dry overnight at room temperature.

**SYPRO Ruby Staining of Proteins**

**Materials and equipment**
SYPRO Ruby protein stain (ready-to-use solution)
Acetic acid
Ethanol
Tween 20
Gel dryer

*Fixing solution*

| | |
|---|---|
| Acetic acid | 10.0 mL |
| Ethanol | 40.0 mL |
| Deionized water to | 100.0 mL |

*Destaining solution*

| | |
|---|---|
| Tween 20 | 0.1 mL |
| Deionized water to | 100.0 mL |

**Procedure**
- Fix a polyacrylamide gel with separated proteins in the fixing solution for 20 min.
- Place the gel into a plastic dish containing fluorescent staining solution. Cover the gel with aluminum foil to protect from light.
- Agitate gently at room temperature for 10–60 min.
- Stain 1D and 2D gels for a minimum of 3 h. Stain IEF gels overnight.
- Destain gel background by incubating overnight in the destaining solution to minimize fluorescence.
- Incubate the gel in a 2 mL/dL glycerol for 30 min.
- Dry the gel using a gel dryer.

**Staining of Lipoproteins with Sudan Black B**

**Materials and equipment**
Sudan black B
Ethanol
Sodium chloride

*Sudan black solution*

| | |
|---|---|
| Sudan black B | 2.0 g |
| Ethanol | 100.0 mL |

### Sudan black staining mixture

Sudan black solution  10.0 mL
2 g/dL NaCl        75.0 mL
Ethanol          90.0 mL

### Destaining solution

2 g/dL NaCl        55.0 mL
Ethanol to        100.0 mL

### Procedure

– Dry a gel with separated lipoproteins at 60 °C.
– Stain in the freshly prepared Sudan black staining mixture for 15 min.
– Destain in the destaining solution for 15 min.
– Wash in deionized water for several seconds.
– Dry the gel at 60 °C.

# References

[1]   Salinovich O, Montelaro RC. Anal Biochem, 1986, 156, 341–347.
[2]   Aebersold R, Leavitt J, Saavedra R, Hood L, Kent S. Proc Natl Acad Sci USA, 1987, 84, 6970–6974.
[3]   Montelaro R. Electrophoresis, 1987, 8, 432–438.
[4]   Gershoni JM, Palade GE. Anal Biochem, 1983, 131, 1–15.
[5]   Soutar AK, Wade DR, Creighton TE eds. Protein Function: A Practical Approach. Wiley–Liss, New York, 1989, 55.
[6]   Towbin H et al. Proc Natl Acad Sci USA, 1979, 76, 4350–4354.
[7]   Fazekas De St. Groth S, Webster RG, Datyner A. Biochim Biophys Acta, 1963, 71, 377–391.
[8]   Meyer TS, Lambert BL. Biochim Biophys Acta, 1965, 107, 144–145.
[9]   Burnette WN. Anal Biochem, 1981, 112, 195–203.
[10]  Diezel W, Kopperschläger G, Hoffmann E. Anal Biochem, 1972, 48, 617–620.
[11]  Shiba KS et al. Clin Chem, 1986, 32, 2209.
[12]  Hiratsuka N et al. Biol Pharm Bull, 1994, 17, 1355–1357.
[13]  Thinakaran N et al. J Hazard Mater, 2008, 151, 316–322.
[14]  Fazekas De St Groth S, Webster RG, Datyner A. Biochim Biophys Acta, 1963, 71, 377–391.
[15]  Meyer TS, Lambert BL. Biochim Biophys Acta, 1965, 107, 144–145.
[16]  Chrambach A, Reisfeld RA, Wyckoff M, Zaccari J. Anal Biochem, 1967, 20, 150–154.
[17]  Zehr BD, Savin TJ, Hall RE. Anal Biochem, 1989, 182, 157–159.
[18]  Chrambach A, Reisfeld RA, Wyckhoff M, Zaccari J. Anal Biochem, 1967, 20, 150–154.
[19]  Diezel W, Kopperschläger G, Hofmann E. Anal Biochem, 1972, 48, 617–620.
[20]  Neuhoff V, Stamm R, Eibl H. Electrophoresis, 1985, 6, 427–448.
[21]  Neuhoff V, Arold N, Taube D, Ehrhardt W. Electrophoresis, 1988, 9, 255–262.
[22]  Anderson NL, Esquer-Blasco R, Richardson F, Foxworthy R, Eacho R. Toxicol Appl Pharmacol, 1996, 137, 75–89.
[23]  Hancock K, Tsang V. Anal Biochem, 1983, 133, 157–162.
[24]  Li K, Geraerts W, Van Elk R, Joosse J. Anal Biochem, 1988, 174, 97–100.
[25]  Fazekas De St Groth S, Webster RG, Datyner A. Biochim Biophys Acta, 1963, 71, 377–391.

[26] Choi JK, Tak KH, Jin LT, Hwang SY, Kwon TI, Yoo GS. Electrophoresis, 2002, 23, 4053–4059.

[27] Merril CR, Switzer RC, Van Keuren ML. Proc Natl Acad Sci USA, 1979, 76, 4335–4339.

[28] Switzer RC, Merril CR, Shifrin S. Anal Biochem, 1979, 98, 231–237.

[29] Graves RR, Haystead TA. Microbiol Mol Biol Rev, 2002, 66, 39–63.

[30] Morrissey JH. Anal Biochem, 1981, 117, 307.

[31] De Moreno MR, Smith JR, Smith RV. Anal Biochem, 1985, 757, 466–470.

[32] Merril CR. Adv Electrophoresis, 1987, 1, 111.

[33] Heukeshoven J, Dernick R. J Chromatogr, 1985, 326, 91–101.

[34] Merril CR. Acta Histochem Cytochem, 1966, 19, 655–667.

[35] Patton W, Lam L, Su Q, Lui M, Erdjument-Bromage H, Tempst P. Anal Biochem, 1994, 220, 324–335.

[36] Castellanos-Serra L, Proenza W, Huerta V, Moritz RL, Simpson RJ. Electrophoresis, 1999, 20, 732–737.

[37] Castellanos-Serra L, Hardy E. Electrophoresis, 2001, 22, 864–873.

[38] Hardy E, Castellanos-Serra LR. Anal Biochem, 2004, 7, 1–13.

[39] Hardy E, Ramon J, Saez V, Beez R, Tejeda Y, Ruiz A. Electrophoresis, 2008, 29, 2363–2371.

[40] Schiff H. Justus Liebigs Ann Chemie, 1866, 140, 92–137.

[41] Hart C, Schulenberg B, Steinberg TH, Leung WY, Patton WF. Electrophoresis, 2003, 24, 588–598.

[42] Pickering CE, Pickering RS. Arch Toxicol, 1978, 39, 249–266.

[43] Hodson AW, Skillen AW. Anal Biochem, 1988, 169, 253–261.

[44] Vulkanen KH, Goldmann D. Electrophoresis, 1988, 9, 132–135.

[45] Peisker K. Nahrung, 1986, 30, 1051–1053.

[46] Kadofuku T, Sato T, Manabe T, Okuyama T. Electrophoresis, 1983, 4, 427–431.

[47] Peisker K. Nahrung, 1986, 30, 463–465.

[48] Biely P, Markovic O, Mislovicova D. Anal Biochem, 1985, 144, 147–151.

[49] Karpetsky TP, Brown CE, McFarland E, Brady ST, Roth W, Rahman A, Jewett P. Biochem J, 1984, 219, 553–561.

[50] Kinzkofer A, Radola BJ. Electrophoresis, 1983, 4, 408–417.

[51] Mukasa H. In Hamada *et al.* (eds.). Molecular Microbiology and Immunobiology of Streptococcus Mutants, Elsevier, Amsterdam, 1986, 121–132.

[52] Mukasa H, Tsumori H, Shimamura A. Electrophoresis, 1987, 8, 29–34.

[53] Ogata M, Suzuki K, Satoh Y. Electrophoresis, 1988, 9, 128–131.

[54] Schiefer S, Teifel W, Kindl H. Hoppe-Seyler´s Z Physiol Chem, 1976, 357, 163–175.

[55] Williams J, Marshall T. Electrophoresis, 1987, 8, 536–537.

[56] Pretsch W, Charles DJ, Narayanan KR. Electrophoresis, 1982, 3, 142–145.

[57] Soltis DE, Haufler CH, Darrow DC, Tastony GJ. Am Fern J, 1983, 73, 9–27.

[58] Schrauwen J. J Chromatogr, 1966, 23, 177–180.

[59] Butler MJ, Lachance MA. Anal Biochem, 1987, 162, 443–445.

[60] Kilias H, Gelfi C, Righetti PG. Electrophoresis, 1988, 9, 187–191.

[61] Plow EF. Biochim Biophys Acta, 1980, 630, 47–56.

[62] Kelleher PJ, Juliano RL. Anal Biochem, 1984, 136, 470–475.

[63] Günther S, Postel W, Wiering H, Görg A. Electrophoresis, 1988, 9, 618–620.

[64] Allen RC, Budowle B, Saravis CA, Lack PM. Acta Histochem Cytochem, 1986, 19, 637–645.

[65] Ohlsson BG, Weström BR, Karlsson BW. Electrophoresis, 1987, 8, 415–420.

[66] Laskey RA, Mills AD. Eur J Biochem, 1975, 56, 335–341.

[67] Shou M, Smith AD, Shackman JG, Peris J, Kennedy RT. J Neurosci Meth, 2004, 138, 189–197.

[68] Chen RF, Scott C, Trepman E. Biochim Biophys Acta, 1979, 576, 440–455.

[69] Stobaugh JF, Repta AJ, Sternson LA, Garren KW. Anal Biochem, 1983, 735, 495–504.

[70] Simons SS Jr, Johnson DF. Anal Biochem, 1978, 90, 705–725.

[71]   Simons SS Jr, Johnson DF. Anal Biochem, 1977, 82, 250–254.
[72]   Fan LY, Cheng YQ, Chen HL, Liu LH *et al*. Electrophoresis, 2004, 25, 3163–3167.
[73]   Keck K, Grossberg AL, Pressmann D. Eur J Immunol, 1973, 3, 99–102.
[74]   Chamberlain JP. Anal Biochem, 1979, 98, 132–135.
[75]   Falk BW, Elliot C. Anal Biochem, 1985, 144, 537–541.
[76]   Horowitz PM, Bowman S. Anal Biochem, 1987, 165, 430–434.
[77]   Berggren K, Chernokalskaya E, Steinberg TH, Kemper C, Lopez MF, Diwu Z, Haugland RP, Patton WF. Electrophoresis, 2000, 21, 2509–2521.
[78]   Lopez MF, Berggren K, Chernokalskaya E, Lazarev A, Robinson M, Patton WF. Electrophoresis, 2000, 21, 3673–3683.
[79]   Flatley B, Malone P, Cramer R. Biochim Biophys Acta, 2014, 1844, 940–949.
[80]   Cantu GR, Nelson JW. BioTechniques, 1994, 16, 322–327.
[81]   Lambert JH. Photometria Sive De Mensura Et Gradibus Luminis, Colorum Et Umbrae (Photometry or on the Measure and Gradations of Light, Colors, and Shade). Eberhardt Klett E, Augsburg, Germany, 1760, 391.
[82]   Beer A. Ann Physik Chemie, 1852, 86, 78–88.
[83]   Manabe T, Okuyama T. J Chromatogr, 1983, 264, 435–443.
[84]   Prehm J, Jungblut P, Klose J. Electrophoresis, 1987, 8, 562–572.
[85]   Anderson NB, Anderson L. Clin Chem, 1982, 28, 739–748.
[86]   Kuick RD, Hanash SM, Strahler JR. Electrophoresis, 1988, 9, 192–198.
[87]   Health Physics Division Annual Progress Report. Oak Ridge National Laboratory. Health Physics Division, Union Carbide Corporation, U.S. National bureau of standards, 1968, 101.
[88]   Dainty JC, Shaw R. Image Science. Academic Press, New York, 1974.
[89]   James TH. The Theory of the Photographic Process. Eastman Kodak, Rochester, 1977.
[90]   Duarte FJ. Tunable Laser Optics. Elsevier, New York, 2003Chapter 10.
[91]   Duarte FJ. *Electro-optical Interferometric Microdensitometer System*, US Patent 5255069, 1993.
[92]   Duarte FJ. Tunable Laser Applications. CRC, New York, 2009 Chapter 12.

# 2.16 Precast gels for protein electrophoresis. Rehydratable gels

2.16.1    Precast agarose gels —— 309
2.16.2    Precast polyacrylamide gels —— 309
2.16.3    Rehydratable polyacrylamide gels —— 311
2.16.4    Protocols —— 311
          Silanization of Glass Plates —— 311
          Casting Agarose Gels on Support Film by Capillary Technique —— 311
          Casting Thin PAA Gels by Cassette Technique —— 312
          Casting Ultrathin PAA Gels by Flap Technique —— 314
          References —— 315

Agarose and polyacrylamide (PAA) gels are most widespread precast gels.

## 2.16.1 Precast agarose gels

The precast agarose gels can be used for months, if stored at 4 °C. They should not be frozen because they shrink. Their thickness is usually 0.5 mm. The precast agarose gels are most commonly used in the clinical diagnosis: they are suitable for zone electrophoresis of proteins or for immunofixation and affinity electrophoresis.

The *precast agarose gels for zone electrophoresis* of proteins are used for separation of serum, cerebrospinal fluid, and urine proteins. Besides, they can be used for separation of hemoglobins and isoenzymes of creatine kinase, lactate dehydrogenase, and alkaline phosphatase (ALP). The *precast gels for immunofixation* of proteins are used for analyzing serum proteins, immunoglobulins G, A, and M, and their light chains (κ and λ). The *precast gels for affinity electrophoresis* of proteins contain lectin. They are used for separation of the isoenzymes of alkaline phosphatase.

## 2.16.2 Precast polyacrylamide gels

The precast PAA gels on a support film or a net are usually 0.5 mm thick. They are covered with a thin polyester film and are packed in polyethylene envelopes. The precast PAA gels can be stored at 4 °C for months, regardless of whether they are intended for disc-electrophoresis (discontinuous electrophoresis), SDS disc-electrophoresis, isoelectric focusing with carrier ampholytes, isoelectric focusing with immobilines, or for rehydratable PAA gels.

The *precast PAA gels for disc-electrophoresis* of native electrophoresis consist of a stacking gel and a resolving gel, or contain no stacking gel. The resolving gel has a PAA concentration (*T*) from 8 to 18 g/dL or even more. They are supplied with gel

https://doi.org/10.1515/9783110761641-026

or paper strip electrodes. The precast PAA gels for disc-electrophoresis are cast on a film or a net of different sizes.

The *precast gels for SDS disc-electrophoresis of proteins* can contain the Laemmli or other buffer system. The gels from ELPHO, Nuremberg, are 0.5 mm thick. They are cast on a film or net [1], and are covered with a polyester film. Prior to electrophoresis, the cover film should be removed so that an application template can be placed onto the gel. The net-supported SDS gels are suitable for electroblotting. All gels are available in different sizes: 51 × 82, 100 × 80, 125 × 125, and 250 × 125 mm.

Pharmacia Biotech, Uppsala (Amersham Pharmacia Biotech), also offers homogeneous and gradient gels for SDS electrophoresis. The $T$-concentration of the homogeneous gels is 7.5, 12.5, or 15 g/dL; the $T$-concentration of the gradient gels is 8–18 or 12–14 g/dL. The gradient gels of $T = 8$–18 g/dL are available as *ExcelGels*. They are 0.5 mm thick, contain a TRIS-acetate-TRICINEate buffer system [2], and consist of a stacking part and a resolving part. The electrode contact with the gel is transmitted through gel electrode strips, which contain buffer.

The *precast PAA gels for isoelectric focusing with carrier ampholytes* are 0.15–0.50 mm thick. Their PAA concentration $T$ is 5 g/dL and the cross-linking degree $C$ is 0.03. The precast gels for isoelectric focusing with carrier ampholytes form different pH gradients (Figure 2.16.1). They are cast on a support film or a net. The net gels are suitable for blotting.

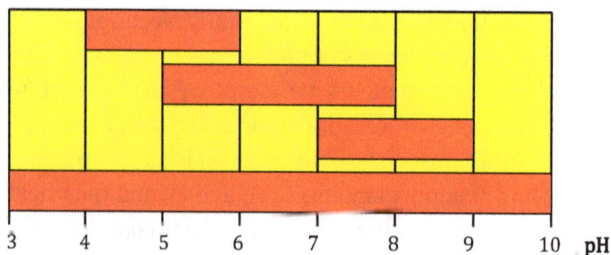

**Figure 2.16.1:** pH gradients, which can be obtained with precast gels.

The *precast PAA gels for isoelectric focusing with immobilines* are usually 0.5 mm thick. They are cast on a film or a net, and are supplied with gel or paper electrode strips containing an anode and a cathode solution. The immobiline precast gels are suitable for pH ranges of 4.0–10.0, 4.0–7.0, 7.0–10.0, and 5.0–6.0, and are available in dimensions of 51 × 82, 100 × 0, 125 × 125, and 250 × 125 mm.

## 2.16.3 Rehydratable polyacrylamide gels

*The rehydratable PAA gels* (CleanGels) represent rehydratable homogeneous PAA gels with a stacking gel of $T = 5$ g/dL and a resolving gel of $T = 10$ g/dL. CleanGels can be rehydrated in native or SDS buffers. They can be stored at −20 °C almost indefinitely. Immobiline DryStrips for pH gradients 3.0–10.0, 3.0–10.5, and 4.0–7.0 are also offered.

## 2.16.4 Protocols

### Silanization of Glass Plates

*The upper glass plate of a casting cassette comes in contact with the gel; therefore, it should be hydrophobic. To do this, it should be coated with Repel-silane (Pharmacia Biotech, Uppsala) or dichlorodimethyl silane (Merck, Darmstadt) (Figure 2.16.2).*

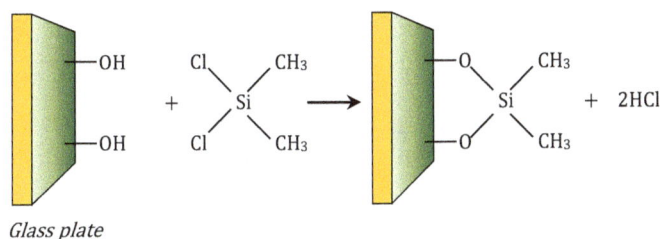

*Glass plate*

**Figure 2.16.2:** Binding of dichlorodimethyl silane (Repel-silane) to a glass surface.

### Materials and equipment
Glass plates
Repel-silane or dichlorodimethyl silane

### Procedure
– Smear a glass plate with 2 mL/dL Repel-silane in chloroform or with 2 mL/dL dichlorodimethyl silane in 1,1,1-trichloroethane.
– Dry the plate until the solution evaporates.
– Wash the plate with running and deionized water and dry. So hydrochloric acid, gained at silanization, is removed.

### Casting Agarose Gels on Support Film by Capillary Technique

*The capillary technique is used for casting thin (0.5 mm) agarose gels.*

**Materials and equipment**
Agarose
Support film
Glass plates
Repel-silane
Silicone spacers
Clamps
Syringe or pipette

**Procedure**
- Place some drops of water onto a glass plate.
- Lay a support film down on the glass plate and roll over so that any air bubbles are removed.
- Hydrophobilize a second glass plate with a repellent solution and mount 0.5 mm silicone spacer on its long sides.
- Fix the glass plates together with clamps to make a cassette.
- Heat the cassette at 60 °C in a drying oven.
- Place the sandwich horizontal onto a flat surface.
- Let the agarose solution penetrate into the cassette using a preheated syringe or pipette.
- Let the cast solution harden at room temperature for 30 min and then at 4 °C for additional 30 min.
- Disassemble the cassette and remove the gel from the cassette. Use the gel or cover it with a polyethylene film and store in a refrigerator at 4 °C for weeks.

### Casting Thin PAA Gels by Cassette Technique

*The cassette technique [3] is used for casting 0.5–1.0 mm (Figure 2.16.3).*

**Materials and equipment**
Glass plates
Repel-silane
Support film for PAG
Photo roller
U-form silicone gasket
Clamps
Monomeric solution

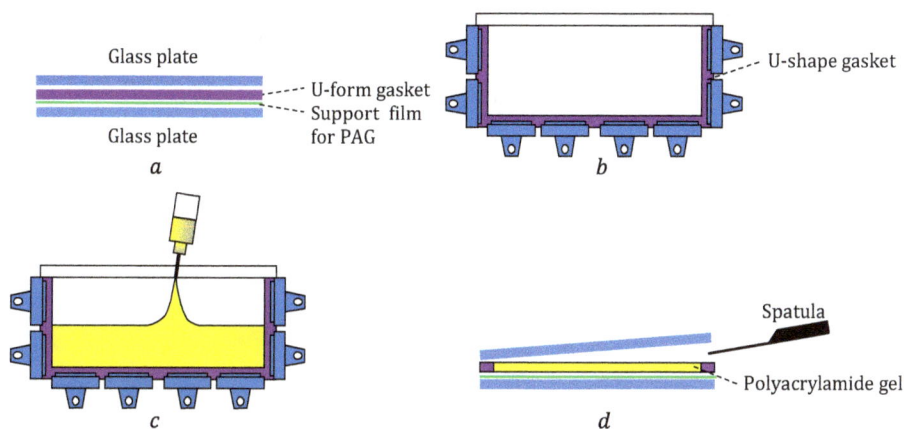

**Figure 2.16.3:** Steps (*a*, *b*, *c*, and *d*) of casting polyacrylamide gel by a cassette technique.

## Procedure

– Put a few drops of water onto a glass plate, place a support film over it, with its hydrophilic side up, and roll the film with a photo roller to remove the air bubbles. Thus, a thin film of water is created between the glass plate and the support film, which holds them together (Figure 2.16.4).

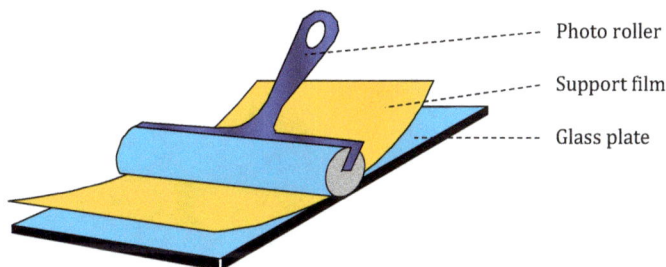

**Figure 2.16.4:** Rolling a support film.

– Place a 0.5–1.0 mm thick U-shaped silicone gasket on the margins of the support film and press it slightly down.
– Smear a second glass plate of same dimensions (cover glass) with a repelling solution.
– Gather the two glasses in a cassette, fix it with clamps, and set it vertically.
– Cast the monomeric solution into the cassette using a syringe or a pipette. After 1 h, remove the clamps from the cassette, disassemble it, and take the gel off. The gel can be used immediately for electrophoresis or can be stored in a plastic envelope for weeks in the refrigerator at 4 °C.

## Casting Ultrathin PAA Gels by Flap Technique

*The flap technique [4] is suitable for production of ultrathin (0.05–0.30 mm) PAA gels, even with dimensions of 40 × 20 cm (Figure 2.16.5).*

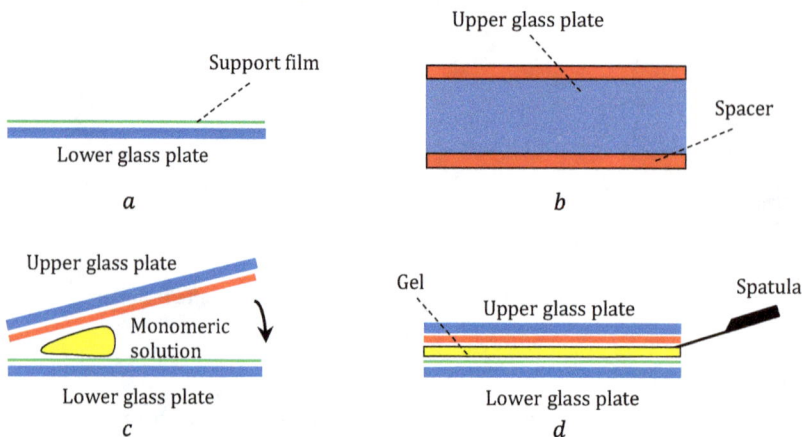

**Figure 2.16.5:** Steps (*a*, *b*, *c*, and *d*) of polyacrylamide gel casting by flap technique.

## Materials and equipment
Glass plates
Repel-silane
Support film for PAG
Silicone spacers
Monomeric solution
Spatula

## Procedure
– Put a few drops of water onto a horizontal glass plate (lower glass plate) and place over a support film with its hydrophobic side down. Roll the support film with a photo roller to remove the air bubbles.
– Smear over a second glass plate (upper glass plate) using a fluff-free cloth soaked with repelling solution to make the glass surface hydrophobic.
– Mount an adhesive tape onto the longer sides of the upper glass plate as spacers, using one or more layers depending on the thickness of the gel required (one layer of the tape provides 0.15 mm thickness).
– Apply a degassed monomeric solution onto the middle of the support film.
– Place slowly the upper glass plate onto the support film, starting at a short end of the lower plate, and avoiding air bubbles.

- Let the monomeric solution polymerize in the resulting sandwich for 60 min at room temperature. Remove the gel with the support film from the glass plates with the help of a spatula and use it, or envelope it into a plastic bag and store it in the refrigerator at 4 °C for up to several weeks.

## References

[1] Michov BM. GIT Fachz Lab, 1992, 36, 746–749.
[2] Schägger H, Jagow G. Anal Biochem, 1987, 166, 368–379.
[3] Görg A, Postel W, Westermeier R. Anal Biochem, 1978, 89, 60–70.
[4] Radola BJ. Electrophoresis, 1980, 1, 43–56.

# 3 Electrophoresis of nucleic acids

Buffers for electrophoresis of nucleic acids —— 317
Polymerase chain reaction —— 318
Procedure —— 318
Applications —— 320
Surface electrophoresis —— 321
References —— 323

The nucleic acid electrophoresis as well as the protein electrophoresis can take place in continuous or discontinuous buffer systems. The electrophoresis in continuous buffers is referred to as zone electrophoresis, whereas the electrophoresis in discontinuous buffer systems is referred to as disc-electrophoresis.

Double-stranded DNA fragments naturally look like rods, whose migration through a gel is a function of their size and gyration [1, 2]. Single-stranded DNA or RNA tends to fold up into structures with complex shapes, which migrate through a gel in a complicated manner because of their structure. Agents that disrupt hydrogen bonds, such as NaOH or formamide, are used to denature the nucleic acids and cause them to resemble long rods [3].

The electrophoretic separation of nucleic acids is carried out in weak acidic, neutral, or alkaline solutions with pH range of 6–10. Under these conditions the nucleic acids behave themselves as polyanions (nucleates) and move in an electrical field to the anode. If the pH value is too low, the bases cytosine and adenine will protonate and obtain positive electric charges; if the pH value is too high, the bases thymine and guanine will deprotonate and gain negative electric charges. In both cases, additional electric charges will cause denaturation of nucleic acids and will change their mobility.

## Buffers for electrophoresis of nucleic acids

For DNA electrophoresis, buffers with pH values in the range of 7–9 are best suited. They should contain electrolytes whose $pK_c$ values are close to the desired pH value. In Table 3.1, the most commonly used buffers for nucleate electrophoresis are given.

Borate ion in the TRIS-**b**orate-**E**DTA (TBE) buffer is a strong inhibitor for many enzymes. As a result, a DNA sample in a TBE buffer can better keep its integrity and, therefore, the size of DNA nucleates can be better analyzed.

TRIS-**a**cetate-**E**DTA (TAE) buffer is used as a running buffer in denaturing gel electrophoresis methods for mutation analysis. High concentration of salts in the sample, for instance, NaCl, retards the DNA mobility.

https://doi.org/10.1515/9783110761641-027

**Table 3.1:** Most commonly used buffers for nucleate electrophoresis.

| Buffer's name | Description | Composition |
|---|---|---|
| TBE (TRIS-borate-EDTA) buffer pH = 8.3 | Low ionic strength buffer which can be used for both preparative and analytical agarose gel electrophoresis. Most commonly used for polyacrylamide gel electrophoresis. Rarely used for sequencing electrophoresis. | **For 1 L 5× buffer** TRIS 54.0 g Boric acid 27.5 g 0.5 mol/L EDTA 20.0 mL *Adjust pH to 8.3 with HCl* Deionized water to 1,000.0 mL |
| TAE (TRIS-acetate-EDTA) buffer pH = 8.0 | Used for analytical and preparative agarose gel electrophoresis when DNA is purified by glass beads. Sometimes it is used in pulse-field electrophoresis. | **For 1 L 50× buffer** TRIS 242.0 g Glacial acetic acid 57.1 mL 0.5 mol/L EDTA 100.0 mL Deionized water to 1,000.0 mL |

## Polymerase chain reaction

The **p**olymerase **c**hain **r**eaction (PCR) was developed in 1983 by Kary Mullis [4, 5], who was awarded the Nobel Prize in Chemistry in 1993. It is a technique to amplify DNA segments *in vitro* [6, 7]. PCR mimics the DNA replication, but confines it to DNA sequences of interest. It requires about 500-fold less DNA than Southern blotting and is less time-consuming.

There are two PCR methods for simultaneous detection and quantification of DNA. The first method uses fluorescent dyes that are retained nonspecifically in between the double DNA strands. The second method uses probes that are fluorescently labeled. The **q**uantitative **PCR** (qPCR) techniques are applied to determine quantitatively the levels of gene expression. An interesting method is the combination of reverse transcription and qPCR. With the help of this technique, mRNA is converted to cDNA, which is further quantified using qPCR. This method lowers the possibility of error in PCR [8] increasing the detection of alterations of gene expression in microbes, tumors, or other diseases [9].

### Procedure

A PCR setup requires several components [10]:
- A *buffer* providing a suitable pH value and ions necessary for the DNA polymerase optimum activity. The following buffer is often used: 10–50 mmol/L TRIS-HCl with pH = 8.3–9.0 containing up to 50 mmol/L KCl and 0.5–5 mmol/L $MgCl_2$. Instead of $Mg^{2+}$, $Mn^{2+}$ can also be used.

– A *DNA template* that contains the DNA target region to be amplified.
– Two DNA *primers* that are complementary to the 3'-ends of each of the sense and anti-sense strands of the DNA target. They are short synthetic oligonucleotides, which contain from 14 to 40 mononucleotides. The optimal concentration of the primers is 0.1–1 μmol/L.
– *Deoxynucleoside (deoxynucleotide) triphosphates* dATP, dGTP, dTTP, and dCTP. They take place in the synthesis of new DNA strands with the help of DNA polymerase according to the principle of complementarity. The four dTTPs should be present at equimolar concentrations in the reaction mixture. Their optimal concentration is 20–200 μmol/L.
– *DNA polymerase* – an enzyme that catalyzes the synthesis of new DNA strands. It binds to a double-stranded region of DNA and uses the primers. The heat-resistant Taq polymerase is commonly used, as it retains its activity at 94 °C for 45–60 min; however, its optimal temperature is 72 °C. The Taq polymerase was first isolated by Chien *et al.* from the microorganism **Thermophylus aquaticus** (Taq) [11].

The PCR is carried out in the following sequence: First, the DNA molecule in the sample is denatured with heat giving two DNA strands. Then two DNA primers are added to occupy the 3'- and 5'-ends of the two DNA strands. Finally, bacterial thermostable Taq polymerase catalyzes the elongation of each primer using four **d**eoxy-**n**ucleoside **t**riphos**p**hates (dNTP).

Every PCR cycle consists of several temperature steps. The temperature in each cycle conforms to the enzyme used for DNA synthesis, the concentration of bivalent ions, dNTPs in the reaction, and the melting temperature of the primers [12]. The cycle steps are as follows:
– *Initialization:* This step is used for thermostable DNA polymerases that require heat activation [13]. During this step, the reaction chamber is heated up to 94–98 °C for 1–10 min.
– *Denaturation:* During this step, the reaction chamber is heated to 94–98 °C for 20–30 s. This causes DNA denaturation (melting) of the double-stranded DNA template by breaking hydrogen bonds between complementary bases, yielding two single-stranded DNA molecules (Figure 3.1).
– *Annealing:* The reaction temperature is lowered to 50–65 °C for 20–40 s. At this temperature, the two primers are annealing to each of the single-stranded DNA templates on the 3'-end of each strand.
– *Elongation:* The temperature for this step depends on the DNA polymerase used. The optimum temperature for the thermostable Taq polymerase is approximately 75–80 °C [14, 15]. The DNA polymerase synthesizes a new DNA strand complementary to the DNA template strand by adding free dNTP from the reaction mixture in the 5'- to 3'-direction. DNA polymerases polymerize a thousand bases per minute. Under optimal conditions, at each elongation step the number of DNA target sequences is doubled. The original template strands plus all newly generated strands

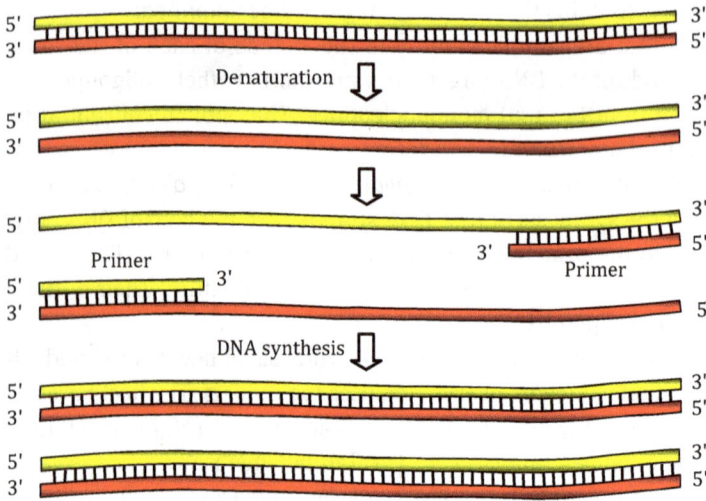

**Figure 3.1:** Steps of polymerase chain reaction.
The strands of double DNA are separated from each other by heating (denaturation), and primers are joined complementarily to their ends. DNA polymerase extends the primers in direction 5′ → 3′ and synthesizes two new chains, which are complementary to the template chains. After multiple cycles of heating and cooling, many copies of the DNA segment are produced.

become template strands for the next round of elongation, leading to exponential amplification of the specific DNA target region.
- *Final elongation:* This step is performed at 70–74 °C for 5–15 min after the last PCR cycle to ensure that any remaining single-stranded DNA is elongated.
- *Final hold:* The reaction chamber is cooled to 4–15 °C.

In automating the process, the time for one cycle is a few minutes. For 20 cycles, the amount of starting DNA is increased $10^6$ times and for 30 cycles it increased $10^5$ times. Most PCR methods amplify DNA fragments between 0.1 and 10 **kilobase pairs** (kbp), although some techniques allow for amplification of up to 40 kbp [16]. These DNA copies are sufficient in quantity to be used in hybridization techniques.

## Applications

PCR has been applied in many areas of molecular genetics:
- Known segments of DNA can easily be produced with the help of PCR.
- PCR is used for studying the gene expression.
- Using PCR, deletions, insertions, translocations, or inversions of DNA can be analyzed.
- PCR generates probes for Southern or northern blot hybridization.

- Although DNA breaks down over time, PCR has been successfully applied to clone ancient genes. For example, the PCR techniques have been successfully used on animals, such as a 40,000-year-old mammoth, and also on men, ranging from the Egyptian mummies to the Russian tsar relatives.

PCR has also been applied to a large number of medical procedures:
- PCR is used for genetic testing. DNA samples for prenatal investigation can be obtained by amniocentesis and chorionic villus sampling, or fetal cells circulating in the mother's bloodstream can be used.
- PCR permits early diagnosis of malignant diseases such as leukemia and lymphomas.
- PCR can also be used as a sensitive test for tissue typing, which is vital for organ transplantation. There is a proposal to replace the antibody-based tests for blood type with PCR-based tests [17].
- The PCR methods are widely performed to detect the presence of a viral infection soon after infection and before the onset of disease [18]. This is very important because the presence of antibodies to the virus circulating in the bloodstream do not appear until weeks after infection. For this purpose, the primers used should be specific to the targeted sequences in the viral DNA.
- Forensic DNA typing is an effective way of identifying or exonerating criminal suspects due to the analysis of criminal evidence discovered. A single human hair with attached hair follicle has enough DNA to conduct the analysis. Similarly, a few sperm, skin samples from under the fingernails, or a small amount of blood can provide enough DNA for conclusive analysis.
- DNA fingerprinting can help in paternity testing, where an individual DNA from unidentified human can be compared with that from possible parents, siblings, or children. The biological father of a newborn can also be confirmed.
- PCR can also be used to determine the sex of not only ancient specimens but also of suspects in crimes [19].

## Surface electrophoresis

A new DNA electrophoresis, called surface electrophoresis (SE), was emerged by Han and Craighead [20]. It involves electrostatic interactions between nucleates and surfaces [21, 22].

Compared with the existing methods, SE has advantages, such as simple device fabrication, low electric field strength (<10 V/cm), and a broad molecular size separation in the range of at least five orders of magnitude [23]. In addition, the resolution of this technique is independent from the bp number. However, the resolution of SE is poor because of the strong diffusion.

SE is carried out in a buffer on the surface of a flat silicon wafer. The separation of DNA nucleates is a result between their moving, from one side, and the friction of the surface and buffer, from other side. The friction can be controlled by coating the silicon surface with silane monolayer film, which permits the separation of different DNA sizes [24].

A single DNA chain on a surface is composed of loops, trains, and tails. Its length correlates with the ionic strength: higher ionic strength leads to a smaller flexible DNA chain. According to Li *et al.* [25], when migrating on a surface, a DNA chain experiences three different forces: driving force, buffer friction force, and surface friction force (Figure 3.2).

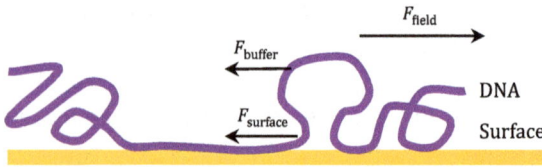

**Figure 3.2:** Forces experienced by a single DNA chain migrating in buffer on surface under electric field.
$F_{field}$ – driving force applied by the electric field; $F_{buffer}$ – friction force created by the buffer; $F_{surface}$ – friction force created by the surface

Let us define $E$ as the electric field strength, $Q$ as the electric charge per bp, $N_{bp}$ as the total number of bp, $f_b$ as the friction coefficient per bp between the DNA chain and the buffer, $v$ as the velocity of the DNA chain, $f_s$ as the friction coefficient per bp between the DNA chain and the surface, $\varepsilon$ as the interaction strength between 1 bp and the surface, and $\alpha$ as the train ratio. Then, the driving force

$$F_{field} = EQN_{bp} \tag{3.1}$$

the buffer friction force

$$F_{buffer} = f_b v N_{bp} \tag{3.2}$$

and the surface friction force

$$F_{surface} = f_s \varepsilon \alpha N_{bp} \tag{3.3}$$

If these three forces are in balance, then

$$EQN_{bp} = f_b v N_{bp} + f_s \varepsilon \alpha N_{bp} \tag{3.4}$$

Taking into account that the mobility

$$\mu = \frac{v}{E} \tag{3.5}$$

it follows from eqs. (3.4) and (3.5) that

$$\mu = \mu_0 \left(1 - \frac{f_s \varepsilon \alpha}{QE}\right) \tag{3.6}$$

where

$$\mu_0 = \frac{Q}{f_b} \tag{3.7}$$

is the free-buffer mobility when $\alpha$ equals zero. Since the mobility should be larger than zero, the threshold in the electric field strength is

$$E_{\min} = \frac{f_s \varepsilon \alpha}{Q} \tag{3.8}$$

According to eq. (3.6), when $\alpha = 0$, i.e., when there is no adsorption of DNA on the surface, $\mu$ is the free-buffer mobility. When $\alpha = 1$, i.e., when there is full adsorption of DNA on the surface, $\mu$ has a fixed value. Therefore, the length-dependent mobility is a function of $\alpha$, which is a function of the chain length. Also according to eq. (3.6), $\mu$ increases with the electric field strength and would approach the free-buffer mobility, when the electric field strength overwhelms the surface friction force.

A device for SE contains separation cell, Pt electrode, silicon substrate, TBE buffer, laser source, dichroic mirror, bandpass filter, and computer. Lee and Kuo [26] fabricated a gel-free microchannel electrophoresis device for surface separation of DNA fragments ranging from 3.5 to 21.2 kbp. Another gel-free microchannel device for SE of DNA was fabricated with polydimethylsiloxane [27].

# References

[1]   Grosberg AY, Khokhlov AR. In Atanov YA (ed.). Statistical Physics of Macromolecules. American institute of physics press, New York, 1994, Vol. 3, 217–220.
[2]   Flory PJ. Principles of Polymer Chemistry. Cornell university, 1953, 428–429.
[3]   Troubleshooting DNA Agarose Gel Electrophoresis, Focus, 1997, 19, 66.
[4]   Bartlett JMS, Stirling D. PCR Protocols. Methods in Molecular Biology, 2nd edn. 2003, Vol. 226, 3–6.
[5]   Mullis KB et al. Process for Amplifying, Detecting, and/or Cloning Nucleic Acid Sequences. US Patent 4683195.
[6]   Saiki R, Scharf S, Faloona F, Mullis K, Horn G, Erlich H, Arnheim N. Science, 1985, 230, 1350–1354.

[7]   Saiki R, Gelfand D, Stoffel S, Scharf S, Higuchi R, Horn G, Mullis K, Erlich H. Science, 1988, 239, 487–491.
[8]   Garibyan A. J Invest Dermatol, 2013, 133, 1–4.
[9]   Garibyan L, Avashia N. J Invest Dermatol, 2013, 133, 1–4.
[10]  Sambrook J, Russel DW. Molecular Cloning: A Laboratory Manual, 3rd edn. Cold Spring Harbor Laboratory Press, NY, 2001, Chapter 8.
[11]  Chien A, Edgar DB, Trela JM. J Bact, 1976, 127, 1550–1557.
[12]  Rychlik W, Spencer WJ, Rhoads RE. Nucl Acids Res, 1990, 18, 6409–6412.
[13]  Sharkey DJ, Scalice ER, Christy KG, Atwood SM, Daiss JL. Bio/Technol, 1994, 12, 506–509.
[14]  Chien A, Edgar DB, Trela JM. J Bacteriol, 1976, 127, 1550–1557.
[15]  Lawyer F, Stoffel S, Saiki R, Chang S, Landre P, Abramson R, Gelfand D. PCR Methods App, 1993, 2, 275–287.
[16]  Cheng S, Fockler C, Barnes WM, Higuchi R. Proc Natl Acad Sci USA, 1994, 91, 5695–5699.
[17]  Quill E. Science, 2008, 319, 1478–1479.
[18]  Cai H, Caswell JL, Prescott JF. Vet Pathol, 2014, 51, 341–350.
[19]  Alonso A. Forensic Sci Intern, 2004, 139, 141–149.
[20]  Han J, Craighead HG. Science, 2000, 288, 1026–1029.
[21]  Seo YS, Samuilov VA, Sokolov J, Rafailovich M, Tinland B, Kim J, Chu B. Electrophoresis, 2002, 23, 2618–2625.
[22]  Seo YS, Luo H, Samuilov VA, Rafailovich MH, Sokolov J, Gersappe D, Chu B. Nano Lett, 2004, 4, 659–664.
[23]  Luo H, Gersappe D. Electrophoresis, 2002, 23, 2690–2696.
[24]  Pernodet N, Samuilov V, Shin K, Sokolov J, Rafailovich MH, Gersappe D, Chu B. Phys Rev Lett, 2000, 85, 5651–5654.
[25]  Li B, Fang X, Luo H, Petersen E, Seo YS, Samuilov V, Rafailovich M, Sokolov J, Gersappe D, Chu B. Electrophoresis, 2006, 27, 1312–1321.
[26]  Lee HH, Kuo Y. Jpn J Appl Phys, 2008, 47, 2300–2305.
[27]  Ghosh A, Patra TK, Kant R, Singh RK, Singh JK, Bhattacharya S. Appl Phys Lett, 2011, 98, 164102.

# 3.1 Agarose gel electrophoresis of nucleic acids

3.1.1      Zone polyacrylamide gel electrophoresis of native nucleic acids —— 326
3.1.1.1   Theory of sieving migration —— 326
           Factors affecting migration of nucleic acids —— 326
3.1.1.2   Practice of zone agarose gel electrophoresis of native nucleic acids —— 327
           Submarine electrophoresis of nucleic acids —— 327
           Restriction fragment length polymorphism —— 328
           Variable number tandem repeat —— 329
           DNA profiling —— 329
           Paternity or maternity testing for child —— 330
           Interpretation of DNA test results —— 331
3.1.2      Zone agarose gel electrophoresis of denatured nucleic acids —— 332
3.1.3      DNA sequencing —— 332
3.1.3.1   Maxam–Gilbert sequencing method —— 332
3.1.3.2   Sanger sequencing method —— 333
3.1.3.3   Single-strand conformation polymorphism method —— 334
3.1.3.4   DNase footprinting assay —— 336
3.1.3.5   Nuclease protection assay —— 337
           $S_1$-nuclease protection assay —— 337
3.1.4      RNA separation —— 338
3.1.4.1   Primer extension assay —— 339
3.1.5      Protocols —— 339
           Native DNA Electrophoresis on Agarose Gels —— 339
           Sanger Sequencing Reactions Using *Taq* DNA Polymerase —— 340
           References —— 341

Native agarose gel electrophoresis of nucleic acids is an analytical technique for separating DNA or RNA fragments by size. Nucleates move in an electric field to the anode because their phosphoric residues are deprotonated and, as a result, negatively charged. Longer DNA nucleates migrate more slowly because they undergo more resistance within the gel; smaller fragments migrate faster.

The nucleic acids to be separated can be prepared in several ways. When the DNA nucleates are very large, they should be cut into smaller fragments using DNA restriction endonucleases. Xylene cyanol or Bromophenol blue is added often to DNA samples to build a moving front. These dyes run with DNA fragments that are 5,000 bp or 300 bp in length, respectively. Other progress markers are Cresol red and Orange G, which run as fast as 125 bp DNA and 50 bp DNA, respectively.

The electrophoretic separations of nucleic acids are carried out in a pH range from 6 to 10. If the pH value is very low, the bases cytosine and adenine bind protons and are positively charged; if the pH value is too high, the bases thymine and guanine split protons and are negatively charged. In both cases, the additional electric charges cause denaturing of nucleic acids and change their mobilities.

DNA is commonly visualized after reacting with ethidium bromide and viewed under UV light. Dyes such as SYBR green, GelRed, methylene blue, and crystal

https://doi.org/10.1515/9783110761641-028

violet are also available. The measurement and analysis of the resolved nucleates are mostly carried out using specialized software.

## 3.1.1 Zone polyacrylamide gel electrophoresis of native nucleic acids

The zone polyacrylamide gel electrophoresis of native nucleic acids is carried out in one buffer.

### 3.1.1.1 Theory of sieving migration

The widely accepted Ogston model treats the gel matrix as a sieve (Ogston sieving) consisting of randomly distributed network of interconnected pores [1, 2]: A random coil DNA moves almost free through the connected pores, but the movement of larger molecules is impeded and slowed down, which leads to separation of nucleates of different sizes. The more the radius of nucleate gyration approximates the size of a gel pore, the more likely the nucleate will pass through the pore.

The Ogston model, however, breaks down for very large molecules moving through pores significantly smaller than the nucleate size. For DNA molecules greater than 1 kb, a *reptation model* is most commonly used [3]. This model assumes that DNA polyions crawl in a snake-like fashion (reptation) through the pores. At higher electric field strength, the leading end of the molecule becomes strongly biased in the forward direction, and pulls the rest of the molecule along.

Double-stranded DNA fragments behave as long rods. Therefore, their migration through a gel is relative to their size or, for cyclic DNA fragments, their radius of gyration. Single-stranded DNA or RNA tends to fold up into different tertiary structure molecules and migrate through the gel in a complicated manner. Therefore, agents as sodium hydroxide or formamide that disrupt the hydrogen bonds are used to denature the nucleic acids and cause them to behave as long rods.

### Factors affecting migration of nucleic acids

A few factors that affect the migration of nucleic acids are: ionic strength of buffer, gel concentration, electric field strength, and concentration of ethidium bromide, if used during electrophoresis [4].

*Ionic strength of buffer.* The mobility of nucleic acids decreases when the buffer ionic strength increases [5]. This can be explained with the concept of the geometric and electrokinetic radius [6]. When the ionic strength increases, the electrokinetic radius of a nucleate, $a_{pi}$, grows, and its $\zeta$-potential decreases. The reverse takes place when the ionic strength decreases – then the electrokinetic radius decreases and the $\zeta$-potential increases.

*Gel concentration.* The gel concentration determines the size of the gel pores, which in turn affects the migration of DNA. Increasing gel concentration reduces the migration speed and improves the separation of smaller DNA nucleates, while lowering gel concentration permits large DNA nucleates to be separated.

*Electric field strength.* The mobility of high-molecular-mass DNA fragments increases when the electric field strength increases whereas the effective range of separation decreases. However, the voltage is limited because it heats the gel, which may cause DNA to melt.

*Concentration of ethidium bromide.* Ethidium bromide in the gel more strongly affects circular DNA than linear DNA. It intercalates into circular DNA, which has different charge, length, and superhelicity. Increasing ethidium bromide concentration can change DNA from a negatively supercoiled through a relaxed form to positively coiled superhelix [7].

### 3.1.1.2 Practice of zone agarose gel electrophoresis of native nucleic acids

Zone agarose gel electrophoresis is used in the following techniques: submarine electrophoresis of nucleic acids, restriction fragment length polymorphism, variable number of tandem repeat, DNA profiling, and more.

### Submarine electrophoresis of nucleic acids

The agarose gel electrophoresis is commonly carried out horizontally in a submarine mode whereby the slab gel is completely submerged in a buffer [8]. The submarine electrophoresis is a standard method for nucleic acid analyzing. Using it, DNA and RNA fragments can be separated and identified, and restriction fragments length polymorphism (RFLP) analysis [9] can be performed (see below).

The unit for submarine electrophoresis contains two buffer tanks, connected with each other by a flat bridge. The agarose gel, cast on a support film, is placed on the bridge [10] (Figure 3.1.1).

In practice, 0.8 g/dL agarose gels are used. Highly concentrated agarose gels (2–4 g/dL) are applied for analysis of PCR products. Higher ionic strength increases the resolution [11], but generates a lot of heat. This may destroy the helix structure of DNA.

The most common buffer for submarine zone electrophoresis is the TRIS-borate-EDTA buffer (TBE), which was introduced by Peacock and Dingman [12]. It has a pH value of 8.3 and TRIS concentration of 50–100 mmol/L. Michov [13] has established that the TRIS-borate buffers contain a complex compound formed by a condensation

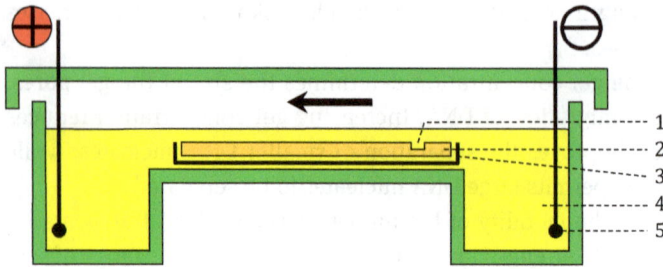

**Figure 3.1.1:** Unit for submarine electrophoresis.
1. Wells for samples; 2. agarose gel; 3. gel dish; 4. electrode buffer; 5. electrode wire.

reaction between boric acid and TRIS. The complex compound, which was referred to as *TRIS-boric acid*, has a zwitterionic structure. Its existence was reproved 30 years later [14].

According to Michov, another buffer, which can be used for nucleic acid separation, is the **TRIS-taurinate-EDTA** buffer (TTE) [15]. Its recipe is as follows:

| TRIS-taurinate-EDTA buffer, pH = 8.5, $I$ = 0.10 mol/L | |
| --- | --- |
| TRIS | 34.16 g |
| Taurine | 48.93 g |
| $Na_2EDTA \cdot 2H_2O$ | 0.37 g |
| Deionized water to | 1,000.00 mL |

## Restriction fragment length polymorphism

RFLP is a laboratory technique, with which differences between samples of homologous DNA molecules can be illustrated. It begins with fragmenting a sample double-helical DNA by restriction enzymes (restrictases) in sequences of lengths of 4–6 base pairs, in a process known as restriction digest. The restriction fragments are separated by length using agarose gel electrophoresis, and are transferred onto a membrane (Southern blotting). Hybridization between them and labeled DNA probe determines their length. *Hybridization* is a process of recombination of complementary chains of different origin (between different DNA molecules, DNA and RNA, or different RNA molecules). The test DNA strand is referred to as a *matrix*, and the DNA or RNA probe of known nucleotide sequence and size is referred to as a *probe*. The probes are labeled either with an isotope ($^{32}P$, $^{35}S$, or $^3H$) or with fluorescent ligands. A probe will hybridize with a matrix, if their chains are complementary to each other and can form a sufficient number of hydrogen bonds.

Restriction fragment length polymorphism is a useful method for genome mapping and genetic diseases analyzing. It plays an important role in the paternity testing, forensics, and DNA fingerprinting.

## Variable number tandem repeat

A **v**ariable **n**umber **t**andem **r**epeat (VNTR) is the location in a genome where a short nucleotide sequence is organized as a tandem repeat. It exists on many chromosomes, and often shows variations in length (number of repeats) among individuals. VNTR is useful in genetics and biology research, forensics, and DNA fingerprinting.

Let us explain the topic with an example: The probe and restriction enzyme could detect a region of the genome that includes a VNTR segment (the boxes on Figure 3.1.2). Allele *a* has four VNTR repeats, and the probe detects a longer fragment between the two restriction sites. In allele *b*, there are only two VNTR repeats, and the probe detects a shorter fragment between the same restriction sites. Insertions, deletions, translocations, and inversions of genes can also lead to restriction fragment length polymorphism.

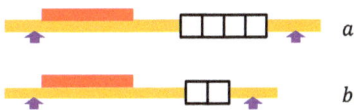

**Figure 3.1.2:** Schematic diagram of restriction fragment length polymorphism by variable number tandem repeat length variation.
*a* and *b* – alleles

VNTR analysis is used to study genetic diversity.

## DNA profiling

DNA profiling (DNA fingerprinting, DNA testing, or DNA typing) is a process of determining individual's DNA characteristics, which are as unique to individuals as the fingerprints. DNA profiling was developed in 1984 by Sir Alec Jeffreys while working in the Department of Genetics at the University of Leicester [16–18]. It is most commonly used as a forensic technique to identify an unidentified person, to place a person at a crime scene, or to eliminate a person from consideration [19]. DNA profiling is also used to help clarify paternity [20] and in genealogical or medical research.

DNA profiling uses repetitive sequences that are highly variable (VNTR). VNTR are similar between closely related individuals, but unrelated individuals are extremely unlikely to have the same VNTR. The DNA testing is carried out after collecting buccal cells using a swab. The swab has wooden or plastic stick handle with cotton on synthetic tip. The swab should rub the inside of a cheek to collect as many buccal cells as possible. When this is not available, a sample of drop of blood, saliva, sperm, vaginal lubrication, hair or other appropriate fluid or tissue from personal items (e.g., a toothbrush or razor) can be used. Next, DNA in the sample should be amplified, which is achieved by the polymerase chain reaction (see there).

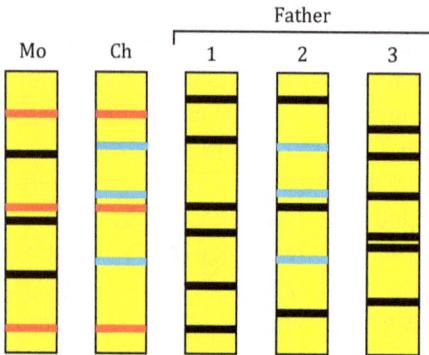

Figure 3.1.3: Example of DNA profiling in order to determine the father of a child. Child's DNA sample (Ch) should contain DNA bands of both parents: mother (Mo) and suspected father (Fa). In this case, man #2 is the father.

For paternity DNA profiling, samples obtained from mother, child, and possible fathers are analyzed. The child's DNA is a composite of its parent DNA. Electrophoretic bands in the child's DNA fingerprint that are not present in the mother's DNA should be contributed by the father (Figure 3.1.3).

DNA profiling may also be used to determine evolutionary relationships among organisms [21].

### Paternity or maternity testing for child

The genetic material of an individual is derived from the genetic material of both parents in equal amounts. It is known as individual genome or nuclear DNA, because it is contained in the nucleus. Beside the nuclear DNA, the mitochondria in the cells also have their own genetic material termed mitochondrial DNA. The mitochondrial DNA comes only from the mother, without any shuffling. Comparison of the mitochondrial genome is easier than that based on the nuclear genome. However, testing the mitochondrial genome can prove only, if two individuals are related through maternal lines. It could not be used to test for paternity.

In testing the paternity of a male child, comparison of the Y chromosome can be used, since it passes directly from father to son.

## Interpretation of DNA test results

A paternity or maternity test report lists the genetic profiles of each tested party. It also lists the relationship index (RI) for each marker – a statistical measure of how powerful a match is at a particular locus. Then RI are multiplied with each other to produce the combined paternity index (CPI), which is the genetic odds that the tested man is more likely to be the biological father. It is generally accepted by government agencies that a CRI of 100 and a probability of 99% or higher is strong proof that the tested man is the biological father.

The recombinant DNA technologies have great importance for forensic medicine. Let us illustrate this by the following example: A girl was found dead and raped. A police doctor collected sperm from the vagina and blood under the girl's nails. The sperm and blood were supposed to be from the rapist and murderer. DNA from the samples was amplified by PCR. Meanwhile, the police detained three men who were with the victim in the night of the murder. Their DNA was compared with the victim's DNA and with the sperm DNA from the vagina. The DNA profiling proved that DNA from the second suspect and sperm DNA had the same profile (Figure 3.1.4).

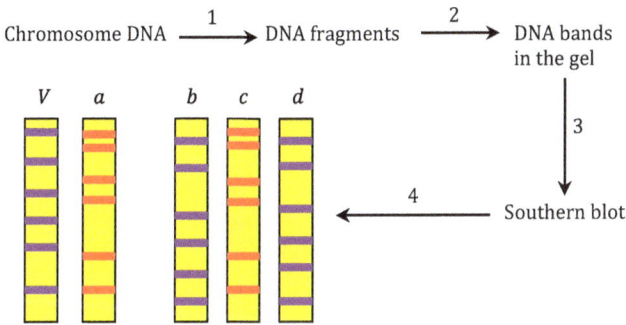

**Figure 3.1.4:** Comparative analysis of victim's DNA and DNA from three suspected men. The test proves that the murder is the suspected man *c*.
1 – DNA restriction with restriction enzymes; 2 – Gel-electrophoresis of DNA fragments;
3 – Southern blotting; 4 – Hybridization with radioactive sample, and autoradiography
*V* – Victim's DNA; *a* – Foreign DNA isolated from victim's body; *b, c, d* – DNA from suspicious men.

Many criminal cases were solved using the DNA profiling. For example, in 1994, Anna Anderson, who claimed to be Grand Duchess Anastasia Nikolaevna of Russia, was tested after her death using samples of her tissue that had been stored at Virginia hospital. The tissue was tested using DNA fingerprinting, which showed that she bore no relation to the Romanovs [22].

## 3.1.2 Zone agarose gel electrophoresis of denatured nucleic acids

The separation of DNA strands from each other is most often achieved by thermal denaturing (melting). Double-stranded DNA that contains more G–C base pairs (with three hydrogen bonds between the bases) requires higher temperature than double-stranded DNA that contains predominantly A–T base pairs (with two hydrogen bonds between the bases). Higher temperatures are also needed when DNA chains are very long and, as a result, contain more hydrogen bonds.

Denaturing agarose gel electrophoresis is used for single-stranded DNA or RNA, which consist of 500 to 20,000 mononucleotides, e.g. for **messenger RNA** (mRNA). RNA samples are most commonly separated on agarose gels containing formaldehyde as a denaturing agent to save the RNA secondary structure [23]. Another method for separation of high-molecular-mass RNA such as ribosomal RNA and their precursors by agarose-formaldehyde gel electrophoresis was described [24]. It can be used for subsequent analysis of RNA by Northern blotting.

## 3.1.3 DNA sequencing

Two DNA sequencing methods are known: Maxam–Gilbert method and Sanger method. In addition, other methods as **single-strand conformation polymorphism** (SSCP) analysis, **deoxyribonuclease** (DNase) footprinting assay, and nuclease protection assay are used for detailed DNA analysis.

### 3.1.3.1 Maxam–Gilbert sequencing method

The Maxam–Gilbert sequencing method, called chemical mode, is based on nucleobase-specific partial chemical modification of DNA and subsequent cleavage of the DNA backbone at sites adjacent to the modified nucleotides [25].

A DNA nucleate should be labeled at its 3'- or 5'-end. In the rule, a $\lambda$-$^{32}$P-ATP is linked at the 5'-end with the help of the enzyme *polynucleotide kinase*. Then, the labeled strand is cut in parts with the help of a *restriction endonuclease* and the obtained fragments are separated in an agarose or a polyacrylamide gel. Labeling at the 3'-end can be achieved with a terminal *transferase* and an $\alpha$-$^{32}$P-ATP followed by alkaline hydrolysis to remove all but the first adenylic acid residue, or by "filling in" the complementary strand of a 5'-single-stranded protruding end created by digestion with the restriction endonuclease. Then the labeled fragments are separated by gel electrophoresis and eluted from the gel.

The chemical treatment generates breaks at a small proportion of one or two of the four nucleotide bases in each of the four reactions (G, A + G, C, and C + T).

Thereafter, the DNA sample is incubated in 1 mol/L piperidine at 90 °C, where phosphate sugar backbones carrying modified bases are destroyed. In this way, a random population of fragments is obtained, whose different lengths correspond to the specific nucleotide position. The fragments in the four reactions are electrophoresed side by side in denaturing gels (Figure 3.1.5).

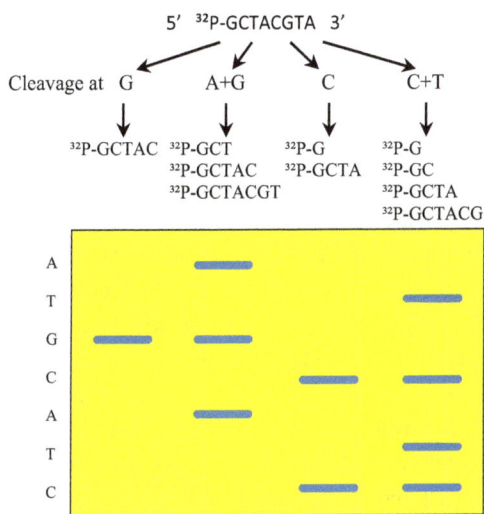

**Figure 3.1.5:** Maxam–Gilbert sequencing reaction.
Cleaving a tagged segment of DNA at different points yields tagged fragments of different size. The fragments are separated by gel electrophoresis.

To visualize the separated bands, the gel is exposed to X-ray film for autoradiography, yielding a series of dark bands, each showing the location of identical radiolabeled DNA nucleates. From the presence or absence of certain fragments, the sequence may be inferred. An automated Maxam–Gilbert sequencing protocol was developed by Boland *et al.* [26].

## 3.1.3.2 Sanger sequencing method

Sanger sequencing (chain termination) method is based on the selective incorporation of chain-terminating dideoxynucleotides by DNA polymerase during DNA replication [27, 28]. It is the most widely used sequencing method.

The normal DNA nucleotides wear no –OH group at 2'-position of ribose, while the –OH group at 3'-position is coupled to the next nucleotide. The **di**deoxynucleoside **tri**phosphates (ddNTP): ddATP, ddTTP, ddGTP, or ddCTP have no –OH groups, neither at their 2'- nor at their 3'-carbon atoms. So they lack a 3'-OH group required for the formation of a phosphodiester bond between two nucleotides, causing DNA polymerase to cease extension of DNA when a modified ddNTP is incorporated.

The DNA sample is divided into four separate reactions, containing all four deoxynucleotides (dATP, dGTP, dCTP and dTTP) and the DNA polymerase. The nucleotides are labeled radioactively or with fluorescent substances whereby a fluorescent substance binds to the 5'-end of the sequencing primer. To each reaction, only one of the four dideoxynucleotides is added. The concentration of each dideoxynucleotide is approximately 100-fold lower than the concentration of the corresponding deoxynucleotide (e.g., 0.005 mmol/L ddATP to 0.5 mmol/L dATP). This is necessary for enough DNA nucleotides to be produced.

If a specific oligonucleotide primer and the four dNTP are present in a mixture, the DNA polymerase synthesizes a 5'→3' nucleate, which is complementary to the DNA template. When a ddNTP is added to the growing nucleotide chain, the polymerase stops the polynucleotide chain elongation and DNA polymerization breaks off. The result is a fragment, whose length corresponds to the position of a particular mononucleotide in the DNA template. Thereafter, the DNA fragments are resolved by denaturing polyacrylamide gel electrophoresis and the DNA sequence is read out by a computer [29]. A fixed laser beam penetrates through the polyacrylamide gel and reaches a photocell, which is located below the resolving track. If a band hits the laser beam, the fluorescent DNA fragment emits light signals [30]. Thus, the migrating bands of each separating track are registered in their order, i.e., the sequence of the separated DNA, in a computer (Figure 3.1.6).

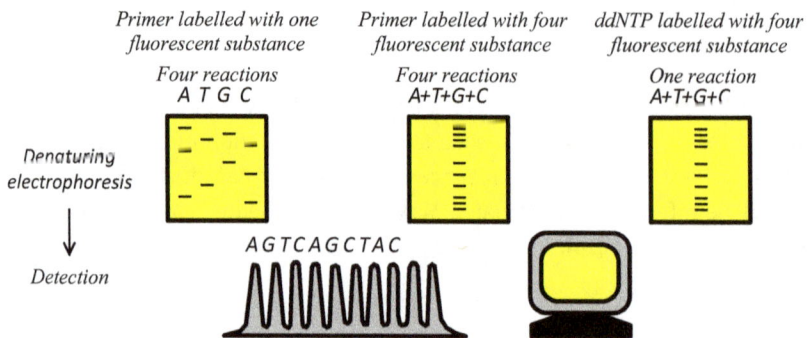

**Figure 3.1.6:** Principle of DNA sequencing according to Sanger.

### 3.1.3.3 Single-strand conformation polymorphism method

SSCP analysis is based on the conformational difference of single-stranded nucleotide sequences of identical length induced by differences in the sequences under certain experimental conditions. This property allows sequences to be distinguished using gel electrophoresis, which separates fragments according to their different conformations [31].

If DNA fragments are denatured by heat or formamide and are rapidly cooled, intramolecular base pairs are formed and single-strands obtain stable conformation. They get different three-dimensional conformations, because the two DNA strands differ in their base sequences, with the exception of the repeat sequence. The difference in shape between two single-stranded DNA strands with different sequences can cause them to migrate differently in an electrophoresis gel, even though the number of nucleotides is the same. The different structures of a DNA strand have different mobilities in a native polyacrylamide gel, so they can be seen as two bands [32] (Figure 3.1.7).

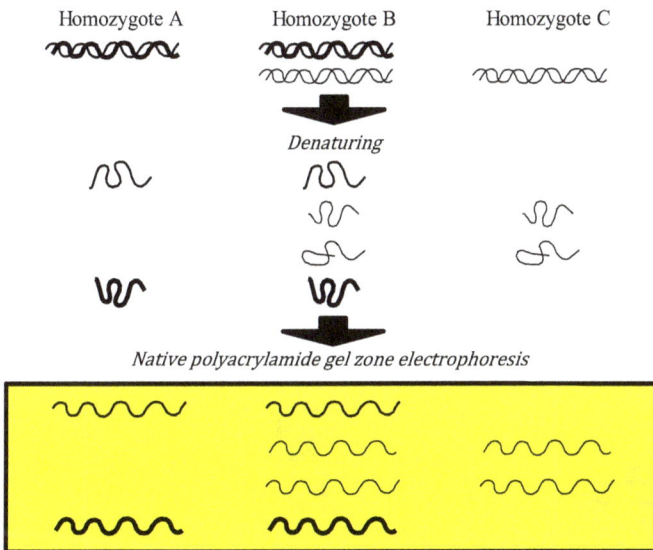

**Figure 3.1.7:** Scheme of SSCP analysis.

SSCP analysis is used in molecular biology, for example in genotyping to detect homozygous individuals of different allelic states, as well as heterozygous individuals [33]. Sequence variations such as small deletions or insertions and even single base substitutions can change the DNA conformation; therefore, they can result in a band shift in PAGE. For example, SSCP is widely used in virology to detect variations in different strains of a virus, if both strains have undergone changes due to mutation, which will cause the two particles to assume different conformations [34].

Single-stranded DNA weakly binds ethidium bromide. Therefore, PCR is needed to amplify 100–500 bp genomic DNA or complementary **DNA** (cDNA) before SSCP analysis. This can be made with the help of reverse transcription on an mRNA.

### 3.1.3.4 DNase footprinting assay

DNase footprinting assay [35, 36] uses the fact that a protein bound to DNA will protect DNA from enzymatic cleavage. With the help of this technique, a protein binding site on a DNA molecule can be located.

First, the enzyme DNase cuts a radioactively end-labeled DNA fragments. Then, the DNA fragments are resolved by gel electrophoresis. The gel is then exposed to a special photographic film. When a protein binds DNA, the binding site of DNA is protected from enzymatic cleavage. This protection will result in a clear area on the gel, which is referred to as "footprint". If DNA is marked at its ends, a marked and an unmarked single strand will be obtained (Figure 3.1.8). The unlabeled strand is not visible on the autoradiogram.

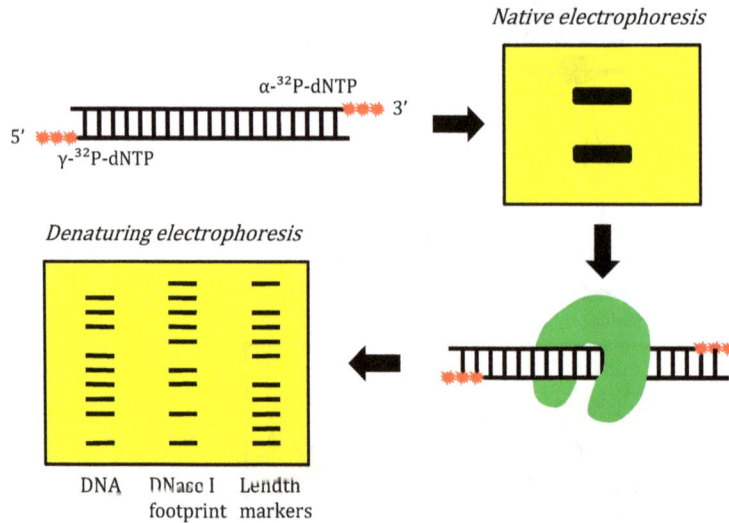

**Figure 3.1.8:** Principle of DNase footprinting assay.

Using DNase footprinting assay, the transcriptional control of MUC1 gene was investigated [37]. It encodes DF3 glycoprotein that is highly expressed in human breast cancer cells.

For analysis of RNA–protein interactions, similar "toeprinting" assay is used [38–40]. In this method, the primer extension reaction is used to examine the binding of protein to RNA, from its 3'- (toe) to the other 5'- (heel) side.

### 3.1.3.5 Nuclease protection assay

The nuclease protection assay is a technique used to identify RNA molecules of known sequence in a heterogeneous RNA sample [41, 42]. RNA is first mixed with DNA or RNA probes that are complementary to the sequence of interest. They form hybridized double-stranded DNA-RNA or RNA-RNA hybrids. Then, the mixture is exposed to ribonucleases that cleave the rest of probe and single-stranded RNA to individual mononucleotides but have no activity against the hybrids. They are analyzed by electrophoresis in a denaturing polyacrylamide gel, and are detected autoradiographically (Figure 3.1.9).

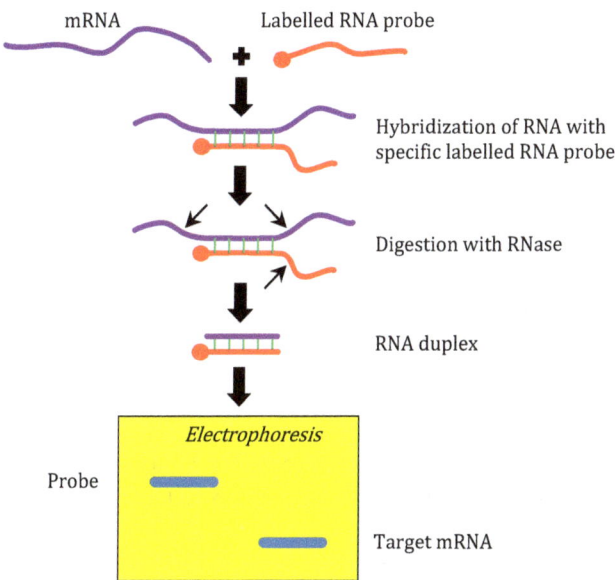

**Figure 3.1.9:** Principle of the nuclease protection assay: mRNA and probes, which are not complementary to each other, are degraded by nucleases. Only hybrids between mRNA and their probes remain to be separated by denaturing electrophoresis.

If the probe is labeled radioactively, the gel should be dried after electrophoresis to be prepared for autoradiography.

### $S_1$-nuclease protection assay

In order to find the 5'-end of an mRNA, $S_1$-nuclease protection assay is used. During the process, labeled DNA probes hybridize with mRNA sample. Then, the single-stranded DNA probe is destroyed by the $S_1$-nuclease and the hybrids are run on a denaturing polyacrylamide gel together with size markers differing in length by only one mononucleotide. So, the distance between the end of the probes and the 5'-end of the mRNA can be determined (Figure 3.1.10).

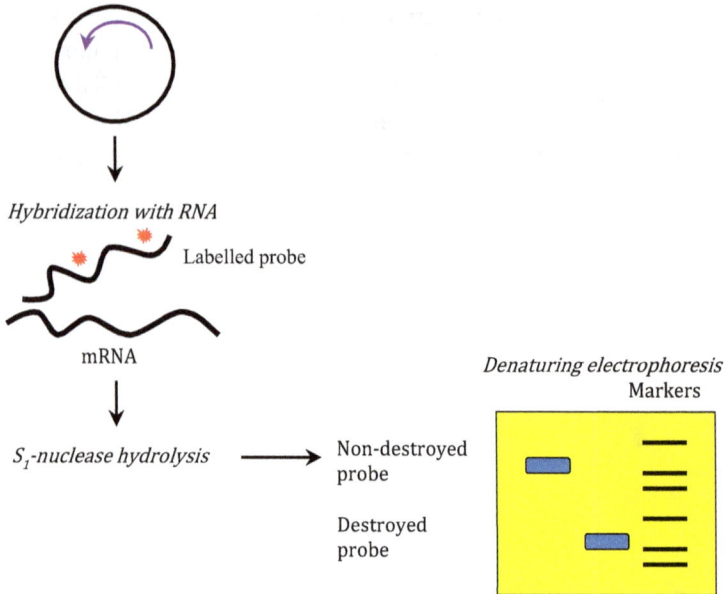

**Figure 3.1.10:** Principle of $S_1$-nuclease protection assay.

Using the $S_1$-nuclease protection assay, the promoter region of mouse gene δ/YY1 has been described [43], which is responsible for a transcription factor.

## 3.1.4 RNA separation

The separation of fragmented RNA finds place on polyacrylamide gels containing urea [44]. An RNA ladder is often run alongside the samples to show the size of fragments obtained. The ribosomal subunits can also act as size markers. Since the large ribosomal subunit is 28 S (approximately 5 kb) and the small ribosomal subunit is 18 S (approximately 2 kb), two prominent bands appear in the gel [45].

A method for resolving RNA fragments [46] is based on the migration of RNA fragments on a polycationic polyacrylamide gel, produced by incorporating positively charged immobilines into a neutral polyacrylamide backbone. The separation is carried out in a 0–10 mmol/L immobiline (p$K$ = 10.3) gradient under denaturing conditions (6 mol/L urea). Using it, good separations of single-stranded RNA of 21–23 mononucleotides in length, which appear to regulate the gene expression, are obtained.

### 3.1.4.1 Primer extension assay

In the **p**rimer **e**xtension **a**ssay (PEA), the transcription start site for a gene is determined experimentally by identifying the 5′-end of the encoded (mRNA. The process begins with a primer, usually a synthetic oligonucleotide of about 20 residues, which is complementary to an mRNA sequence of 50–150 nucleotides downstream of the anticipated 5′ end. The primer is 5′-end-labeled using γ-$^{32}$P-ATP and T4 polynucleotide kinase and is annealed to the specific mRNA molecules within a RNA sample.

Reverse **t**ranscriptase (RT), deoxyribonucleoside triphosphates, and appropriate buffer components are added to the primer-mRNA hybrids to catalyze elongation of the primer to the 5′-end of the mRNA. So, the 5′-end of mRNA is reached and the distance to the transcription start site can be determined. The resulting radiolabeled cDNA products are analyzed by denaturing polyacrylamide gel electrophoresis, followed by autoradiography. If the labeled cDNA products are within the resolution range of the gel, the transcription start site can be determined with an accuracy of plus or minus one nucleotide.

A disadvantage of the PEA is that it can be difficult to find a primer that works for a new gene. Moreover, background bands caused by premature termination of reverse transcription (because of the RNA secondary structure) often appear which makes difficult to find the start site location. Because of these and other limitations, other methods, such as RNase protection assay, are recommended.

## 3.1.5 Protocols

### Native DNA Electrophoresis on Agarose Gels

#### Materials and equipment
TRIS-**b**orate-EDTA (TBE) buffer, pH = 8.0
DNA samples and markers
10 µg/mL ethidium bromide
Electrophoresis cell with gel bridge
DC power supply
Longwave UV lamp

#### Procedure
- Blot the upper surface of an agarose gel near the cathode with a thin filter paper.
- Place an application template onto the gel in front of the cathode.
- Press the template carefully against the gel to remove the air bubbles.
- Pipette 5 µL each of the DNA samples and DNA markers into the slots of the application template and wait for 5 min to diffuse into the gel.

- Blot the remained sample volumes with a thin filter paper and remove it together with the application template.
- Hang the gel with the gel side below on the gel bridge.
- Full the tanks of the electrophoresis cell with TBE buffer and put the bridge with the gel into the electrophoresis cell to obtain direct contact between gel and electrode buffer.
- Close the electrophoresis cell with its lid and switch on the power supply.
- Run electrophoresis at a constant current and room temperature according to the following electrophoresis program until the tracking dye reaches the anode.

| Gel size (mm) | Current (mA) | Voltage (V) |
|---|---|---|
| 51 × 82 | 10 | 80 |
| 260 × 125 | 78 | 120 |

- Soak the gel after the electrophoresis in 0.5 µg/mL ethidium bromide at room temperature for 45 min and dry it.
- Visualize the DNA bands under a UV lamp as shadows.

**Sanger Sequencing Reactions Using *Taq* DNA Polymerase**

**Materials and equipment**
10x *Taq* polymerase buffer (500 mmol/L TRIS-HCl, pH = 9.0)
*Taq* Sanger mixtures A, C, G, and T
*Taq* DNA polymerase
dNTP chase
50 75 °C water bath

**Procedure**
- Anneal primer to template in 10x *Taq* polymerase buffer.
- Add 3 µL of *Taq* Sanger mixtures A, C, G, and T to the bottom of labeled A, C, G, and T tubes, respectively.
  *Keep tubes closed to prevent evaporation.*
- Prior to use, dilute *Taq* DNA polymerase to 2.5 U/µL in the *Taq* polymerase buffer, and keep on ice.
- Add
  2 µL 10 mCi/mL α-$^{35}$S-dATP and
  1 µL diluted *Taq* DNA polymerase
  to the annealed primer and template (total 13 µL) and mix by pipetting up and down.
- Add 2.5 µL of primer/template mixture to each tube containing *Taq* Sanger mixture.

- Incubate at 50 to 75 °C for 10 min.
- Add 1.0 μL dNTP chase to each reaction, mix, and incubate at 50 to 75 °C for 10 min.
- Electrophorese.

# References

[1]   Ogston AC. Trans Faraday Soc, 1958, 54, 1754–1757.
[2]   Zhu L, Wang H (Tian WC, Finehout E (eds.)). Microfluidics for Biological Applications. Springer, 2009, 125.
[3]   Slater GW. Electrophoresis, 2009, 30, Suppl 1, 181–187.
[4]   Lucotte G, Baneyx F. Introduction to Molecular Cloning Techniques. Wiley-Blackwell, 1993, 32.
[5]   Bahga SS, Bercovici M, Santiago JG. Electrophoresis, 2010, 31, 910–919.
[6]   Michov BM. Electrochim Acta, 2013, 108, 79–85.
[7]   Voet D, Voet JG. Biochemistry, 2nd edn. John Wiley & Sons, 1995, 877–878.
[8]   Freifelder D. Physical Biochemistry: Applications to Biochemistry and Molecular Biology, 2nd edn. WH, Freeman, 1982, 292–293.
[9]   Rickwood D, Hames BD. Gel Electrophoresis of Nucleic Acid. IRL Press, 1982.
[10]  Davis AR, Nayak DP, Ueda M, Hitti AL, Dowbenko D, Kleid DG. Proc Natl Acad Sci USA, 1981, 78, 5376–5380.
[11]  Budowle B, Chakraborty R, Giusti AM, Eisenberg AJ, Allen RC. Am J Hum Genet, 1991, 48, 137–144.
[12]  Peacock AC, Dingman CW. Biochemistry, 1968, 7, 668–674.
[13]  Michov BM. J Appl Biochem, 1982, 4, 436–440.
[14]  Tournie A, Majerus O, Lefevre G, Rager MN, Walme S, Caurant D, Barboux P. J Colloid Interface Sci, 2013, 400, 161–167.
[15]  Michov BM Protein Separation by SDS Electrophoresis in a Homogeneous Gel Using a TRIS-formate-taurinate Buffer System and Homogeneous Plates Suitable for Electrophoresis. (Proteintrennung Durch SDS-Elektrophorese in Einem Homogenen Gel Unter Verwendung Eines TRIS-Formiat-Taurinat-Puffersystems Und Dafür Geeignete Homogene Elektrophoreseplatten), German Patent 4127546, 1991.
[16]  Tautz D. Nucl Acids Res, 1989, 17, 6463–6471.
[17]  Newton G. Discovering DNA Fingerprinting: Sir Alec Jeffreys Describes Its Development. Wellcome Trust, 2010.
[18]  Jeffreys A. Genetic Fingerprinting. University of Leicester.
[19]  Murphy E. Annu Rev Criminol, 2018, 1, 497–515.
[20]  Petersen JK. Handbook of Surveillance Technologies, 3rd edn. Boca Raton, Fl., CRC Press, 2012, 815.
[21]  Pombert JF, Sistek V, Boissinot M, Frenette M. BMC Microbiol, 2009, 9, 232.
[22]  Gill P, Ivanov PL, Kimpton C, Piercy R, Benson N, Tully G, Evett I, Hagelberg E, Sullivan K. Nat Genet, 1994, 6, 130–135.
[23]  Yamanaka S, Poksay KS, Arnold KS, Innerarity TL. Genes Dev, 1997, 11, 321–333.
[24]  Mansour FH, Pestov DG. Anal Biochem, 2013, 441, 18–20.
[25]  Maxam AM, Gilbert W. Proc Natl Acad Sci USA, 1977, 74, 560–564.
[26]  Boland EJ, Pillai A, Odom MW, Jagadeeswaran P. Biotechniques, 1994, 16, 1088–1092, 1094–1095.
[27]  Sanger F, Coulson AR. J Mol Biol, 1975, 94, 441–448.

[28] Sanger F, Nicklen S, Coulson AR. Proc Natl Acad Sci USA, 1977, 74, 5463–5467.

[29] Connell CR, Fung S, Heiner C et al. Biotechniques, 1987, 5, 342–348.

[30] Ansorge W, Sproat BS, Stegemann J, Schwager C. J Biochem Biophys Methods, 1986, 13, 315–323.

[31] Orita M, Iwahana H, Kanazawa H, Hayashi K, Sekiya T. Proc Natl Acad Sci USA, 1989, 86, 2766–2770.

[32] Orita M, Suzuki Y, Sekiya T, Hayashi K. Genomics, 1989, 5, 874–879.

[33] Oto M, Miyake S, Yuasa Y. Anal Biochem, 1993, 213, 19–22.

[34] Kubo KS, Stuart RM, Freitas-Astua J, Antonioli-Luizon R, Locali-Fabris EC, Coletta-Filho HD, Machado MA, Kitajima EW. Arch Virol, 2009, 154, 1009–1014.

[35] Galas DJ, Schmitz A. Nucl Acids Res, 1978, 5, 3157–3170.

[36] Brenowitz M, Senear DF, Shea MA, Ackers GK. Meth Enzymol, 1986, 130, 132–181.

[37] Abe M, Kufe D. Proc Natl Acad Sci USA, 1993, 90, 282–286.

[38] Hartz D, McPheeters DS, Traut R, Gold L. Meth Enzymol, 1988, 164, 419–425.

[39] Ringquist S, MacDonald M, Gibson T, Gold L. Biochemistry, 1993, 32, 10,254–10,262.

[40] Shirokikh NE, Alkalaeva EZ, Vassilenko KS, Afonina ZA, Alekhina OM, Kisselev LL, Spirin AS. Nucl Acids Res, 2010, 38, e15.

[41] Sandelin A, Carninci P, Lenhard B et al. Nature Rev Genet, 2007, 8, 424–436.

[42] Green M, Sambrook J. Molecular Cloning: A Laboratory Manual, 4th edn. Cold Spring Harbor Laboratory Press, 2012.

[43] Safrany G, Perry RP. Proc Natl Acad Sci USA, 1993, 90, 5559–5563.

[44] Valoczi A, Hornyik C, Varga N, Burgyan J, Kauppinen S, Havelda Z. Nucl Acids Res, 2004, 32, e175.

[45] Gortner G, Pfenninger M, Kahl G, Weising K. Electrophoresis, 1996, 17, 1183–1189.

[46] Zilberstein G, Shlar I, Baskin E, Korol L, Righetti PG, Bukshpan S. Electrophoresis, 2009, 30, 3696–3700.

# 3.2 Pulsed-field gel electrophoresis of nucleic acids

3.2.1    Theory of pulsed-field gel electrophoresis of nucleic acids —— 343
3.2.2    Types of pulsed-field gel electrophoresis —— 344
3.2.3    Mapping the human genome —— 345
3.2.3.1  STR analysis —— 345
3.2.3.2  AmpFLP —— 345
3.2.3.3  DNA family relationship analysis —— 346
3.2.3.4  Y-chromosome analysis —— 346
3.2.3.5  Mitochondrial analysis —— 346
3.2.4    Protocols —— 346
         Pulsed-field Gel Electrophoresis of Chromosomal DNA —— 346
         References —— 347

Agarose gel electrophoresis is the method of choice regarding DNA fragments in size between 1 and 23 kbp. However, it cannot separate DNA nucleates in length of more than 30 kbp because they are moving together through the gel. In 1984, David Schwartz and Charles Cantor resolved larger DNA nucleates by agarose gel electrophoresis in an electric field with changing direction (pulsed electric field) [1]. Now this method is known as **pulsed-field gel electrophoresis** (PFGE). The PFGE allows separation of linear double-stranded DNA nucleates of 20 kbp up to 10 Mbp, for example, of whole chromosomes [2]. This is important for the mapping of human genome, which comprises about 3,000 Mbp.

## 3.2.1 Theory of pulsed-field gel electrophoresis of nucleic acids

PFGE is similar to the standard gel electrophoresis. However, the voltage is periodically switched on and off in three directions: one through the central axis of the gel, and two in directions located at an angle of 60° on both sides. So the long DNA nucleates migrate at first as reptiles, and then unfold their molecules and enlarge their volume (Figure 3.2.1). Whereas smaller nucleates have enough time to fold and unfold, and as a result migrate quickly through the gel, greater nucleates need more time to fold and unfold and as a result move slowly.

The pulse times are equal for each direction. They can be increased for extremely large DNA nucleates (up to 2 Mbp). PFGE takes between 10 h and several days depending on the size of DNA fragments. For folding and unfolding, larger nucleates need longer time compared to smaller nucleates [3].

https://doi.org/10.1515/9783110761641-029

Direction of
first electric field

Direction of
second electric field

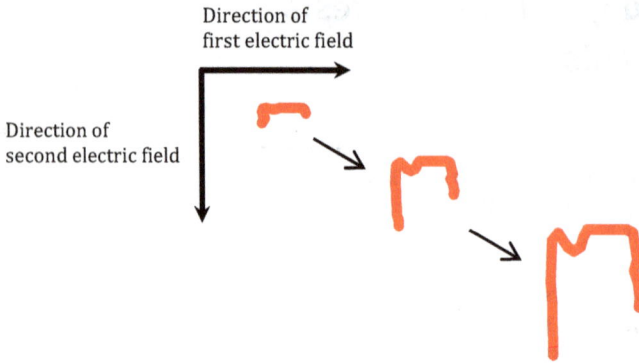

**Figure 3.2.1:** Enlargement of a nucleic acid molecule.

## 3.2.2 Types of pulsed-field gel electrophoresis

PFGE requires a special separation chamber, where the polarity, the angle of separation direction, and the pulse rate are changing alternatively by microprocessor-
controlled point electrodes. Depending on this, following types of PFGE are known:
gel electrophoresis in contour-clamped homogeneous electric fields (CHEF), field
inversion gel electrophoresis (FIGE), transverse alternating field gel electrophoresis (TAFGE), and rotating gel electrophoresis (RGE) in a constant electric field [4]
(Figure 3.2.2).

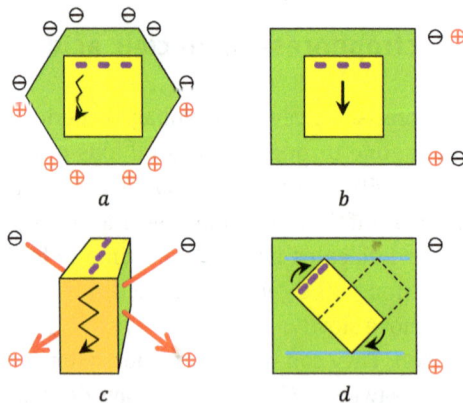

**Figure 3.2.2:** Types of pulsed-field gel electrophoresis.
(*a*) Gel electrophoresis in a contour-clamped homogeneous electric fields, generated by
hexagonally arranged point electrodes; (*b*) field inversion gel electrophoresis; (*c*) transverse
alternating field gel electrophoresis; (*d*) rotating gel electrophoresis in a constant electric field.

CHEF gel electrophoresis [5, 6] is a type of PFGE that enables the resolution of DNA fragments with lengths from 10 kb up to 9 Mb.

In FIGE, the electric field is reversed in certain pulse frequencies [7, 8]. Besides a gel electrophoresis unit and a power supply, a device is needed for periodic inversion of the electric field direction over the electrophoresis. The four-electrode PFGE is reduced to a standard submarine gel electrophoresis with only two electrodes. This configuration generates a uniform electric field across the gel, making the lane-to-lane comparison easy.

In TAFGE [9, 10], two electrodes are located on each side of a vertical agarose gel. The gel is located in a large Plexiglas tank, where the electrode wires stretch.

RGE is a technique for separating large DNA fragments [11–13]. It requires periodic changing of the orientation of the electric field by setting the gel on a rotating platform within a constant homogeneous electric field. The DNA nucleates of different sizes are separated by adjusting the frequency of rotation.

## 3.2.3 Mapping the human genome

Due to PFGE, genome mapping is carried out [14–16]. This was made for mammalian genomes that can be split by restriction endonucleases, as well as for bacterial chromosomes that can be separated in undigested form. PFGE is also used for karyotyping and analysis of chromosomes.

### 3.2.3.1 STR analysis

The people have different numbers of short DNA sequences. The short tandem repeats (STR) consist usually of four bases. Each STR is polymorphic but the number of alleles is very small. Typically, each STR allele is shared by 5–20% of individuals. In the STR analysis, their loci (locations on a chromosome) are targeted with sequence-specific primers and amplified using PCR. Then the resulting DNA fragments are separated by electrophoresis.

### 3.2.3.2 AmpFLP

Amplified fragment length polymorphism (AmpFLP) [17] relies on variable number tandem repeat (VNTR) polymorphisms to distinguish various alleles, which are separated on a gel in the presence of an allelic ladder. The bands obtained can be visualized by silver staining. AmpFLP allows for easy creation of phylogenetic trees based on comparing individual samples of DNA. It can be automated.

### 3.2.3.3 DNA family relationship analysis

All the cells forming the body contain the same DNA – half from the father and half from the mother. The DNA of an individual is the same in all somatic (nonreproductive) cells. As a result, every cell type can be used for testing the family relationship: cells from the cheeks, blood, or other biological samples. In a routine DNA paternity test, the markers used are STR. They are short DNA pieces that characterize the individuals.

### 3.2.3.4 Y-chromosome analysis

The Y chromosome is found only in males. For its analysis in forensics, paternity, and genealogical DNA testing, STR on the **Y**-chromosome (Y-STR) analysis is used. The Y male sex-determining chromosome, as it is inherited only from the father, is almost identical along the patrilineal line. Y-STR analysis can help in the identification of paternally related males. It was performed to determine, if Thomas Jefferson had sired a son with his slave Sally Hemings [18].

### 3.2.3.5 Mitochondrial analysis

Mitochondrial **DNA** (mtDNA) is maternally inherited. Therefore, directly linked maternal relatives can be used as match references. Scientists amplify the HV1 and HV2 regions of mtDNA, and then sequence each region to compare single-nucleotide differences. Heteroplasmy and poly-C differences may throw off straight sequence comparisons. mtDNA can be obtained from the same materials as nuclear DNA: hair shafts, bones or teeth. Using mtDNA analysis, it was determined that Anna Anderson was not Anastasia Romanov – the Russian princess she had claimed to be (see above).

## 3.2.4 Protocols

**Pulsed-field Gel Electrophoresis of Chromosomal DNA**

**Materials and equipment**

Cells
TAE or TBE buffer
Low melting agarose
Molecular mass markers: chromosomes (1.05–3.13 Mbp), *Schizosaccharomyces* (3.5–5.7 Mbp), and yeast chromosome PFG (225–1,900 kbp)

**Procedure**
- Centrifuge cells at 4,000 g for 5 min.
- Resuspend the pellet to concentration of $8.6 \times 10^7$ cells/mL.
- Add 3.1 mL of 37% formaldehyde into 40 mL centrifuge tubes and place them on ice.
- Centrifuge at 4,000 g for 5 min.
- Wash samples by resuspending them in 10 mL of buffer with 50 mmol EDTA and following centrifugation at 4,000 g for 5 min. Repeat the wash step 3 times.
- Resuspend infected cells in 0.5 mL of buffer.
- Add to the cells 0.5 mL of 2% low melting agarose kept at 45 °C, mix well, and pour the mixture into plug molds.
- Place plug molds in refrigerator for 15 min to solidify.
- Remove agarose blocks from the mold and place them into buffer with 1 mg/mL proteinase K.
- Incubate the agarose blocks for 24 h at 50 °C.
- Wash the agarose blocks for 30 min 4 times. Cut the blocks in small pieces to fit gel wells.
- Prepare 1% agarose gel in TAE buffer using a casting stand.
- Load the agarose blocks into the gel wells and seal them with melted 1% low melting agarose in running buffer.
- Separate the chromosomal DNA in an electrophoresis unit with TAE running buffer.
- Run electrophoresis at 3 V/cm with pulse time ramping from 250 to 900 s for 60 h. Change the buffer every 24 h.
- Stain the nucleates with 0.5 mg/L ethidium bromide for 20 min.

# References

[1] Schwartz DC, Cantor CR. Cell, 1984, 37, 67–75.
[2] Ito T, Hohjoh H, Sakaki Y. Electrophoresis, 1993, 14, 278–282.
[3] Viovy JL. Rev Mod Phys, 2000, 72, 813–872.
[4] Gardiner K. Anal Chem, 1991, 63, 658–655.
[5] Vollrath D, Davis RW. Science, 1986, 234, 1582–1585.
[6] O'Brien FG, Udo EE, Grubb WB. Nat Protoc, 2006, 1, 3028–3033.
[7] Carle GF, Frank M, Olson MF. Science, 1986, 232, 65–68.
[8] Carle GF, Carle GF. Meth Mol Biol, 1992, 12, 3–18.
[9] Gardiner K, Patterson D. Electrophoresis, 1989, 10, 296–302.
[10] Gardiner K. Meth Mol Biol, 1992, 12, 51–61.
[11] Serwer P. Electrophoresis, 1987, 8, 301–304.
[12] Southern EM, Anand R, Brown WRA, Flecher DS. Nucl Acids Res, 1987, 15, 5925–5943.
[13] Serwer P, Hayes SJ. App Theor Electrophoresis, 1989, 1, 95–98.
[14] Boultwood J. Meth Mol Biol, 1994, 31, 121–133.

[15]   Maule JC. MRC Human Genetics Unit. Published online, Edinburgh, UK, 2001.
[16]   Gemmill RM, Bolin R, Albertsen H, Tomkins JP, Wing RA. Curr Protoc Hum Genet. 2002, Ch. 5, Unit 5.1.
[17]   Keygene.com, 2013.
[18]   Foster EA, Jobling MA, Taylor PG, Donnely P, De Knijff P, Mieremet R, Zerjal T, Tyler-Smith C. Nature, 1998, 396, 27–28.

# 3.3 Capillary electrophoresis of nucleic acids

3.3.1    Theory of capillary electrophoresis of nucleic acids —— 349
3.3.2    Instrumentation for capillary electrophoresis —— 350
3.3.2.1  Capillaries —— 350
3.3.2.2  Coating polymers —— 350
3.3.2.3  Gels —— 351
3.3.3    Running capillary electrophoresis —— 352
3.3.3.1  Injection —— 352
3.3.3.2  Separation —— 352
3.3.3.3  Detection —— 352
3.3.4    Pulsed-field capillary electrophoresis —— 353
3.3.5    Applications of capillary electrophoresis —— 353
3.3.6    Protocols —— 354
         Quantitative PCR Analysis —— 354
         References —— 354

The capillary electrophoresis (CE) is effective for DNA sequencing, fragment analysis, and DNA typing [1]. In CE methods, nucleates are separated according to their electrophoretic mobility and non-covalent interactions with a phase in capillaries. CE has many advantages over slab gel electrophoresis in terms of speed, resolution, sensitivity, and data handling.

## 3.3.1 Theory of capillary electrophoresis of nucleic acids

In free solutions, DNA nucleates have a constant size-to-charge ratio, and, as a result, their electrophoretic mobilities are independent of size. In an entangled polymer matrix, however, the friction coefficient of DNA nucleates increases with the DNA chain length, which results in decrease of the DNA electrophoretic mobilities (Figure 3.3.1).

A strong denaturant to cleave intramolecular hydrogen bonds in RNA is required for RNA size separation. Sumitomo *et al.* [2] have found that carboxylic acids (for example acetic acid) were strong denaturants for RNA and that RNA separation performance was dramatically improved by CE with a sieving matrix containing acetic acid. The same authors have established that the denaturing ability of 2.0 mol/L acetic acid was stronger than that of either 2.5 mol/L formaldehyde or 7.0 mol/L urea by estimating DNA melting temperature.

https://doi.org/10.1515/9783110761641-030

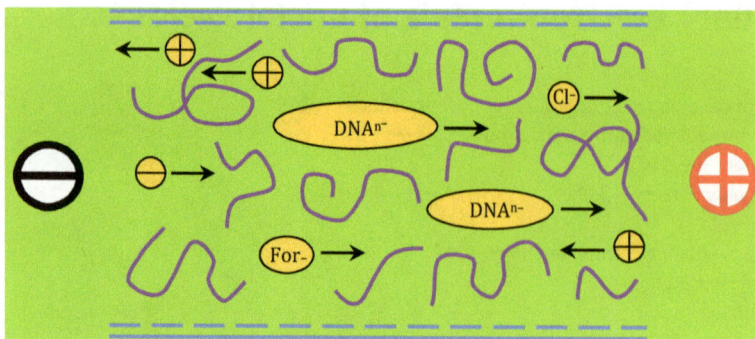

**Figure 3.3.1:** DNA migrates through entangled strands of a polymer twisting and tumbling through the gel. Small ions such as chloride (Cl⁻) and formate (For⁻) have higher electrophoretic mobility.

## 3.3.2 Instrumentation for capillary electrophoresis

The CE instrument contains capillaries, coating polymers, and gels. It also requires a suitable sample injection module, a detector, adequate temperature control, and protection from high voltages is used.

### 3.3.2.1 Capillaries

The capillaries are thin walled, which allows for dissipation of the Joule heating during electrophoresis. This minimizes convective effects that result in band broadening. The fused-silica capillary is coated on the outside with a polyimide layer that eliminates oxidation of its glass. The polyimide sheathing is burned from a small portion of the capillary to expose a clear section of the silica. This section is placed in the light path of a UV or fluorescence detector and becomes the on-column flow cell. As DNA migrates through the capillary, it passes through the detector light and is measured.

### 3.3.2.2 Coating polymers

A capillary should be coated to eliminate charge effects that are contributed by the native silica surface. This can be achieved by cellulose-derived polymers or modified acrylamide polymers, which interact strongly with the inner surface of the bare fused-silica capillary. The synthesis and characterization of a hydrophilic, replaceable, self-coating polymer matrix, **p**oly-*N*-**h**ydroxy**e**thyl**a**crylamide (PHEA), for application in DNA CE was reported [3].

### 3.3.2.3 Gels

An ideal separation medium for DNA separations using CE should possess the following properties: high sieving ability, dynamic coating ability, and relatively low viscosity. A separation medium with these specifications can be used for automation of CE and for further enhancing of its performance [4]. For this purpose, homopolymers, such as linear **p**oly**a**crylamide (LPA) and its derivatives [5–7], **p**oly**v**inyl**p**yrrolidone (PVP) [8], **p**oly**e**thylene **o**xide (PEO) [9], and cellulose and its derivatives [10–12] are used.

LPA is hydrophilic. It forms strongly entangled networks that are excellent for high-quality DNA separations of long sequencing read lengths (up to 1,300 bases) [13]. Doherty *et al.* [14, 15] showed that by adding very small amounts of cross-linking monomers into the polymer solution to form LPA "nanogels," sequencing performance could be improved compared with LPA synthesized in the absence of any cross-linker. However, LPA does not bind well to the silica surface to suppress electroosmotic flow in CE. Therefore, the use of LPA solution needs precoated capillaries.

In addition to linear polyacrylamide, the acrylamide derivative **p**oly(*N, N*-**dim**ethyl**a**crylamide) (pDMA) has also been demonstrated as a high-quality polymer for sequencing DNA [16–18]. Sequencing matrices using pDMA generally have lower viscosities than LPA matrices and are also able to form coatings on the capillary surfaces, eliminating the need to perform chemical reactions on the capillary surface prior to DNA fragment separation. Dynamic coatings are alternatives to covalent coatings because these coatings can be applied relatively quickly (45 min or less) and are extremely stable [19].

Two types of CE gels are employed in DNA separations: cross-linked and non-cross-linked gels.

*Cross-linked gels* are polymerized inside the capillary and can be reused for 30 to 100 separations. They are usually covalently bound to the capillary surface and are not removed from the capillary between runs. These gels are preferred when larger DNA molecules should be resolved – e.g., for fragment analysis and DNA sequencing.

*Non-cross-linked (flowable) gels* are viscous hydrophilic polymer solutions that can be pumped into the capillary. The flowable polymer has the advantage that it can be expelled from the capillary by pressure. These capillaries have lifetimes of several hundred injections, being replaced for loss of the surface coating or mechanical breakage. Flowable polymers can be used for oligonucleotides and **d**ouble-**s**tranded **DNA** (dsDNA) fragment analysis.

## 3.3.3 Running capillary electrophoresis

The CE can be run when a fused-silica capillary, filled with gel or liquid, joins two buffer tanks and is placed at a high voltage (10–30 kV). The resulting electric field drives the nucleic acids along the capillary. Generally, the capillaries are 20–50 cm long and have 50–100 μm internal diameter.

The CE of nucleic acids is carried out in three steps: injection, separation, and detection of the separated nucleic acids.

### 3.3.3.1 Injection

DNA samples are injected electrokinetically or by pressure into the sample tube. Then, they are introduced into the capillary using a voltage of 5–15 kV for a few seconds. The samples should be diluted with formamide or water [20] to reduce the ionic strength.

### 3.3.3.2 Separation

The separation of DNA is carried out in a small-diameter (50–100 μm) quartz capillary under a high voltage in a few minutes up to an hour [21–23]. The separation is performed in TBE buffer with alkaline pH value. Urea is often included as a denaturant when analyzing single-stranded DNA (ssDNA). ssDNA interacts efficiently with the sieving polymer, and its size is proportional to its length. Denaturants prevent also the complementary sequences within ssDNA molecules to bind to each other. This could create loops and hairpins. The temperature should be controlled, as it strongly influences DNA mobility and the calculated allele size [24]. The separation times range from 10 to 45 min at 1–10 kV. In the presence of appropriate standards with known electrophoretic mobility, the physical properties of DNA can be examined [25].

### 3.3.3.3 Detection

CE uses laser-induced fluorescence to detect DNA. In many CE systems, a 488 nm argon ion laser is focused onto a window burned in the polyimide coating of the capillary, providing efficient access to the sample.

The laser universally excites dyes added to the 5'-end of each DNA fragment. Once excited, the dyes emit fluorescent light at different wavelengths. The emitted light is separated into its constituent wavelengths and is captured by a photoarray detector. Each dye has a characteristic emission spectrum.

DNA fragments can be detected also in the UV spectrum (at 260 nm) in the presence or absence of ethidium bromide. Sensitivity can be increased by at least two orders of magnitude by using fluorescence detection. Since DNA possesses no native fluorescence, intercalating dyes such as cyanine derivatives [26] or rhodamine derivatives should be covalently attached to the DNA prior to electrophoresis or added to the electrophoresis buffer.

## 3.3.4 Pulsed-field capillary electrophoresis

Another type of CE is **p**ulsed-**f**ield **c**apillary **e**lectrophoresis (PFCE) [27]. It can be used not only for resolution of DNA chains containing above 1 kbp [28, 29], but also for resolution of DNA chains into the Mbp range. PFCE can be carried out in a gel or in an ultradilute polymer solution. It is believed that DNA interacts mechanically with the ultradilute polymer solutions and the polymer leads to a transient entity, which has lower mobility than random-coil DNA.

Visualization experiments in 0.01 g/dL DNA has a size-dependent reorientation time, which depends on its length and the electric field strength. At constant field, a DNA fragment collides with **h**ydroxy**e**thyl **c**ellulose (HEC), becomes elongated, and then collapses back to the random coil configuration. Long-chain DNA fragments are more stretched than short-chain fragments. As a result, they migrate together. The PFCE in ultradilute polymer solutions containing buffer is used to separate long-chain dsDNA of below 10 kbp and greater than 1.5 Mbp in less than 4 min.

## 3.3.5 Applications of capillary electrophoresis

Schwartz *et al.* [30] demonstrated that 0.5 g/dL hydroxypropyl-methyl-cellulose having $M_r$ = 90,000 allowed for adequate separation in the 72–1,353 bp range and a 3.0-bp peak capacity in the 100–200 bp size range. With this approach, Del Principe *et al.* [31] showed that CE is an effective method for genomic analysis. They indicated that it is suitable for prenatal diagnosis of X-linked recessive disorders.

Optimized CE and fluorescence-labeled PCR appear to be the methods of choice in **H**untington **d**isease (HD) and **s**ickle **c**ell **d**isease (SCD) diagnostics, often combined with Southern blotting.

HD is a neurodegenerative genetic disorder that affects muscle coordination and leads to mental decline and behavioral symptoms. With the time, the uncoordinated body movements become more apparent and the mental deficiency turns into dementia [32]. In 1993, a research group, using genetic linkage analysis [33], isolated the causal gene at 4p16.3 [34].

SCD, is a group of genetically blood disorders. It is characterized by a hemoglobin abnormality. This leads to a sickle-like erythrocytes under certain circumstances

[35]. Temperature changes, dehydration, high altitude and more may develop pain attacks (sickle cell crises), anemia, bacterial infections, and stroke [36]. SCD occurs when a person inherits two abnormal hemoglobin genes, one from each parent. A person with a single abnormal gene does not have symptoms and is referred to as a carrier [37].

The analysis of dsDNA fragments, such as those found in polymerase chain reaction products and DNA restriction digests [38, 39], has led to the development of a complete map of the human genome.

## 3.3.6 Protocols

### Quantitative PCR Analysis

*The quantitative PCR analysis can be used in conjunction with CE separation to amplify and quantitate DNA target sequence by conjecting an intercalating dye with the samples or by using a covalently modified and fluorescently labeled oligonucleotide primer.*

**Materials and equipment**
TBE buffer
Resolving gel
65 cm 100 µm i.d. coated capillary
Standard 174 DNA ladder (in a concentration of 10 µg/mL and stored at −20 °C)
CE instrument with fluorescent detection

**Procedure**
– Prepare gel buffer mixture and add intercalating dye. Store at 4 °C up to 30 days.
– Reverse standard polarity of CE instrument electrodes.
– Rinse with buffer at high pressure for 10 min, if the capillary is new.
– Load standard DNA ladder into the capillary at low pressure for 10 s.
– Electrophorese at 9.4 kV and 25 °C for 30 min.
  *Replace inlet gel reservoir after 30 injections.*

## References

[1]  Cohen AS, Najarian AR, Paulus A, Guttman JA, Smith BL, Karger BL. Proc Natl Acad Sci USA, 1988, 85, 9660–9663.
[2]  Sumitomo K, Sasaki M, Yamaguchi Y. Electrophoresis, 2009, 30, 1538–1543.
[3]  Albarghouthi MN, Buchholz BA, Huiberts PJ et al. Electrophoresis, 2002, 23, 1429–1440.
[4]  Zhang J, Wang YM, Liang DH, Ying QC, Chu B. Macromolecules, 2005, 38, 1936–1943.
[5]  Salas-Solano O, Carrilho E, Kotler L, Miller AW et al. Anal Chem, 1998, 70, 3996–4003.

[6]    Heller C. Electrophoresis, 1999, 20, 1962–1977.
[7]    Zhou H, Miller AW, Sosie Z, Buchholz B *et al*. Anal Chem, 2000, 72, 1045–1052.
[8]    Gao Q, Yeung ES. Anal Chem, 1998, 70, 1382–1388.
[9]    Iki N, Yeung ES. J Chromatogr A, 1996, 731, 273–282.
[10]   Hammond RW, Oana H, Schwinefus JJ, Bonadio J *et al*. Anal Chem, 1997, 69, 1192–1196.
[11]   Atha DH. Electrophoresis, 1998, 19, 1428–1435.
[12]   Raucci G, Maggi CA, Parente D. Anal Chem, 2000, 72, 821–826.
[13]   Zhou HH, Miller AW, Sosie Z, Buchholz B *et al*. Anal Chem, 2000, 72, 1045–1052.
[14]   Doherty EAS, Kan CW, Barron AE. Electrophoresis, 2003, 24, 4170–4180.
[15]   Doherty EAS, Kan CW, Paegel BM, Yeung SHI *et al*. Anal Chem, 2004, 76, 5249–5256.
[16]   Chiari M, Riva S, Gelain A, Vitale A, Turati E. J Chromatogr A, 1997, 781, 347–355.
[17]   Heller C. Electrophoresis, 1998, 79, 3114–3127.
[18]   Madabhushi RS. Electrophoresis, 1998, 79, 224–230.
[19]   Doherty EAS, Berglund KD, Buchholz BA, Kourkine IV *et al*. Electrophoresis, 2002, 23, 2766–2776.
[20]   Crivellente F, McCord BR. J Capill Electrophoresis, 2002, 7, 73–80.
[21]   Hjerten S. Chromat Rev, 1967, 9, 122–219.
[22]   Jorgensen JW, Lukacs KD. Science, 1983, 222, 266–272.
[23]   Heller C. Electrophoresis, 2001, 22, 629–643.
[24]   Nock T, Dove J, McCord BR, Mao D. Electrophoresis, 2001, 22, 755–762.
[25]   Stellwagen N, Celfi C, Righetti PG. Electrophoresis, 2002, 23, 167–175.
[26]   Zhu Y, Takeda T, Nasmyth K, Jones N. Genes Dev, 1994, 8, 885–898.
[27]   Kim Y, Morris MD. Electrophoresis, 1996, 17, 152–160.
[28]   Demana T, Lanan M, Morris MD. Anal Chem, 1991, 63, 2795–2797.
[29]   Heiger DN, Cohen AS, Karger BL. J Chromatogr, 1990, 516, 33–48.
[30]   Schwartz HE, Ulfelder KJ, Sunzeri FJ, Busch MP, Brownlee RG. J Chromatogr, 1991, 559, 267–283.
[31]   Del Principe D, Iampieri MP, Germani D, Menchelli A, Novelli G, Dallapiccola B. J Chromatogr, 1993, 638, 277–281.
[32]   Hammond K, Tatum B. The Behavioral Symptoms of Huntington's Disease. Huntington's outreach project for education at Stanford, 2010.
[33]   Bertram L, Tanzi RE. J Clin Inv, 2005, 115, 1449–1457.
[34]   MacDonald ME, Ambrose CM, Duyao MP *et al* .Cell, 1993, 72, 971–983.
[35]   What Is Sickle Cell Disease?. National heart, lung, and blood institute, NIH, 2015.
[36]   What are the Signs and Symptoms of Sickle Cell Disease?. National heart, lung, and blood institute, NIH, 2015.
[37]   Sickle Cell Disease and Other Hemoglobin Disorders. Fact sheet N°308, 2011.
[38]   Ulfelder KJ, Schwartz HE, Hall JM, Sunzeri FJ. Anal Biochem, 1992, 200, 260–267.
[39]   Schwartz HE, Ulfelder KJ, Sunzeri FJ, Busch MP, Brownlee RG. J Chromatogr, 1991, 559, 267–283.

# 3.4 Polyacrylamide gel electrophoresis of nucleic acids

3.4.1    Zone polyacrylamide gel electrophoresis of native nucleic acids —— 357
3.4.1.1  Practice of zone polyacrylamide gel electrophoresis of native nucleic acids —— 358
3.4.2    Disc-electrophoresis of native nucleic acids —— 360
3.4.2.1  Theory of disc-electrophoresis of native nucleic acids —— 360
3.4.2.2  Running disc-electrophoresis of native nucleic acids —— 360
3.4.2.3  Disc-electrophoresis of double-stranded PCR products —— 361
3.4.3    Electrophoretic mobility shift assay —— 363
3.4.4    Clinical applications —— 364
3.4.5    Protocols —— 364
         DNA Disc-electrophoresis in TRIS-taurinate Buffer at Two pH Values According to
         Michov —— 364
         Separation of PCR Products by Polyacrylamide Gel Disc-Electrophoresis According
         to Michov —— 366
         Electrophoretic Mobility Shift Assay —— 367
         Electroelution of Small DNA Fragments from Polyacrylamide Gel —— 368
         References —— 368

If small double-stranded DNA fragments are to be resolved on agarose gels, 2 g/dL agarose gels should be used. However, these gels are opalescent and, as a result, the sensitivity of the ethidium bromide proving is reduced. In addition, the zone electrophoresis on agarose gels runs for several hours. Therefore, low-molecular-mass DNA fragments are separated in native polyacrylamide gels.

The native polyacrylamide gel electrophoresis is used for the separation of double-stranded DNA with a length of 15–300 bp fragments, for example, PCR fragments, and single-stranded DNA with the same length. It takes place in continuous or discontinuous buffer systems [1, 2]. The electrophoresis should be run at pH range of 6.0 to 10.0. If the pH value is too low, the bases cytosine and adenine protonate and become positively charged; if the pH value is too high, the bases thymine and guanine deprotonate and become negatively charged. In both cases, additional electric charges cause denaturation of nucleic acids and change their mobility.

## 3.4.1 Zone polyacrylamide gel electrophoresis of native nucleic acids

The native polyacrylamide gel electrophoresis of nucleic acids is a zone electrophoresis carried out in a buffer. It is characterized by continuous strength of the electric field during electrophoresis. With the help of the native zone electrophoresis on polyacrylamide gel, DNA nucleates, oligonucleotides, RNA nucleates, and viroids are resolved

https://doi.org/10.1515/9783110761641-031

according to their molecular masses. DNA fragments may be separated if they differ from each other with 6 bp in a length of 500 to 1,200 bp ($M_r$ up to 800,000).

### 3.4.1.1 Practice of zone polyacrylamide gel electrophoresis of native nucleic acids

*Preparation of polyacrylamide gels.* Polyacrylamide gels for zone electrophoresis of nucleic acids are cast in the same way as any other polyacrylamide gels: A support film is placed with its hydrophobic side down onto a glass plate covered with drops of water and is rolled by a roller. Then, it is framed with a U-shaped spacer and covered with another glass plate. The resulting cassette is clamped together and placed upright. A buffer, for example TRIS-borate-EDTA buffer, is mixed with a monomeric solution, TMEDA, and APS. The mixture is cast into the cassette and covered with deionized water to prevent air–oxygen diffusion into the mixture.

If the monomeric concentration $T$ is 50 g/dL, the polyacrylamide gels for zone electrophoresis can be prepared as follows (Table 3.4.1):

**Table 3.4.1:** Producing polyacrylamide gels.

| Reagents | Total monomeric concentration $T$, g/dL | | | | |
|---|---|---|---|---|---|
| | 3.5 | 5.0 | 8.0 | 12.0 | 20.0 |
| TRIS-borate-EDTA buffer, 10x | 10.00 mL | 10.00 mL | 10.00 mL | 10.00 mL | 10.00 mL |
| Monomeric solution ($T$ = 50 g/dL, $C$ = 0.03) | 7.00 mL | 10.00 mL | 16.00 mL | 24.00 mL | 40.00 mL |
| 10 g/dL TMEDA | 0.46 mL | 0.46 mL | 0.46 mL | 0.46 mL | 0.46 mL |
| 10 g/dL APS | 0.97 mL | 0.68 mL | 0.43 mL | 0.38 mL | 0.17 mL |
| Deionized water to | 100.00 mL | 100.00 mL | 100.00 mL | 100.00 mL | 100.00 mL |

After 1 h of polymerization at room temperature, the polyacrylamide gel should be removed from the cassette: the cassette is opened and the cover glass plate is raised up using a spatula. Thereafter, the spacer is removed and the gel is covered with a hydrophobic polyester film, or packed in a wrap.

Prior to electrophoresis, a thermostat is set to 10 °C and the cooling plate of a separation unit is covered with 0.5–1 mL of kerosene or silicone oil DC-200 as a contact fluid. The gel, with its support film down, is placed onto the cooling plate so that the bubbles between the plate and support film are removed. Then, the cover film is removed from the gel and paper electrodes strips, impregnated with electrode buffer, are placed onto the cathode and anode ends of the gel. Before the cathode strip, at a distance of 1 cm, a silicone template is placed and lightly pressed against the gel. Afterward, 0.1 mg/mL DNA containing 0.025 g/dL of Na salt of Xylene cyanol or Bromophenol blue is applied in the template slots.

*Electrophoresis conditions.* The electrophoresis is carried out at a constant electric current of 50 mA, limiting voltage of 500 V, and limiting power of 25 W. It is terminated after the dye front has run the opposite end of the gel.

On principle, native double-stranded DNA nucleates move faster than denatured (single-stranded) DNA with the same masses. Moreover, smaller DNA fragments move together with Bromophenol blue, while larger DNA fragments follow Xylene cyanol [3]. This can be explained by the structures of Xylene cyanol and Bromophenol blue (Figure 3.4.1): The molecular mass of Xylene cyanol Na salt ($M_r = 538.61$) represents approximately 3/4 of the molecular mass of Bromophenol blue Na salt ($M_r = 691.94$), but Xylene cyanol includes four additional methyl groups, which diminishes its hydrophilicity. Therefore, Xylene cyanol moves slower than Bromophenol blue.

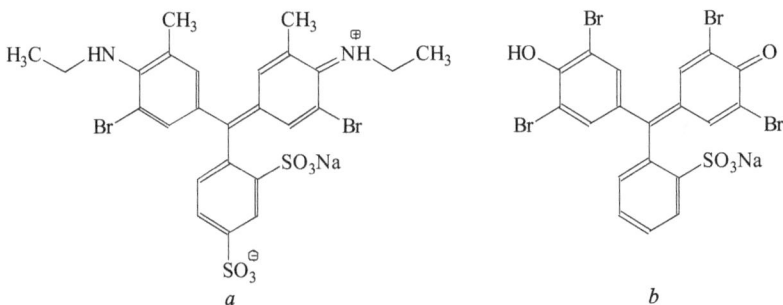

**Figure 3.4.1:** Chemical structures of the sodium salts of (*a*) Xylene cyanol and (*b*) Bromophenol blue.

The relationship between the polyacrylamide concentration and the masses of nucleic acids to be resolved can be seen in Table 3.4.2. It shows that there are differences between the behavior of native and denatured DNA nucleates during electrophoresis.

After run, the nucleic acid bands can be stained with ethidium bromide. If the DNA concentration is too low, a silver staining can be applied or radioactive labeling.

With the help of native zone electrophoresis on polyacrylamide gel, DNA nucleates, oligonucleotides, RNA nucleates, and viroids are resolved in accordance with their molecular masses. DNA fragments may be separated, if they differ from each other with 6 bp in a length of 500 to 1,200 bp ($M_r$ up to 800,000).

Table 3.4.2: Polyacrylamide concentrations used for different sizes (lengths) of DNA nucleates.

| Polyacrylamide concentration $T$, in g/dL | Double-stranded DNA, in bp, moving with Xylene cyanol | Double-stranded DNA, in bp, moving with Bromophenol blue | Polyacrylamide concentration $T$, in g/dL | Single-stranded DNA, in b, moving with Xylene cyanol | Single-stranded DNA, in b, moving with Bromophenol blue |
|---|---|---|---|---|---|
| 3.5 | 460 | 100 | 5.0 | 130 | 35 |
| 5.0 | 260 | 65 | 6.0 | 106 | 26 |
| 8.0 | 160 | 45 | 8.0 | 76 | 19 |
| 12.0 | 70 | 20 | 10.0 | 55 | 12 |
| 15.0 | 60 | 15 | 20.0 | 28 | 8 |
| 20.0 | 45 | 12 | | | |

## 3.4.2 Disc-electrophoresis of native nucleic acids

The disc-electrophoresis is used for separation of native 20 to 60 bp double-stranded DNA [4–6], and RNA [7]. It gives sharper bands than the zone (continuous) electrophoresis [8] in TRIS-borate-EDTA buffer [9, 10].

### 3.4.2.1 Theory of disc-electrophoresis of native nucleic acids

The negative charges of phosphate backbones cause DNA nucleates to move toward the positively charged anode. The migration in a solution, in the absence of gel, is independent of molecular masses, i.e., no separation by size happens [11, 12]. During a native disc-electrophoresis in a gel, however, the nucleic acids are concentrated in an ionic boundary between the leading and trailing ion, and in the resolving gel, they are overtaken by the ionic boundary. Thereafter, the electrophoresis runs as a zone electrophoresis. If the concentration of the leading ion is enough high, nucleates run slower but the resolution of electrophoresis is higher [13].

### 3.4.2.2 Running disc-electrophoresis of native nucleic acids

Prior to electrophoresis, the nucleic acid to be separated can be prepared in several ways: In the case of large DNA molecules, DNA is cut by DNA restriction endonuclease into smaller fragments. In PCR-amplified samples, the enzymes which are present in the sample are to be removed prior to analysis. Once the nucleic acid is properly prepared, the samples are placed into wells of the gel or into slots and a voltage is applied through the gel.

The DNA fragments of different lengths are visualized using ethidium bromide, a specific DNA fluorescent dye. Under UV light, these gels show bands corresponding to nucleic acid bands.

After electrophoresis, DNA nucleates can be eluted. For this purpose, their bands can be cut out from the gel with a scalpel or razor blade and transferred into a centrifuge tube containing 200 µL of sterile deionized water. After several hours or overnight extraction at 80 °C, the suspension is centrifuged for 5 min in a microcentrifuge at high rotation. The supernatant is filled with sterile deionized water to 300 µL, and diluted in 30 µL of 3 mol/L sodium acetate with pH = 4.8, and 750 µL of ethanol. After 30 min at −70 °C, the precipitated DNA is centrifuged and washed with 1 mL of 70 mL/dL ethanol. Then, DNA is dried briefly in vacuum and dissolved in a small volume of sterile TRIS-borate-EDTA or TRIS-taurinate-EDTA buffer. The measurement and analysis are done with specialized gel analysis software.

### 3.4.2.3 Disc-electrophoresis of double-stranded PCR products

Nucleates of greater length cannot be separated by native polyacrylamide gel electrophoresis. Therefore, they should be broken by endonucleases, and the nucleate fragments should be multiply in the **p**olymerase **c**hain **r**eaction (PCR) [14]. The products of the PCR can be later separated in horizontal polyacrylamide gels using, for example, disc-electrophoresis in a TRIS-formate-borate buffer system [15], or TRIS-formate-taurinate buffer system [16].

The TRIS-formate-taurinate buffer system consists of a TRIS-formate buffer as leading buffer (the anode buffer), and a TRIS-taurinate buffer as trailing buffer (the cathode buffer). The TRIS-formate buffer is in the polyacrylamide gel, and the TRIS-taurinate buffer is in the anode strip. During the electrophoresis, a moving ionic boundary is formed between the formate ion in the gel and the taurinate ion in the cathode strip, which concentrates the PCR products and overtakes them later (Figure 3.4.2).

The recipes for the TRIS-formate-taurinate buffer system and the preparation of polyacrylamide gels for disc-electrophoresis are presented in Tables 3.4.3 and 3.4.4, and some results are shown on Figure 3.4.3.

A simple gel electrophoresis apparatus for size-selective separations of DNA fragments in both polyacrylamide and agarose gels was described [17]. It employs a microslab gel format with a novel gel casting technique that eliminates the need for combs to define sample loading wells. Real time fluorescence detection of the migrating DNA fragments is accomplished using a digital microscope that connects to any PC with a USB interface. The microscope is adaptable for this application by replacing its white light source with a blue **l**ight-**e**mitting **d**iode (LED) and using an appropriate emission filter.

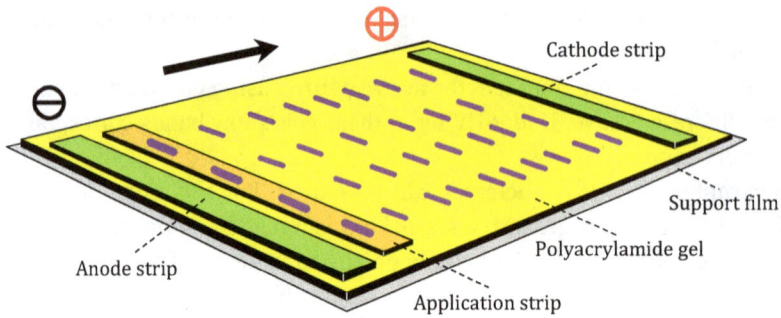

**Figure 3.4.2:** Disc-electrophoresis of PCR products.

**Table 3.4.3:** Recipes for the TRIS-formate-taurinate buffer system for disc-electrophoresis of nucleic acids.

| | Anode buffer, pH = 7.4 | Cathode buffer pH = 8.3 | Monomeric solution $T = 50$ g/dL, $C = 0.03$ |
|---|---|---|---|
| TRIS, g | 16.32 | 2.23 | – |
| 99 mL/dL Formic acid, mL | 4.57 | – | – |
| Taurine, g | – | 6.02 | – |
| Acrylamide, g | – | – | 48.50 |
| BIS, g | – | – | 1.50 |
| 87% Glycerol, mL | 10.00 | 10.00 | – |
| $Na_2EDTA \cdot 2H_2O$, g | 0.45 | 0.22 | – |
| $NaN_3$, g | 0.06 | 0.03 | – |
| Deionized water to, mL | 100.00 | 100.00 | 100.00 |

**Table 3.4.4:** Recipes for polyacrylamide gels containing TRIS-formate-taurinate buffer system for disc-electrophoresis of nucleic acids.

| | $T = 5$ g/dL pH = 7.4 | $T = 7$ g/dL pH = 7.4 | $T = 9$ g/dL pH = 7.4 | $T = 11$ g/dL pH = 7.4 | $T = 13$ g/dL pH = 7.4 |
|---|---|---|---|---|---|
| Anode buffer, mL | 16.67 | 16.67 | 16.67 | 16.67 | 16.67 |
| Monomeric solution, mL | 10.00 | 14.00 | 18.00 | 22.00 | 26.00 |
| 87% glycerol, mL | 20.00 | 20.00 | 20.00 | 20.00 | 20.00 |
| 10 g/dL TMEDA, mL 10 g/dL APS, mL | 0.46 0.68 | 0.46 0.49 | 0.46 0.38 | 0.46 0.31 | 0.46 0.26 |
| Deionized water to, mL | 100.00 | 100.00 | 100.00 | 100.00 | 100.00 |

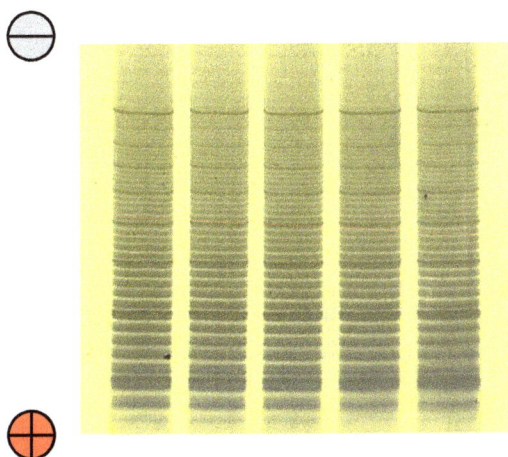

**Figure 3.4.3:** Disc-electrophoresis of 20 to 100 bp DNA fragments in a *T7* polyacrylamide gel. Silver staining.

After electrophoresis, the PCR products can be detected by various methods. The most common among them is the silver staining [18], which is more sensitive than the ethidium bromide method [19]. The PCR products can also be labeled with fluorescent substances [20, 21], and the labeled PCR products can be visualized by the chemoluminescence digoxigenin system [22].

## 3.4.3 Electrophoretic mobility shift assay

The **e**lectrophoretic **m**obility **s**hift **a**ssay (EMSA), referred also as mobility shift electrophoresis, band shift assay, gel mobility shift assay, or gel retardation assay, is based on methods described by Garner and Revzin [23], and Fried and Crothers [24, 25]. It is used to deduce the affinities of a protein for one or more DNA or RNA sites or to compare the affinities of different proteins for same nucleate site.

The proteins should be in solutions with physiological ionic strength and physiological pH value to recognize specific sequences of DNA/RNA. On the other hand, DNA should be native and double-stranded to interact with them. Therefore, EMSA is carried out in buffers.

If a double-stranded DNA fragment contains a detection sequence for some protein, the protein is bound to this DNA fragment during electrophoresis. The resulting DNA–protein complexes migrate slower through the gel and, therefore, displace from the DNA band – *band shift* (Figure 3.4.4). When the DNA fragment binds more than one protein polyion, the mobility of the complex reduces further.

To visualize EMSA, the nucleic acid fragments in the gel are usually labeled radioactively or fluorescently, or are bound to biotin. While the isotopically $^{32}$P-labeled

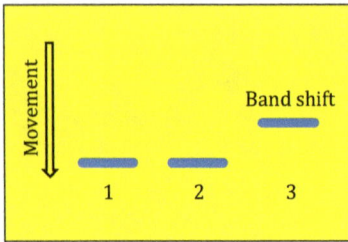

**Figure 3.4.4:** Scheme of the electrophoretic mobility shift assay.
*Lane 1* contains only genetic material. *Lane 2* contains protein and DNA fragments, which do not interact each with other. *Lane 3* contains protein and DNA fragments that have reacted with each other to build a slower moving complex.

DNA has less or no effect on the protein binding affinity, the non-isotopic labels including fluorophores or biotin can alter the affinity and the stoichiometry of the protein–DNA interaction. The biotin label uses streptavidin, conjugated to an enzyme, such as horseradish peroxidase.

If the starting concentrations of a protein and its DNA probe, as well as the stoichiometry of the complex are known, the apparent affinity between the protein and nucleic acid sequence can be determined [26]. If the protein concentration is not known but the complex stoichiometry is known, the protein concentration can be determined by increasing the concentration of DNA probe until all the protein is bound. Then, the number of protein molecules bound to the DNA fragment can be calculated.

## 3.4.4 Clinical applications

EMSA seems well suited to detect many types of genetic differences, including single-base substitutions and larger sequence alterations such as insertions, deletions, and inversions, and could be used for the efficient screening of samples for sequence alterations [27]. It was applied also to study the MUC1 promoter in breast cancer cells [28]. The MUC1 gene encodes the DF3 protein, which represents a high-molecular-mass glycoprotein. Membrane proteins can also be identified by the EMSA using charged detergents.

## 3.4.5 Protocols

**DNA Disc-electrophoresis in TRIS-taurinate Buffer at Two pH Values According to Michov**

**Materials and equipment**
TRIS
Taurine
Acrylamide
BIS

87% Glycerin
**B**romo**p**henol **b**lue (BPB) Na salt
TMEDA
**A**mmonium **p**eroxydisulfate (APS)
Casting cassette

***Buffer 1(Anode buffer), pH = 8.6, I = 2×0.20 mol/L***
| | |
|---|---|
| TRIS | 14.52 g (1.2 mol/L) |
| Taurine | 13.01 g (1.04 mol/L) |
| Deionized water to | 100.00 mL |

***Buffer 2 (Cathode buffer), pH = 7.6, I = 2×0.02 mol/L***
| | |
|---|---|
| TRIS | 6.93 g (0.06 mol/L) |
| Taurine | 13.01 g (0.10 mol/L) |
| Deionized water to | 1,000.00 mL |

***Monomeric solution, T = 50 g/dL, C = 0.03***
| | |
|---|---|
| Acrylamide | 48.50 g |
| BIS | 1.50 g |
| Deionized water to | 100.00 mL |

The gel solutions should be prepared according to Table 3.4.5.

**Table 3.4.5:** Preparing the stacking and resolving gels for DNA electrophoresis in TRIS-taurinate buffer at 2 pH values according to Michov.

| | Stack. gel $T = 6$ g/dL, pH = 7.6, $I = 0.02$ mol/L | Resolv. gel $T = 9$ g/dL | Resolv. gel $T = 11$ g/dL | Resolv. gel $T = 13$ g/dL | Resolv. gel $T = 15$ g/dL | Resolv. gel $T = 17$ g/dL |
|---|---|---|---|---|---|---|
| | | pH = 8.6, $I = 0.20$ mol/L | | | | |
| Buffer 1, mL | – | 50.00 | 50.00 | 50.00 | 50.00 | 50.00 |
| Buffer 2, mL | 50.00 | – | – | – | – | – |
| Monomeric solution, mL | 12.00 | 18.00 | 22.00 | 26.00 | 30.00 | 34.00 |
| 87% Glycerin, mL | 30.00 | 10.00 | 10.00 | 10.00 | 10.00 | 10.00 |
| 10 mL/dL TMEDA, mL | 0.45 | 0.45 | 0.45 | 0.45 | 0.45 | 0.45 |
| 10 g/dL APS, mL | 0.36 | 0.20 | 0.16 | 0.14 | 0.12 | 0.11 |
| Deionized water, mL, to | 100.00 | 100.00 | 100.00 | 100.00 | 100.00 | 100.00 |

**Procedure**
- Place some drops of water onto a glass plate.
- Lay a support film down on the glass plate and roll over so that any air bubbles are removed.
- Place on the film 0.5 mm U-form silicone gasket, put over a second hydrophobilized glass plate, and fix the cassette by clamps.
- Pour stacking solution into the vertical casting cassette until one-third of it is filled.
- Overlay the stacking solution with some deionized water.
- After the stacking solution has gelled, pour out the deionized water and overlay the stacking gel with resolving solution.
- Overlay the resolving solution with some deionized water.
- Allow the resolving solution to gel.
- Run the electrophoresis at 20 mA for 120 min.
  *Use paper electrode strips with undiluted anode buffer and cathode buffer, respectively.*

**Separation of PCR Products by Polyacrylamide Gel Disc-Electrophoresis According to Michov**

**Materials and equipment**
TRIS-formate buffer, pH = 8.3
TRIS-taurinate buffer, pH = 7.4
Monomeric solution ($T$ = 50 g/dL, $C$ = 0.03)
10 g/dL TMEDA
10 g/dL APS
DNA samples
DNA ladder
Gel electrophoresis cell
Glass plates, spacers and combs
DC power supply
Ethidium bromide or SYBR Green EMSA nucleic acid gel stain

**Procedure**
- Cast a gel on support film as above.
- Put 0.5–1.0 mL of contact fluid, for example, kerosene or silicone oil DC 200, onto the cooling plate of the electrophoretic cell.
- Place the film-supported gel, together with its cover film, onto the cooling plate.
- Roll gently on the cover film with a photo roller so that the contact fluid is evenly distributed between the cooling plate and support film.

- Remove the cover film and place onto the gel ends paper electrode strips according to their polarity.
- Lay on an application template (preferably of silicone) in front of the cathode strip.
- Apply 5–8 µL of nucleic acid samples with a concentration of 10 mg/mL into the slots of the application template.
- Place the electrodes on the corresponding paper electrode strips.
- Run disc-electrophoresis at constant electric current and a temperature of 10 °C until the dye front reaches the opposite strip.
- Stain the DNA bands with ethidium bromide or SYBR Green EMSA nucleic acid gel stain.

**Electrophoretic Mobility Shift Assay**

**Materials and equipment**
TRIS-**b**orate-EDTA buffer (TBE)
10,000x Concentrate in dimethylsulfoxide
**Tri**chloro**a**cetic acid (TCA)
7 mL/dL Acetic acid
10 mL/dL Methanol
DNA sample
DNA ladder
Orange G
SYBR Green EMSA nucleic acid gel stain
SYPRO Ruby EMSA protein gel stain
Staining tray

*0.5x TBE running buffer*

| | |
|---|---|
| TRIS | 0.54 g (44.5 mmol/L) |
| Boric acid | 0.28 g (44.5 mmol/L) |
| $Na_2EDTA$ | 0.04 g (1 mmol/L) |
| Deionized water to | 100.0 mL |

**Procedure**
- Peel the tape from the bottom of casting cassette.
- Pull the comb from the gel.
- Fill the upper buffer chamber with TBE running buffer. The buffer level should exceed the level of the wells.
- Load DNA samples and DNA ladder.
- Fill the lower buffer chamber also with TBE running buffer.
- Run electrophoresis at 100 V for 90 min.
- When the Orange G dye front reaches the bottom, shut off the power.

### Gel staining

- Warm the SYBR Green EMSA gel stain concentrate to room temperature.
- Dilute 5 µL of 10,000x SYBR green EMSA gel stain concentrate into 50 mL TBE buffer and pour into gel staining tray.
- Transfer the gel into the staining tray.
- Incubate for 20 min on orbital shaker, protected from light.
- Add SYPRO Ruby EMSA protein gel stain solution with TCA.
- Incubate for 3 h on orbital shaker at 50 rpm, protected from light.
- Wash the gel with deionized water for 10 s.
- Destain the gel in 10 mL/dL methanol and 7 mL/dL acetic acid for 60 min.
- Wash the gel with deionized water for 10 s.


## Electroelution of Small DNA Fragments from Polyacrylamide Gel

### Materials and equipment
TBE buffer, pH = 8.0
3 mol/L Sodium acetate, pH = 5.2
Ethanol
Samples with DNA fragments
Dialysis bag
Horizontal gel electrophoresis cell

### Procedure
- Run gel electrophoresis of DNA.
- Cut out the desired DNA band with a scalpel or razor blade.
- Transfer the gel piece in an appropriate dialysis bag and add enough TBE buffer.
- Place the bag in a horizontal gel electrophoresis cell containing TBE buffer and run electrophoresis at 4 V/cm for 2 h, if DNA < 300 bp; or for 6 h, if DNA are longer.
- Reverse the polarity of the electrophoresis cell for 1 min to free the bound DNA.
- Rinse the gel piece and the inside of the dialysis bag to obtain all DNA.
- Precipitate DNA molecules with 3 mol/L sodium acetate, pH = 5.2, and ethanol.


## References

[1]    Michov BM. Elektrophorese. Theorie Und Praxis. Walter de Gruyter, Berlin – New York, 1996.
[2]    Martin R. Electrophoresis of Nucleic Acids (Elektrophorese Von Nucleinsäuren). Spektrum, Heidelberg – Berlin – Oxford, 1996.
[3]    Maniatis TA, Jeffrey A, Kleid DG. Proc Natl Acad Sci USA, 1975, 72, 1184–1188.

[4]    Doktycz MJ, Gibson WA, Arlinghaus HF, Jacobson KB. Appl Theoret Electrophoresis, 1993, 3, 157–162.
[5]    Allen RC, Budowle B, Reeder DJ. Appl Theoret Electrophoresis, 1993, 3, 173–181.
[6]    Vasudevan J, Ayyappan A, Perkovic M, Bulliard Y, Cichutek K, Trono D, Häussinger D, Münk C. J Virol, 2013, 87, 9030–9040.
[7]    Richards EG, Coll JA, Gratzer WB. Anal Biochem, 1965, 12, 452–471.
[8]    Doktycz MJ. Anal Biochem, 1990, 213, 400–406.
[9]    Allen RC, Graves GM. BioTechnology, 1990, 8, 1288–1290.
[10]   Orban L, Chrambach A. Electrophoresis, 1991, 12, 233–240.
[11]   Old RW, Primrose SB. Principle of Gene Manipulation: An Introduction to Genetic Engineering, 5th edn, Blackwell Science, Oxford, England, 1989, 9.
[12]   Zimm BH, Levene SD. Q Rev Biophys, 1992, 25, 171–204.
[13]   Allen RC, Graves GM, Budowle B. BioTechniques, 1989, 7, 736–744.
[14]   Saiki RK, Scharf S, Faloona F, Mullis KB, Horn GT, Ehrlich HA, Arnheim N. Science, 1985, 230, 1350–1354.
[15]   Haas H, Budowle B, Weiler G. Electrophoresis, 1994, 15, 153–158.
[16]   Michov BM Protein Separation by SDS Electrophoresis in a Homogeneous Gel Using a TRIS-formate-taurinate Buffer System and Suitable Homogeneous Electrophoresis Plates [Proteintrennung Durch SDS-Elektrophorese in Einem Homogenen Gel Unter Verwendung Eines TRIS-Formiat-Taurinat-Puffersystems Und Dafür Geeignete Homogene Elektrophoreseplatten]. German Patent 4127546, 1991.
[17]   Chen X, Ugaz VM. Electrophoresis, 2006, 27, 387–393.
[18]   Allen RC, Budowle B. Gel Electrophoresis of Proteins and Nucleic Acids. Walter de Gruyter, Berlin – New York, 1994, 262.
[19]   Ansorge W. Stathakos D (ed.). Electrophoresis ´82. Walter de Gruyter, Berlin – New York, 1983, 235.
[20]   Mayrand PE, Corcoran KP, Ziegle JS, Robertson JM, Hoff LB, Kronick MN. Appl Theoret Electrophoresis, 1992, 3, 1.
[21]   Robertson J, Ziegle J, Kronick M, Madden D, Budowle B. Burke T, Dolt G, Jeffries AJ, Wolf R (eds.). Fingerprinting. Approaches and Applications. Birkhäuser, Basel, 1991, 391.
[22]   Rüger R, Höltke HJ, Reischl U, Sagner G, Kessler C. J Clin Chem Clin Biochem, 1990, 28, 566.
[23]   Garner MM, Revzin A. Nucl Acids Res, 1981, 9, 3047–3060.
[24]   Fried M, Crothers DM. Nucl Acids Res, 1981, 9, 6505–6525.
[25]   Fried MG. Electrophoresis, 1989, 10, 366–376.
[26]   Fried MG. Electrophoresis, 1989, 10, 366–376.
[27]   Hestekin CN, Barron AE. Electrophoresis, 2006, 27, 3805–3815.
[28]   Abe M, Kufe D. Proc Natl Acad Sci USA, 1993, 90, 282–286.

# 3.5 Microchip electrophoresis of nucleic acids

3.5.1    Theory of microchip electrophoresis of nucleic acids —— 371
3.5.2    Construction of a microchip for electrophoresis of nucleic acids —— 371
3.5.2.1  Polymers —— 372
3.5.3    Running microchip electrophoresis of nucleic acids —— 373
3.5.4    Applications of DNA microchip electrophoresis —— 373
         References —— 373

(Micro)chip electrophoresis (ME, microfluidic electrophoresis, lab-on-a-chip) is one of the most powerful analytical tools in recent years [1]. Since its introduction in the early 1990s by Manz *et al.* [2] and Harrison *et al.* [3], it has become an important method for analyzing DNA fragments because of speed, miniaturization, small reagent volumes, high resolution, and automation [4–6]. Using this method, DNA can be sized, sequenced, and genotyped [7].

## 3.5.1 Theory of microchip electrophoresis of nucleic acids

Under influence of an electric field, the negatively charged nucleates migrate through a buffer with velocity $v$, in m/s, which can be expressed by the product of the electrophoretic mobility $\mu$ and the electric field strength $E$ [8, 9]:

$$v = \mu E \tag{3.5.1}$$

Since $E$ is equal to the ratio between the voltage $U$ and the length $l$ between the electrodes, eq. (3.5.1) can be transformed to give

$$v = \mu \frac{U}{l} \tag{3.5.2}$$

As the length between the electrodes is a constant, the last equation shows that only the voltage controls the velocity of nucleates.

## 3.5.2 Construction of a microchip for electrophoresis of nucleic acids

Usually, a chip for electrophoresis of nucleic acids consists of separation channel, injection channel, and reservoirs. The separation channel can be 85 mm long, 50 μm wide and 20 μm deep. It begins with a buffer inlet reservoir and ends with buffer outlet reservoir (buffer waste). The injection channel can be 8.0 mm long. It

https://doi.org/10.1515/9783110761641-032

represent a double-T channel with a 100 µm offset, beginning with sample inlet and ending with sample outlet (sample waste). All reservoirs can have 2.0 mm diameter and can be 1 mm deep (Figure 3.5.1) [10].

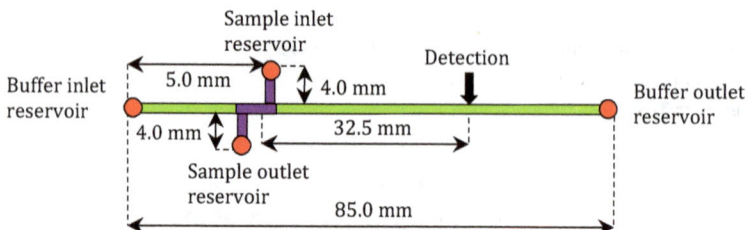

**Figure 3.5.1:** Schematic diagram of a microfluidic chip.

All microchips are fabricated by hot embossing method with a master fabricated on soda-lime glass using photolithography techniques. The reservoirs are created by drilling. The sequencing chip has a four-layer construction, consisting of three 100 mm diameter glass wafers and a **polydim**ethylsiloxane (PDMS) membrane. Prior to experiments, the microchip should be rinsed with 1 mol/L NaOH and then with deionized water.

### 3.5.2.1 Polymers

The polymer solutions used as DNA sieving matrices define the quality of the separation [11, 12]. Among them, the N-alkoxyalkylacrylamide polymers – **poly(N-m**ethoxy **ethylacrylamide) (**pNMEA) and **poly(N-e**thoxyethylacrylamide) (pNEEA) were investigated [13]. The polymers pNMEA and pNEEA were synthesized *via* free-radical polymerization from the monomers **N-m**ethoxyethylacrylamide (NMEA) and **N-e**thoxy ethylacrylamide (NEEA), respectively (Figure 3.5.2). The concentrations of the monomers are presented by their mass parts in the total monomeric mixture; e.g., NMEA90-NEEA10 means 90% NMEA monomer and 10% NEEA monomer in the total mixture.

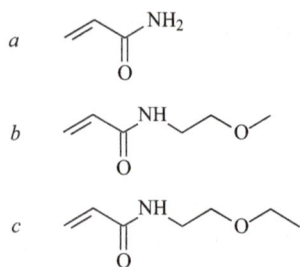

**Figure 3.5.2:** Monomer structures.
(*a*) Acrylamide; (*b*) NMEA; (*c*) NEEA.

### 3.5.3 Running microchip electrophoresis of nucleic acids

Various preconcentration techniques were developed and applied for ME to increase the sensitivity, including high salt stacking [14], field-amplified sample stacking [15–17], sweeping [18], base stacking [19], porous membrane stacking [20], and transient isotachophoresis TITP [21–23].

Kang et al. [24] proposed a TRIS-borate-EDTA (TBE) buffer (pH = 8.3) with 0.5 µg/mL ethidium bromide (EtBr) for the ME run. The dynamic coating matrix was made by dissolving 0.5 g/dL of $M_r$ = 1,000,000 polyvinylpyrrolidone (PVP), and the sieving matrix was made by dissolving 0.5 g/dL of $M_r$ = 8,000,000 polyethylene oxide (PEO). The DNA sample was injected into the injection region by applying a voltage of 480 V.

### 3.5.4 Applications of DNA microchip electrophoresis

Lin et al. [25] separated dsDNA by a PMMA electrophoresis microchip using polymer solutions containing gold nanoparticles. Later, Xu et al. [26, 27] have developed a convenient single-step quantitation technique for separating specific dsDNA fragments in PCR products based on PMMA electrophoresis microchip with UV or fluorescence detection.

The analysis of single-base mutations in genomic DNA was carried out using a PMMA chip for detection of the allele-specific ligation products using a universal array platform. Sung et al. [28] have used PMMA ME for analyzing PCR products. Lee et al. [29] have developed a hot-embossed PMMA-PCR chip for amplifying a human cancer tumor-suppressing DNA sequence. Liu et al. [30] have reported a method based on PMMA microchips for separation of multiplex PCR products.

A fast diagnosis by ME was evaluated, using programmed field strength gradients (PFSG) in a conventional glass double-T microfluidic chip [31]. Compared to ME with a constantly applied electric field, ME-PFSG achieved 15-fold faster analysis.

Highly sensitive and rapid analysis of the methylated p16 gene in plasma and tissue DNA in cancer patients was realized by Zhou et al. [32] in a PMMA microchip. Hashimoto et al. [33] have performed PCR, ligase detection reaction, and hybridization assays using flow-through PMMA microchip for the detection of low-abundant DNA point mutations.

### References

[1]   Liu CN, Toriello NM, Richard A, Mathies RA. Anal Chem, 2006, 78, 5474–5479.
[2]   Manz A, Graber N, Widmer HM. Sens Actuators B, 1990, 1, 244–248.
[3]   Harrison DJ, Manz A, Fan Z, Lüdi H, Widmer HM. Anal Chem, 1992, 64, 1926–1932.
[4]   Woolley AT, Mathies RA. Anal Chem, 1995, 67, 3676–3680.

[5]    Shi YN, Anderson RC. Electrophoresis, 2003, 24, 3371–3377.

[6]    Kan CW, Fredlake CP, Doherty EAS, Barron AE. Electrophoresis, 2004, 25, 3564–3588.

[7]    Verpoorte E. Electrophoresis, 2002, 23, 677–712.

[8]    Guttma A, Wanders B, Cooke N. Anal Chem, 1992, 64, 2348–2351.

[9]    Baker DR. Capillary Electrophoresis. John Wiley & Sons, New York, 1995, 19–52.

[10]   Kang SH, Park M, Cho K. Electrophoresis, 2005, 26, 3179–3184.

[11]   Chiesl TN, Putz KW, Babu M, Mathias P et al. Anal Chem, 2006, 78, 4409–4415.

[12]   Chiesl TN, Forster RE, Root BE, Larkin M, Barron AE. Anal Chem, 2007, 79, 7740–7747.

[13]   Kan CW, Doherty EAS, Barron AE. Electrophoresis, 2003, 24, 4161–4169.

[14]   Chen CC, Yen SH, Makamba H, Tsai ML, Chen SH. Anal Chem, 2007, 79, 195–201.

[15]   Yang H, Chien RL. J Chromatogr A, 2001, 924, 155–163.

[16]   Beard NP, Zhang CX, deMello AJ. Electrophoresis, 2003, 24, 732–739.

[17]   Jung B, Bharadwaj R, Santiago JG. Electrophoresis, 2003, 24, 3476–3483.

[18]   Seram Y, Matsubara N, Otsuka NK, Terabe S. Electrophoresis, 2001, 22, 3509–3513.

[19]   Kim DK, Kang SH. J Chromatogr A, 2005, 1064, 121–127.

[20]   Song S, Singh AK, Kirby BJ. Anal Chem, 2004, 76, 4589–4592.

[21]   Wainright A, Williams SJ, Ciambrone G, Xue Q, Wei J. J Chromatogr A, 2002, 979, 69–80.

[22]   Jeong Y, Choi K, Kang MK, Chun K, Chung DS. Sens Actuators B, 2005, 104, 269–275.

[23]   Chen L, Prest JE, Fielden PR, Goddard NJ et al. Lab Chip, 2006, 6, 474–487.

[24]   Kang SH, Park M, Cho K. Electrophoresis, 2005, 26, 3179–3184.

[25]   Lin YW, Huang MJ, Chang HT. J Chromatogr A, 2003, 1014, 47–55.

[26]   Xu F, Jabasini M, Zhu BM, Ying L et al. J Chromatogr A, 2004, 1051, 147–153.

[27]   Xu R, Jabasini M, Baba Y. Electrophoresis, 2005, 26, 3013–3020.

[28]   Sung WC, Lee GB, Tzeng CC, Chen SH. Electrophoresis, 2001, 22, 1188–1193.

[29]   Lee DS, Park SH, Chung KH, Yoon TH et al. Lab Chip, 2004, 4, 401–407.

[30]   Liu DY, Zhou XM, Zhong RT, Ye NN et al. Tatanta, 2006, 68, 616–622.

[31]   Kang SH, Park M, Cho K. Electrophoresis, 2005, 26, 3179–3184.

[32]   Zhou XM, Shao SJ, Xu GD, Zhong RT et al. J Chromatogr B, 2005, 816, 145–151.

[33]   Hashimoto M, Barany F, Soper SA. Biosens Bioelect, 2006, 21, 1915–1923.

# 3.6 Blotting of nucleic acids

3.6.1     Blotting principles —— 375
3.6.1.1   Transfer —— 375
          Diffusion transfer —— 376
          Capillary transfer —— 376
          Vacuum transfer —— 376
          Electrotransfer —— 377
3.6.1.2   Blocking —— 378
3.6.1.3   Detection by probes, dyes, and autoradiography —— 378
3.6.2     Southern blotting —— 379
3.6.3     Northern blotting —— 380
3.6.4     Middle-Eastern blotting —— 381
3.6.5     Protocols —— 382
          Southern Blotting by Downward Capillary Transfer —— 382
          Electroblotting of Polyacrylamide Gel onto Nylon Membrane —— 383
          References —— 384

Blotting of nucleic acids is used to understand the organization, sequences, and hybridization of nucleic acids, as well as the gene expression, inherited diseases and infectious agents.

## 3.6.1 Blotting principles

As with proteins, nucleic acids blotting runs in three steps: transfer of nucleate bands onto a blot membrane, blocking the free (unoccupied) binding sites of the blot membrane by blocking agents, and detection of the blotted nucleates by specific high molecular ligands (probes).

### 3.6.1.1 Transfer

The transfer of electrophoretically separated nucleates onto a membrane is necessary because nucleates in a gel have difficultly reacting with probes. Polyacrylamide gels should be blotted in a transfer buffer of low ionic strength. There are a few transfer methods: diffusion transfer, capillary transfer, vacuum transfer, and electrotransfer. Among them, capillary and vacuum transfer are most commonly used, whereas the diffusion transfer and electrotransfer are limited used.

https://doi.org/10.1515/9783110761641-033

## Diffusion transfer

In the diffusion transfer (blotting) [1], the blot membrane is placed onto the gel surface and the nucleate transfer is carried out by diffusion (Figure 3.6.1). The blotting time takes a long time – up to 36–48 h. The gel could be also blotted between two blot membranes, if it is not cast on a support film. The transfer can be accelerated by increasing the temperature, what is referred to as thermotransfer (thermoblotting). In practice, this method is not more used as it takes too much time.

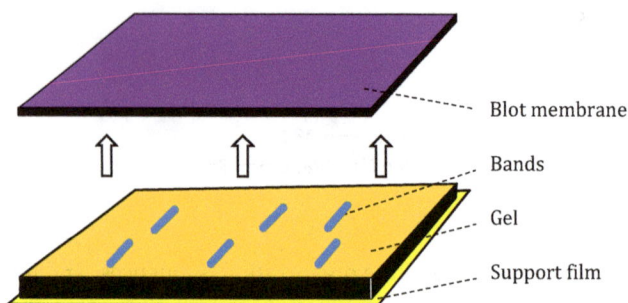

Blot membrane

Bands

Gel

Support film

**Figure 3.6.1:** Diffusion blotting.

## Capillary transfer

In the capillary transfer (blotting) [2], a buffer is sucked up into a stack of dry papers, which are placed onto a blot membrane covering a gel with the resolved nucleates. The paper stack is usually covered with a glass plate and the extra ballast. The paper stack acts with its capillary force on the gel, sucks the gel buffer upstairs and, as a result, the resolved nucleates stop on the blot membrane. The transfer is carried out for a long time, usually overnight (Figure 3.6.2).

In addition to the upward capillary transfer, a downward capillary transfer is known (see *Protocols*).

## Vacuum transfer

The vacuum transfer (blotting) [3, 4] requires a vacuum blotter that creates negative pressure (incorrect named vacuum) of 200–400 Pa (20–40 cm of water) (Figure 3.6.3). For this purpose, a water-jet pump can be used. The vacuum transfer should be made carefully, since the gel could be broken. It takes place in 30–40 min.

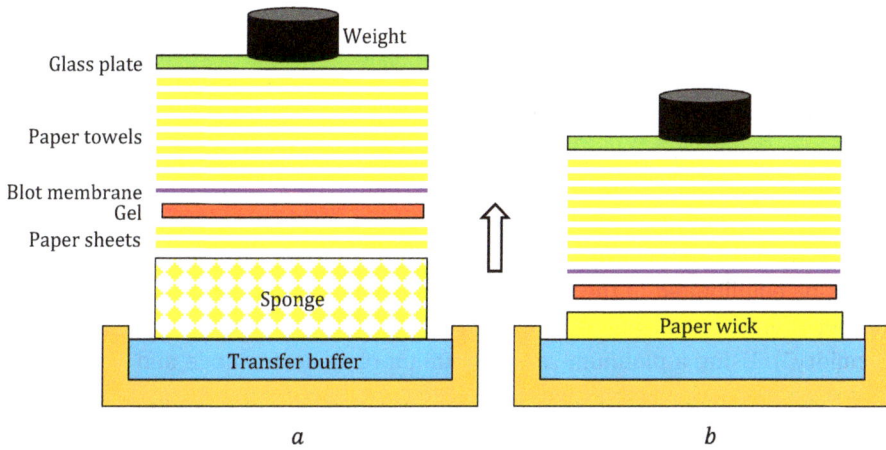

**Figure 3.6.2:** Upward capillary transfer.
(*a*) Sponge method; (*b*) Wick method.

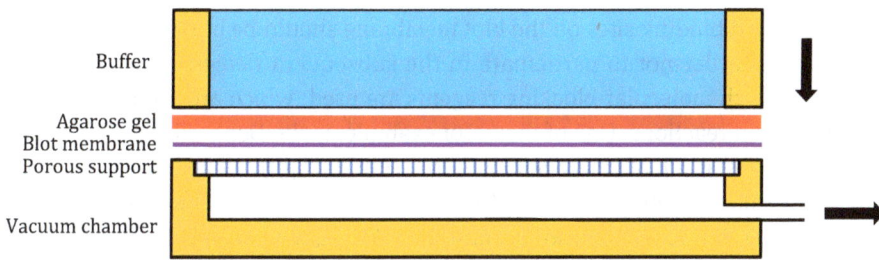

**Figure 3.6.3:** Vacuum blotting.

## Electrotransfer

The electrophoretic transfer (electrotransfer, electroblotting) of nucleic acids is carried out when the gel, containing electrophoretically separated DNA or RNA, is clamped to a blotting membrane in a cassette, and is put between two plate electrodes creating an electric field. It is faster than the capillary transfer and takes place in a few minutes to 24 h. If the gel is connected to a support film, it should be freed from it [5, 6]. Meanwhile, net-supported gels [7, 8] are available, which are suitable for electroblotting.

The electrotransfer has two versions: tank transfer (blotting) and semidry transfer (blotting).

The *tank transfer* [9] of nucleates is preferred in comparison with the semidry transfer. Prior to it, the gel with the resolved nucleates and the blot membrane are placed between filter papers soaked with buffer. Then, the blot sandwich is introduced into a grid cassette and put vertically between two platinum electrode wires that are in a tank containing a buffer. The buffer should be cooled so that the blot sandwich is not heated. The transfer is carried out in a few hours to overnight.

In the *semidry transfer* [10, 11], the electrodes have a direct contact to the gel – nitrocellulose membrane sandwich to make a fast transfer. The polyacrylamide gels should be equilibrated in the transfer buffer to remove electrophoresis buffer salts and detergents. The nitrocellulose membranes and filter papers are wetted scarce ("semidry"). Using a platinum-coated titanium plate as an anode and a stainless-steel plate as a cathode, the transfer is carried out in horizontal location without a buffer tank or gel cassettes.

### 3.6.1.2 Blocking

The unoccupied binding sites on the blot membrane should be blocked with certain substances in order not to participate in the subsequent detection reactions. To cover them, high molecular-blocking reagents are used, which are not taking place in the detection reactions. One of them, for example, is the Denhardt's buffer [12]. It contains 10 to 50 g/mL of heterogeneous DNA with 0.02 g/dL of bovine serum albumin, 0.02 g/dL of ficoll, 0.02 g/dL of polyvinyl pyrrolidone, 1 mmol/L $Na_2EDTA$, and 50 mmol/L NaCl in a 10 mmol/L TRIS-chloride buffer with pH = 7.0.

### 3.6.1.3 Detection by probes, dyes, and autoradiography

After transfer and blocking, the nucleate bands can be made visible using specific high molecular ligands (probes) or dyes. Besides, radiolabeled DNA or RNA probes bind (hybridize) with complementary DNA or RNA. To prevent radioactivity, the same probes can be labeled with the biotin-avidin system or digoxigenin.

If radioactivity is incorporated in the probes, the position of the blotted nucleates on the membrane can be determined by autoradiography: The blot membrane is placed in contact with photographic film and left to expose (hours to weeks) in the dark and cold. If required, the gel can be overlaid on the negative and the radioactive spots obtained can be cut up, eluted, and scintillated.

There are different blotting methods for nucleic acids: Southern blotting, Northern blotting, reverse Northern blotting, Middle-eastern blotting, Eastern-western blotting, Eastern blotting, Far-eastern blotting, and other (Table 3.6.1).

**Table 3.6.1:** The famous blotting methods for nucleates.

| Blotting methods | Goals |
| --- | --- |
| Southern blotting | DNA |
| Northern blotting | RNA |
| Reverse Northern blotting | RNA |
| Middle-eastern blotting | polyA RNA |

## 3.6.2 Southern blotting

Southern blotting is created by the British biologist Edwin Southern [13]. Later, in reference to Southern's name, other blotting methods were named (i.e., Western blotting, Northern blotting, Eastern blotting, Southwestern blotting, *etc.*), because they employ similar principles, but different electrophoretically resolved polyions.

Southern blotting is an analytical technique used to detect DNA sequence. After a gel electrophoresis, the separated DNA bands are transferred onto a membrane to become visible by hybridization with specific probes. The method is fulfilled in the following way:

– Using restriction endonucleases, high-molecular-mass DNA strands are cut into smaller fragments.
– The DNA fragments are electrophoresed in an agarose gel.
– The gel is placed into an alkaline solution to denature the double-stranded DNA.
– A nitrocellulose membrane is placed onto the gel. Then, a stack of filter papers and a weight are put above to ensure good contact between gel and membrane. As a result of originated capillary forces, the buffer moves from the gel into the filter papers and transfer the DNA bands from the gel onto the membrane. The transferred DNA bands bind to the membrane due to their negative charges and the positive charges of the membrane.
– The membrane is baked in a vacuum or regular oven at 80 °C for 2 h, or exposed to ultraviolet radiation to permanently attach the transferred DNA to the membrane.
– Then, the membrane is exposed to a DNA hybridization probe. In some cases, RNA can also be used as a hybridization probe rather than DNA.
– After hybridization, the excess probe is washed from the membrane, and the pattern of hybridization is visualized on X-ray film by autoradiography, if a radioactive or fluorescent probe is used; or by color reaction on the membrane, if a chromogenic method is used.

Hybridization of the probe to a specific DNA fragment on the membrane indicates that the DNA fragment contains a complementary DNA sequence. So, Southern blots obtained with restriction enzyme-digested genomic DNA may be used to determine the number of sequences (e.g., gene copies) in a genome. A probe that hybridizes only to a single DNA segment produces a single band, whereas multiple bands are observed when the probe hybridizes several similar sequences. Sequences that hybridize with the probe are further analyzed to obtain the full-length sequence of the targeted gene. Southern blotting can also be used to identify methylated sites in genes.

Southern blotting allows also detection of larger gene alterations such as deletions, insertions or rearrangements. Hereditary motor and sensory neuropathy type I, for example, are characterized by progressive distal muscle weakness and wasting between the ages of 10 and 30 years. Usually, the disorder is a result of a duplication of 1.5 Mb of genomic DNA on the chromosome 17 short arm, including the gene for the **p**eripheral **m**yelin **p**rotein **22** [14].

## 3.6.3 Northern blotting

Northern blotting is a method, which is used to study RNA sequences with labeled RNA probes. It was developed by Alwine, Kemp, and Stark in 1977 at Stanford University [15]. The name of this technique comes from the similarity between it and the Southern blotting. The difference between two methods is that in Northern blotting RNA, rather than DNA, is analyzed (Figure 3.6.4).

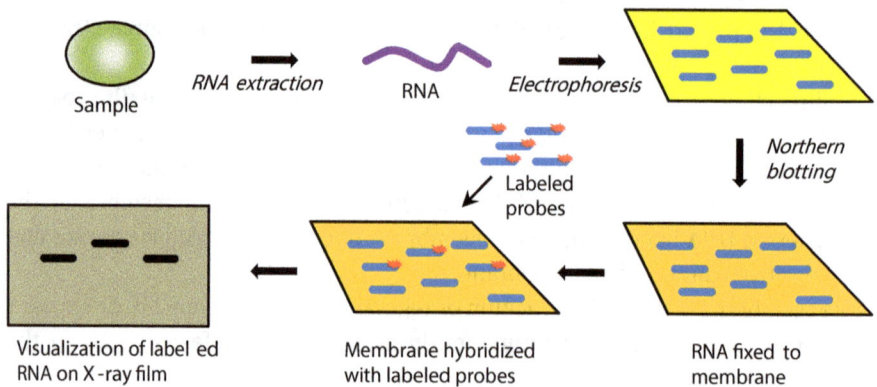

**Figure 3.6.4:** RNA detection by Northern blotting.

The procedure of Northern blotting starts with the extraction of total RNA from a homogenized tissue or cells. mRNA can be isolated from the total RNA using oligo-(dT)-cellulose chromatography to maintain only RNAs with poly(A) tails [16, 17].

Then, electrophoresis in an agarose gel is carried out, during which RNA are resolved by size. Since the agarose gels are fragile and the probes cannot enter them, the RNA samples are transferred onto a nylon membrane using a capillary or vacuum transfer system. The nylon membrane is most effective since it is positive charged while RNA carries negative charges. The transfer buffer contains formamide that lowers the annealing temperature of the probe-RNA interaction preventing the RNA degradation at high temperatures [18].

After RNA is transferred on a membrane, it is immobilized through covalent linkage to the membrane by UV light or heat. Then, a labeled probe is hybridized to the bound RNA. At the end, the membrane is washed to remain only the complexes between RNA and the probe. The hybrid signals are detected by an X-ray film and are quantified by densitometry.

The probes are labeled either with radioactive isotopes ($^{32}$P) or with chemiluminescent substances. The chemiluminescent labelling can occur in two ways: either the probe is attached to **al**kaline **p**hosphatase (ALP) or **h**orse**r**adish **p**eroxidase (HRP), or is labeled with biotin, whose ligand (avidin or streptavidin) is attached to the enzyme (e.g. HRP). An X-ray film can detect both the radioactive and chemiluminescent signals. Many researchers prefer chemiluminescent signals because they are faster, more sensitive, and reduce the health hazards that go along with radioactive labels [19].

Northern blotting is used to determine gene expression rates during differentiation and morphogenesis, as well as in abnormal or diseased conditions [20]. It also allows observing the gene expression in tissues, organs, developmental stages, pathogen infection, and over the course of treatment [21, 22]. This technique has also been used to explore the overexpression of oncogenes and downregulation of tumor-suppressor genes in cancerous cells, as well as the gene expression in the rejection of transplanted organs. Besides, the Northern blotting can show which region of RNA is missing [23].

A variant of the Northern blotting is the reverse Northern blotting. In it, DNA rather than RNA is first immobilized on a blot membrane and then is detected with labeled RNA probes. The Northern blotting is used to determine the levels of gene expression in particular tissues. In comparison to Northern blotting, it is able to probe a large number of transcripts at once [24]. However, the reverse Northern blotting is limited by its ability to probe only with one mRNA at a time, while q-PCR requires transcripts to be long enough to generate primers for the sequence and probes can be costly.

## 3.6.4 Middle-Eastern blotting

Middle-eastern blotting was described in 1984. It combines a blotting of polyA RNA, resolved in an agarose gel, and its linkage with DNA probes [25].

## 3.6.5 Protocols

**Southern Blotting by Downward Capillary Transfer**

**Materials and equipment**
TRIS-borate-EDTA (TBE) buffer
Saline-sodium citrate (SSC) buffer
Nondenaturing or denaturing polyacrylamide gel
Chromatographically clean filter paper

**Procedure**
- Run electrophoresis of DNA samples in a nondenaturing or denaturing poly-acrylamide gel.
- Soak two pads in TBE buffer and remove air pockets by squeezing.
- Cut pieces of chromatographically clean filter paper of the same size as the gel and soak them also in TBE-buffer for 15 min.
- Wet a nylon membrane with the dimensions of the gel in water for 5 min.
- Make a 2–3 cm stack of paper towels in a glass dish, followed by a few pieces of the same paper, and then a chromatographically clean filter paper prewetted with transfer buffer.
- Wet the membrane and place it onto the filter paper, followed by the gel (Figure 3.6.5).

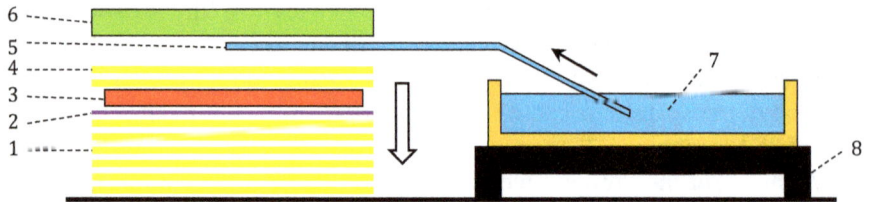

**Figure 3.6.5:** Downward capillary transfer.
1. Paper towels; 2. blot membrane; 3. gel; 4. paper sheets; 5. tubing; 6. glass plate; 7. transfer buffer; 8. support

- Place two pieces of soaked paper onto the top so that one end is submerged in a glass dish containing transfer buffer. Cover the sandwich with a glass plate to reduce evaporation, and transfer for 1 h.
- Remove paper towels and filter papers. Mark with a pencil the positions of the wells on the membrane. Indicate up–down and back–front orientations by marking or cutting off one corner of the membrane.
- Rinse in SSC buffer, place on a sheet of chromatographically clean filter paper, and dry at room temperature for 30 min.

- For nylon membrane, wrap in UV-transparent plastic wrap, place DNA side down on a UV transilluminator or UV light box, and irradiate. For nitrocellulose membrane, place between two sheets of filter paper and bake under vacuum at 80 °C for 2 h.
- Store between sheets of chromatographically clean filter paper or in a desiccator at room temperature.

### Electroblotting of Polyacrylamide Gel onto Nylon Membrane

#### Materials and equipment
TRIS-borate-EDTA (TBE) buffer
Saline-sodium citrate (SSC) buffer
0.4 mol/L NaOH (for nondenaturing gels)
Chromatographically clean filter paper
Electroblotting cell with cooling coil and pads

#### Procedure
- Run electrophoresis of DNA samples in a polyacrylamide gel, stain and photograph it (if nondenaturing).
- Soak the gel in TBE buffer for 20 min.
- Cut a few pieces of chromatographically clean filter paper to the same size as the gel.
- Wet a piece of a nylon membrane with dimensions of the gel in water for 5 min.
- Cover the gel with the dry piece of chromatographically clean filter paper and lift the gel off peeling away the filter paper.
- Place one of the saturated pads onto a panel and cover it with three soaked filter papers, carefully removing trapped bubbles by rolling a glass pipette over the surface.
- Flood the filter paper carrying the gel with TBE-buffer and place on top of the stack, gel side up. Flood the gel with TBE-buffer and place the prewetted membrane on top. Flood the surface with 0.5 × TBE buffer and place a few sheets of saturated filter paper on top, followed by the second saturated pad.
- Transfer at 30 V (~125 mA) for 4 h using precooled at 4 °C buffer, and a recirculating bath.
- Mark wells and orientation of the membrane after disassembly.
- If the gel is nondenaturing, denature the membrane for 10 min by placing DNA-side-up on three pieces of chromatographically clean filter paper soaked in 0.4 mol/L NaOH.
- Rinse in SSC buffer, place onto a sheet of filter paper, and allow drying completely. Immobilize the DNA and store the membrane.

- Place DNA side down on a UV transilluminator or UV light box, and irradiate. For nitrocellulose, place between two sheets of chromatographically clean filter paper and bake under vacuum at 80 °C for 2 h.
- Store between sheets of chromatographically clean filter paper or in a desiccator at room temperature.

# References

[1] Bowen B, Steinberg J, Laemmli UK, Weintraub H. Nuc Acids Res, 1980, 8, 1–20.
[2] Olsson BG, Weström BR, Karlsson BW. Electrophoresis, 1987, 8, 377–464.
[3] Medveczky P, Chang CW, Oste C, Mulder C. BioTechniques, 1987, 5, 242–246.
[4] Olszewska E, Jones K. Trends Gen, 1988, 4, 92–94.
[5] Westermeier R. Elektrophorese-Praktikum. VCH, Weinheim, 1990.
[6] Jägersten C, Edström A, Olsson B, Jacobson G. Electrophoresis, 1988, 9, 662–665.
[7] Nishizawa H, Murakami A, Hayashi N, Iida M, Abe Y. Electrophoresis, 1985, 6, 349–350.
[8] Kinzkofer-Peresch A, Patestos NP, Fauth M, Kögel F, Zok R, Radola BJ. Electrophoresis, 1988, 9, 497–511.
[9] Bittner M, Kupferer P, Morris CF. Anal Biochem, 1980, 102, 459–471.
[10] Trnovsky J. BioTechniques, 1992, 13, 800–804.
[11] Michov BM. GIT Fachz Lab, 1992, 36, 746–749.
[12] Denhardt D. Biochem Biophys Res Commun, 1966, 20, 641–646.
[13] Southern EM. J Mol Biol, 1975, 98, 503–517.
[14] Pareyson D, Taroni F. Curr Opin Neurol, 1996, 9, 348–354.
[15] Alwine JC, Kemp DJ, Stark GR. Proc Natl Acad Sci USA, 1977, 74, 5350–5354.
[16] Durand GM, Zukin RS. J Neurochem, 1993, 61, 2239–2246.
[17] Mori H, Takeda-Yoshikawa Y, Hara-Nishimura I, Nishimura M. Eur J Biochem, 1991, 197, 331–336.
[18] Yang H, McLeese J, Weisbart M, Dionne JL, Lemaire I, Aubin RA. Nuc Acids Res, 1993, 21, 3337–3338.
[19] Yang H, McLeese J, Weisbart M, Dionne JL, Lemaire I, Aubin RA. Nuc Acids Res, 1993, 21, 3337–3338.
[20] Kevil CG, Walsh L, Laroux FS, Kalogeris T, Grisham MB, Alexander JS. Biochem Biophys Res Comm, 1997, 238, 277–279.
[21] Mori H, Takeda-Yoshikawa Y, Hara-Nishimura I, Nishimura M. Eur J Biochem, 1991, 197, 331–336.
[22] Liang P, Pardee AB. Current Opinion Immunol, 1995, 7, 274–280.
[23] Alberts B, Johnson A, Lewis J, Raff M, Roberts K, Walter P. Molecular Biology of the Cell, 5th edn. Garland Science, Taylor & Francis Group, NY, 2008, 538–539.
[24] Primrose SB, Twyman R. Principles of Genome aAnalysis and Genomics. John Wiley & Sons, 2009.
[25] Wreschner DH, Herzberg M. Nucl Acids Res, 1984, 12, 1349–1359.

# 3.7 Evaluation of nucleic acid pherograms

3.7.1    Counter-ion dye staining of nucleic acids —— 385
3.7.2    Silver staining of nucleic acids —— 385
3.7.3    Fluorescence methods for detecting nucleic acids —— 386
3.7.4    Autoradiography of nucleic acids —— 389
3.7.5    Labeling of nucleic acids with proteins —— 390
3.7.6    Absorption spectroscopy of nucleic acids —— 391
3.7.7    Protocols —— 391
         Detection of DNA and RNA in Gel Using Ethidium Bromide —— 391
         Fast and Sensitive Silver Staining of DNA Bands According to Han *et al.* [34] —— 392
         Autoradiography of Radiolabeled DNA in Gels and Blots —— 392
         References —— 394

The nucleic acids, which are colorless, should be made visible after electrophoresis. This can be realized by counter-ion dye staining, silver staining, fluorescent substances, autoradiography, or protein probes. The results can be evaluated quantitatively by densitometry or photography.

## 3.7.1 Counter-ion dye staining of nucleic acids

The counter-ion dye staining of nucleic acids in agarose and polyacrylamide gels [1, 2] is a result of the interaction between negatively charged phosphate groups on DNA or RNA and cationic dyes, which intercalate between the base pairs of nucleic acids.

The main counter-ion dyes for staining of DNA in gels are **Methylene blue** (MB) [3, 4], **Brilliant cresyl blue** (BB) [5], **Crystal violet** (CV) [6], **Nile blue** (NB) [7, 8], and **Iodine blue** (IB) [9]. It is simple, but its sensitivity is inferior to ethidium bromide (EB) staining (see below).

More sensitive is the staining with a mixture of IB and methyl orange. IB has a positively charged group and six aromatic rings while methyl orange has a negatively charged group. The staining mixture has pH = 4.7, consists of 0.008 g/dL Indoin blue and 0.002 g/dL Methyl orange, and contains 10 mL/dL ethanol and 0.2 mol/L sodium acetate. It can detect 5–10 ng of DNA. The staining is fast and does not need a destaining of the gel background.

## 3.7.2 Silver staining of nucleic acids

Silver staining of nucleic acids in agarose and polyacrylamide gels [10–12] enables the detection of 0.03 ng DNA or RNA per $mm^2$ [13, 14], which corresponds to less than 50 pg nucleic acids per band. It provides higher sensitivity than EB fluorescence

https://doi.org/10.1515/9783110761641-034

method, however, requires multiple steps, preparation of prior to use solutions, and the toxic formaldehyde [15, 16].

The silver staining of nucleic acids, as the silver staining of proteins, passes in the same steps: fixing, sensitizing, developing, stopping, and drying the gel.

The *fixing* causes precipitation of the nucleic acids by denaturation. For this purpose, weak acids, such as acetic acid, are used, with or without alcohol. The low pH value of these solutions avoids the protons dissociation from the nucleic acids. As a result, the nucleic acids lose their negative charges and precipitate.

During the *sensitizing* (impregnation), the precipitated nucleic acids bind silver ions. This process is usually carried out in a neutral solution of silver nitrate and allows the building of silver nuclei.

The *developing* (visualizing) requires a strong reducing agent, such as formaldehyde, which reduces the silver ions to metallic silver. The reaction takes place faster in the vicinity of silver nuclei than in the gel, since the silver nuclei catalyze it. During the developing, the nucleic bands transform in dark brown to black bands on a slightly yellowish to colorless background. The developing takes place at higher pH values, for example, in a sodium carbonate solution [17, 18].

The developing should be quickly *stopped*. Otherwise, all silver ions will be reduced to metallic silver, resulting in a black background. The stopping is caused by changing the pH value.

The *drying* is carried out under a hair dryer or in a dryer. If the polyacrylamide concentration of the gel is high, the wet gel should be placed with the support film down onto a glass plate, and should be covered free of air bubbles with a water-swollen cellophane membrane. The protruding edges of the cellophane membrane should be folded over the underside of the glass plate and the gel should be dried. The ultrathin polyacrylamide gels and agarose gels do not need to be covered by cellophane.

According to the fast method of Han *et al.* [19], nitric acid and ethanol in the impregnation step eliminate the need for prior treatment of polyacrylamide gels with a fixing solution, followed by water rinsing. The procedure can be completed within 10 min. The sensitivity of this method is significantly improved by the silver-ion sensitizer **e**riochrome **b**lack **T** (EBT), which reduces the gel background staining and increases the sensitivity of the method [20]. Compared to the conventional silver staining methods, the method of Han *et al.* saves time and displays high sensitivity (Table 3.7.1).

## 3.7.3 Fluorescence methods for detecting nucleic acids

The localization of nucleic acids in gels is most often carried out by fluorescence detection methods. For this purpose usually EB is used.

**Table 3.7.1:** Procedures and comparison of different methods for DNA silver staining.

| | Method of Bassam et al. [21] | Method of Carlos et al. [22] | Method of Qu et al. [23] | Method of Han et al. [24] |
|---|---|---|---|---|
| Fixing | 10 mL/dL ethanol and 0.5 mL/dL acetic acid for 3 min | 10 mL/dL acetic acid for 20 min | – | – |
| Rinsing | $H_2O$ for 2 min, twice | $H_2O$ for 2 min, thrice | – | – |
| Impregnating | 10 mL/dL ethanol, 0.5 mL/dL acetic acid and 0.2 g/dL $AgNO_3$ for 5 min | 0.1 g/dL $AgNO_3$ and 1.5 mL 37 g/dL HCHO for 30 min | 25 mL/dL ethanol, 1 g/dL $HNO_3$ and 0.2 g/dL $AgNO_3$ for 2–5 min | 1 g/dL $HNO_3$ and 0.1 g/dL $AgNO_3$ for 5 min |
| Rinsing | $H_2O$ for 2 min, twice | $H_2O$ for 20 s, optional | $H_2O$ for 2 min | $H_2O$ for 5 s, twice |
| Developing | 3 g/dL NaOH and 0.1 g/dL HCHO for 5 min | 3 g/dL $Na_2CO_3$, 1.5 mL 37 g/dL HCHO and 2 mg/L $Na_2S_2O_3$ for 2–5 min | 3 g/dL $Na_2CO_3$ and 0.2 g/dL HCHO for 2–5 min | 2 g/dL NaOH, 0.04 g/dL $Na_2CO_3$, 2 mL 0.0025 g/dL EBT and 1.5 mL 37 g/dL HCHO for 5 min |
| Stopping | 10 mL/dL ethanol and 0.5 mL/dL acetic acid for 5 min, then in $H_2O$ for 10 min | 10 mL/dL acetic acid for 5 min | 10 mL/dL acetic acid for 2–5 min | 2.5 g/dL ampicillin for 5 s, then $H_2O$ |
| Sensitivity denat. DNA | 3.50 ng | 0.44 ng | 0.44 ng | 0.11 ng |
| Sensitivity native DNA | 14 ng | 14 ng | 3.5 ng | 1.75 ng |
| Background | lightest | dark | light | golden yellow |
| Used time | ~40 min | ~60 min | ~20 min | ~10 min |

EB is the most popular and commonly used reagent to detect nucleic acids (DNA and RNA). Its molecule has four flat atom rings, which intercalate between adjacent purine and pyrimidine rings of DNA or RNA.

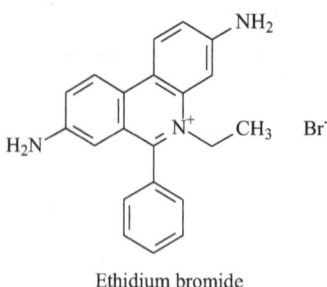

Ethidium bromide

In intercalated state, nucleates absorb **u**ltraviolet (UV) light and emit it again in the visible range. As a result, the nucleic acid-EB complexes are seen in agarose gels as orange-red fluorescent bands on a dark background [25] (Figure 3.7.1).

UV lamp

Fluorescent plate

Gel

DNA shadows

Support film

**Figure 3.7.1:** Visualizing the separated nucleic acids by ethidium bromide under UV light.

EB can be added to a sample prior to or after electrophoresis. If EB is added prior to electrophoresis, it combines with the double-chain nucleates and its positive charges neutralize partially the negative charges of DNA. As a result, the DNA-EB complexes move by about 15% slower than the "uncolored" DNA. If EB is added after the electrophoresis, the agarose gel should be kept in a TRIS-borate-EDTA buffer containing 0.5 μg/mL EB at room temperature for 45 min. The nucleates, "colored" with EB, are usually photographed under UV light.

EB is a strong mutagen. The UV irradiation may also damage both the DNA in the samples and the experimenters [26–28]. To avoid the mutagenic effect of EB, a fluorescent method using **b**erberine/**M**ordant Yellow **3 R** (BB/M3R) has been introduced [29].

In addition to EB, three other high sensitivity fluorescent substances (SYTO 13, TOTO-1, and YOYO-1) are used for staining of DNA and RNA [30, 31]. TOTO-1 and YOYO-1 represent dimers of the compounds TO-PRO-1 (benzothiazolium-4-quinolinolium) and YO-PRO-1 (benzoxazolium-4-quinolinolium). DNA can be "colored" with these substances prior to electrophoresis because they dissociate difficultly from nucleic acids.

In recent years, it was found that the nucleic acid dye GelRed™ is the most sensitive and safest dye under UV light and blue light excitation [32].


## 3.7.4 Autoradiography of nucleic acids

Autoradiography is the most sensitive detection method for nucleic acids. It requires isotope labeled nucleic acids. After electrophoresis, the resolving gel should be dried before contacting with an X-ray film. Under these conditions, a photon of light, or a β-particle, or γ-ray released from a radioactive molecule "activates" silver bromide crystals on the film emulsion. This reduces the silver ions to atomic (metal) silver in form of grains, which blacken on the X-ray film (Figure 3.7.2). Then, the film is scanned by a special detector, which transmits the information into a computer.

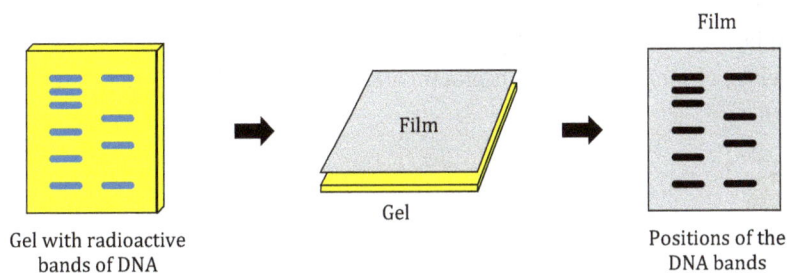

Gel with radioactive bands of DNA

Positions of the DNA bands

**Figure 3.7.2:** Autoradiogram of electrophoretically separated deoxynucleates.

In autoradiography, three isotopes: $^{32}P$, $^{33}P$, and $^{35}S$ are commonly used. The isotopes $^{32}P$ and $^{33}P$ can be incorporated into the α-phosphate residues at the 5′-end of DNA or RNA strands. The sulfur atom is similar to the oxygen atom. Therefore, oxygen atoms in the α-phosphate residues of nucleoside triphosphates can be replaced with $^{35}S$ atoms and later be incorporated by DNA or RNA polymerases in nucleotide chains.

The isotopes $^{32}P$, $^{33}P$, and $^{35}S$ release high-energy electrons named β-particles (Table 3.7.2). The isotope $^{32}P$ radiates highest energy whereas the isotope $^{33}P$ radiates much weaker energy. The energy of the isotope $^{35}S$ is less than that of the isotope $^{33}P$.

**Table 3.7.2:** Characteristics of isotopes used in autoradiography.

| Isotope | Strength of ß-radiation, MeV | Half-life, days | Maximum specific activity, Ci/mmol |
|---|---|---|---|
| $^{32}P$ | 1.71 | 14.3 | 6,000 |
| $^{33}P$ | 0.249 | 25.4 | 3,000 |
| $^{35}S$ | 0.167 | 87.4 | 1,500 |

## 3.7.5 Labeling of nucleic acids with proteins

The nucleic acids can be labeled (coupled) with enzymes. Typically **h**orseradish **p**eroxidase (HRP) and **al**kaline **p**hosphatase (ALP) are used, which are often referred to as reporter enzymes. These enzymes can be detected by colorimetric or chemiluminescent substances. In the chemical luminescence, a chemical reaction produces visible light, which can last for several hours. It can be recorded on an X-ray film, similarly as in the autoradiography.

The labeling of nucleic acids with proteins (enzymes) may be indirect or direct (Figure 3.7.3).

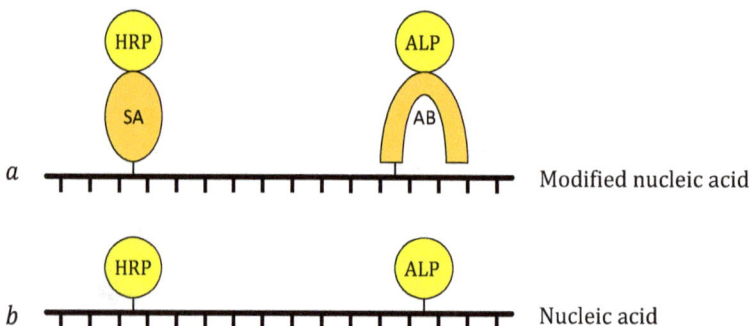

**Figure 3.7.3:** Labeling of nucleic acids with enzymes.
Horseradish peroxidase (HRP) or alkaline phosphatase (ALP) can be coupled to nucleic acids indirectly (*a*) or directly (*b*). SA – streptavidin; AB – antibody.

The *indirect coupling* of proteins with nucleic acids is based on modified nucleotides of DNA or RNA. Under these circumstances special proteins, for example, **s**treptavidin (SA) or **a**nti**b**odies (AB) can be bound to the nucleotide chain, which in turn can fix HRP or ALP. For chemical modifications of nucleic acids, three substances are used: biotin, fluorescein, and digoxigenin. If nucleic acids are labeled with biotin, they can bind the bacterial protein SA. Biotin should be at the end of the nucleic acid; only then, biotin can bind to the large SA molecule. So, the steric hindrance is avoided and the SA molecule can come undisturbed closely to the biotin ligand. SA conjugates with the modified nucleic acid alone or bound to a reporter enzyme.

The *direct coupling* of proteins on nucleic acids is used also for the reporter enzymes HRP and ALP [33]. The direct coupling changes the DNA mobility. The advantage of the direct DNA–protein coupling is that it is made quickly, because the nucleotides should not be labeled with SA or AB. A protein should be coupled directly to a nucleic acid, only if it is used as a probe during hybridization of nucleic acids – in Southern or Northern blotting.

## 3.7.6 Absorption spectroscopy of nucleic acids

The absorption of nucleic acids at 260 or 280 nm wavelengths can be used to calculate the nucleic acid concentrations. Absorption at 325 nm indicates particles in the solution or dirty cuvettes. Contaminants containing peptide bonds or phenol absorb at 230 nm.

## 3.7.7 Protocols

### Detection of DNA and RNA in Gel Using Ethidium Bromide

**Materials and equipment**

Agarose gel
DNA or RNA
Ethidium bromide
  *Ethidium bromide is a mutagen. Handle with gloves.*

*Tris-Borate-EDTA (TBE) buffer, 10x, pH = 8.3*

| | |
|---|---|
| TRIS | 1.08 g (0.089 mol/L) |
| Boric acid | 0.55 g (0.089 mol/L) |
| $Na_2EDTA$ | 0.07 g (0.002 mol/L) |
| Deionized water to | 100.00 mL |

  *Instead TBE, TRIS-acetate-EDTA (TAE) buffer can be used.*

*Ethidium bromide stock*

| | |
|---|---|
| Ethidium bromide | 1.0 g |
| Deionized water to | 100.0 mL |

  *The solution is stable for 1–2 months at room temperature in the dark.*

*Buffer preparation, pH = 8.3*

| | |
|---|---|
| TBE buffer, 10x | 10.00 mL |
| Ethidium bromide stock | 0.05 mL |
| Deionized water to | 100.00 mL |

### Gel preparation
- Dissolve and heat agarose in buffer.
- Allow gel solution to cool to 60–70 °C.
- Add ethidium bromide stock to 0.5 µg/mL final concentration.
- Pour agarose solution into a cassette to gel.

### Procedure
- After electrophoresis run, place the agarose gel into a plastic wrap on a UV light box. Bands will appear bright orange on a faint orange background.
  *This method will detect approximately 5 ng of DNA.*

## Fast and Sensitive Silver Staining of DNA Bands According to Han *et al.* [34]

### Materials and equipment
Polyacrylamide slab gel
$HNO_3$
$AgNO_3$
NaOH
$Na_2CO_3$
Eriochrome black T (EBT)
HCHO
Ampicillin

### Procedure
- Impregnate the gel in a mixture of 0.1 g/dL $AgNO_3$ and 1 g/dL $HNO_3$ at room temperature for 5 min.
- Rinse with deionized water for 5 s, twice.
- Develop in a mixture of 2 g/dL NaOH and 0.04 g/dL $Na_2CO_3$, containing 2 mL 0.0025 g/dL EBT and 1.5 mL 37 g/dL HCHO at room temperature for about 5 min until the gel color turns to yellow and all the bands of DNA fragments have appeared distinctly.
- Stop in 2.5 g/dL ampicillin for 5 s, twice.
- Rinse with deionized water.
- Dry in the air at room temperature.

## Autoradiography of Radiolabeled DNA in Gels and Blots

### Materials and equipment
Dried gel or membrane (e.g., after immunoblotting)

Developer
Fixer
Film or paper cassette with particle-board supports and metal binder clips
Plastic wrap
X-ray film

## Procedure

- Place the sample (dried gel or membrane) in the film cassette in a darkroom illuminated with a safelight. Cover the sample with a plastic wrap to prevent it from sticking to the film and contaminating the cassette with radioactivity (Figure 3.7.4).

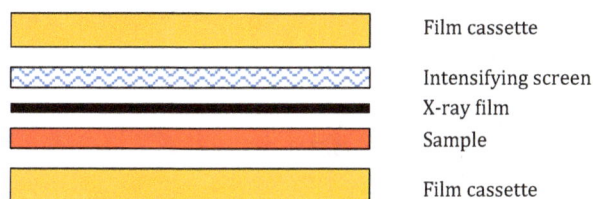

Film cassette
Intensifying screen
X-ray film
Sample
Film cassette

**Figure 3.7.4:** Autoradiography setup.

- Place an X-ray film on top of the sample, and put the sample in a film cassette.
- Expose the film at appropriate temperature for desired length of time. The sensitivity of the film can be improved using intensifying screens (Table 3.7.3). Time of exposure depends on the strength of radioactivity in the sample. A Geiger counter can be used to detect the relative amount of radioactivity in the sample.

**Table 3.7.3:** Film choice and exposure temperature for autoradiography.

| Isotope | Enhancement method | Film | Exposure temperature |
|---|---|---|---|
| $^3H$ | Fluorography | Double-coated | −70 °C |
| $^{35}S$, $^{14}C$, $^{32}P$ | None | Single-coated | Room temperature |
| $^{35}S$, $^{14}C$, $^{32}P$ | Fluorography | Double-coated | −70 °C |
| $^{32}P$, $^{125}I$ | Intensifying screens | Double-coated | −70 °C |

- After exposure, return the cassette into the darkroom and remove the film for developing.
- Immerse the film in 20 °C developer for 5 min, and wash in running water for 1 min.
- Immerse the film in 20 °C fixer for 5 min, and wash in running water for 15 min.
- Hang the film to dry.

# References

[1]   Jung DW, Yoo GS, Choi JK. Anal Biochem, 1999, 272, 254–256.
[2]   Yang YL, Jung DW, Bai DG, Yoo GS, Choi JK. Electrophoresis, 2001, 22, 855–859.
[3]   Daru YS, Ramesh K. Methods Cell Mol Biol, 1989, 1, 183–187.
[4]   Flores N, Valle F, Bolivar F, Merino E. BioTechniques, 1992, 13, 203–205.
[5]   Torres JS, Noyala R. Tech Tips, 1993, 9, 40.
[6]   Yang Y, Jung DW, Bai DG, Yoo GS, Choi JK. Electrophoresis, 2001, 22, 855–859.
[7]   Adkins S, Burmeister M. Anal Biochem, 1996, 240, 17–23.
[8]   Yang YI, Hong HY, Lee IS, Bai DG, Yoo GS, Choi JK. Anal Biochem, 2000, 280, 322–324.
[9]   Hwang SY, Jin LT, Yoo GS, Choi JK. Electrophoresis, 2006, 27, 1739–1743.
[10]  Beidler JL, Hilliard PR, Rill RL. Anal Biochem, 1982, 126, 374–380.
[11]  Blum H, Beier H, Gross HJ. Electrophoresis, 1987, 8, 93–99.
[12]  Merril CR. Nature, 1990, 343, 779–780.
[13]  Goldman D, Merril CR. Electrophoresis, 1982, 3, 24–26.
[14]  Guillemette JG, Lewis PN. Electrophoresis, 1983, 4, 92–94.
[15]  Gottlieb M, Chavko M. Anal Biochem, 1987, 165, 33–37.
[16]  Datar RH, Bhisey AN. Ind J Biochem Biophys, 1988, 25, 373–375.
[17]  Merril CR, Goldman D, Van Keuren ML. Electrophoresis, 1982, 3, 17–23.
[18]  Nielsen BL, Brown LR. Anal Biochem, 1984, 144, 311–315.
[19]  Han YC, Teng CZ, Hu ZL, Song YC. Electrophoresis, 2008, 29, 1355–1358.
[20]  Hwang SY, Jin LT, Yoo GS, Choi JK. Electrophoresis, 2006, 27, 1744–1748.
[21]  Bassam BJ, Caetano-Anolles G, Gresshoff PM. Anal Biochem, 1991, 796, 80–83.
[22]  Carlos JS, Emmanuel DN, Andrew FG. BioTechniques, 1994, 17, 907–914.
[23]  Qu LJ, Li X, Wu G, Yang Y. Electrophoresis, 2005, 26, 99–101.
[24]  Han YC, Teng CZ, Hu ZL, Song YC. Electrophoresis, 2008, 29, 1355–1358.
[25]  Sharp PA, Sugden B, Sambrook J. Biochemistry, 1973, 12, 3055–3062.
[26]  Kantor GJ, Hull DR. Biophys J, 1979, 27, 359–370.
[27]  Marks R. Cancer, 1995, 75, 607–612.
[28]  Ohta T, Tokishita SI, Yamagata H. Mutat Res, 2001, 492, 91–97.
[29]  Jung DW, Yoo GS, Choi JK. Anal Biochem, 1999, 272, 254–256.
[30]  Guindulain T, Comas J, Vives-Rego J, Appl Environ Microbiol, 1997, 63, 4608–4611.
[31]  Ullal AJ, Pisetsky DS, Reich CFIII. Cytometry A, 2010, 77, 294–301.
[32]  Haines AM, Tobe SS, Kobus HJ, Linacre A. Electrophoresis, 2015, 36, 941–944.
[33]  DIG Application Manual for Filter Hybridization. Roche diagnostics, 2008.
[34]  Han YC, Teng CZ, Hu ZL, Song YC. Electrophoresis, 2008, 29, 1355–1358.

# 3.8 Precast gels for nucleic acid electrophoresis

3.8.1    Precast agarose gels —— 395
3.8.2    Precast polyacrylamide gels —— 395
3.8.3    Protocols —— 396
         Casting Mini- and Midi-agarose Gels —— 396

The precast gels for nucleic acid electrophoresis are as important, as the precast gels for protein electrophoresis. They can contain agarose or polyacrylamide.

## 3.8.1 Precast agarose gels

The precast agarose gels on film or net for native or denaturing zone electrophoresis of nucleic acids contain buffers. In principle, they are 0.5 mm thick. The precast gels are available in different dimensions: 51 × 82, 80 × 100, 125 × 125, 125 × 200, 125 × 250 mm or more. The concentration of the agarose can be 0.60, 0.80, 1.00 g/dL or other.

The precast agarose gels are located between two polyester films. The cover film is loosely bound to the gel. It should be removed prior to the application of a template on the gel. The bottom film is chemically bound to the gel.

## 3.8.2 Precast polyacrylamide gels

The precast polyacrylamide gels on film or net for native or denaturing disc-electrophoresis of nucleic acids are usually 0.5 mm thick. They are supplied together with paper strips. The gels and paper strips contain buffers: the gel and anode strip contain, for example, TRIS-formate-EDTA buffer, and the cathode strip contains, for example, TRIS-taurinate-EDTA buffer.

The polyacrylamide concentration $T$ of the precast gels may be 5, 7, 9, 11, or 13 g/dL, at dimensions corresponding to the dimensions of the precast agarose gels. Film-supported gels are used for electrophoresis and diffusion blotting, whereas net-supported gels are suitable for electroblotting.

Similarly to the precast agarose gels, the precast polyacrylamide gels are located between two polyester films. The cover film should be removed prior to the application of a template on the gel. The lower film of a net-supported gel, which is loosely bound to the gel, should be removed after electrophoresis for electroblotting.

Precast polyacrylamide gels should be stored at 4–8 °C in the refrigerator.

https://doi.org/10.1515/9783110761641-035

## 3.8.3 Protocols

### Casting Mini- and Midi-agarose Gels

**Materials and equipment**
Appropriate buffer
Agarose
Casting cassette for vertical mini- or midi-agarose gels

**Procedure**
- Fulfill the casting cassette (for example, with inner dimensions of 80 × 50 × 2 mm) with agarose solution (10 mL).
- Place a comb into the solution.
- After gelling, remove carefully and under buffer the comb from the gel.

# 4 Iontophoresis

4.1      Theory of iontophoresis —— 397
4.2      Factors affecting iontophoresis —— 398
4.2.1    Physicochemical factors —— 398
4.2.2    Biological factors —— 399
4.3      Calculating the iontophoretic current —— 399
4.4      Iontophoresis device —— 399
4.5      Diagnostic iontophoresis —— 400
4.6      Therapeutic iontophoresis —— 400
4.6.1    Transdermal iontophoresis —— 400
         Hyperhidrosis iontophoresis —— 401
         Other transdermal applications —— 401
4.6.2    Ocular iontophoresis —— 402
         References —— 403

The term *iontophoresis* stems from Greek words "ion" and "phorein" (to carry) and is referred to as inserting ions into the body driven by electric field.

The advantages of iontophoresis are: painless drug delivering in the body, without injections; reduced risk of infection because the skin and outer mucosae remain unhurt; direct drug delivery into the ill place; and no risks of fibrosis in contrast to the injection drug delivering. Iontophoresis has, however, some disadvantages. The most common is that the iontophoretic procedures make the skin dry. Skin irritation, blistering, and peeling might also occur.

*Reverse iontophoresis* is also known. It is a technique for removing ions or neutral molecules from the body. Reverse iontophoresis can be caused when the skin is negatively charged, which makes it permeable for cations (sodium and potassium ions). They, in turn, cause electroosmosis that pushes solvent toward the anode. The reverse iontophoresis can be applied to extract glucose across the skin in diabetic patients to monitor the blood glucose concentration [1, 2]. This is used in the GlucoWatch device [3].

## 4.1 Theory of iontophoresis

In iontophoresis, charged drugs are inserted into skin or mucosae for an appropriate length of time [4]. This is carried out with the help of electric field [5]. The electric current should be direct and below the level of patient's pain threshold. Besides, the drugs should have relatively small molecules ($M_r < 8,000$).

The ions to be driven into a tissue should be placed under the electrode with the same polarity. If so, the anode repels cations, and the cathode repels anions (Figure 4.1). The ground electrode might be placed elsewhere on the body surface. In principal, the negative electrode (cathode) is larger than the positive electrode

https://doi.org/10.1515/9783110761641-036

(anode) to avoid skin irritation. For example, the surface of the cathode can be twice larger than the surface of the anode.

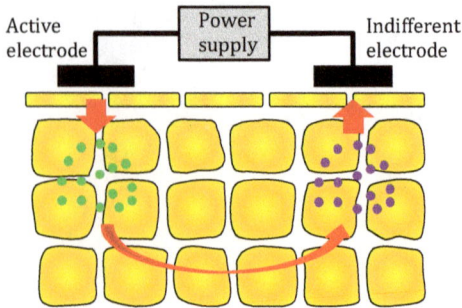

**Figure 4.1:** Ions of the polarity as the polarity of stimulating electrode are repelled into the skin, and on the contrary.
Active ion; Counter-ion

## 4.2 Factors affecting iontophoresis

Factors affecting iontophoresis have physicochemical and biological origin.

### 4.2.1 Physicochemical factors

Physicochemical factors that affect iontophoresis are: concentration, pH value and ionic strength of the buffer, solution viscosity, and electric current.

*Concentration.* The drug should be water soluble and ionizable. Smaller molecules are more mobile.

*Buffer pH.* The pH value of the buffer containing the drug should have such value, at which the drug molecules are ionized [6]. This pH value is called pH optimum of the drug.

*Buffer ionic strength.* If a drug is dissolved in a buffer, the buffer ions compete with the drug ions in the electric field. This process has a prominent role in iontophoresis because the buffer ions are generally smaller than the drug ions; therefore, they are faster than the drug ions.

*Solution viscosity.* The migration of a drug is inversely related to the viscosity of the solution to be introduced.

*Electric current.* The electric current can be direct, alternate or pulsed. The alternating electric current (AC) iontophoresis shows better results than the conventional direct electric current (DC) iontophoresis [7].

### 4.2.2 Biological factors

The biological factors, which influence iontophoresis, are the thickness and permeability of skin or mucosa. Sweat glands in the skin are in principal the path for inserting charged ions. This fact was demonstrated by Papa and Kligman, who proved this using Methylene blue [8].

## 4.3 Calculating the iontophoretic current

The electric current density $J$ at the cathode should not be above 0.5 mA/cm$^2$, and the electric current density at the anode should not be above 1.0 mA/cm$^2$. If an electric current $I$ of 2 mA should be delivered through an electrode surface $S$ of 6 cm$^2$, then the electric current density at either the anode or cathode should be

$$J = \frac{I}{S} = \frac{2}{6} = 0.33 \, \text{mA/cm}^2 \tag{4.1}$$

On the contrary, it is possible to calculate the electric current that should be applied. If the surface $S$ of the active electrode is 6 cm$^2$, and the maximum electric current density is 0.5 mA/cm$^2$, then, according to eq. (4.1), the maximum electric current $I$ should be

$$I = J \times S = 0.5 \times 6 = 3 \, \text{mA} \tag{4.2}$$

## 4.4 Iontophoresis device

An iontophoresis device consists of a microprocessor-controlled battery, two electrodes, and a reservoir.

The iontophoresis *battery* creates low voltage – usually 9 V.

The *electrodes* of an iontophoresis device are active and passive. They are made of metal or gel and possess special wells or receptacle areas for drugs (Figure 4.2).

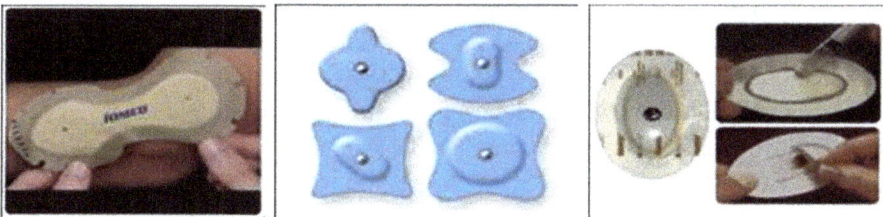

**Figure 4.2:** Electrodes for iontophoresis.

The *reservoir* contains a gauze or gel pad, to which a drug solution is applied or injected through.

Two types of iontophoresis are run: diagnostic and therapeutic iontophoresis.

## 4.5 Diagnostic iontophoresis

For diagnostic purposes, pilocarpine iontophoresis is often used to stimulate sweat secretion, as part of the diagnosis of *cystic fibrosis* [9–11].

Acetylcholine iontophoresis is used for testing the endothelium by stimulating endothelium-dependent synthesis of NO and subsequent *microvascular vasodilation* [12]. Acetylcholine ions are positively charged; therefore, they are placed in the anode chamber.

Iontophoresis can be used for testing *contact eczema* [13]: The test substances are administered through epidermis into dermis.

Iontophoresis of epinephrine for assessment of neuromuscular synapsis response improves the diagnosis of *myasthenia gravis*. For this purpose, an improved method called iontophoresis of epinephrine with **r**epetitive **n**erve **s**timulation (RNS) was successfully used [14].

## 4.6 Therapeutic iontophoresis

Usually, iontophoresis for disease treatment is carried out with drugs in low concentrations – up to 5 g/dL, and under electric current up to 5 mA. Treatment times are in the 20–40 min range.

Most significant types of therapeutic iontophoresis are transdermal and ocular iontophoresis.

### 4.6.1 Transdermal iontophoresis

Iontophoresis for transdermal delivery of drugs has been used for more than a century [15–17]. It is applicable to small ions, and polyions of $M_r$ up to a few thousand. The ions are driven into the skin *via* the sweat gland ducts, which have low resistance, and hair follicles.

In transdermal iontophoresis, low voltage is applied between an iontophoretic chamber (containing a same-polar drug) on the skin and another electrode chamber also on the skin (Figure 4.3).

The transdermal drug delivery can be controlled by a microprocessor [18]. The most treated disease is hyperhidrosis.

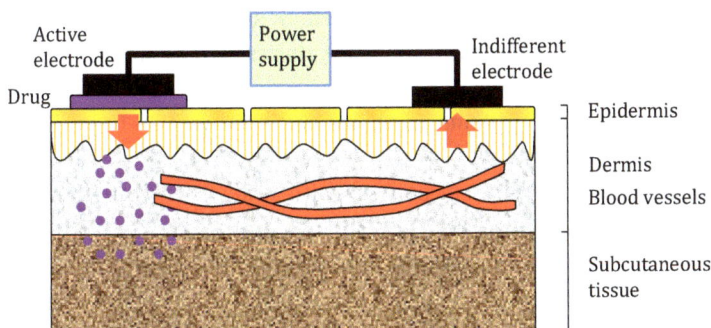

**Figure 4.3:** Transdermal drug delivery by iontophoresis.

## Hyperhidrosis iontophoresis

*Hyperhidrosis* is characterized with excessive sweating under no reason. This disease affects about 4% of the population. Hyperhidrosis can occur anywhere on the body (hands, feet, underarms, and face) and can affect all genders and ages, but is most common in teenagers and young adults.

Iontophoresis blocks temporarily the sweat glands and stops the sweating. Patients with hyperhidrosis should undergo several iontophoresis applications per week, lasting from 20 to 40 min [19]. In serious cases of hyperhidrosis, water solutions with cholinergic inhibitors, such as methyl sulfate [20], atropine [21], or glycopyrronium bromide [22–24], can be used.

## Other transdermal applications

*Myasthenia gravis* is another disease that can be treated iontophoretically. It is a result of a genetic defect in the neuromuscular synapsis [25]. In most cases, myasthenia gravis is caused by circulating antibodies that block acetylcholine receptors on the postsynaptic membrane, inhibiting so the neurotransmitter acetylcholine. Acetylcholine ion is positively charged; therefore, it can be forced out from the presynaptic terminal by a microcathode [26].

Iontophoresis with histamine, 2 g/dL tolazoline hydrochloride or 20 g/dL ZnO (ointment) on the positive pole of the iontophoretic device is carried out in treatment of *trophic ulcers* [27–29].

Iontophoresis is also used for treatment of *plantar fasciitis*, *bursitis*, lateral and medial *epicondylitis* (commonly referred to as tennis elbow and golfers elbow, respectively), and other diseases [30].

Some *viral infections* are also positively influenced by iontophoresis. Iontophoresis can be an alternative for delivering of acyclovir to treat *herpes labialis* [31].

*Planar warts* can be successfully treated with sodium salicylate iontophoresis [32]. Iontophoretic application of idoxuridine is effective in *herpes simplex* treatment [33].

*Hyperkeratosis* with fissuring of palms and soles is treated with 5–10 g/dL sodium salicylate and shows improvement within 3–4 weeks [34].

Iontophoresis with 1 g/dL meladine at patients with *vitiligo* shows marked repigmentation [35].

Iontophoresis with hyaluronidase leads to temporary increased skin softness and flexibility and decreases the cold sensitivity in patient with *scleroderma* [36].

Methyl prednisolone iontophoresis is used for healing of erosive *lichen planus* [37].

Iontophoresis can be used in the treatment of *lymphedema* of the limbs [38].

For *muscle and joint pain* 2–3 g/dL sodium salicylate or 10 g/dL trolamine salicylate ointment on the cathode is used. Salicylate ion inhibits the synthesis of prostaglandins, which take part in inflammation and pain.

*Muscle spasm* is treated with 2–5 g/dL $CaCl_2$ on the negative pole. Calcium ions stabilize membranes, and decrease excitability threshold of peripheral nerves and skeletal muscle. Also 2 g/dL $MgSO_4$ (ointment) on the same pole can be used, because magnesium ions push out calcium ions. As a result, muscles relax.

Iontophoretic patches enable patients to administer a bolus of fentanyl for *pain relief* [39]. Soft tissue pain can be treated with 4–5 g/dL lidocaine (ointment) on anode.

Dentists have used it for anesthesia of the oral mucosa [40, 41]. The anesthesia of the skin can be achieved by local administration of lidocaine and epinephrine [42, 43].

### 4.6.2 Ocular iontophoresis

Iontophoresis is also used for delivery of therapeutic drugs into the eye. It has the capacity to provide high drug concentration safely, while only 1% of the same drugs delivered *per os* reach the eye. Two types of ocular iontophoresis are known: transcorneal and transscleral.

*Transcorneal iontophoresis* can deliver drugs to the anterior segment of the eye (cornea, aqueous humor, ciliary body, iris, and lens) for treatment of anterior segment diseases. It is used for the treatment of *glaucoma, dry eyes, keratitis, corneal ulcers,* and *ocular inflammations*. However, the transcorneal iontophoresis cannot insert drugs into the posterior segment of the eye [44].

The lens-iris diaphragm is the main barrier of local drugs to pass through the choroid in the posterior eye tissues such as vitreous and retina. *Transscleral iontophoresis* overcomes this barrier and delivers drugs to the vitreous and retina. This treatment is useful for posterior ocular disorders, such as *retinitis, optic nerve atrophy, uveitis, endophthalmitis, pediatric retinoblastoma,* and **age-related macular degeneration** (AMD).

# References

[1]   Rao G, Guy RH, Glikfeld P, LaCourse WR, Leung L, Tamada J, Potts RO, Azimi N. Pharm Res, 1995, 12, 1869–1873.

[2]   Tamada JA, Garg SK, Jovanovic L, Pitzer KR, Fermi S, Potts RO. JAMA, 1999, 282, 1839–1844.

[3]   Sieg A, Guy RH, Delgado-Charro MB. Clin Chem, 2004, 50, 1383–1390.

[4]   Peter ME, Jui-Chen T, Gopinathan KM (Bolognia JL, Jorrizo JL, Rapini RP (eds.)). Dermatology. Mosby, London, Vol. 2003, 1969–1978.

[5]   Crumay MH. Goldsmith H (ed.). Physical Modalities in Dermatologic Therapy, 1st edn. Springer Verlag, New York, 1978, 190–196.

[6]   Gajjar V, Gupta SK, Singhvi IJ et al. Asian J Pharm Sci Clin Res, 2011, 1, 41–54.

[7]   Zhu H, Li SK, Peck KD, Miller DJ, Higuchi WI. J Control Release, 2002, 82, 249–261.

[8]   Papa CM, Kligman AM. J Invest Dermatol, 1966, 47, 1–9.

[9]   Sawyer CJ, Scott AV, Summer GK. South Med J, 1966, 59, 197–202.

[10]  Beauchamp M, Lands LC. Pediatr Pulmonol, 2005, 39, 507–511.

[11]  Sam AH, James THT. Rapid Medicine, 2nd edn. Wiley-Blackwell, 2010.

[12]  Cooke JP, Tsao PS. Arterioscler Thromb, 1994, 14, 653–655.

[13]  Walberg JE. Acta Derm Venerol (Stockh), 1970, 50, 255–262.

[14]  Talebian S, Abolfazli R, Olyaei GR, Hajizadeh S. Acta Med Iran, 2005, 43, 323–330.

[15]  Banga AK. Electricly-Assisted Transdermal and Topical Drug Delivery. Taylor & Francis, London, 1998.

[16]  Pikal MJ. Adv Drug Deliv Rev, 2001, 46, 281–305.

[17]  Kalia YN, Naik A, Garrison J, Guy RH. Adv Drug Deliv Rev, 2004, 56, 619–658.

[18]  Subramony JA, Sharma A, Phipps JB. Int J Pharmaceutics, 2006, 317, 1–6.

[19]  Grice K. Physiother, 1980, 66, 43–44.

[20]  Grice K, Sattar H, Baker H. Br J Dermatol, 1972, 86, 72–78.

[21]  Gibinski K, Giec L, Zmudzinski J, Dosiak J, Waclawczyk J. J Appl Physiol, 1973, 34, 850–852.

[22]  Abell E, Morgan K. Br J Dermatol, 1974, 91, 87–91.

[23]  Solish N, Bertucci V, Dansereau A, Hong HCH, Lynde C, Lupin M, Smith KC, Storwick G. Derm Surg, 2007, 33, 908–923.

[24]  Walling HW, Swick BL. Am J Clin Derm, 2011, 12, 285–295.

[25]  Kandel E, Schwartz J, Jessel T, Siegelbaum S, Hudspeth A. Principles of Neural Science, 5th edn. McGraw–Hill, New York, 2012, 318–319.

[26]  Byrne JH. Neuroscience online. Chapter 4: Synaptic Transmission and the Skeletal Neuromuscular Junction. The University of Texas Medical School at Houston.

[27]  Abramson DI, Tuck S, Chu LS, Buso E. Arch Phys Med Rehabil, 1967, 48, 583–591.

[28]  Cornwall MW. Phys Ther, 1981, 61, 359–360.

[29]  Gherardini G, Gürlek A, Gregory RD, Lundeberg T. Plastic Reconstruct Sur, 1998, 101, 90–93.

[30]  Caufield TG Tim Caufield PhD, 2013.

[31]  Morrel EM, Spruance SL, Goldberg DI. Clin Infect Dis, 2006, 43, 460–467.

[32]  Gordon NH, Weinstein MV. Phys Ther, 1969, 49, 869–870.

[33]  Gangarosa LP Sr, Merchant HW, Park NH, Hill JM. Methods Find Exp Clin Pharmacol, 1979, 1, 105–109.

[34]  Mukherjee S, Gupta AB, Malakar S, Haldar B. Indian J Dermatol Venereol Leprol, 1989, 55, 22–24.

[35]  Moawad MB. Dermatol Monatsschr, 1969, 155, 388–394.

[36]  Popkin RJ. J Invest Dermatol, 1951, 16, 97–102.

[37]  Gangarosa LP Sr.. Iontophoresis in Dental Practice. Quintessence Publ. Co., Chicago, 1983, 40–52.

[38]  Schwartz HS. Arch Intern Med, 1955, 95, 662–668.

[39]  Mayes S, Ferrone M. Ann Pharmacother, 2006, 40, 2178–2186.

[40]  Comeau M, Brummett R, Vernon J. Arch Otolaryngol, 1973, 98, 114–120.

[41]  Gangarosa LP Sr.. J Am Dent Assoc, 1974, 88, 125–128.

[42]  Gangarosa LP Sr.. Methods Find Exp Clin Pharmacol, 1981, 3, 83–94.

[43]  Zempsky WT, Sullivan J, Paulson DM, Hoath SB. Clin Ther, 2004, 26, 1110–1119.

[44]  Fishman PH, Jay WM, Rissing JP, Hill JM, Shockley RK. Invest Ophthalmol Vis Sci, 1984, 25, 343–345.

# 5 History of electrophoresis and iontophoresis

5.1        History of electrophoresis —— **405**
5.1.1    Discovery of electrophoresis —— **405**
5.1.2    Zone electrophoresis of Tiselius —— **406**
5.1.3    Paper and cellulose acetate electrophoresis —— **406**
5.1.4    Gel electrophoresis —— **406**
            Agarose gel electrophoresis —— **406**
            Polyacryalamide gel electrophoresis —— **407**
5.1.5    Isoelectric focusing —— **407**
5.1.6    Two-dimensional electrophoresis —— **407**
5.1.7    Blotting —— **408**
            Blotting of proteins —— **408**
            Blotting of nucleic acids —— **409**
5.1.8    Staining methods —— **409**
5.1.9    Outline history of electrophoresis —— **409**
5.2        History of iontophoresis —— **413**
5.2.1    Outline history of iontophoresis —— **414**
            References —— **415**

The history of electrophoresis and iontophoresis continues a few centuries.

## 5.1 History of electrophoresis

### 5.1.1 Discovery of electrophoresis

The movement of charged particles in an electric field was first observed as long ago as 1807 by the Russian chemists Pyotr Ivanovich Strakhov (1757–1813) and Ferdinand Frederic Reuss (1778–1852) at the Moscow State University [1]. Using a microscope, they noticed that the application of a constant electric field caused dispersed in water clay particles to migrate. Reuss discovered also the opposite flow of water (electroosmosis).

G. Quincke [2, 3] showed that the migration speed of a charged particle in an electric field is linearly related to the potential gradient and is affected by the solution pH value.

In 1892, H. Picton and S. Linder [4] observed that hemoglobin, a colored protein, was moving in a U-tube, filled with an electrolyte solution and placed in an electric field. In 1899, Hardy reported the movement of serum globulins in an electric field and showed that the electric charge of a protein could be changed from positive to negative when pH value was varied [5, 6].

In 1909, L. Michaelis introduced the term *electrophoresis*. This term is composed of the Greek words *elektron* (amber), which was connected in the ancient times with the electricity, and *phorein* (to carry).

https://doi.org/10.1515/9783110761641-037

In 1985, B. Michov drew up the parameter of ionic mobility, which can be used for calculating the ionic mobilities at different ionic strengths [7]. In 1988, he simplified the Henry's function [8].

### 5.1.2 Zone electrophoresis of Tiselius

The history of modern electrophoresis began with the works of Arne Tiselius in Sweden. In 1937, he published a new method for analyzing bio-colloids, and developed the "Tiselius apparatus" for moving boundary electrophoresis [9]. Tiselius detected separated but colorless protein bands using a Schlieren-scanning system.

Over time, the electrophoresis in solutions was replaced by electrophoresis in solid media. At first, filter paper was used, then cellulose acetate membrane, and finally gels.

### 5.1.3 Paper and cellulose acetate electrophoresis

Paper electrophoresis was introduced by P. König in 1937 [10]. He soaked paper in a buffer for zone electrophoresis and suggested UV detection. Using the paper electrophoresis, Linus Pauling and colleagues [11] proved in 1949 that patients with sickle cell disease contained an abnormal hemoglobin.

In 1957, J. Kohn introduced cellulose acetate membranes as a solid support for hemoglobin electrophoresis [12].

### 5.1.4 Gel electrophoresis

In 1955, O. Smithies introduced partially hydrolyzed starch gels for electrophoretic separation of serum proteins [13, 14]. Later, agarose, polyacrylamide, and other gels enabled efficient separation of proteins and nucleic acids.

### Agarose gel electrophoresis

Agarose, a purified form of agar, was introduced in 1961 by Stellan Hjerten [15] as separation medium for electrophoresis. It continues to be applied for separation of proteins and nucleic acids. In 1967, S. Hjerten [16] used 3 mm i.d. tubes for capillary electrophoresis. Later, in 1981, J. Jorgenson and K. Lukacs [17] demonstrated high-performance capillary electrophoresis in 75 µm i.d. capillaries. During 1988 to 1990, Karger's group [18, 19, 20] separated single stranded DNA in gel-filled capillaries.

## Polyacryalamide gel electrophoresis

Polyacrylamide gel has been introduced as electrophoresis medium in 1954 by G. Oster [21]; however, credit for this development is usually given to S. Raymond and L. Weintraub [22] who published their work in 1959. Later S. Hjerten gave detailed formulae and recipes for preparation of polyacrylamide gels with sieving characteristics [23].

In 1967, A. Shapiro *et al.* [24] showed that the detergent **s**odium **d**odecyl **s**ulfate (SDS) improved the resolution of polyacrylamide gel electrophoresis. In 1969, K. Weber and M. Osborn [25] optimized this method. The explanation of the processes in SDS electrophoresis was presented by J. Reynolds and C. Tanford [26].

The discovery of disc-electrophoresis is a great progress in the electrophoresis history. Its theoretical basis was developed by L. Ornstein [27]; its practical implementation was improved by B. Davis [28]. Later, U. Laemmli [29] used the Ornstein buffer system for SDS disc-electrophoresis of proteins.

During 1982 to 1986, B. Michov established the complex compound TRIS-boric acid in the TRIS-borate buffers and calculated the mobility of TRIS-borate ion [30, 31, 32]. In 1989, he described a disc-electrophoresis in one buffer at two pH values [33].

## 5.1.5 Isoelectric focusing

The isoelectric focusing of proteins was applied by Alexander Kolin [34, 35]. He developed it in a column containing a sucrose gradient where two buffer solutions were allowed to diffuse into one another to generate a pH gradient.

A few years later, H. Svensson (H. Rilbe) created the theoretical basis of the isoelectric focusing and gave the name of carrier ampholytes [36]. He obtained the first carrier ampholyte solution (a polypeptide mixture obtained from partially hydrolyzed blood) and separated there hemoglobins [37].

Next, O. Vesterberg [38, 39] patented a method of creating synthetic carrier ampholytes. They were mixtures of aliphatic aminocarboxylic acids, which formed a continuous pH gradient when placed in an electric field.

## 5.1.6 Two-dimensional electrophoresis

In the two-dimensional electrophoresis, proteins are separated by their pI values in a pH gradient in a first dimension, and then are separated due to their molecular sizes using SDS-PAGE in a second dimension. The method was introduced in 1975 simultaneously by three independent groups of researchers, namely P. O'Farrell [40], J. Klose [41], and G. Scheele [42], although the credit usually goes to O'Farrell.

### 5.1.7 Blotting

The blotting of polyions, resolved by electrophoresis, was discovered by E. Southern [43]. Later, many scientists created similar techniques for different substances.

#### Blotting of proteins

Many forms of protein blotting are used: Western, Southwestern, Northwestern, Far-Western, Eastern, and Far-Eastern.

*Western blotting* (called also "protein blotting" or "protein immunoblot") was developed in 1979 by three groups of researchers in Stanford, Basel, and Seattle, which worked independently.

George Stark's group at Stanford University, including also J. Renart and J. Reiser [44], published in July 1979 this method. They used passive transfer of proteins, then $^{125}$I-labeled protein A for detection.

Harry Towbin's group in Basel, Switzerland, including also T. Staehelin (at Hoffman–La Roche) and J. Gordon (at Friedrich-Miescher-Institut) [45] published in September 1979 a similar technique. They transferred proteins from a SDS-PAGE gel to blot membrane using electric current and secondary antibodies for detection.

W. Neal Burnette, working in Robert Nowinski's laboratory at the Fred Hutchinson Cancer Research Center in Seattle, submitted in 1979 a similar paper, which was at first rejected. This paper was published in 1981 [46] after his friends repeated the process [47]. W. Neal Burnette named the technique "Western blotting" as a nod to Southern blotting and because his laboratory was on the west coast [48,49].

So, Stark's group published first; Towbin's group developed the most common method, including the electrophoretic transfer, as well as the use of secondary antibodies; and Burnette popularized the technique under the name Western blotting.

*Southwestern blotting* is a technique, which was first described by B. Bowen and J. Steinberg *et al.* in 1980 [50]. It is used to characterize proteins that bind to specific DNA probes.

*Northwestern blotting*, also known as *Northwestern assay*, is a hybrid technique of Western blotting and Northern blotting. It detects interactions between RNA and proteins [51].

*Far-Western blotting* is based on the technique of Western blotting. While Western blotting uses an antibody to detect a protein of interest, Far-Western blotting uses a non-antibody protein for the same purpose. So Far-Western blotting is employed to detect protein: protein interactions.

*Eastern blotting* was created by H. Tanaka *et al.* [52]. It is a technique used to analyze protein post-translational modifications and to detect carbohydrate epitopes.

*Far-Eastern blotting* was named in 2000 by D. Ishikawa and T. Taki [53]. It is based on antibody or lectin proving lipids transferred to PVDF membranes.

**Blotting of nucleic acids**

Two main types of nucleic acid blotting exist: Southern and Northern blotting. Beside them, Middle-Eastern and Eastern-Western blotting are known.

*Southern blotting* of DNA was discovered in 1975 by the British biologist E. Southern at University of Edinburgh [54] and carries his name.

*Northern blotting* of RNA was developed in 1977 by George Stark's group at Stanford University [55].

*Middle-Eastern blotting* was described in 1984 [56]. It was developed for blotting and immobilizing of polyA-RNA using DNA probes.

*Eastern-Western blotting* was introduced in 1996 by M. Bogdanov *et al.* [57]. It is based on a method of Towbin, created in 1984 [58], and a work by T. Taki *et al.* in 1994 [59]. During the Eastern-Western blotting, phospholipids are blotted on PVDF or nitrocellulose membrane, and then proteins are transferred by Western blotting onto the same membrane where they are probed with specific antibodies.

## 5.1.8 Staining methods

The detection of resolved proteins relies on dyes such as Bromophenol blue [60], Azocarmine B [61], and Amido black 10B [62]. A better staining assay is that of M. Bradford [63] who bound the dye **C**oomassie **b**rilliant **b**lue (CBB) R-250 to proteins. This dye was developed as an acid wool dye. Its binding to proteins was first studied by Fazekas de St Groth *et al.* [64]. CBB R-250 was employed for staining of protein bands in polyacrylamide gels by A. Chrambach *et al.* in 1967 [65].

In 1979, C. Merril *et al.* [66] and R. Switzer *et al.* [67] applied silver staining, which is 10–100 times more sensitive than CBB R-250, for protein detection in pherograms.

## 5.1.9 Outline history of electrophoresis

| Year | Researcher(s) | Development |
|------|---------------|-------------|
| 1807–1809 | Strakhov and Reuss [68] | Observed movement of dispersed clay particles in an electric field – the discovery of electrophoresis in Moscow State University |
| 1859 | Quincke [69, 70] | Studied the electroosmosis and discovered the electric double layer |

(continued)

| Year | Researcher(s) | Development |
|------|---------------|-------------|
| 1879 | Helmholtz [71] | Proposed the first model of the electric double layer |
| 1887 | Arrhenius [72] | Developed the theory of electrolyte dissociation |
| 1892 | Picton and Lindner [73] | Invented the boundary electrophoresis |
| 1896 | Hardy [74] | Observed the movement of globulin in a U-tube by electric current |
| 1897 | Kohlrausch [75] | Described the "persistent" function in a system of two or more electrolyte solutions |
| 1903 | Smoluchowski [76, 77] | Derived the first equation of ion mobility |
| 1904 | Perrin [78] | Used the Helmholtz model for dissolved particles |
| 1910–1913 | Gouy [79] and Chapman [80] | Developed, independently from each other, the theory of the diffuse counter-ionic atmosphere of the electric double layer |
| 1923 | Debye and Hückel [81] | Created the theory of electrolyte solutions and the theory of electric double layer |
| 1923 | Brönsted [82] | Developed the proteolytic theory of acids and bases |
| 1924 | Hückel [83] | Derived the second equation of ion mobility |
| 1924 | Stern [84] | Developed the current theory of the electric double layer |
| 1926 | Onsager [85, 86] | Derived the equation of conductivity of an ionic solution |
| 1931 | Henry [87, 88] | Combined the equations of Smoluchowski and Hückel in a mathematical function |
| 1937 | Tiselius [89, 90] | Carried out the free electrophoresis of serum proteins and improved an apparatus for their detection |
| 1937–1948 | König [91, 92], Wieland and Fischer [93] | Introduced the paper electrophoresis and an apparatus for it |
| 1942 | Martin [94] | Developed the isotachophoresis |
| 1949 | Pauling *et al.* [95] | Proved, using electrophoresis, that an abnormal hemoglobin caused sickle cell disease |
| 1950 | Grassmann and Hannig [96, 97] | Developed the free-flow electrophoresis |

(continued)

| Year | Researcher(s) | Development |
|------|---------------|-------------|
| 1953 | Grabar and Williams [98] | Developed the immunoelectrophoresis |
| 1954 | Kolin [99, 100] | Introduced artificial pH gradients for IEF |
| 1959 | Robinson and Stokes [101] | Completed the Onsager equation |
| 1955 | Smithies [102, 103] | Described electrophoresis in a starch gel |
| 1957 | Kohn [104, 105] | Introduced cellulose acetate as an electrophoretic medium |
| 1957 | Poulik [106] | Used the TRIS-citrate-borate buffer system as a discontinuous buffer system for electrophoresis |
| 1959 | Raymond and Weintraub [107] | Introduced the polyacrylamide gel as an electrophoresis medium |
| 1961 | Hjerten [108] | Used agarose gel for electrophoresis |
| 1961 | Svensson [109, 110, 111] | Developed the theory of isoelectric focusing |
| 1962 | Hierten [112] | Gave detailed formulae for preparation of gels with specific sieving characteristics |
| 1963 | Fazekas *et al.* [113] | Introduced Coomassie brilliant blue as a highly sensitive dye for detection of electrophoretically separated proteins |
| 1964 | Ornstein [114] and Davis [115] | Invented the theory and practice of disc-electrophoresis in a TRIS-chloride-glycinate buffer system |
| 1965 | Hjerten [116] | Described the concentration of polyions on the boundary to a solution with higher ionic strength – Hjerten's effect |
| 1965–1967 | Laurell [117, 118] and Clarke and Freeman [119] | Described the rocket immunoelectrophoresis |
| 1966 | Thorne [120] | Used agar as separation medium |
| 1966 | Vesterberg [121, 122] | Synthesized carrier ampholytes and introduced the isoelectric focusing of proteins in the practice |
| 1967–1969 | Shapiro *et al.* [123] and Maizel [124] | Developed SDS-PAGE in continuous buffers for determination of molecular masses of proteins |
| 1967–1971 | Margolis and Kenrick [125, 126], Kopperschläger *et al.* [127], Slater [128, 129] and Rodbard *et al.* [130] | Introduced gradient gels in the electrophoretic practice and analyzed theoretically the gradient electrophoresis |

(continued)

| Year | Researcher(s) | Development |
|---|---|---|
| 1969 | Weber and Osborn [131] | Optimized the SDS electrophoresis for separation of protein subunits |
| 1969 | Allen *et al.* [132, 133] | Introduced the disc-electrophoresis in step-gradient gels |
| 1970 | Laemmli [134, 135] | Used SDS in the discontinuous buffer system of Ornstein and Davis, and created so the SDS disc-electrophoresis |
| 1972 | Aaij and Borst [136] | Introduced agarose gels with ethidium bromide for DNA electrophoresis |
| 1973 | Jovin and Chrambach [137, 138] | Calculated mobilities of ions used in electrophoresis |
| 1975 | O'Farrell [139], Klose [140] and Scheele [141] | Introduced independently the two-dimensional electrophoresis, which combines isoelectric focusing with SDS electrophoresis |
| 1975 | Southern [142] | Carrid out blotting of DNA |
| 1975–1982 | Gasparic *et al.* [143] and Bjellquist *et al.* [144] | Used immobilized pH gradients for isoelectric focusing |
| 1976 | Lambin *et al.* [145, 146] and Lasky [147] | Used gradient gels for SDS electrophoresis |
| 1978 | Hannig [148, 149] | Invented the free-flow electrophoresis |
| 1979–1987 | Mikkers *et al* [150], Jorgensen and DeArman [151], Tsuda *et al.* [152], Hjerten [153], Gebauer *et al.* [154], Terabe *et al.* [155], Green and Jorgenson [156], Hjerten *et al.* [157, 158] | Developed the capillary electrophoresis |
| 1979 | Rosen *et al.* [159] | Used agarose gels of low electroosmosis as a medium for isoelectric focusing |
| 1979–1981 | Towbin *et al.* [160], Renart *et al.* [161] and Burnette [162] | Invented Western blotting |
| 1979–1981 | Switzer *et al.* [163] and Merril *et al.* [164] | Introduced the silver method as a highly sensitive staining method for electrophoretically resolved bands |
| 1981 | Jorgenson and Lukacs [165] | Demonstrated high performance capillary electrophoresis separations in 75 μm i.d. capillaries |

(continued)

| Year | Researcher(s) | Development |
|------|--------------|-------------|
| 1982–1986 | Michov [166, 167, 168] | Established the complex compound TRIS-boric acid in the TRIS-borate buffers, and calculated the mobility of TRIS-borate ion. |
| 1983 | Mullis [169] | Invented the polymerase chain reaction (PCR) |
| 1984 | Schwartz and Cantor [170] | Introduced the pulsed-field gel electrophoresis for separation of large DNA nucleates |
| 1985 | Michov [171] | Drew up the parameter of ionic mobility, which can be used for calculating the ionic mobilities at different ionic strengths |
| 1987 | Schägger and Jagow [172] | Published the TRIS-acetate-TRICINEate buffer system for SDS-electrophoresis in gradient gels |
| 1988 | Michov [173] | Simplified the Henry's function |
| 1988–1990 | Karger's group [174, 175, 176] | Separated single stranded DNA oligonucleotides in gel-filled capillaries |
| 1989 | Booth [177] | Separated chromosomes in agarose gels by pulsed-field electrophoresis |
| 1989 | Michov [178] | Described a disc-electrophoresis in one buffer at two pH values |
| 1992 | Harrison *et al.* [179] | Introduced the microchip-based capillary electrophoresis (MCE) |
| 2004 | Kastenholz [180] | Improved electrophoresis of metaloproteins |
| 2013 | Michov [181, 182] | Proved that every ion or polyion has two radii and two electric potentials: geometric and electrokinetic, and established the dependence of the electrokinetic radius on the ionic strength |

## 5.2 History of iontophoresis

Iontophoresis exists as a medical method after the early 1900s when French biologist Stephane Leduc (1853–1939), professor at the Ecole de Medecine de Nantes, published his works [183, 184, 185]. Fritz Frankenhäuser (born 1868) is said to have introduced the term "iontophoresis."

R. Wirtz employed iontophoresis in ophthalmology in 1908. He passed electric current through electrolyte-saturated cotton sponges, placed over the eye globe, for treatment of corneal ulcers, keratitis, and episcleritis [186].

In 1942, S. Witzel *et al.* applied iontophoresis for delivery of a variety of antibiotics such as tetracyclines, chloramphenicol, penicillin, streptomycin, neomycin, and bacitracin in rabbits [187].

In 1959, L. Gibson and R. Cooke demonstrated that sweating could be induced by pilocarpine iontophoresis [188].

In 1973, M. Corneau *et al.* used iontophoresis for local anesthesia of the ear [189].

In 1986, L. Gangarosa and J. Hill applied iontophoresis of vidarabine monophosphate for treatment of herpes orolabialis [190].

In 1992, W. Rigano *et al.* demonstrated treating of burned ears by gentamicin iontophoresis [191].

In 2013, M. Patane *et al.* showed that repeated transscleral iontophoresis with dexamethasone phosphate in the rabbit eye could be used in the treatment of inflammatory disorders of the orbit [192].

## 5.2.1 Outline history of iontophoresis

| Year | Researcher(s) | Development |
|------|---------------|-------------|
| 1900 | Leduc [193, 194, 195] | Published works on the physiological effects of electric current on the body |
| 1908 | Wirtz [196] | Employed iontophoresis in ophthalmology |
| 1942 | Witzel *et al.* [197] | Applied iontophoresis for delivery of antibiotics in rabbits |
| 1959 | Gibson and Cooke [198] | Demonstrated that sweating could be induced by pilocarpine iontophoresis |
| 1973 | Corneau and Brummett [199] | Used iontophoresis for local anesthesia of the ear |
| 1986 | Gangarosa and Hill [200] | Applied iontophoresis of vidarabine monophosphate for treatment of herpes orolabialis |
| 1992 | Rigano *et al.* [201] | Demonstrated treating of burned ears by gentamicin iontophoresis |
| 2003 | Monti *et al.* [202] | Studied the iontophoresis on the permeation of β-blocking agents across rabbit corneas *in vitro* |
| 2013 | Patane *et al.* [203] | Treated inflammatory disorders of the orbit with repeated transscleral iontophoresis of dexamethasone phosphate in the rabbit eye |

# References

[1] Reuss FF. Sur Un Nouvel Effet De L'électricité Glavanique. Mémoires De La Société Impériale Des Naturalistes De Moscou, 1809, 2, 327–336.

[2] Quincke G. Pogg Ann, 1859, 107, 1–47.

[3] Quincke G. Ann Phys, 1861, 113, 513–598.

[4] Picton H, Linder SE. J Chem Soc, 1892, 61, 148–172.

[5] Hardy WB. J Physiol, 1899, 24, 288–304.

[6] Whetham WCD. Phil Mag, 1899, 48, 474–477.

[7] Michov BM. Electrophoresis, 1985, 6, 471–474.

[8] Michov BM. Electrophoresis, 1988, 9, 199–200.

[9] Tiselius A. Trans Far Soc, 1937, 33, 524–531.

[10] König P. Acta E Terceiro Congresso Sud-americano De Chimica Rio De Janeiro E Sao Paulo, 1937, 2, 334–336.

[11] Pauling L, Itano HA, Singer SJ, Wells IC. Science, 1949, 110, 543–548.

[12] Kohn J. Biochem J, 1957, 65, 9P.

[13] Smithies O. Biochem J, 1955, 61, 629–641.

[14] Smithies O. Nature, 1955, 775, 307–308.

[15] Hjerten S. Biochem Biophys Acta, 1961, 53, 514–517.

[16] Hjerten S. Chromatogr Rev, 1967, 9, 122–219.

[17] Jorgenson JW, Lukacs KD. Clin Chem, 1981, 27, 1551–1553.

[18] Cohen AS, Najarian DR, Paulus A, Guttman A, Smith JA, Karger BL. Proc Natl Acad Sci USA, 1988, 85, 9660–9663.

[19] Karger BL. Nature, 1989, 339, 641–642.

[20] Heiger DN, Cohen AS, Karger BL. J Chromatogr, 1990, 516, 33–48.

[21] Oster G. Nature, 1954, 173, 300–301.

[22] Raymond S, Weintraub LS. Science, 1959, 130, 711–721.

[23] Hjerten S. Arch Biochem Biophys, 1962, Suppl 1, 147–151.

[24] Shapiro AL, Vinuela E, Maizel JV. Biochem Biophys Res Commun, 1967, 28, 815–820.

[25] Weber K, Osborn M. J Biol Chem, 1969, 244, 4406–4412.

[26] Reynolds JA, Tanford C. J Biol Chem, 1970, 245, 5161–5165.

[27] Ornstein L. Ann N Y Acad Sci, 1964, 121, 321–349.

[28] Davis BJ. Ann N Y Acad Sci, 1964, 121, 404–427.

[29] Laemmli UK. Nature, 1970, 227, 680–685.

[30] Michov BM. J Appl Biochem, 1982, 4, 436–440.

[31] Michov BM. Electrophoresis, 1984, 5, 171.

[32] Michov BM. Electrophoresis, 1986, 7, 150–151.

[33] Michov BM. Electrophoresis, 1989, 10, 686–689.

[34] Kolin A. J Chem Phys, 1954, 22, 1628–1629.

[35] Kolin A. Methods Biochem Anal, 1958, 6, 259–288.

[36] Svensson H. Acta Chem Scand, 1961, 15, 325–341.

[37] Svensson H. Arch Biophys Biochem, 1962, Suppl 1, 132–140.

[38] Vesterberg O, Svensson H. Acta Chem Scand, 1966, 20, 820–834.

[39] Vesterberg O. Acta Chem Scand, 1969, 23, 2653–2666.

[40] O'Farrell PH. J Biol Chem, 1975, 250, 4007–4021.

[41] Klose J. Humangenetik, 1975, 26, 231–243.

[42] Scheele GA. J Biol Chem, 1975, 250, 5375–5385.

[43] Southern EM. J Mol Biol, 1975, 98, 503–517.

[44] Renart J, Reiser J, Stark GR. Proc Natl Acad Sci USA, 1979, 76, 3116–3120.

[45] Towbin H, Staehelin T, Gordon J. Proc Natl Acad Sci USA, 1979, 76, 4350–4354.

[46] Burnette WN. Anal Biochem, 1981, 112, 195–203.

[47] Mukhopadhyay RW. The American Society for Biochemistry and Molecular Biology. Rockville, MD, 2012, 17–19.

[48] Burnette WN. In Kurien BT, Scofield RH (eds.). Methods in Molecular Biology. Springer Verlag, New York, 2009, Vol. 536, 5–8.

[49] Burnette WN. Clin Chem, 2011, 57, 132–133.

[50] Bowen B, Steinberg J, Laemmli UK, Weintraub H. Nucl Acids Res, 1980, 8, 1–20.

[51] Bagga PS, Wilusz J. Meth Mol Biol, 1999, 118, 245–256.

[52] Tanaka H, Fukuda N, Shoyama Y et al. J Agr Food Chem, 2009, 31, 296–303.

[53] Ishikawa D, Taki T. Meth Enzymol, 2000, 312, 145–157.

[54] Southern EM. J Mol Biol, 1975, 98, 503–517.

[55] Alwine J, Kemp D, Stark G. Proc Natl Acad Sci USA, 1977, 74, 5350–5354.

[56] Wreschner DH, Herzberg M. Nucl Acids Res, 1984, 12, 1349–1359.

[57] Bogdanov M, Sun J, Kaback HR, Dowhan WA. J Biol Chem, 1996, 271, 11615–11618.

[58] Towbin H, Schoenenberger C, Ball R, Braun DG, Rosenfelder G et al. J Immunol Meth, 1984, 72, 471–479.

[59] Taki T, Handa S, Ishikawa D. Anal Biochem, 1994, 221, 312–316.

[60] Durrum EL. Chem Eng News, 1949, 27, 601–610.

[61] Durrum EL. J Am Chem Soc, 1950, 72, 2943–2948.

[62] Grassmann W. Naturwissenschaften, 1951, 38, 200–206.

[63] Bradford M. Anal Biochem, 1976, 72, 248–254.

[64] Fazekas De St Groth S, Webster RG, Datyner A. Biochim Biophys Acta, 1963, 71, 377–391.

[65] Chrambach A, Reisfeld RA, Wyckoff M, Zaccari J. Anal Biochem, 1967, 20, 150–154.

[66] Merril CR, Robert C, Switzer RC, Van Keuren ML. Proc Natl Acad Sci USA, 1979, 76, 4335–4339.

[67] Switzer RC, Merril CR, Shifrin S. Anal Biochem, 1979, 98, 231–237.

[68] Reuss FF. Sur Un Nouvel Effet De L'électricité Glavanique. Mémoires De La Société Impériale Des Naturalistes De Moscou, 1809, 2, 327–336.

[69] Quincke G. Pogg Ann, 1859, 107, 1–47.

[70] Quincke G. Pogg Ann, 1861, 113, 513–598.

[71] Helmholtz HL. Wiedemann's Ann Phys Chem, 1879, 7, 337–382.

[72] Arrhenius S. Z Phys Chem, 1987, 1, 631–648.

[73] Picton H, Under SE. J Chem Soc, 1892, 61, 148–172.

[74] Hardy WB, Whetham WCD. J Physiol, 1896, 24, 288–304.

[75] Kohlrausch F. Ann Phys Chem, 1897, 62, 209–239.

[76] Smoluchowski M. Bull Acad Sci Cracovie, 1903, 182–200.

[77] Smoluchowski M. In Graetz L (ed.). Handbuch Der Elektrizität Und Des Magnetismus. Barth, Leipzig, 1921, Vol. 2, 366–428.

[78] Perrin J. J Chim Phys, 1904, 2, 601–651.

[79] Gouy G. J Phys, 1910, 9, 457–468.

[80] Chapman DL. Phil Mag, 1913, 25, 475–481.

[81] Debye P, Hückel E. Phys Z, 1923, 24, 185–206.

[82] Brönsted JN. Rec Trav Chim, 1923, 42, 718–728.

[83] Hückel E. Phys Z, 1924, 25, 204–210.

[84] Stern O. Z Elektrochem, 1924, 30, 508–516.

[85] Onsager L. Phys Z, 1926, 27, 388–392.

[86] Onsager L. Chem Rev, 1933, 13, 73–89.

[87] Henry DC. Proc Roy Soc, 1931, A 133, 106–129.

[88]   Henry DC. Trans Faraday Soc, 1948, 44, 1021–1026.
[89]   Tiselius A. Biochem J, 1937, 31, 313–317.
[90]   Tiselius A. Trans Far Soc, 1937, 33, 524–531.
[91]   König P. Acta E Terceiro Congresso Sud-americano De Chimica Rio De Janeiro E Sao Paulo, 1937, 2, 334–336.
[92]   Von Klobusitzky D, König P. Arch Exp Pathol Pharmacol, 1939, 192, 271–275.
[93]   Wieland T, Fischer E. Naturwissenschaften, 1948, 35, 29–30.
[94]   Martin AJP, Everaerts FM. Proc Roy Soc Ser A, 1970, 316, 493–514.
[95]   Pauling L, Itano HA, Singer SJ, Wells IC. Science, 1949, 110, 543–548.
[96]   Grassmann W, Hannig K. Naturwissenschaften, 1950, 37, 397–402.
[97]   Hannig K. Fresenius Zeitschr Anal Chem, 1961, 181, 244–254.
[98]   Grabar P, Williams CA. Biochim Biophys Acta, 1953, 10, 193–194.
[99]   Kolin A. J Chem Phys, 1954, 22, 1628–1629.
[100]  Kolin A. Proc Natl Acad Sci USA, 1955, 41, 101–110.
[101]  Robinson RA, Stokes RH. Electrolyte Solutions. Butterworths, London, 1959.
[102]  Smithies O. Biochem J, 1955, 61, 629–641.
[103]  Smithies O. Adv Protein Chem, 1959, 14, 65–113.
[104]  Kohn J. Nature, 1957, 180, 986–987.
[105]  Kohn J. Biochem J, 1957, 65, 9P.
[106]  Poulik MD. Nature, 1957, 180, 1477–1479.
[107]  Raymond S, Weintraub L. Science, 1959, 130, 711–721.
[108]  Hjerten S. Biochim Biophys Acta, 1961, 53, 514–517.
[109]  Svensson H. Acta Chem Scand, 1961, 15, 325–341.
[110]  Svensson H. Acta Chem Scand, 1962, 16, 456–466.
[111]  Svensson H. Acla Chem Scand, 1966, 20, 820–834.
[112]  Hjerten S. Arch Biochem Biophys, 1962, Suppl 1, 147–151.
[113]  Fazekas De St Groth S, Webster RG, Datyner A. Biochim Biophys Acta, 1963, 71, 377–391.
[114]  Ornstein L. Ann N Y Acad Sci, 1964, 121, 321–349.
[115]  Davis BJ. Ann N Y Acad Sci, 1964, 121, 404–427.
[116]  Hjerten S, Jerstedt S, Tiselius A. Anal Biochem, 1965, 11, 219–223.
[117]  Laurell CB. Anal Biochem, 1965, 10, 358–361.
[118]  Laurell CB. Anal Biochem, 1966, 15, 45–52.
[119]  Clarke HGM, Freeman T. In Peeters H (ed.). Prot Biol Fluids. Elsevier, Amsterdam, 1967, Vol. **14**, 503–522.
[120]  Thorne HV. Virology, 1966, 29, 234–239.
[121]  Vesterberg O, Svensson H. Acta Chem Scand, 1966, 20, 820–834.
[122]  Vesterberg O. Acta Chem Scand, 1969, 23, 2653–2666.
[123]  Shapiro AL, Vinuela E, Maizel JV. Biochem Biophys Res Commun, 1967, 28, 815–820.
[124]  Maizel JV Jr. In Habel K, Salzman NP (eds.). Fundamental Techniques in Virology. Acad Press, New York, 1969, 334–362.
[125]  Margolis J, Kenrick KG. Nature, 1967, 214, 1334–1336.
[126]  Margolis J, Kenrick KG. Anal Biochem, 1968, 25, 347–362.
[127]  Kopperschläger B, Diezel W, Bierwagen B, Hofmann E. FEBS Lett, 1969, 5, 221–224.
[128]  Slater GG. Anal Biochem, 1968, 24, 215–217.
[129]  Slater GG. Anal Chem, 1969, 41, 1039–1041.
[130]  Rodbard D, Kapadia G, Chrambach A. Anal Biochem, 1971, 40, 135–157.
[131]  Weber K, Osborn M. J Biol Chem, 1969, 244, 4406–4412.
[132]  Allen RC, Moore DJ, Dilworth RH. J Histochem Cytochem, 1969, 17, 189–190.

[133] Allen RC, Budowle B. Gel Electophoresis of Proteins and Nucleic Acids. Walter de Gruyter, Berlin – New York, 1994.
[134] Laemmli UK. Nature, 1970, 27, 680–685.
[135] Laemmli UK, Favre M. J Mol Biol, 1973, 80, 575–599.
[136] Aaij C, Borst P. Biochim Biophys Acta, 1972, 269, 192–200.
[137] Jovin TM. Ann N Y Sci, 1973, 209, 477–496.
[138] Chrambach A, Jovin TM, Svendsen PJ, Rodbard D. In Catsimpoolas N (ed.). Methods of Protein Separation. Plenum Press, New York, 1979, 27–144.
[139] O'Farrell PH. J Biol Chem, 1975, 250, 4007–4021.
[140] Klose J. Humangenetik, 1975, 26, 231–243.
[141] Scheele GA. J Biol Chem, 1975, 250, 5375–5385.
[142] Southern EM. J Mol Biol, 1975, 98, 503–517.
[143] Gasparic V, Biellquist B, Rosengren A. *Swedish patent No 14049-1*, 1975.
[144] Biellquist B, Ek K, Righetti PG *et al.* J Biochem Biophys Methods, 1982, 6, 317–339.
[145] Lambin P, Rochu D, Fine JM. Anal Biochem, 1976, 74, 567–575.
[146] Lambin P. Anal Biochem, 1978, 85, 114–125.
[147] Lasky M. In Catsimpoolas N (ed.). Electrophoresis'78. Elsevier, Amsterdam, 1976, 195–210.
[148] Hannig K. In Catsimpoolas N (ed.). Electrophoresis '78. Elsevier, Amsterdam, 1978, 69–76.
[149] Hannig K. Electrophoresis, 1982, 3, 235–243.
[150] Mikkers FEP, Everaerts FM, Verheggen TPEM. J Chromatogr, 1979, 169, 11–20.
[151] Jorgensen JW, DeArman L. J Chromatogr, 1981, 218, 209–216.
[152] Tsuda T, Nomura K, Nakagawa G. J Chromatogr, 1982, 248, 241–247.
[153] Hjerten S. J Chromatogr, 1983, 270, 1–6.
[154] Gebauer P, Deml M, Bocek P, Janak J. J Chromatogr, 1983, 267, 455–457.
[155] Terabe S, Otsuka K, Ichikawa K, Tsuchiya A, Andn T. Anal Chem, 1984, 56, 111–113.
[156] Green JS, Jorgenson JW. J High Res Chrom, 1984, 7, 529–531.
[157] Hjerten S, Zhu MD. J Chromatogr, 1985, 327, 157–162.
[158] Hjerten S, Elenbring K, Kilar F, Liao JL, Chen AJ, Siebert CJ, Zhu MD. J Chromatogr, 1987, 403, 47–61.
[159] Rosen A, Ek K, Aman P, Vesterberg O. In Peeters H (ed.). Protides of Biological Fluids. Pergamon Press, Oxford, 1979, 707–710.
[160] Towbin H, Staehelin T, Gordon J. Proc Natl Acad Sci USA, 1979, 76, 4350–4354.
[161] Renart J, Reiser J, Stark GR. Proc Natl Acad Sci USA, 1979, 76, 3116–3120.
[162] Burnette WN. Anal Biochem, 1981, 112, 195–203.
[163] Switzer RC, Merril CR, Shifrin S. Anal Biochem, 1979, 98, 231–237.
[164] Merril CM, Goldman D, Sedman SA, Ebert MH. Science, 1981, 211, 1437–1438.
[165] Jorgenson JW, Lukacs KD. Clin Chem, 1981, 27, 1551–1553.
[166] Michov BM. J Appl Biochem, 1982, 4, 436–440.
[167] Michov BM. Electrophoresis, 1984, 5, 171.
[168] Michov BM. Electrophoresis, 1986, 7, 150–151.
[169] Mullis KB. Nobel Lecture: The Polymerase Chain Reaction. 1993.
[170] Schwartz DC, Cantor CR. Cell, 1984, 37, 67–75.
[171] Michov BM. Electrophoresis, 1985, 6, 471–474.
[172] Schägger H, Jagow G. Anal Biochem, 1987, 166, 368–379.
[173] Michov BM. Electrophoresis, 1988, 9, 199–200.
[174] Cohen AS, Najarian DR, Paulus A, Guttman A, Smith JA, Karger BL. Proc Natl Acad Sci USA, 1988, 85, 9660–9663.
[175] Karger BL. Nature, 1989, 339, 641–642.
[176] Heiger DN, Cohen AS, Karger BL. J Chromatogr, 1990, 516, 33–48.

[177] Boots S. Anal Chem, 1989, 61, 551A–553A.

[178] Michov BM. Electrophoresis, 1989, 10, 686–689.

[179] Harrison DJ, Manz A, Fan Z, Ludi H, Widmer HM. Anal Chem, 1992, 64, 1926–1932.

[180] Kastenholz B. Anal Letters, 2004, 37, 657–665.

[181] Michov BM. Electrophoresis, 1985, 6, 470–471.

[182] Michov BM. Electrochim Acta, 2013, 108, 79–85.

[183] Leduc S. Ann D'electrobiologie, 1900, 3, 545–560.

[184] Leduc S. Les Bases Physiques De La Vie Et La Biogenese. Masson, Paris, 1906.

[185] Leduc S. La biologie synthétique. In: Poinat A ed. Étude De Biophysique. Vol. 2, Peiresc, Paris, 2012.

[186] Wirtz R. Klin Monatsbl Augenheilkd, 1908, 46, 543–579.

[187] Witzel SH, Fielding IZ, Ormsby HL. Am J Ophthalmol, 1956, 42, 89–95.

[188] Gibson LE, Cooke RE. Pediatrics, 1959, 23, 545–549.

[189] Corneau M, Brummett R, Vernon J. Arch Otolaryngol, 1973, 98, 114–120.

[190] Gangarosa LP, Hill JM . J Infect Dis, 1986, 154, 930–934.

[191] Rigano W, Yanik M, Barone FA et al. J Burn Care Rehabil, 1992, 13, 407–409.

[192] Patane MA, Schubert W, Sanford T et al. J Ocul Pharmacol Ther, 2013, 29, 760–769.

[193] Leduc S. Ann D'electrobiologie, 1900, 3, 545–560.

[194] Leduc S. Les Bases Physiques De La Vie Et La Biogenese. Masson, Paris, 1906.

[195] Leduc S. La biologie synthétique. In: Poinat A ed. Étude De Biophysique. Vol. 2, Peiresc, Paris, 2012.

[196] Wirtz R. Klin Monatsbl Augenheilkd, 1908, 46, 543–579.

[197] Witzel SH, Fielding IZ, Ormsby HL. Am J Ophthalmol, 1956, 42, 89–95.

[198] Gibson LE, Cooke RE. Pediatrics, 1959, 23, 545–549.

[199] Corneau M, Brummett R, Vernon J. Arch Otolaryngol, 1973, 98, 114–120.

[200] Gangarosa LP, Hill JM. J Infect Dis, 1986, 154, 930–934.

[201] Rigano W, Yanik M, Barone FA et al. J Burn Care Rehabil, 1992, 13, 407–409.

[202] Monti D, Saccomani L, Chetoni P et al. Int J Pharm, 2003, 250, 423–429.

[203] Patane MA, Schubert W, Sanford T et al. J Ocul Pharmacol Ther, 2013, 29, 760–769.

# 6 Troubleshooting

6.1    Protein electrophoresis troubleshooting —— 421
6.2    IEF troubleshooting —— 425
6.3    Nucleic acid electrophoresis troubleshooting —— 427
6.4    Blotting troubleshooting —— 431
6.5    Iontophoresis troubleshooting —— 433

## 6.1 Protein electrophoresis troubleshooting

| Problem | Cause | Solution |
|---|---|---|
| **CA electrophoresis** | | |
| The surface of CA membrane is spotty. | The drying of cellulose acetate membrane has taken too long. | Shorten the drying. |
| No protein bands on CA membrane. | The proteins have left CA membrane. | Monitor the running of dye front. |
| | The proteins have precipitated on application site. | Let electrophoresis run in another buffer with a different pH value. |
| **Agarose electrophoresis** | | |
| The agarose gel is too soft. | The agarose concentration is too low. | Check the recipe. Increase the agarose concentration. |
| | The gelling time was too short. | The gelling time should be at least 60 min at room temperature. It is better, if the agarose gel is stored overnight at 4 °C. |
| | Urea changed the agarose structure. | Use rehydratable agarose gels. |
| Air bubbles in the agarose gel. | The support film or glass plates of casting cassette were unclean. | Do not touch the hydrophilic side of support film with fingers. Wash the glass plates before use. |
| | The glass plates or agarose solution were too cold. | Heat the glass plates to 60–70 °C before casting. |
| The agarose gel separates from support film. | An incorrect support film was used. | Do not exchange the support films for agarose and polyacrylamide gels. |
| | The gel was cast on the wrong side of the support film. | Cast the gel on the hydrophilic side of the support film. Check the sides with water drops. |

https://doi.org/10.1515/9783110761641-038

(continued)

| Agarose electrophoresis | | |
| --- | --- | --- |
| The agarose gel does not stick to the support film, but to the glass plate. | The glass plate was too hydrophilic. | Clean the glass plate and coat it with Repel-silane. |
| The gel "sweats" (is covered with water drops) during electrophoresis run. | Strong electroosmosis. | Add 20 g/dL glycerol or 10 g/dL sorbitol to the agarose gel. |

### Polyacrylamide electrophoresis

#### Prior to electrophoresis

| | | |
| --- | --- | --- |
| The monomeric solution polymerizes too slowly or does not polymerize. | The concentration of TMEDA or APS in the monomeric solution is too low. | Increase the concentration of TMEDA or APS in the monomeric solution. |
| | The APS solution is too old or is stored improperly. | Use a new APS solution. Store the APS solution in the refrigerator. |
| | The gelling temperature is too low. | Cast gels at 20–25 °C. |
| | Too much oxygen in the monomeric solution. | Vent the monomeric solution with a water-jet pump. |
| | The pH value for the gel polymerization is not optimal. | Increase the concentration of TMEDA, if the pH value of the monomeric solution is too low. |
| The monomeric solution polymerizes too quickly. | The concentration of TMEDA or APS in the monomeric solution is too high. | Check the recipe for gel casting. Reduce the concentrations of TMEDA or APS. |
| | The polymerization temperature is too high. | Cast gels at 20–25 °C. |
| The polyacrylamide gel is too soft. | The concentration of monomeric solution was too low. | Increase the concentration of monomeric solution. |
| | The APS solution was too old or stored improperly. | Use a new APS solution. Store the APS solution in the refrigerator. |
| The polyacrylamide gel is sticky. | The concentration of acrylamide or BIS in the monomeric solution is too low. | Check the composition of the monomeric solution. |
| | The APS solution is too old or stored improperly. | The usable life of an APS solution is a week. Store the APS solution in the refrigerator. |
| | The atmospheric oxygen inhibited the polymerization of gel surface. | Degas the monomeric solution with a water-jet pump and overlay the monomeric solution with deionized water after casting. |

(continued)

| Prior to electrophoresis | | |
| --- | --- | --- |
| Air bubbles in the polyacrylamide gel. | The glass plate was dirty. | Wash the glass plate with detergent and ethanol before use. |
| | The support film was unclean. | Do not touch the hydrophilic side of the support film with fingers. |
| | There were bubbles in the outlet tube of the gradient maker or in the casting cassette. | Keep the outlet tube clean and dry. Remove the air bubbles in the casting cassette by vigorous knocking on it. |
| The gel separates from the support film. | An incorrect support film was used. | Do not exchange the support film for polyacrylamide and agarose gels. |
| | The gel was poured onto the wrong side of the support film. | Cast monomeric solution on the hydrophilic side of the support film; check first the side with water drops. |
| | The support film was stored incorrectly or too long. | Store the support film dry and in the dark at room temperature. |
| The polyacrylamide gel does not stick to the support film, but to the glass plate. | The glass plate is too hydrophilic. | Coat the glass plate with Repel-silane. |
| **During electrophoresis** | | |
| The electric current does not flow or flows too little during the electrophoresis run. | Some of the power connections has no or poor contact. | Check all power connections. |
| The gel "evaporates" (the lid of the separation chamber is covered with water condensation). | The electric power is too high. The cooling is insufficient. | Reduce the electric power. Check the coolant temperature (10–15 °C are recommended). The cooling plate should be made of glass, metal or best of ceramic. |
| The gel sparks and burns. | Thin areas in the gel, because the support film was not fixed on the glass plate during gel casting. | Roll strongly the support film on the glass plate of the casting cassette. |
| | Poor contact between the gel and the electrode bridges. | Weight the electrode bridges on the gel with a glass plate. |

(continued)

| After electrophoresis | | |
| --- | --- | --- |
| No protein bands in the gel. | Proteins have left the gel. The protein concentration in the sample was very low. The cathode and anode were exchanged. The detection sensitivity of the staining method is too low. | Monitor the movement of dye front. Apply more sample volume or concentrate the sample. Check the connections to the power supply. Use another staining method, e.g., silver staining. |
| Precipitates on the place of sample application. | The proteins were not dissolved totally. The protein or salt concentration in the sample was too high. | Use higher urea concentration (9 mol/L). Add non-ionic detergent (Nonidet NP-40) to the sample. Dilute the sample and desalt. |
| The protein bands are pale. | The protein concentration or sample volume was low. The sample was not completely dissolved. The sensitivity of the staining method is low. | Concentrate the sample or apply more volume of the sample. Treat the sample with ultrasound; centrifuge, if there is opacity. Stain more time. Use another staining method, for example, silver staining. |
| The protein bands have formed tails. | Precipitates or solid components in the sample. The sample was not dissolved. The sample was old. | Centrifuge the sample before applying. Treat the sample with ultrasound. Use fresh sample. |
| The samples from adjacent tracks have run into each other. | The sample volumes were too large. The application template did not laid tight on the gel. | Concentrate the samples and apply fewer volumes. Press lightly the application template toward the gel to remove any air between the template and gel. |
| The protein bands are blurred. | Too much protein in the sample. The separated proteins diffused in the gel. | Apply smaller volume or dilute the sample. Fix the proteins immediately after electrophoresis. |
| The gel rolls on during drying. | The gel contracts. | Add 5 g/dL glycerol in the final wash solution. |
| The gel dissolves from the support film during staining. | The bonds between the support film and the gel were partially hydrolyzed by strong acids (TCA) in some staining solutions. | The concentrated gels ($T > 10$ g/dL) should have cross-linking degree $C = 0.02$. |

(continued)

| After electrophoresis | | |
| --- | --- | --- |
| The Coomassie color has insufficient intensity. | SDS was not completely removed from the proteins. | Clean the gels more time to remove SDS, and color longer time. |
| | The alcohol concentration in the destaining solution was too high. | Reduce the concentration of ethanol or methanol in the destaining solution, or use a colloidal staining method. |

## 6.2 IEF troubleshooting

| Problem | Cause | Solution |
| --- | --- | --- |
| **Prior to isoelectric focusing** | | |
| Polyacrylamide gel has not polymerized or is sticky. | The concentration of monomerics was too low. | Increase monomeric concentration. |
| | The concentration of TMEDA or APS in the monomeric solution was too low. | Use 5 μL of 10 g/dL TMEDA and 5 μL of 10 g/dL APS per 1 mL of monomeric solution. |
| | The APS solution was too old or improperly stored. | Use new APS solution. Store the APS solution in the refrigerator at 4–8 °C for a week. |
| | The concentration of air oxygen in the monomeric solution was too high. | Degas the monomeric solution with a water-jet pump. |
| | The high carrier ampholytes concentration (more than 3 g/dL) has retarded the polymerization. | Decrease the carrier ampholyte concentration or polymerize the gel solution in the absence of carrier ampholytes and later add them. |
| Monomeric solution has polymerized too fast. | The concentration of TMEDA or APS in the monomeric solution was too high. | Check the recipe for gel casting. |
| | The environment temperature was too high. | Cast gels at 20–25 °C. |
| Polyacrylamide gel separates from the support film. | A wrong support film was used. | Do not exchange the support films for polyacrylamide and agarose gels. |
| | The gel was cast on the wrong side of the support film. | Cast gel solution only on the hydrophilic side of the support film. Check the film sides with drops of water. |
| | The monomeric solution contained high concentrations of the non-ionic detergent Triton X-100 or Nonidet NP-40. | Polymerize the gel solution in absence of detergents. The detergents can be brought later. |

(continued)

| Problem | Cause | Solution |
|---|---|---|
| **Prior to isoelectric focusing** | | |
| The gel does not adhere to the support film, but to the glass plate. | The glass plate is hydrophilic. | Clean the glass plate and treat it with Repel-silane. |
| Air bubbles in the gel. | The glass plate that comes in contact with the monomeric solution was dirty. | Wash the glass plate before use. |
| | The support film was unclean. | Do not touch the hydrophilic side of the support film with fingers. |
| **During isoelectric focusing** | | |
| It does not flow or flow too little electric current. | One of the connectors has no or poor contact. | Check all connections. The default setting for IPG gel is 5,000 V, 1.0 mA, and 5.0 W. |
| | There is a bad contact with the electrode strips. | Put a glass plate onto the gel. |
| The electric current intensity increases during IEF. | The electrode solutions are changed. | Use basic solution for the cathode; and acidic solution for the anode. |
| Water condensation on the lid of the separation cell. | The electric power is too high. | The electric power should not exceed 2.5 W/mL gel. |
| | The cooling is insufficient. | Check the cooling temperature. The cooling block should be made of glass, metal, or best of ceramic. Give kerosene or silicone oil DC 200 between the cooling block and support film. |
| The gel "sweats" (is covered with drops of water). | No water-binding additives in the gel. | Add 20 g/dL glycerol, 10 g/dL sucrose, or 10 g/dL sorbitol to the monomeric solution. |
| The gel sparks and burns. | The support film was not fixed on the glass plate during the gel casting. | Prior to gel casting, roll strongly the support film on the glass plate of the casting cassette. |
| | Poor cooling of the gel. | Check the cooling temperature. The cooling block should be made of glass, metal, or best of ceramic. |
| Urea crystallizes in IPG gels. | The IEF temperature is too low. | Focus at 10–15 °C. |
| | The surface dries out. | Add 0.5 mL/dL Nonidet NP-40 to the rehydration solution. |

(continued)

| During isoelectric focusing | | |
| --- | --- | --- |
| Iso-pH-lines in the gel. | The salt concentration in the sample is too high. | Desalinate the samples. |
| | The concentration of APS was too high. | Reduce the APS concentration or use rehydratable gels. |

| After isoelectric focusing | | |
| --- | --- | --- |
| The protein bands are not sharp. | The concentration of the carrier ampholytes was too low. | The concentration of a carrier ampholytes in the gel should be at least 2 g/dL. |
| | The electrofocusing time was too short. | Prolong the electrofocusing time. |
| The bands pull tails. | There were precipitates or particles in the samples. | Centrifuge the sample. |
| The bands are crooked. | The salt concentration in the samples was too high. | Desalinate the samples. |
| The samples from adjacent tracks run into each other. | The sample volumes were too large. | Concentrate the samples and apply less volumes. |
| | The application template was not placed tight on the gel. | Press gently the application template toward the gel to remove any air bubbles between the template and gel. |
| The background remains blue after Coomassie staining. | Basic immobiline groups bind to Coomassie dyes. | Use colloidal or silver staining method. |

# 6.3 Nucleic acid electrophoresis troubleshooting

| Problem | Cause | Solution |
| --- | --- | --- |
| **Prior to electrophoresis** | | |
| The consistency of agarose gel is insufficient. | The agarose concentration is too low. | Check the recipe. Keep agarose dry because it is hygroscopic. |
| | The gelling time of agarose was too short. | The gelling of agarose solution should continue at least 60 min at room temperature or better overnight. |

(continued)

| Problem | Cause | Solution |
|---|---|---|
| **Prior to electrophoresis** | | |
| Air bubbles in agarose gels. | The support film or glass plates that came into contact with the agarose solution were dirty.<br><br>The agarose solution or glass plates were cold – the agarose solution gelled during casting. | Do not touch the hydrophilic side of the support film or glass plate with fingers. Prior to use wash the glass plates with dishwashing detergent.<br>Preheat the agarose solution and glass plates at 70 °C. |
| The agarose gel separates from the support film. | An incorrect support film was used.<br>The gel was cast on the wrong side of the support film. | Do not exchange the support films for agarose and polyacrylamide gels.<br>Cast gel only on the hydrophilic side of support film. Prior to casting, check the film side with water drops. |
| The monomeric solution does not polymerize or polymerizes to slow. | The concentration of TMEDA or APS in the monomeric solution was too low.<br>The APS solution was too old or stored improperly.<br>The room temperature was too low. | Check the composition of the monomeric solution. Increase the concentration of TMEDA or APS.<br>Use new APS solution. Store the APS solution in the refrigerator.<br>Cast gels at 20–25 °C. |
| The polyacrylamide gel is too soft and sticky. | The concentration of acrylamide or BIS was too low.<br>The concentration of TMEDA or APS in the monomeric solution was too low.<br>The APS solution was old.<br><br>There was too much oxygen in the monomeric solution. | Increase the concentration of acrylamide or BIS.<br>Use 5 µL of 10 g/dL TMEDA and 5 µL of 10 g/dL APS per 1 mL of monomeric solution.<br>The maximum storage for an APS solution in the refrigerator is a week.<br>Degas the monomeric solution using a water-jet pump. |
| Air bubbles in the polyacrylamide gel. | The support film was unclean.<br><br>The glass plate that came in contact with the monomeric solution was dirty. | Do not touch the hydrophilic side of support film with fingers.<br>Wash the glass plate with dishwashing detergent or ethanol before use. |

(continued)

| Problem | Cause | Solution |
|---|---|---|
| **Prior to electrophoresis** | | |
| The polyacrylamide gel separates from support film. | An improper support film was used. The gel was cast onto the wrong side of the support film. | Do not exchange the support film for agarose and polyacrylamide gels. Cast gel only on the hydrophilic side of the support film, first checking the side with water drops. |
| | The support film was stored incorrectly or too long. The glass plate that came into contact with the monomeric solution was unclean. | Store the support film at room temperature in the dark. Prior to use, wash the glass plate with dishwashing detergent or ethanol. |
| The gel does not adhere to the support film, but to the glass plate. | The glass plate is too hydrophilic. | Treat the glass plate with Repel-silane. |
| **During electrophoresis** | | |
| It does not flow or flows too little electric current. | One of the connectors has no or poor contact. | Check all connections. |
| | Poor contact between the electrode strips and gel. | Put a glass plate onto the electrodes on the gel. |
| The gel "sweats" – is covered with water drops. | No water binding additives in the gel. | Add 20 g/dL glycerol or 10 g/dL sucrose to the monomeric solution. |
| The gel "evaporates" – the separation chamber lid is covered with condensed water. | The voltage is too high. The cooling is insufficient. | Reduce the voltage. Check the coolant temperature. The cooling block should be made of glass, metal or best of ceramic. |
| The gel sparks and burns. | The polyacrylamide gel contains thin spots because the support film was not fixed on the glass plate of casting cassette. | Prior to casting, roll the support film on a wet glass plate to get an even contact between the film and plate. |

**After electrophoresis**

| | | |
|---|---|---|
| No nucleic acid bands on the gel. | The cathode and anode were changed. | Check the connections to the power supply. |
| | The nucleic acids have left the gel. | Monitor the passage of the dye front. |
| | The mass of the nucleic acid was too large – the nucleic acid could not enter the gel. | Use agarose gel or reduce the total concentration $T$ of the polyacrylamide gel. |
| | Insufficient quantity of DNA on the gel. | Increase the amount of DNA; don't exceed 50 ng/band. |
| | DNA was degraded. | Avoid nuclease contamination. |
| | Improper light source was used for visualization of ethidium bromide-stained DNA. | Use a short wavelength (254 nm) light. |
| The nucleic acid bands are very weak. | The concentration of nucleic acids or the sample volume was too small. | Concentrate the sample, apply more volume, and use more sensitive detection methods. |
| The DNA separation is bad. | Wrong agarose concentration. | Higher agarose concentration will help to resolve smaller DNA molecules, while lower agarose concentration will help to resolve larger DNA molecules |
| The bands are fuzzy. | Excess salt in the samples. | Check the salt concentrations in the gel and electrode buffers. |
| DNA bands are smeared. | DNA was degraded. | Avoid nuclease contamination. |
| | Too much DNA was loaded on the gel. | Decrease the DNA concentration. |
| | Too much salt in the DNA sample. | Remove excess salts using ethanol precipitation. |
| The nucleic acid bands form tails. | The concentration of nucleic acids in the sample was too high. | Apply small sample volume or dilute the sample. |
| | Diffusion after the separation. | Shorten the time between the end of electrophoresis and detection procedure. |
| The samples from adjacent tracks ran into each other. | The sample volumes were too much. | Concentrate the samples and apply smaller volumes. |
| | The application template was not tight on the gel. | Press lightly the application template toward the gel to push away air bubbles between the template and gel. |

(continued)

| After electrophoresis | | |
|---|---|---|
| The DNA fragments are arranged in a curved manner – "smiling" effect. | The temperature in the central part of the gel was higher than in the gel margins. | Reduce the voltage. Apply the samples at some distance from the gel margins. |
| Autoradiogram is too light. | Old label<br>Not enough primer<br>Not enough template DNA<br>Low enzyme activity | Use new label.<br>Check the primer concentration.<br>Check the template concentration.<br>Use ~0.5 pmol template DNA per reaction.<br>Use fresh enzyme. |
| Amplified DNA concentration is lower than expected. | Amplification cycle setting is low.<br>Low MgCl$_2$ concentration.<br>Low sample concentration.<br>Inhibitors in the template.<br>PCR reagents are contaminated or expired.<br>Degraded primers. | Add three to five cycles.<br><br>Increase the MgCl$_2$ concentration.<br>Increase the sample concentration.<br>Purify the template.<br>Use fresh PCR reagents.<br><br>Store unused primers at –20 °C. |
| PCR inhibition. | Sample contains PCR inhibitors: hemoglobin, heparin, and polysaccharides. | Dilute the sample before amplification to reduce the concentration of PCR inhibitors. |
| Intensive background. | Impure reagents were used.<br>The water quality was poor. | Use reagents p.a.<br>Use only deionized water. |

# 6.4  Blotting troubleshooting

| Prior to blotting | | |
|---|---|---|
| Air bubbles between the filter papers of blot sandwich. | The blot sandwich was not assembled under a buffer. | Roll the blot sandwich with a photo roller under a buffer to push out air bubbles. |
| Nitrocellulose membrane is prewet unevenly. | The nitrocellulose membrane is old or contaminated. | Use new nitrocellulose membrane. |
| Nitrocellulose membrane is getting fragile after baking. | The nitrocellulose membrane is overbaked. | Check oven temperature. |

(continued)

| During blotting | | |
| --- | --- | --- |
| No electric current or too small electric current. | One of the connectors had no or poor contact. | Check all connections. |
| Electric power is too high. | Electric current flows around the blot sandwich. | Cut filter papers and blot membrane according to gel size. |
| Voltage increases during blotting. | Gas bubbles between the electrode plates and filter papers. | Weight the cathode plate with 1 kg to push the gas bubbles laterally. |
| Blotter becomes hot during blotting. | Electric current intensity is very high. | Blot at 0.8–1.0 mA/cm$^2$. |

| After blotting | | |
| --- | --- | --- |
| No transfer on the blot membrane. | The direction of blotting was wrong. | Blot toward the anode in an alkaline buffer. |
| Transfer is incomplete. | The transfer time was too short. | Extend the transfer time. |
| | The electric current passed around the blot sandwich. | Cut the blot membrane and filter papers according to the gel size. |
| Transfer is unregularly. | The electric current flow was irregular because the electrode plates were dry. | Prior to blotting wet the electrode plates with deionized water. |
| Loss of protein bands. | Low molecular mass proteins ($M_r < 20,000$) may be lost during the post transfer washes. | Use glutardialdehyde fixation and smaller pore size nitrocellulose membranes (0.2 μm). |
| Proteins have left the gel, but are not found on the blot membrane. | Low-molecular peptides have been washed during the detection procedure. | When using a nylon membrane, the fixative should contain glutardialdehyde. |
| Strong background on the blot membrane. | Blocking was ineffective. | Block longer or use higher temperature (37 °C). |
| | Cross-reactions with the blocking agent. | Use a different blocking agent, i.e. fish gelatin or skim milk. |

(continued)

| After blotting | | |
| --- | --- | --- |
| Low sensitivity of hybridization. | Low concentration of RNA. | Increase the RNA concentration up to 30 mg in each line. Use poly(A) RNA instead of total RNA; 10 mg of poly(A) RNA is ≈300–350 mg of total RNA (3–5%). Use optimal hybridization temperature. Use freshly synthesized probes. Increase exposure time. |
| More bands after RNA hybridization instead of one expected. | Cross-hybridization. | The probe concentration was too high. |
| | The hybridization temperature was very low. | Increase the hybridization temperature up to 52 °C. |
| Specks and splotches. | Contaminated membrane. Bubbles in hybridization solution. | Use new membrane. Avoid bubbles, mix during hybridization. |

## 6.5 Iontophoresis troubleshooting

| Problem | Cause | Solution |
| --- | --- | --- |
| **Prior to isoelectric focusing** | | |
| Device does not power up. | Battery Battery contact | Check the battery. If necessary, change it. Check both contacts between the battery and biological tissue. |
| Iontophoresis device is shutting down. | Treatment site has high resistance. | Leave the electric current to 0.5 mA. |
| | Treatment site is callused. | Wet the callused site. |
| Battery indicator lights when device is turned on. | Battery voltage is too low. | Replace the battery. |
| Battery indicator lights during treatment. | Battery voltage is decreasing during treatment. | Replace the battery. |
| Electric current stops. | Treatment has paused by decreasing electric current to zero mA. | Restart the treatment. |
| Resistance limits. | Skin resistance at the electrode site is too high. | Set the electric current level as high as possible and prolong the time to deliver desired dose. |

# Problems

1    Fundamentals of electrophoresis
2    Electrophoresis of proteins
3    Electrophoresis of nucleic acids

## 1 Fundamentals of electrophoresis

1.1   p$K$ value of dihydrogen phosphate ($H_2PO_4^- \leftrightarrow H^+ + HPO_4^{2-}$) is equal to 7.20 at 25 °C. Calculate its p$K_c$ value at $I = 0.1$ mol/L.

1.2   Calculate the ionic strength of a solution containing 0.1 mol/L $CH_3COONa$ and 0.2 mol/L NaCl.

1.3   A buffer contains 0.1 mol/L $CH_3COOH$ and 0.1 mol/L $CH_3COONa$ at 25 °C. Calculate its pH value, if 0.01 mol/L HCl or 0.01 mol/L NaOH is added. The p$K_c$ value of $CH_3COOH$ is 4.64.

1.4   Calculate the buffer capacity of a buffer with pH = 5.21 at 25 °C, which contains 0.1 mol/L pyridine and 0.1 mol/L HCl. The p$K_c$ value of pyridinium ion is 5.21.

1.5   Calculate the mobility of TRIS ion $\mu_{HT^+}$ at 0 °C and at 25 °C, at ionic strengths of 0.01 and 0.07 mol/L. The absolute mobility of TRIS ion is $12.75 \times 10^{-9}$ m$^2$/(s V) at 0 °C and $27.86 \times 10^{-9}$ m$^2$/(s V) at 25 °C. Use the ionic mobility parameter of the linear equation for calculation.

1.6   Using the equation $\mu = \dfrac{dl}{tU}$, calculate the mobility of albumin polyion $\mu_{albumin^{n-}}$, if it has run 25 mm ($d$) on a cellulose strip in 1 h ($t$), and if the distance ($l$) between the anode and cathode ends of the strip was 100 mm and the potential difference was 250 V ($U$).

1.7   Which is the most important buffer in the cell?
      A) Hydrogen carbonate buffer
      B) Hemoglobinate buffer
      C) Proteinate buffer
      D) Hydrogen phosphate buffer

1.8   Which is the p$K_b$ value of ammonia, if the p$K$ value of ammonium ion is 9.25?
      A) 3.25
      B) 2.15
      C) 2.25
      D) 4.75

https://doi.org/10.1515/9783110761641-039

1.9 Which buffer is most significant for the blood plasma?
A) Proteinate buffer
B) Hydrogen carbonate buffer
C) Hemoglobinate buffer
D) Hydrogen phosphate buffer

1.10 How high will be the pH value of a solution, if 10 mL of 0.1 mol/L weak acid ($K_a = 10^{-4}$ mol/L) are mixed with 10 mL of 0.1 mol/L sodium salt of the same acid?
A) 5
B) 3
C) 4
D) 6

1.11 What will be the ratio between the concentrations of the charged and uncharged amino groups in a solution with pH = 6.5, if $pK$ value of amino group is equal to 7.5?
A) 1:1
B) 100:1
C) 10:1
D) 1:10

1.12 How high will be the pH value of a solution obtained, if 10 mL of 0.1 mol/L sodium lactate are added to 10 mL of 0.1 mol/L lactic acid ($pK_a = 3.86$)?
A) 3.86
B) 4.86
C) 2.86
D) 1.86

# 2 Electrophoresis of proteins

2.1 How can be prepared a monomeric solution of acrylamide and BIS with a total monomeric concentration $T = 50$ g/dL and a degree of crosslinking $C = 0.03$?

2.2 The "persistent" function of Kohlrausch $F_k$ is equal to 0.8. What should be the concentration of the trailing electrolyte, if the concentration of the leading electrolyte $c_{HA}$ is equal to 0.10 mol/L?

2.3 Which concentration will reach the protein $H_nP$, if the concentration of the strong leading acid $c_{HA} = 0.10$ mol/L and the Kohlrausch function $F_k = 0.02$?

2.4 Calculate the electrophoretic mobility of negatively charged albumin $\mu_{alb}$, if it has run the distance $d$ of 38 mm for one hour at voltage $U$ of 250 V between the electrodes, which are located from each other at a distance $l$ of 10 cm.

2.5 At what pH value borate, taurinate, and glycinate ions will follow certain proteins, if the slowest of them has a velocity of $-13 \times 10^{-9}$ m²/(s V) and is running in a solution with $I$ of 01 mol/L at 25 °C? At 25 °C the absolute mobilities of these ions are equal to $-33.63 \times 10^{-9}$, $-34.11 \times 10^{-9}$, and $38.45 \times 10^{-9}$ m²/(s V), respectively. The p$K$ values of the corresponding acids at the same temperature and $I$ = 0.1 mol/L are 9.02, 8.84, and 9.56, respectively.

2.6 Prepare a TRIS-glycinate buffer with pH = 8.30 and an ionic strength of 0.10 mol/L at 25 °C. The p$K$ values of TRIS-ion p$K_{\text{TRIS}+}$ and glycine p$K_{\text{HG}}$ at 25 °C are 8.07 and 9.78, respectively.

2.7 Which bonds are most important for the alpha-helix structure of polypeptide chains?
A) Hydrogen bonds, which are parallel to the axis of the polypeptide chain
B) Disulfide bonds
C) Hydrogen bonds that are chaotically scattered along an alpha-helix
D) Hydrogen bonds, which are perpendicular to the axis of the polypeptide chain

2.8 Which amino acid forms disulfide bridges in proteins?
A) Methionine
B) Serine
C) Cysteine
D) Threonine

2.9 The isoelectric point pI (pH(I)) of pepsin (an enzyme in gastric juice) is approximately equal to 1. Which amino acids are present in it?
A) Tryptophan and tyrosine
B) Serine and alanine
C) Aspartic and glutamic acid
D) Lysine and arginine

2.10 What happens in agarose gel electrophoresis?
A) Proteins separate from each other according to their electric charges
B) The separation of DNA fragments does not depend on the concentration of agarose
C) Proteins move with same speed in buffers of different pH values
D) Proteins are separated according to their molecular masses

2.11 Blotting techniques are used to identify
   A) Unique proteins
   B) Nucleic acid sequences
   C) Both A and B
   D) None of the above

2.12 Western blotting is a method for testing the presence of
   A) DNA
   B) RNA
   C) Protein
   D) None of the above

## 3 Electrophoresis of nucleic acids

3.1 How many base pairs have a double-helical DNA with $M_r$ of $2 \times 10^6$ and $2 \times 10^7$?

3.2 How many electric charges have single-stranded RNAs with $M_r$ of $4 \times 10^4$ and $4 \times 10^5$, respectively?

3.3 Why DNA which consists of AT pairs melts at 70 °C, and why DNA which consists of GC pairs melts at 90 °C?

3.4 Which equation expresses the relationship between the mobility of DNA and agarose concentration?

3.5 Why the migration of DNA fragments diminishes in a gel containing ethidium bromide?

3.6 Calculate the electrophoretic mobility $\mu$ of DNA, which has run a distance $d$ of 127 mm in an electric field with a distance between the electrodes of 20 cm at voltage $U$ of 250 V in one hour electrophoresis.

3.7 Which bonds bind the mononucleotide residues in polynucleotide chains?
   A) Peptide bonds
   B) N-glycosidic bonds
   C) O-glycosidic bonds
   D) 3',5'-Phosphodiester bonds

3.8 Which are the bonds between neighboring bases in a DNA strand?
   A) Ester bonds
   B) Covalent bonds
   C) Hydrogen bonds
   D) Hydrophobic interactions

3.8   Which bonds stabilize the double helix structure of DNA?
   A) Hydrogen bonds and hydrophobic interactions between the bases
   B) Ionic bonds between the phosphate residues
   C) Hydrogen bonds between the deoxyriboses and bases
   D) Disulfide bridges

3.10  What is true?
   A) DNA contains uracil, and RNA contains thymine
   B) DNA has a greater mass and volume than RNA
   C) Nucleases attack DNA, but not RNA
   D) Phosphodiester bonds in DNA are stable against alkalis, on the contrary
      in RNA

3.11  Which bonds between both DNA strands are broken at denaturation?
   A) 3′,5′-Phosphodiester bonds
   B) Covalent bonds
   C) Hydrogen bonds
   D) Ionic bonds

3.12  Southern blotting is a method for checking the presence of
   A) DNA
   B) RNA
   C) Protein
   D) None of the above

# Solution of problems

1   Fundamentals of electrophoresis
2   Electrophoresis of proteins
3   Electrophoresis of nucleic acids

## 1 Fundamentals of electrophoresis

1.1   It follows from the relationship between $pK$ and the ionic strength that

$$pK_c = 7.20 - \frac{2 \times 0.1^{1/2}}{1 + 0.1^{1/2}} = 7.20 - 0.48 = 6.72$$

1.2   Since $[Na^+] = 0.1 + 0.2 = 0.3$ mol/L, $[CH_3COO^-] = 0.1$ mol/L, and $[Cl^-] = 0.2$ mol/L, the ionic strength

$$I = \frac{1}{2}\left[0.3 \times 1^2 + 0.1(-1)^2 + 0.2(-1)^2\right] = 0.3 \, mol/L$$

1.3   If 0.01 mol/L HCl is added to a buffer,

$$pH = 4.64 + \log\frac{0.1 - 0.01}{0.1 + 0.01} = 4.55$$

If 0.01 mol/L NaOH is added to a buffer,

$$pH = 4.64 + \log\frac{0.1 + 0.01}{0.1 - 0.01} = 4.73$$

1.4   The pH value of the buffer is equal to the value of the $pK_c$ of pyridinium ion. Therefore, $\alpha = 0.5$, i.e., the equilibrium concentrations of pyridine and pyridinium ion are equal. Then it follows that

$$\beta = \ln 10 \times 0.5(1 - 0.5) \times 0.1 = 0.0576 \, mol/L$$

1.5   At 0 °C (273.15 K), the values of $\eta$ and $\varepsilon_r$ of water are equal to $1.787 \times 10^{-3}$ Pa s, and 87.74, respectively, i.e., the parameters of the ionic mobility $p_i = 15.45 \times 10^{-9} \, z_i I^{1/2}$ $m^2/(s \, V)$. At 25 °C (298.15 K), the values of $\eta$ and $\varepsilon_r$ of water are equal to $0.8904 \times 10^{-3}$ Pa s, and 78.30, respectively, i.e., the parameter of the ionic mobility $p_i = 31.42 \times 10^{-9} \, z_i I^{1/2} \, m^2/(s \, V)$. Then, at 0 °C and $I = 0.01$ mol/L

https://doi.org/10.1515/9783110761641-040

$$\mu_{HTRIS^+} = \left(12.75 - 15.45 \times 0.01^{1/2}\right) \times 10^{-9} = 11.20 \times 10^{-9} \text{ m}^2/(\text{s V})$$

and at the same temperature but at $I = 0.07$ mol/L

$$\mu_{HTRIS^+} = \left(12.75 - 15.45 \times 0.07^{1/2}\right) \times 10^{-9} = 8.66 \times 10^{-9} \text{ m}^2/(\text{s V})$$

At 25 °C and $I = 0.01$ mol/L

$$\mu_{HTRIS^+} = \left(27.86 - 31.42 \times 0.01^{1/2}\right) \times 10^{-9} = 24.72 \times 10^{-9} \text{ m}^2/(\text{s V})$$

and at the same temperature but at $= 0.07$ mol/L

$$\mu_{HTRIS^+} = \left(27.86 - 31.42 \times 0.07^{1/2}\right) \times 10^{-9} = 19.55 \times 10^{-9} \text{ m}^2/(\text{s V})$$

1.6   Taking into account that 25 mm = $25 \times 10^{-3}$, 100 mm = $100 \times 10^{-3}$, and 1 h = 3,600 s, according to the ionic mobility equation,

$$\mu_{albumin^{n-}} = \frac{25 \times 10^{-3} \times 100 \times 10^{-3}}{3,600 \times 250} = 2.8 \times 10^{-9} \text{ m}^2/(\text{s V})$$

1.7   D

1.8   D

1.9   B

1.10  C

1.11  C

1.12  A

# 2 Electrophoresis of proteins

2.1   The total monomeric concentration and the degree of crosslinking of a polyacryl-amide gel are given by the equations

$$T = a + b$$

and

$$C = \frac{b}{a + b}$$

respectively, where $a$ and $b$ are the concentrations of acrylamide and BIS, in g/dL. According to the problem, these equations can be transformed into

$$a + b = 50$$

and

$$0.03(a + b) = b$$

where from follows that 48.5 g of acrylamide and 1.5 g of BIS should be resolved in 100.0 mL water to obtain the desired $T$ and $C$ values.

2.2 The "persistent" function of Kohlrausch $F_k$ is calculated according to the equation

$$F_k = \frac{c_{HB}}{c_{HA}}$$

Hence

$$c_{HB} = F_k \times c_{HA} = 0.8 \times 0.10 = 0.08 \, \text{mol/L}$$

2.3 For the protein $H_nP$, the "persistent" function of Kohlrausch $F_k$ can be calculated according to equation

$$F_k = \frac{c_{H_nP}}{c_{HA}}$$

Hence

$$c_{H_nP} = F_k \times c_{HA} = 0.02 \times 0.10 = 0.002 \, \text{mol/L}$$

2.4 The mobility of albumin is

$$\mu_{alb} = \frac{v_{alb}}{E} = \frac{d/t}{U/l} = \frac{dl}{tU}$$

where $v_{alb}$ (in m/s) is the velocity of albumin; $E$ (in V/m) is the electric field strength; $d$ (in m) is the distance between the electrodes; $t$ (in s) is the time; $U$ (in V) is the voltage, and $l$ (in m) is the run distance. Since $d = -0.038$ m, $l = 0.1$ m, $t = 3,600$ s, and $U = 250$ V, it follows that

$$\mu_{alb} = -\frac{0.038 \times 0.1}{3,600 \times 250} = -4.2 \times 10^{-9} \, \text{m}^2/(\text{s V})$$

2.5 From the equation of the ionic mobility parameter follows that at $I = 0.1$ mol/L and 25 °C, the mobility of borate, taurinate, and glycinate ion is $-28.51 \times 10^{-9}$, $-24.17 \times 10^{-9}$, and $-23.69 \times 10^{-9}$ m²/(s V), respectively. Therefore, the dissociation degrees of the corresponding acids

$$\alpha_{\text{boric acid}} = \frac{-13 \times 10^{-9}}{-23.69 \times 10^{-9}} = 0.55$$

$$\alpha_{\text{taurine}} = \frac{-13 \times 10^{-9}}{-24.17 \times 10^{-9}} = 0.54$$

and
$$\alpha_{\text{glycine}} = \frac{-13 \times 10^{-9}}{-28.51 \times 10^{-9}} = 0.46$$

Hence, according to the Henderson–Hasselbalch equation, glycine, taurine, and boric acid should be dissolve at pH = 9.49, 8.91, and 9.11, respectively.

2.6 From the Debye–Hückel equation follows that

$$pK_{c\,\text{HTRIS}+} = 8.07 + \frac{0.01^{1/2}}{1 + 0.01^{1/2}} = 8.16$$

and
$$pK_{c\,\text{HG}} = 9.78 + \frac{0.01^{1/2}}{1 + 0.01^{1/2}} = 9.69$$

Then, according to Henderson–Hasselbalch equation,

$$\alpha_{\text{TRIS}} = \left(1 + 10^{pH - pK_c}\right)^{-1} = \left(1 + 10^{8.30 - 8.16}\right)^{-1} = 0.420$$

and
$$\alpha_{\text{HG}} = \left(1 + 10^{pK_c - pH}\right)^{-1} = \left(1 + 10^{9.69 - 8.30}\right)^{-1} = 0.039$$

Hence,
$$c_{\text{TRIS}} = \frac{0.01}{0.420} = 0.024 \text{ mol/L}$$

and
$$c_{\text{glycine}} = \frac{0.01}{0.039} = 0.256 \text{ mol/L}$$

2.7 A

2.8 C

2.9 C

2.10 A

2.11 C

2.12 C

# 3 Electrophoresis of nucleic acids

3.1 The relative molecular mass $M_r$ of the mononucleotide residues is 314, approximately 330. This means that two mononucleotide residues (a base pair) have a molecular mass of approx. $2 \times 330$. Therefore, a double-helical DNA with $M_r$ of $2 \times 10^6$ contains

$$\frac{2 \times 10^6}{2 \times 330} \approx 3,000$$

base pairs, and a double-helical DNA with $M_r$ of $2 \times 10^7$ contains

$$\frac{2 \times 10^7}{2 \times 330} \approx 30,\ 000$$

base pairs.

3.2 A mononucleotide residue carries a negative electric charge. The relative molecular mass of a mononucleotide residue is approximately 330. Therefore, a single-stranded RNA with $M_r$ of $4 \times 10^4$ has

$$\frac{4 \times 10^4}{330} \approx 120$$

negative charges, and a single-stranded RNA with $M_r$ of $4 \times 10^5$ has

$$\frac{4 \times 10^5}{330} \approx 1,200$$

negative charges.

3.3 If a DNA contains predominantly AT pairs, it melts at a lower temperature (70 °C) because two hydrogen bonds exist between A and T; when a DNA contains predominantly GC pairs, it melts at a higher temperature (90 °C) because three hydrogen bonds exist between G and C.

3.4 The equation, which expresses the linear relationship between the logarithm of DNA mobility $\mu$ and the agarose concentration $c$, is:

$$\frac{\mu}{\mu_0} = 10^{-K_R c}$$

or

$$\log\mu = \log\mu_0 - K_R c$$

where $\mu_0$ is its electrophoretic DNA mobility in an agarose gel of concentration of 0 g/dL, and the slope $K_R$ is the retardation coefficient. The value of $K_R$ depends on the ionic strength of the buffer and the properties of the gel. This equation becomes invalid when the agarose concentration is over 0.9 g/dL.

3.5 The migration of DNA fragments slows down in a gel in presence of ethidium bromide by about 15% because ethidium bromide intercalates between the bases and thus increases the mass of DNA.

3.6 The mobility of DNA

$$\mu_{DNA} = \frac{v_{DNA}}{E} = \frac{d/t}{U/l} = \frac{dl}{tU}$$

where $v_{DNA}$ (in m/s) is the DNA velocity; $E$ (in V/m) is the strength of the electric field; $d$ (in m) is the distance between the electrodes; $t$ (in s) is the time; $U$ (in V) is the voltage; and $l$ (in m) is the run distance. In our example, $d$ = 0.127 m, $l$ = 0.2 m, $t$ = 3,600 s, and $U$ = 250 V. Therefore,

$$\mu_{DNA} = \frac{0.127 \times 0.2}{3,600 \times 250} = 28.2 \times 10^{-9} \ m^2/(s \ V)$$

3.7 D

3.8 D

3.9 A

3.10 B

3.11 C

3.12 A

# Reagents for electrophoresis

https://doi.org/10.1515/9783110761641-041

| Reagent | Molecular formula | $M_r$ | Density, $d$ | $pK_{25C}$ | Useful pH range |
|---|---|---|---|---|---|
| ACES | | 182.19 | | 6.78 | 6.1–7.5 |
| Acetic acid | | 60.05 | 1.05 | 4.76 | 4.1–5.5 |
| Acrylamide | | 71.08 | | | |
| Agarose | | 630.55 | | | |
| L-Alanine | | 89.09 | | 9.69 | 9.0–10.4 |
| β-Alanine | | 89.09 | | 10.19 | 9.5–10.9 |

(continued)

(continued)

| Reagent | Molecular formula | $M_r$ | Density, $d$ | $pK_{25C}$ | Useful pH range |
|---|---|---|---|---|---|
| Amido black 10B | | 616.49 | | | |
| 4-Aminobutyric acid (GABA) | | 103.12 | | 10.43 | 9.7–11.1 |
| Ammediol (2-amino-2-methyl-1,3-propanediol) | | 105.14 | | | |
| Ammonium hydroxide | | 35.05 | | | |
| Ammonium sulfate | | 132.13 | | | |

| Name | Structure | Molecular weight | pKa | pH range |
|---|---|---|---|---|
| APS (Ammonium peroxydisulfate) | | 228.19 | | |
| L-Arginine | | 174.20 | | |
| Asparagine | | 132.12 | 8.86 | 8.2–9.6 |
| L-Aspartic acid | | 133.10 | | |
| Barbital (Veronal) | | 184.20 | 7.98 | 7.3–8.7 |
| BES [N,N-Bis(2-hydroxyethyl)-2-aminoethanesulfonic acid] | | 213.25 | 7.09 | 6.4–7.8 |
| BICINE | | 163.17 | 8.26 | 7.6–9.0 |

(continued)

(continued)

| Reagent | Molecular formula | $M_r$ | Density, $d$ | $pK_{25C}$ | Useful pH range |
|---|---|---|---|---|---|
| BIS (Bisacrylamide, Diacrylamide) | | 127.13 | | | |
| BISTRIS | | 209.24 | | 6.46 | 5.8–7.2 |
| Boric acid | | 61.83 | | 9.24 | 8.5–9.9 |
| Bromophenol blue Na salt | | 691.94 | | | |

| | | | |
|---|---|---|---|
| Calcium lactate | | 218.22 | |
| Carrier ampholytes, pH = 3-10 | | | |
| CHES | | 207.29 | 9.30  8.6–10.0 |
| Coomassie brilliant blue R-250 | | 825.97 | |
| Disodium hydrogen phosphate | | 141.96 | |
| 1,4-**Dithiothreitol** (DTT) | | 154.24 | |

(continued)

(continued)

| Reagent | Molecular formula | $M_r$ | Density, $d$ | $pK_{25C}$ | Useful pH range |
|---|---|---|---|---|---|
| EDTA.Na₂ | | 336.21 | | | |
| Ethanol | | 46.07 | 0.79 | | |
| Ethanolamine | | 62.08 | 1.02 | 9.50 | 8.8–10.2 |
| Ethidium bromide | | 394.32 | | | |
| Formaldehyde 37% | | 30.03 | 1.09 | | |
| Formic acid | | 46.03 | 1.22 | 3.75 | |
| L-Glutamic acid | | 147.13 | | | |

| Name | MW | pKa | pH range |
| --- | --- | --- | --- |
| Glutardialdehyde 25% | 100.12 | 1.06 | |
| Glycerol 87% | 92.09 | 1.23 | |
| Glycine | 75.07 | 9.78 | 9.1–10.5 |
| Glycylglycine | 132.12 | 8.25 | 7.6–9.0 |
| HEPES | 238.30 | 7.48 | 6.8–8.2 |
| HEPPS | 252.33 | 8.00 | 7.3–8.7 |
| L-Histidine | 155.16 | 8.97 | 8.3–9.7 |
| Hydrochloric acid 37% | 36.46 | 1.18 | |
| Imidazole | 68.08 | 6.95 | 6.2–7.8 |
| Immobilines, pH = 4-10 | | | |
| 2-Iodoacetamide | 184.96 | | |

(continued)

(continued)

| Reagent | Molecular formula | $M_r$ | Density, $d$ | $pK_{25C}$ | Useful pH range |
|---|---|---|---|---|---|
| Lactic acid | | 90.08 | | 3.84 | |
| MES | | 195.23 | | 6.16 | 5.5–6.9 |
| Methanol | | 32.04 | 0.79 | | |
| Methylparaben (Nipagin, methyl 4-hydroxybenzoate) | | 152.15 | | | |
| MOPS | | 209.26 | | 7.20 | 6.5–7.9 |
| Naphthol blue black – see Amido black 10B | | | | | |
| Nitric acid | | 63.01 | 1.40 | | |
| Phosphoric acid | | 98.00 | 1.69 | 2.15 | |

| Name | Structure | M | | |
|---|---|---|---|---|
| PIPES | | 302.36 | 6.76 | 6.1–7.5 |
| Potassium dihydrogen phosphate | | 136.08 | 7.20 | 6.5–7.9 |
| Potassium hydroxide | | 56.11 | | |
| L-Proline | | 115.13 | | |
| Pyridine | | 79.10 | 5.21 | 4.5–5.9 |
| Pyronin Y (G) | | 302.80 | | |
| SDS | | 288.38 | | |
| Silver nitrate | | 169.87 | | |

(continued)

(continued)

| Reagent | Molecular formula | $M_r$ | Density, $d$ | $pK_{25C}$ | Useful pH range |
|---|---|---|---|---|---|
| Sodium acetate | | 82.03 | | | |
| Sodium carbonate | | 105.99 | | | |
| Sodium dihydrogen phosphate | | 119.98 | | | |
| Sodium hydroxide | | 40.00 | | | |
| Sodium thiosulfate | | 158.10 | | | |
| D-Sorbitol | | 182.17 | | | |
| 5-Sulfosalicylic acid | | 254.21 | | | |

| Name | Structure | MW | pKa | pH range |
|---|---|---|---|---|
| Sulfuric acid | | 98.07 | | |
| Taurine | | 125.14 | 9.06 | 8.4–9.8 |
| Trichloroacetic acid | | 163.38 | 1.62 | |
| TES | | 229.25 | 7.55 | 6.9–8.3 |
| TMEDA | | 116.21 | | |
| TRICINE | | 179.17 | 8.09 | 7.4–8.8 |
| Triethanolamine | | 149.19 | 7.74 | |
| TRIS | | 121.14 | 8.07 | 7.4–8.8 |

(continued)

(continued)

| Reagent | Molecular formula | $M_r$ | Density, $d$ | $pK_{25C}$ | Useful pH range |
|---|---|---|---|---|---|
| Triton X-100 | | 250.38 | 1.07 | | |
| Tween 20 (Polysorbate 20) | | 1227.54 | | | |
| Urea | | 60.06 | | | |
| Water | | 18.02 | 1.00 | 15.74 | |

# Recipes for electrophoresis solutions

**Table R-1:** Molarities and relative density of concentrated acids and bases.

| Protolyte | $M_r$ | % by mass | Relative density, d |
|---|---|---|---|
| *Acids* | | | |
| HCOOH | 46.03 | 90 | 1.21 |
| | | 98 | 1.22 |
| $CH_3COOH$ | 60.05 | 99.6 | 1.05 |
| $H_3PO_4$ | 98.00 | 85 | 1.70 |
| HCl | 36.46 | 36 | 1.18 |
| $H_2SO_4$ | 98.07 | 98 | 1.835 |
| $HNO_3$ | 63.01 | 70 | 1.42 |
| *Bases* | | | |
| $NH_4OH$ | 35.0 | 28 | 0.90 |
| NaOH | 40.0 | 50 | 1.53 |
| KOH | 56.11 | 45 | 1.45 |
| | 56.11 | 50 | 1.51 |

## Buffers

*Electrophoresis buffer, pH = 3.5*

| | |
|---|---|
| Pyridine | 0.50 mL |
| Acetic acid | 5.00 mL |
| Deionized water to | 100.00 mL |

Check pH. Store at room temperature.

*Electrophoresis buffer, pH = 4.7*

| | |
|---|---|
| Pyridine | 2.50 mL |
| Acetic acid | 2.50 mL |
| *n*-Butanol | 5.00 mL |
| Deionized water to | 100.00 mL |

Check pH. Store at room temperature.

*Sodium-acetate buffer, 4x, pH = 5.0*

| | |
|---|---|
| Sodium acetate·$3H_2O$ | 2.72 g (0.2 mol/L) |
| Acetic acid | 1.16 mL (0.2 mol/L) |
| Deionized water to | 100.00 mL |

https://doi.org/10.1515/9783110761641-042

*Sodium acetate (SA) buffer, 3 mol/L, pH = 6.0*
Na·acetate·H$_2$O                    40.82 g (3 mol/L Na·acetate)
   *Titrate with 3 mol/L acetic acid [17.24 mL acetic acid to 100 mL*
   *deionized water (3 mol/L)] to pH = 6.0*
Deionized water to                    100.00 mL
   *Store at room temperature up to 6 months.*

*Electrophoresis buffer, pH = 6.5*
Pyridine                              10.00 mL
Acetic acid                            0.40 mL
Deionized water to                    100.00 mL
   *Check pH. Store at room temperature.*

*Hydrogen phosphate buffer, 0.01 mol/L*
Prepare 800 mL of deionized water in a suitable container.
Add sodium phosphate dibasic heptahydrate (Na$_2$HPO$_4$·7H$_2$O, $M_r$ = 268.07).
Add sodium phosphate monobasic monohydrate (NaH$_2$PO$_4$·H$_2$O, $M_r$ = 137.99).
Adjust solution to desired pH using HCl or NaOH.
Add deionized water to 1,000 mL.

**pH = 6.6**
Na$_2$HPO$_4$·7H$_2$O                    10.76 g (0.04 mol/L)
NaH$_2$PO$_4$·H$_2$O                      8.26 g (0.06 mol/L)
Deionized water to                    1,000.00 mL

**pH = 7.0**
Na$_2$HPO$_4$·7H$_2$O                    15.48 g (0.06 mol/L)
NaH$_2$PO$_4$·H$_2$O                      5.83 g (0.04 mol/L)
Deionized water to                    1,000.00 mL

**pH = 7.4**
Na$_2$HPO$_4$·7H$_2$O                    20.21 g (0.08 mol/L)
NaH$_2$PO$_4$·H$_2$O                      3.39 g (0.02 mol/L)
Deionized water to                    1,000.00 mL

*Hydrogen phosphate buffer, 0.5 mol/L, pH = 6.8*
Na$_2$HPO$_4$                            3.55 g
NaH$_2$PO$_4$                            3.45 g
Deionized water to                    100.00 mL

*Sodium chloride-sodium citrate (SSC) buffer, 20x, pH = 7.0*
Na$_2$citrate·2H$_2$O                    8.80 g (0.3 mol/L)
*Adjust pH to 7.0 with 1 mol/L HCl.*
NaCl                                  17.50 g (3 mol/L)
Deionized water to                    100.00 mL

*Phosphate-**b**uffered **s**aline (PBS), 10x, pH = 7.3*

| | |
|---|---|
| $Na_2HPO_4 \cdot 7H_2O$ | 1.15 g (4.3 mmol/L) |
| $KH_2PO_4$ | 0.20 g (1.4 mmol/L) |
| NaCl | 8.00 g (137 mmol/L) |
| KCl | 0.20 g (2.7 mmol/L) |
| Deionized water to | 100.00 mL |

*PBS-Tween*

| | |
|---|---|
| PBS (see there) | |
| Tween-20 | 0.50 mL |
| Deionized water to | 100.00 mL |

*Store at room temperature.*

*T**RIS-E**DTA-**Na** (TEN) buffer, pH = 7.4*

| | |
|---|---|
| TRIS | 0.61 g (0.05 mol/L) |

*Add 1 mol/L HCl to pH = 7.4.*

| | |
|---|---|
| NaCl | 0.88 g (0.15 mol/L) |
| $Na_2EDTA$ | 3.36 g (0.01 mol/L) |
| Deionized water to | 100.00 mL |

*Store at room temperature up to 6 months.*

*T**RIS-E**DTA-**S**DS (TES) buffer, pH = 7.5*

| | |
|---|---|
| TRIS | 0.12 g (0.01 mol/L) |

*Add 1 mol/L HCl to pH = 7.5.*

| | |
|---|---|
| $Na_2EDTA$ | 3.36 g (0.01 mol/L) |
| SDS | 0.50 g (0.017 mol/L) |
| Deionized water to | 100.00 mL |

*Store at room temperature.*

*Low-**s**alt **b**uffer (LSB), pH = 7.5*

| | |
|---|---|
| HEPES | 0.48 g (0.02 mol/L) |
| NaCl | 0.58 g (0.10 mol/L) |

*Store at 4 °C up to several weeks. Before use add:*

| | |
|---|---|
| 2-Mercaptoethanol | 0.01 mL (1.0 mmol/L) |
| PMSF | 0.09 g (0.5 mmol/L) |
| Deionized water to | 100.00 mL |

*T**RIS-b**uffered **s**aline (TBS), pH = 7.5*

| | |
|---|---|
| TRIS | 1.21 g (0.10 mol/L) |

*Add 1 mol/L HCl to pH = 7.5.*

| | |
|---|---|
| NaCl | 0.88 g (0.15 mol/L) |
| Deionized water to | 100.00 mL |

*Store at 4 °C up to several months.*

*Sodium chloride-TRIS-EDTA (STE) buffer, pH = 8.0*

| | |
|---|---|
| TRIS | 0.12 g (10 mmol/L) |
| *Add 1 mol/L HCl to pH = 8.0.* | |
| $Na_2EDTA$ | 0.03 g (1 mmol/L) |
| NaCl | 0.58 g (0.1 mol/L) |
| Deionized water to | 100.00 mL |

*TRIS-borate-EDTA (TBE) buffer, 10x, pH = 8.0*

| | |
|---|---|
| TRIS | 108.00 g (0.89 mol/L) |
| Boric acid | 55.00 g (0.89 mol/L) |
| *Add 1 mol/L HCl to pH = 8.0.* | |
| $Na_2EDTA$ | 6.72 g (0.02 mol/L) |
| Deionized water to | 1,000.00 mL |

*TRIS-buffered saline with Triton X-100 (TBST), pH = 8.0*

| | |
|---|---|
| TRIS | 0.12 g (0.01 mol/L) |
| *Add 1 mol/L HCl to pH = 8.0.* | |
| NaCl | 0.88 g (0.15 mol/L) |
| Triton X-100 | 0.05 mL |
| Deionized water to | 100.00 mL |

*Store at room temperature up to 6 months.*

*TRIS-EDTA (TE) buffer, pH = 8.0*

| | |
|---|---|
| TRIS | 0.12 g (0.01 mol/L) |
| *Bring with HCl to pH = 8.0.* | |
| $Na_2EDTA$ | 0.03 g (1 mmol/L) |
| Deionized water to | 100.00 mL |

*Triethanolamine (TEA) buffer, pH = 8.0*

| | |
|---|---|
| Triethanolamine | 1.33 g (0.1 mol/L) |
| Deionized water | 80.00 mL |
| *Adjust pH to 8.0 with 1 mol/L HCl* | |
| Deionized water to | 100.00 mL |

*Prepare fresh daily.*

*TRIS-borate-EDTA (TBE) buffer, 5x, pH = 8.3*

| | |
|---|---|
| TRIS | 5.45 g (0.45 mol/L) |
| Boric acid | 2.78 g (0.45 mol/L) |
| $Na_2EDTA$ | 0.17 g (0.005 mol/L) |
| Deionized water to | 100.00 mL |

*Store at room temperature up to several weeks.*

*TRIS-acetate-EDTA (TAE) buffer, 50x, pH = 8.5*

| | |
|---|---|
| TRIS | 24.20 g (0.2 mol/L) |
| Acetic acid | 5.72 mL (0.1 mol/L) |
| $Na_2EDTA$ | 3.36 g (0.1 mol/L) |
| Deionized water to | 100.00 mL |

## Solutions for agarose gel electrophoresis

*Agarose gel, 1 g/dL*

| | |
|---|---|
| Agarose | 1.00 g |
| Ethidium bromide | 0.05 g |
| Gel buffer to | 100.00 mL |

*Melt for several minutes.*

*Sodium-citrate buffer for acidic agarose electrophoresis of hemoglobins, 10x, pH = 6.0,*
*I = 10 x 0.08 mol/L*

**Solution A (0.2 mol/L Na₃ citrate)**

| | |
|---|---|
| $Na_3$·citrate | 5.88 g (0.23 mol/L) |
| $NaN_3$ | 0.02 g (0.003 mol/L) |
| Deionized water to | 100.00 mL |

**Solution B (0.2 mol/L citric acid)**

| | |
|---|---|
| Citric acid | 4.20 g (0.22 mol/L) |
| $NaN_3$ | 0.02 g (0.003 mol/L) |
| Deionized water to | 100.00 mL |

| | |
|---|---|
| Solution A | 41.50 mL |
| Solution B | 9.50 mL |
| $NaN_3$ | 0.02 g (0.003 mol/L) |
| Deionized water to | 100.00 mL |

*TRIS-borate-EDTA buffer for agarose electrophoresis, 10x, pH = 8.6, I = 10 x 0.012 mol/L*

| | |
|---|---|
| TRIS | 5.40 g (0.045 mol/L) |
| $H_3BO_3$ | 1.55 g (0.025 mol/L) |
| $Na_2EDTA \cdot 2H_2O$ | 0.29 g (0.008 mol/L) |
| $NaN_3$ | 0.10 g (0.015 mol/L) |
| Deionized water to | 100.00 mL |

*Electrode buffer for alkaline agarose electrophoresis of hemoglobins and lipoproteins, 10x, pH = 8.8, I = 10 x 0.04 mol/L*

| TRIS | 3.15 g (0.3 mol/L) |
|---|---|
| NaOH | 1.48 g (0.4 mol/L) |
| Taurine | 11.60 g (0.9 mol/L) |
| $NaN_3$ | 0.65 g (0.10 mol/L) |
| Deionized water to | 100.00 mL |

*TRIS-barbitalate buffer for agarose electrophoiresis, 10x, pH = 8.9*

| TRIS | 6.06 g (0.50 mol/L) |
|---|---|
| Barbital | 2.58 g (0.14 mol/L) |
| Sodium barbitalate | 10.31 g (0.50 mol/L) |
| Deionized water to | 100.00 mL |

*Electrode buffer for agarose electrophoresis, 10x, pH = 9.0, I = 10 x 0.09 mol/L*

| Ammediol | 18.34 g (1.74 mol/L) |
|---|---|
| BICINE | 17.40 g (1.07 mol/L) |
| $NaN_3$ | 0.65 g (0.10 mol/L) |
| Deionized water to | 100.00 mL |

*TRIS-taurinate-NaOH buffer for agarose electrophoresis, 10x, pH = 9.1, I = 10 x 0.06 mol/L*

| TRIS | 1.82 g (0.15 mol/L) |
|---|---|
| NaOH | 2.33 g (0.58 mol/L) |
| Taurine | 12.17 g (0.49 mol/L) |
| $NaN_3$ | 0.65 g (0.10 mol/L) |
| Deionized water to | 100.00 mL |

*TRIS-barbitalate-NaOH buffer for agarose electrophoresis, 10x, pH = 9.2, I = 10 x 0.06 mol/L (with $Na_2EDTA$)*

| TRIS | 7.20 g (0.60 mol/L) |
|---|---|
| Sodium barbitalate | 10.30 g (0.25 mol/L) |
| [8.46 Barbital + 1.84 NaOH] | |
| Barbital | 1.84 g (0.10 mol/L) |
| $Na_2EDTA$ | 3.72 g (0.10 mol/L) |
| $NaN_3$ | 1.00 g (0.03 mol/L) |
| Deionized water to | 100.00 mL |

## Solutions for affinity electrophoresis

***Al****kaline **p**hosphatase (ALP) buffer, pH = 9.5*
| | |
|---|---|
| TRIS | 1.21 g (0.1 mol/L) |

*Add 1 mol/L HCl to pH = 9.5.*
| | |
|---|---|
| NaCl | 0.58 g (0.1 mol/L) |
| $MgCl_2$ | 0.48 g (0.05 mol/L) |
| Tween 20 | 0.10 mL |
| Deionized water to | 100.00 mL |

*Store at 4 °C up to 3 months.*

***B****ES-**b**uffered **s**olution (BBS), 2x, pH = 6.95*
| | |
|---|---|
| BES | 1.07 g (0.05 mol/L) |
| $Na_2HPO_4$ | 0.02 g (0.0015 mol/L) |
| NaCl | 1.64 g (0.28 mol/L) |
| Deionized water | 80.00 mL |

*Adjust pH to 6.95 with 1 mol/L NaOH*
| | |
|---|---|
| Deionized water to | 100.00 mL |

*Store at −20 °C.*

*5-**B**romo-4-**c**hloro-3-**i**ndolyl **p**hosphate (BCIP) solution*
| | |
|---|---|
| BCIP | 0.50 g |
| Dimethylformamide | 10.00 mL |

*Store at −20 °C in aliquots.*

*4-**N**itro **b**lue **t**etrazolium (NBT) buffer, pH = 9.2*
| | |
|---|---|
| TRIS | 0.12 g (0.01 mol/L) |

*Add 80 mL deionized water and adjust pH to 9.2 with 1 mol/L HCl.*
| | |
|---|---|
| NBT chloride | 0.41 g (0.005 mol/L) |
| BCIP | 0.20 g (0.006 mol/L) |
| $MgCl_2$ | 0.56 g (0.059 mol/L) |
| Deionized water to | 100.00 mL |

*NBT-BCIP substrate buffer, pH = 9.5*
| | |
|---|---|
| ALP buffer (see there) | 50.00 mL |
| NBT buffer (see there) | 0.22 mL |
| BCIP solution (see there) | 0.17 µL |

*Prepare fresh.*

*Developing solution*
| | |
|---|---|
| NBT stock (see there) | 66.00 µL |
| ALP buffer (see there) | 10.00 mL |
| BCIP stock (see there) | 33.00 µL |

## Solutions for native disc-electrophoresis

*Acrylamide/bisacrylamide (T30, C0.03)*

| | |
|---|---|
| Acrylamide | 29.00 g |
| Bisacrylamide | 1.00 g |
| Deionized water to | 100.00 mL |

    *Wear gloves when working.*

*Acrylamide-BIS solution (T50, C0.03)*

| | |
|---|---|
| Acrylamide | 48.50 g |
| BIS | 1.50 g |
| Deionized water to | 100.00 mL |

*Buffers for disc-electrophoresis according to Michov*

*Resolving buffer, 10x, pH = 6.5*
**[Electrode buffer *a* (+)]**

| | |
|---|---|
| TRIS | 19.64 g (1.62 mol/L) |
| Formic acid | 6.04 mL (1.60 mol/L) |
| 87% Glycerol | 10.00 mL |
| NaN$_3$ | 0.10 g (0.015 mol/L) |
| Deionized water to | 100.00 mL |

*Electrode buffer, 10x, pH = 7.4*
**[Electrode buffer *b* (−)]**

| | |
|---|---|
| TRIS | 2.61 g (0.22 mol/L) |
| BICINE | 21.98 g (1.35 mol/L) |
| 87% Glycerol | 10.00 mL |
| 0.1 g/dL NaBPB | 0.69 mL |
| NaN$_3$ | 0.10 g (0.015 mol/L) |
| Deionized water to | 100.00 mL |

*Sample buffer, 2x, pH = 6.5, 2 x 0.20 mol/L*

| | |
|---|---|
| Electrode buffer *a* | 25.00 mL |
| Na$_2$EDTA | 0.07 g (0.002 mol/L) |
| 1 g/l NaBPB | 2.77 mL (0.04 mmol/L) |
| 87% Glycerol | 20.00 mL |
| Deionized water to | 100.00 mL |

## Solutions for SDS disc-electrophoresis

*Nonreducing sample buffer for SDS electrophoresis, pH = 6.5, 0.20 mol/I*

| | |
|---|---|
| Electrode buffer *a* | 12.50 mL (0.20 mol/L) |
| SDS | 1.13 g (0.04 mol/L) |
| 1 g/dL NaBPB | 1.38 mL (0.02 mmol/L) |
| 87% Glycerol | 10.00 mL |
| Deionized water to | 100.00 mL |

*Reducing sample buffer, 2x for SDS electrophoresis, pH = 6.5, 2 x 0.2 mol/L*

| | |
|---|---|
| Electrode buffer *a* | 25.00 mL (0.40 mol/L) |
| SDS | 2.27 g (0.08 mol/L) |
| 87% Glycerol | 20.00 mL |
| 1,4-Dithiothreitol | 0.31 g (0.02 mol/L) |
| 1 g/l NaBPB | 2.77 mL (0.04 mmol/L) |
| Deionized water to | 100.00 mL |

*SDS sample buffer, 2x, pH = 6.8*

| | |
|---|---|
| TRIS | 4.00 mL (0.125 mol/L) |

*Add 1 mol/L HCl to pH = 6.8.*

| | |
|---|---|
| Glycerol | 2.00 mL |
| $Na_2EDTA$ | 0.02 g (0.006 mol/L) |
| SDS | 0.40 g (0.14 mol/L) |
| 2-Mercaptoethanol | 0.40 mL (0.57 mol/L) |

*or*

| | |
|---|---|
| 0.2 mol/L DTT | 0.40 mL (0.008 mol/L) |
| NaBPB | 0.02 g |
| Deionized water to | 10.00 mL |

   *Filter and store at −20 °C up to 6 months.*

*Alkylating solution for SDS electrophoresis, 2x, for SDS-electrophoresis*

| | |
|---|---|
| 2-Iodacetamide | 3.70 g (0.20 mol/L) |
| Deionized water to | 100.00 mL |

*Buffers for SDS electrophoresis according to Michov*

*Resolving buffer, 10x, pH = 6.5*

**[SDS electrode buffer *a* (+)]**

| | |
|---|---|
| TRIS | 17.68 g (1.46 mol/L) |
| Formic acid | 5.44 mL (1.44 mol/L) |
| SDS | 0.23 g (0.008 mol/L) |
| 87% Glycerol | 10.00 mL |
| Deionized water to | 100.00 mL |

*Electrode buffer, 10x, pH = 7.4*

**[SDS electrode buffer *b* (–)]**

| | |
|---|---|
| TRIS | 2.36 g (0.19 mol/L) |
| BICINE | 19.78 g (1.21 mol/L) |
| SDS | 0.23 g (0.008 mol/L) |
| 87% Glycerol | 10.00 mL |
| 0.1 g/dL NaBPB | 0.69 mL |
| Deionized water to | 100.00 mL |

*SDS electrophoresis buffer, 5x, pH = 8.3*

| | |
|---|---|
| TRIS | 1.51 g (0.12 mol/L) |
| Glycine | 7.20 g (0.96 mol/L) |
| SDS | 0.50 g (0.02 mol/L) |
| Deionized water to | 100.00 mL |

*Store at 0° to 4 °C up to 1 month.*

## Solutions for IEF

*Sample solution for native IEF*

| | |
|---|---|
| Carrier ampholytes 3-10 | 1.00 mL |
| Carrier ampholytes 4-9 | 0.25 mL |
| 87% Glycerol | 40.00 mL |
| Amberlite MB-3 | 1.00 g |

*(Amberlite MB-3 deionizes the solution.)*

| | |
|---|---|
| Deionized water to | 100.00 mL |

*Sample solution for denaturing IEF*

| | |
|---|---|
| Carrier ampholytes 3-10 | 1.00 mL |
| Carrier ampholytes 4-9 | 0.25 mL |
| Urea | 48.05 g (8.0 mol/L) |
| 1,4-Dithiothreitol | 0.31 g (0.02 mol/L) |
| 87% Glycerol | 40.00 mL |
| Amberlite MB-3 | 1.00 g |

*(Amberlite MB-3 deionizes the solution.)*

| | |
|---|---|
| Deionized water to | 100.00 mL |

*Anode (+) fluid 3 for IEF*

| | |
|---|---|
| L-aspartic acid | 0.17 g (0.0125 mol/L) |
| L-glutamic acid | 0.18 g (0.0125 mol/L) |
| 87% Glycerol | 50.00 mL |
| NaN₃ | 0.02 g (0.003 mol/L) |
| Deionized water to | 100.00 mL |

*Cathode (−) fluid 10 for IEF*

| | |
|---|---|
| L-arginine | 0.22 g (0.0125 mol/L) |
| L-lysine | 0.18 g (0.0125 mol/L) |
| [0.21 g/dL L-lysine·$H_2O$] | |
| Ethylene diamine | 6.00 mL (0.9 mol/L) |
| 87% Glycerol | 50.00 mL |
| $NaN_3$ | 0.02 g (0.003 mol/L) |
| Deionized water to | 100.00 mL |

*Rehydrating buffer for native PAGE, pH = 9.0*

| | |
|---|---|
| El. buffer, 10x, pH = 9.0 | 0.25 mL |
| Carrier ampholytes | 0.40 mL |
| Carrier ampholytes 4-9 | 0.10 mL |
| Dextran 8 | 5.00 g |
| $NaN_3$ | 0.02 g (0.003 mol/L) |
| Deionized water to | 100.00 mL |

*Rehydrating buffer for denatured PAGE, pH = 9.0*

| | |
|---|---|
| El. buffer, 10x, pH = 9.0 | 0.25 mL |
| Carrier ampholytes | 0.40 mL |
| Carrier ampholytes 4-9 | 0.10 mL |
| Urea | 48.05 g (8.0 mol/L) |
| [x 2/3 = **32.03** mL] | |
| 1,4-Dithiothreitol | 0.62 g (0.04 mol/L) |
| Dextran 8 | 5.00 g |
| $NaN_3$ | 0.02 g (0.003 mol/L) |
| Deionized water to | 100.00 mL |
| *Prepare fresh.* | |

## Blotting solutions

*Transfer buffer, pH = 10.0*

| | |
|---|---|
| CAPS | 5.00 mL (0.50 mol/L) |
| Methanol | 20.00 mL |
| Deionized water to | 100.00 mL |
| *Degas before transfer.* | |

*Blocking buffer, pH = 7.4*

| | |
|---|---|
| TRIS | 0.12 g (0.01 mol/L) |

*Add 1 mol/L HCl to pH = 7.4.*

| | |
|---|---|
| BSA | 5.00 g |
| Hen ovalbumin | 1.00 g |
| NaCl | 0.88 g (0.15 mol/L) |
| $NaN_3$ | 0.02 g (0.003 mol/L) |
| Deionized water to | 100.00 mL |

  *Store at 4 °C for up to 6 months.*

*Blocking solution with BSA, pH = 7.8*

| | |
|---|---|
| TRIS | 1.21 g (0.1 mol/L) |

*Add 1 mol/L HCl to pH = 7.8.*

| | |
|---|---|
| BSA | 5.00 g |
| NaCl | 0.88 g (0.15 mol/L) |
| $NaN_3$ | 0.02 g (0.003 mol/L) |
| Deionized water to | 100.00 mL |

  *Prepare fresh.*

## Fixative solutions

*Fixative TS (**t**richloroacetic acid with 5-**s**ulfosalicylic acid), 10x*

| | |
|---|---|
| TCA | 50.00 g (3.06 mol/L) |
| SSA dihydrate | 10.00 g (0.39 mol/L |
| Deionized water to | 100.00 mL |

*Fixative A (**a**cetic acid)*

| | |
|---|---|
| Acetic acid | 7.00 mL (1.11 mol/l ) |
| Deionized water to | 100.00 mL |

*Fixative GE (**g**lutardialdehyde-**e**thanol), 10x*

| | |
|---|---|
| 25 mL/dL Glutardialdehyde | 10.00 mL (0.5 mol/L) |
| Ethanol | 50.00 mL |
| Deionized water to | 100.00 mL |

*Fixative CE (**c**itric acid-**e**thanol), 2x*

| | |
|---|---|
| Citric acid | 1.00 g (0.05 mol/L) |
| Ethanol | 60.00 mL |
| Deionized water to | 100.00 mL |

*Fixative AE (**a**cetic acid-**e**thanol)*

| | |
|---|---|
| Acetic acid | 10.00 mL (1.75 mol/L) |
| Ethanol | 40.00 mL |
| Deionized water to | 100.00 mL |

## Staining solutions

*Ponceau S staining solution*

| | |
|---|---|
| Ponceau S | 0.30 g |
| Trichloroacetic acid | 3.00 g |
| Deionized water to | 100.00 mL |

*Ponceau S – TCA–sulfosalicylic acid (PTS) staining solution, 10x*

| | |
|---|---|
| Ponceau S | 3.00 g |
| Trichloroacetic acid | 30.00 g |
| Sulfosalicylic acid | 30.00 g |
| Deionized water to | 100.00 mL |

Store at room temperature. Prior to use dilute tenfold with deionized water.

*Coomassie **b**rilliant **b**lue R-250 (CBB) solution*

| | |
|---|---|
| CBB R250 | 0.20 g |
| Methanol | 30.00 mL |
| Acetic acid | 10.00 mL |
| Deionized water | 60.00 mL |

*Coomassie **b**rilliant **v**iolet R-200 (CBV) solution*

| | |
|---|---|
| CBV R-200 | 0.20 g |
| Methanol | 30.00 mL |
| Acetic acid | 10.00 mL |
| Deionized water | 60.00 mL |

*Amido **b**lack 10B (AB) solution*

| | |
|---|---|
| Amido black 10B | 0.20 g |
| Methanol | 30.00 mL |
| Acetic acid | 10.00 mL |
| Deionized water | 60.00 mL |

*Sudan **b**lack B (SB) solution, 10x*

| | |
|---|---|
| Sudan black B | 1.00 g |
| Ethanol to | 100.00 mL |

*Bromophenol blue (BPB) solution*

| | |
|---|---|
| NaBPB· | 0.10 g |
| [0.1 g/dL BPB + 0.01 g/dL NaOH] | |
| Citric acid | 0.50 g |
| Deionized water to | 100.00 mL |

*Filter.*

## Destaining solutions

*Destaining solution A (acetic acid), 10x*

| | |
|---|---|
| $CH_3CO_2H$ | 50.00 mL |
| Deionized water to | 100.00 mL |

*Destaining solution AM (acetic acid–methanol)*

| | |
|---|---|
| Acetic acid | 10.00 mL |
| Methanol | 20.00 mL |
| Deionized water to | 100.00 mL |

*Destaining solution AE (acetic acid–ethanol), 2x*

| | |
|---|---|
| Acetic acid | 20.00 mL |
| Ethanol | 60.00 mL |
| Deionized water to | 100.00 mL |

*Destaining solution C (citric acid), 10x*

| | |
|---|---|
| Citric acid | 5.00 g |
| Deionized water to | 100.00 mL |

*Destaining solution P (phosphoric acid), 10x*

| | |
|---|---|
| Phosphoric acid | 3.00 g |
| Deionized water to | 100.00 mL |

## Silver staining solutions

*Preparing solution for silver staining, 10x*

| | |
|---|---|
| $Na_2S_2O_3\cdot5H_2O$ | 3.00 g (0.12 mol/L) |
| Deionized water to | 100.00 mL |

$[2Na_2S_2O_3 + AgBr \rightarrow Na_3[Ag(S_2O_3)_2] + NaBr]$

*Store in a brown flask.*

*Silver stain solution A (Sensitizer) for silver staining of agarose gels*

| | |
|---|---|
| AgNO$_3$ | 0.20 g (0.012 mol/L) |
| NH$_4$NO$_3$ | 0.20 g (0.024 mol/L) |
| Silicotungstic acid·aq, H$_4$[Si(W$_3$O$_{10}$)$_4$]·aq. | |
| | 1.00 g (0.003 mol/L) |
| 37 g/dL HCHO | 0.89 mL (0.12 mol/L) |
| Deionized water to | 100.00 mL |

*Silver stain solution B (Developer) for silver staining of agarose gels*

| | |
|---|---|
| Na$_2$CO$_3$ | 8.00 g (0.755 mol/L) |
| [2[Ag(NH$_3$)$_2$]OH + HCHO → 2Ag + HCOONH$_4$ + 3NH$_3$ + H$_2$O] | |
| Deionized water to | 100.00 mL |

*Silver stain kit for Paa gels, 10x*

| | |
|---|---|
| AgNO$_3$ | 1.00 g (0.006 mol/L) |
| Deionized water to | 100.00 mL |
| Store in a brown flask. | |

**Developer 1 for silver staining of Paa gels, 10x**

| | |
|---|---|
| Na$_2$CO$_3$ | 29.60 g (2.8 mol/L) |
| Deionized water to | 100.00 mL |

**Developer 2 for silver staining of Paa gels**

| | |
|---|---|
| 37 g/dL HCHO | 10.00 mL |

# Other solutions

*Electrode buffer for CA (cellulose acetate) electrophoresis, pH = 8.6, I = 0.10 mol/L*

| | |
|---|---|
| Sodium barbitalate | 2.06 g (0.10 mol/L) |
| [16.94 Barbital + 3.68 NaOH] | |
| Barbital | 0.40 g (0.02 mol/L) |
| NaN$_3$ | 0.14 g (0.02 mol/L) |
| Deionized water to | 100.00 mL |

*Denhardt's solution, 100x*

| | |
|---|---|
| Ficoll 400 | 10.00 g |
| Polyvinylpyrrolidone | 10.00 g |
| BSA (Fraction V) | 10.00 g |
| Deionized water to | 500.00 mL |
| Filter sterilize and store at −20 °C in aliquots. | |

*Urea solutions*

**9.5 mol/L urea**

| | |
|---|---|
| Urea | 28.50 g |
| Deionized water to | 50.00 mL |

**10 mol/L urea**

| | |
|---|---|
| Urea | 30.03 g |
| Deionized water to | 50.00 mL |

*Sucrose solution, 10 g/dL, pH = 7.5*

| | |
|---|---|
| TRIS | 0.24 g (0.02 mol/L) |
| Deionized water to | 80.00 mL |

*Add 1 mol/L HCl to pH = 7.5.*

| | |
|---|---|
| Sucrose | 10.00 g |
| NaCl | 5.84 g (1 mol/L) |
| EDTA | 0.17 g (0.005 mol/L) |
| Deionized water to | 100.00 mL |

*Lysis buffer, pH = 8.0*

| | |
|---|---|
| TRIS | 0.12 g (0.01 mol/L) |
| Deionized water to | 80.00 mL |

*Add 1 mol/L HCl to pH = 8.0.*

| | |
|---|---|
| NaCl | 0.82 g (0.14 mol/L) |
| $MgCl_2$ | 0.14 g (0.015 mol/L) |
| **N**onidet **P-40** (NP-40) | 0.50 mL |
| Deionized water to | 100.00 mL |

    *Store at room temperature.*

*Elution buffer, pH = 8.0*

| | |
|---|---|
| TRIS | 0.61 g (0.05 mol/L) |
| Deionized water to | 80.00 mL |

*Add 1 mol/L HCl to pH = 8.0.*

| | |
|---|---|
| $Na_2EDTA$ | 0.34 g (0.01 mol/L) |
| SDS | 1.00 g (003 mol/L) |
| Deionized water to | 100.00 mL |

    *Filter sterilize and store at room temperature.*

# SI units and physical constants used in electrophoresis

The International system of units SI (Système International) was introduced in 1960 [1–3]. It contains seven *base units* (Table 1).

**Table 1:** SI base units.

| Quantity | Abbreviation | Name of unit | SI unit |
|---|---|---|---|
| Mass | $m$ | Kilogram | kg |
| Length | $l, b, h, \delta, d, r, s, \lambda$ | Meter | m |
| Time | $t$ | Second | s |
| Thermodynamic temperature | $T$ | Kelvin | K |
| Amount of substance | $n$ | Mole | mol |
| Electric current | $I$ | Ampere | A |
| Luminous intensity | $I_V$ | Candela | cd |

The *mol* is referred to as an amount $n$ of a substance, which contains as much particles as 0.012 kg carbon isotope $^{12}C$. The particles can be of different nature: atoms, molecules, ions, or electrons. The following relationship exists between the mass $m$, in kg, and the amount, in mols, of a homogeneous substance x:

$$m = nM_x = nN_Am_x = nN_AM_rm_u \tag{1}$$

where $M_x = N_Am_x$ is the molar mass of the particle, in kg/mol; $N_A$ (6.022 045 × $10^{23}$ mol$^{-1}$) is Avogadro constant (the number of particles in 1 mol); $m_x$ is the mass of the particle x, in kg; $M_r$ is the relative mass of the particle, dimensionless; and $m_u$ is the atomic mass unit, in kg, equal to 1/12 of the mass of one atom of carbon isotope $^{12}C$.

The derived units of the SI base units and the physical constants, which are used in the electrophoresis methods, are listed in Tables 2 and 3, respectively.

https://doi.org/10.1515/9783110761641-043

**Table 2:** Derived units of the SI base units.

| Derived quantity | Abbreviation and equation | SI unit |
| --- | --- | --- |
| **Space and time** | | |
| Area | $A, S = l^2$ | $m^2$ |
| Volume | $V = l^3$ | $m^3$, L, l (liter) |
| Volume flow rate | $q_v = dV/dt$ | $m^3/s$ |
| Velocity (rate) of migration | $v = ds/dt$ | $m/s$ |
| Velocity of an ion | $v_i = ds_i/dt$ | $m/s$ |
| Mobility of an ion | $\mu_i = v_i/E$ | $m^2/(s\,V)$ |
| Absolute mobility of an ion | $\mu_{\infty i} = v_{\infty i}/E$ | $m^2/(s\,V)$ |
| Effective mobility of an ion | $\mu'_i = \alpha\mu_i$ | $m^2/(s\,V)$ |
| **Mechanics** | | |
| Density | $\rho = m/V$ | $kg/m^3$ |
| Force | $F = ma$ | N (Newton) = kg $m/s^2$ = $CV/m$ |
| Pressure | $p = F/A$ | Pa = $N/m^2$ = $J/m^3$ |
| Dynamic viscosity | $\eta = \tau\,dz/dv$ | Pa s = kg/(m s) |
| Relative viscosity | $\eta_r = \eta/\eta_s$ | Dimensionless |
| **Thermodynamics** | | |
| Energy, work, heat | $E,\ W = Fl$ | J (Joule) = Nm = Ws = CV |
| Quantity of heat | $Q$ | J |
| Power | $P = dW/dt$ | W (watt) = J/s |
| Celsius temperature | $t$ | °C (degree Celsius) |
| Joule heating | $P = UI$ | W |
| **Electricity** | | |
| Electric charge | $Q$ | C = As |
| Charge number (electrovalence) | $z$ | Dimensionless |

**Table 2** (continued)

| Derived quantity | Abbreviation and equation | SI unit |
|---|---|---|
| Electric current area density | $J = \dfrac{dI}{dS} = F\sum c_i z_i v_i$ | $A/m^2$ |
| Electric field strength (intensity) | $E = F/Q = dV/dr$ | $V/m = N/C$ |
| Electric potential | $\varphi,\ V = A/Q = Q/(4\pi\varepsilon r)$ | V (volt) $= W/A = J/C$ |
| Electrokinetic potential | $\zeta$ | V |
| Electric potential difference (voltage) | $U$ | $V = C/F$ |
| Electric energy | $W = QU$ | J |
| Electric power | $P = UI$ | $W = VA$ |
| Force acting on an electric charge | $F = QE = Q_1 Q_2/(4\pi\varepsilon r^2)$ | N |
| Electric capacity | $C = dQ/dU = \varepsilon S/d$ | F (farad) $= C/V$ |
| (Di)electric permittivity | $\varepsilon = D/E$ | F/m |
| Relative (di)electric permittivity | $\varepsilon_r = \varepsilon/\varepsilon_0$ | Dimensionless |
| Electric resistance | $R = U/I$ | $\Omega$ (Ohm) $= V/A$ |
| Electric conductance | $G = 1/R$ | S (Siemens) $= \Omega^{-1}$ |
| Electrophoretic mobility | $\mu = v/E$ | $m^2/(s\ V)$ |
| Specific conductivity | $\gamma, \sigma = \dfrac{1}{\rho} = F\sum c_i z_i \mu_i$ | S/m |
| Molar conductivity | $\Lambda_m = \dfrac{\gamma}{c} = F(z^+\mu^+ + z^-\mu^-)$ | S $m^2/$(k)mol |
| **Physical chemistry** | | |
| Amount of substance | $n,\ v$ | (k)mol |
| Molar mass | $M = m/n$ | kg/mol |
| Relative molecular mass | $M_r = M_x/m_u$ | Dimensionless |
| Molar volume concentration (molarity) | $c_B = n/V$ | $kmol/m^3$, mol/L |
| Molar mass concentration (molality) | $m_B = n/m$ | (k)mol/kg |
| Mass concentration | $c_m = m/V$ | $kg/m^3$, g/L, g/dL |

**Table 2** (continued)

| Derived quantity | Abbreviation and equation | SI unit |
|---|---|---|
| The equilibrium concentration of substance B | $[B]$ | mol/L |
| Activity coefficient of substance B | $\gamma_B$ | Dimensionless |
| Relative activity of substance B | $a_B = \gamma_B[B]$ | mol/L |
| Diffusion coefficient | $D$ | $m^2/s$ |
| Friction coefficient | $\mu = F/N$ | Dimensionless |
| Mass ionic strength (molal ionic strength) | $I_m = \dfrac{1}{2}\displaystyle\sum_{i=1}^{s} c_i z_i^{\,2}$ | mol/kg |
| Volume ionic strength (concentration ionic strength) | $I_c = \dfrac{1}{2}\displaystyle\sum_{i=1}^{s} c_i z_i^{\,2}$ | mol/L, $kmol/m^3$ |
| Dissociation degree | $\alpha$ | Dimensionless |
| Hydrogen exponent | $pH = -\log\dfrac{[H^+]}{[H^+{}_o]}$ | Dimensionless |
| Isoelectric point | $pI$, $pH(I)$ | Dimensionless |
| Persistent function of Kohlrausch | $\dfrac{v^+}{v^-} = \dfrac{z^+ c^+}{z^- c^-}$ | Dimensionless |

**Table 3:** Physical constants.

| Physical constants | Symbol and equation | Value |
|---|---|---|
| Atomic mass unit | $m_u$ | $1.6605655 \times 10^{-27}$ kg |
| Elementary charge | $e$ | $1.6021892 \times 10^{-19}$ C |
| (Di)electric constant | $\varepsilon_0$ | $8.854187818 \times 10^{-12}$ F/m |
| Avogadro constant | $N_A$ | $6.022045 \times 10^{26}$ $kmol^{-1}$ |
| Faraday constant | $F = N_A e$ | $9.648455461 \times 10^7$ C/kmol |
| Molar gas constant | $R = N_A k$ | 8,314.41 J/(kmol K) |
| Boltzmann constant | $k = R/N_A$ | $1.380662 \times 10^{-23}$ J/K |

# References

[1]     Dybkaer R. Clin Chim Acta, 1979, 96, 157F–183F.

[2]     Siggaard-Andersen O, Durst R, Maas AHJ. Pure Appl Chem, 1984, 56, 567–584.

[3]     Mills I, Cvitas T, Homann K, Kallay N, Kuchitsu K. Quantities, Units and Symbols in Physical Chemistry. Blackwell Scientific Publications, Oxford, 1988.

# Electrophoresis terms

| | |
|---|---|
| *Acid* | Proton donor. An acid gives a proton to a base. |
| *Affinity electrophoresis* | Zone electrophoresis in usually agarose gel containing specific ligands. |
| *Agarose* | Natural polymer composed of D-galactose and 3,6-L-anhydrogalactose. Agarose gel is used as separation medium in different electrophoresis methods, especially in the clinical routine. |
| *Agarose gel electrophoresis* | Zone electrophoresis on agarose gel. |
| *Allele* | Alternative form of a gene. |
| *Ampholyte* | Chemical compound having acidic and basic properties. |
| *Analytical electrophoresis* | Electrophoresis used for resolving of polyions as bands. |
| *Autoradiography* | Blackening on an X-ray film by irradiation of labeled with isotopes polyions, which are separated each from other by gel electrophoresis. |
| *Base* | Proton catcher. The base binds a proton from an acid. |
| *Blotter* | Device where transfer of proteins or nucleic acids onto a blot membrane is carried out. |
| *Blotting* | Method for transferring electrophoretically resolved bands onto a membrane where they can react with non-specific or specific reagents. Blotting is a proofing method, not an electrophoresis method. |
| *Buffer* | Solution of a base and its conjugate acid, which keeps constant pH value. |
| *Buffer capacity* | The capacity of a buffer to maintain constant pH value when the buffer is diluted or bases (acids) are added. |
| *Capillary electrophoresis* | Electrophoresis (zone electrophoresis, isotachophoresis, or isoelectric focusing) in capillaries. |
| *Carrier ampholytes* | Synthesized ampholytes that reach their isoelectric points in an electric field to form a stable pH gradient. |
| *Cellulose acetate* | Natural compound used for production of film-formed electrophoretic medium. |
| *Cellulose acetate electrophoresis* | Zone electrophoresis in cellulose acetate membranes. |
| *Densitometry* | Method for determining the number and concentration of electrophoretically separated bands by measuring their optical density using a light or laser beam. |
| *Deoxyribonucleates* | DNA polyions. |
| *Disc-electrophoresis* | Combination of isotachophoresis and zone electrophoresis, which takes place in appropriate buffer systems. |
| *DNA (deoxyribonucleic acids)* | Polynucleotides that consist of specifically arranged deoxyribonucleotide residues. They are negatively charged in a solution. |
| *DNase (deoxyribonuclease)* | Enzyme that catalyzes the hydrolysis of DNA. |
| *Effect of Hjerten* | Concentrating of polyions on the boundary between solutions with a lower and higher ionic strengths. |

https://doi.org/10.1515/9783110761641-044

(continued)

| | |
|---|---|
| *Effect of molecular sieve* | Separation of polyions in a gel, depending on its pores. |
| *Effective mobility* | Ionic mobility multiplied by the dissociation degree of the electrolyte building the ion. |
| *Electric double layer* | Electric layer formed by a charged surface and its counter-ion(s). |
| *Electroblotting* | Blotting carried out in an electric field. |
| *Electrolyte* | Chemical compound that forms ions in a solution. |
| *Electrophoresis* | Separation of dissolved electrically charged particles in an electric field. |
| *Exons* | mRNA fragments containing genetic information for proteins. In most eukaryotes and in some prokaryotes, the exons are divided by introns. |
| *Fibrillar proteins* | Usually insoluble proteins whose polypeptide chains form filaments. |
| *Fluorography* | Blackening on a photosensitive film, caused by visible light irradiation, emitted by polyions labeled with isotopes. |
| *Free-flow electrophoresis* | Electrophoresis in a flowing buffer. It can be carried out as zone electrophoresis, isotachophoresis, or isoelectric focusing. |
| *Gel* | Porous soft material. |
| *Genome* | Total genetic information contained in the nucleic acids of an organism. |
| *Globular proteins* | Soluble proteins whose polypeptide chains are folded in the space. |
| *Gradient gel* | Gel with linearly or exponentially changeable concentration. |
| *Gradient gel electrophoresis* | Electrophoresis in a gradient gel. |
| *Gradient maker* | Device for casting gradient gels. |
| *Henderson–Hasselbalch equation* | Equation expressing the relationship between the pH value, on the one hand, and the $pK_c$ value and ratio between the base and its conjugated acid, on the other. |
| *Homogeneous gel* | Gel of same concentration in all its parts. |
| *Homogeneous gel electrophoresis* | Electrophoresis in a homogeneous gel. |
| *Immunoelectrophoresis* | Zone electrophoresis on an agarose gel containing or receiving immunoglobulins. As a result, the electrophoretically resolved polyions take part in precipitation reactions. |
| *Introns* | mRNA fragments that are located between exons. |
| *Ionic strength* | Mathematical term that takes into account the concentration and electrovalence of all ions in a solution. |
| *Ionization constant* | Constant that characterizes an electrolytic reaction. It is a function of the equilibrium concentrations of the reactants and products at a certain temperature and pressure. |
| *Isoelectric focusing* | Electrophoresis in a pH gradient. The moving polyions stop at their isoelectric points. |
| *Isoelectric focusing with carrier ampholytes* | Isoelectric focusing in a pH gradient created by carrier ampholytes. |
| *Isoelectric focusing in immobilized pH gradients* | Isoelectric focusing in a pH gradient created by bound immobilines. |

(continued)

| | |
|---|---|
| Isoelectric point | pH value at a certain ionic strength and temperature, where the sum of all electric charges of a polyion is equal to zero. |
| Isotachophoresis | Electrophoresis that is carried out in a system of two or more buffers, which form a moving ionic boundary between them. |
| mRNA (messenger RNA) | Type of RNA. The information of mRNA is translated by ribosomes onto proteins. |
| Net charge | It is less (in absolute value) than the total charge due to the counter-ions, which are bound to the polyion surface, according to Stern model. |
| Oligonucleotides | Short nucleic acids, similar to oligopeptides, which serve as primers. |
| Paper electrophoresis | Zone electrophoresis on paper. |
| Persistent function of Kohlrausch | Electrochemical effect according to which the concentration of an ion depends on its mobility. If certain conditions are met, leading and trailing ions are moving with same velocity forming an ionic boundary. |
| Polyacrylamide | Polymer consisting of acrylamide and bis-acrylamide residues. The polyacrylamide gel has no electroosmosis and is the most widely used separation medium in electrophoresis. |
| Polyacrylamide gel electrophoresis | Zone electrophoresis in homogeneous polyacrylamide gels. |
| Polyion | Large particle, e.g. nucleate or proteinate, which has negative, positive, or both charges. |
| Precast gels | Polyacrylamide or agarose gels that are cast on a support film or support net prior to electrophoresis. |
| Preparative electrophoresis | Quantitative electrophoresis, during which the polyions are separated from each other into separate solutions. |
| Protein | Macromolecule that is composed of one or more polypeptide chains. It has many and different electric charges in a solution. |
| Protolyte | Electrolyte that gives or receives one or more protons in a solution. |
| Pulsed-field electrophoresis | Zone electrophoresis that is used for separation of chromosomes or other large particles in an electric pulsed-field. |
| 3'-Poly (A) | Tail of hundreds of adenosine mononucleotide residues hanging on the 3'-end of most eukaryotic mRNA. The poly(A) tails are involved in the translation control. |
| Rehydratable gel | Dried gel, usually of polyacrylamide, that may be converted into usable wet gel in an appropriate solution. |
| Relaxation | Dislocation of the electric charges of a particle in an electric field whereby its velocity slows down. |
| Restriction endonucleases | Enzymes, which are isolated from different bacterial species. They cut off specific sequences of DNA nucleates. The restriction endonucleases protect the bacteria from virus infections. DNA of the host bacterium is resistant, because it is methylated. |
| Ribonucleates | RNA nucleates. |

(continued)

| | |
|---|---|
| RNA (ribonucleic acid) | Polynucleotides that consist of specifically arranged ribonucleotide residues. They carry negative charges in a solution. |
| RNases (ribonucleases) | Enzymes that degrade RNA. |
| rRNA (ribosomal RNA) | Two types of rRNA are known in eukaryotes: 28S and 18S rRNA. The bacteria contain also two types of rRNA: 23S and 16S rRNA. |
| SDS disc-electrophoresis | Disc-electrophoresis whose buffer system contains the denaturing agent SDS. |
| SDS zone electrophoresis | Zone electrophoresis whose buffer contains the denaturing agent SDS. |
| Splicing process | Process during which intron transcripts are cut off from the RNA copy of a gene, whereby the exon transcripts are joined together. |
| Starch gel | Natural polymer that consists of D-glucose residues. |
| Starch gel electrophoresis | Zone electrophoresis in a starch gel. Today it is replaced by the electrophoresis in polyacrylamide or agarose gels. |
| Submarine electrophoresis | Agarose zone electrophoresis for separation of nucleates that is carried out under a buffer. |
| Support film | Chemically pretreated film, usually of polyester, which is used for chemical fixation of a gel. |
| Support net | Chemically pretreated net, usually of polyester, which is used for chemical fixation of a gel. |
| Total charge | Sum of all the electric charges of a polyion in an infinitely dilute solution. |
| tRNA (transfer RNA) | RNA used for transporting amino acids in the process of protein synthesis. |
| Two-dimensional electrophoresis | Combination between isoelectric focusing with carrier ampholytes or in immobilized pH gradients, and SDS disc-electrophoresis. |
| Zone electrophoresis | Electrophoresis that is run in one buffer, hence at continuous electric field strength. |
| Zwitterion | Ion that carries a positive and a negative charge. |

# Index

absolute mobility  28–29, 32, 34–36
absorption of nucleic acids  391
ACES  447
ACESate ion  34
acetate ion  35
acetic acid  74–75, 290, 293–294, 298–299,
        303, 447
N-acetylneuraminic acid  121
AcrylAide  42
acrylamide  41–43, 45–46, 447
adsorption layer  7, 9–10
affinity electrophoresis  64, 119–120, 122–123,
        125–126
affinity-trap electrophoresis  119, 125
agarobiose  39
agarose  39, 447
agarose gel  77–78, 83, 93–95, 98–106
agarose gel electrophoresis  63, 343
agarose gel electrophoresis of nucleic
        acids  325
β-alanine  447
L-alanine  447
alaninium ion  34
alkaline phosphatase  119–122, 390
Amido black 10B  291, 298, 448, 454
Amido black staining of proteins  291
4-aminobutirate ion (GABAate ion)  34
4-Aminobutyric acid (GABA)  448
ammediol  448
ammediolium ion  34
ammonium hydroxide  448
ammonium sulfate  448
amplified fragment length polymorphism  345
AMPro  122, 128
analbuminemia  80
anaphoresis  1
annealing  319
APS (ammonium peroxydisulfate)  449
L-arginine  449
L-aspartic acid  449
asparagine  449
automated electrophoresis system  59
autoradiography  236, 239, 290, 297, 389, 392

BAC  43
barbital (veronal)  449
barbitalate (veronalate) ion  34

Becher glasses  55
Bence Jones proteins  115
BES  449
BICINE  449
BICINEate ion  35
biological buffers  24
BIS (bisacrylamide)  42, 450
bisalbuminemia  80
BISTRIS  450
BISTRIS ion  34
blocking  271, 276, 279, 281, 286, 375, 378
blocking reagents  275
blot membrane  271–278, 281, 284–285
blotter  55, 59
blotting  271–276, 278–280, 282–283,
        375, 379
blue native electrophoresis  139, 142
bone ALP isoenzyme  121
borate ion  34
boric acid  450
bridge-type elution  243, 245
Bromocresol green  2
Bromophenol blue  2, 450
buffer capacity  20–21, 23
buffer recirculator  58
buffer-gel system  132–133, 155, 173
buffers  19–23, 317–318

CA films  39
calcium ion  34
calcium lactate  451
capillaries  349–352
capillary electrophoresis  62, 217, 224,
        349–350, 352–353
capillary isoelectric focusing  62
capillary isotachophoresis  62
capillary transfer  272, 274, 375–377
carrier ampholytes  65, 187, 189–195,
        199–201, 451
casting cassettes  55, 57
cataphoresis  1
Cellogel  74
cellulose  350–351, 353
cellulose acetate  39
cellulose acetate electrophoresis  63, 73–74
ceramic  56
cerebrospinal fluid  78, 94, 104

Chapman 30
charge free agarose 41
charge shift electrophoresis 126
charged particles 1
chemiluminescence detection 276
CHES 451
chloride ion 35
colorimetric detection 276
combined paternity index 331
computer 55, 57
continuous semidry buffer 274
contour-clamped homogeneous electric
    fields 345
coomassie brilliant blue 3
coomassie brilliant blue (CBB) R-250 292
coomassie brilliant blue G-250 292–293, 298,
    303–304
Coomassie brilliant blue staining of
    proteins 292–293, 302
copolymerization 43–46
counterion dye staining 385
counterionic atmosphere 36
counter-ions 6, 8
covalent binding of proteins 264
creatine kinase 95
critical micelle concentration
    161
cross-linked gels 351
cross-linking degree C 42
cylindrical gels 56

denaturation 319
denatured DNA 16
densitometers 55–56, 301
densitometry 299
destaining 298, 305–306
detection 222–223, 226
detector 350, 352
developing 294–295, 304, 386, 393
DHEBA 42–43
diazobenzyloxymethyl paper 272
diazophenylthioether paper 272
dielectro phoresis 66
diffusion 53
diffusion layer 9
diffusion transfer 375–376
dihydrogen phosphate ion 34
dioxane 74–75
direct coupling 391

disc-electrophoresis 151, 153, 156–157, 173,
    176, 262
disc-electrophoresis of double-stranded PCR
    products 361
disc-electrophoresis of native nucleic
    acids 360
discontinuous buffer system 151, 167, 171
discontinuous conductivity gradient
    elution 243, 245
disc-semidry blotting 274
dissociation degree 152–154, 169
1,4-dithiothreitol (DTT) 451
DNA 1, 3
DNase footprinting assay 332, 336
drying of gels 289, 291, 295,
    299, 386
dynamic capillary coating 220
dynamic viscosity 27, 29, 35
dynamic wall coating 258

Eastern blotting 280
EDTA.Na$_2$ 452
effective mobility 28
electric charge 5–6
electric field strength 327
electroblotting 377
electroelution 243
electrokinetic ion 9–10
electrokinetic potential 6, 9, 11, 28, 30
electrokinetic radius 9–11, 50
electrolyte solutions 28
electron charge 35
electroosmosis 41
electrophoresis 1–4, 27
electrophoresis cell 73–75
electrophoresis conditions 359
electrophoretic cell 55–56
electrophoretic mobility shift assay
    363–364
electrotransfer 272, 375, 377
elongation 319
elution by diffusion 243
elution by gel dissolving 243
EMSA 363, 366–368
equation of Robinson–Stokes 32–33
Erlenmeyer flasks 55
ethanol 452
ethanolamine 452
ethidium bromide 327, 339, 387–388, 391

Far-Western blotting 278–279
Ferguson plots 137
fibrinogen 83–84
field-inversion gel electrophoresis 345
final elongation 320
final hold 320
first direction 231
fixing 289–291, 294, 303–305, 386
fixing of proteins 290
fluorescent detection 276–277
fluorescent substances 385, 388
fluorography 236, 290, 298
footprint 336
formaldehyde 17, 452
formamide 17
formate ion 35
formic acid 452
free-flow electrophoresis 62, 209–210,
    213–215
free-flow isoelectric focusing 210, 212, 214
free-flow isotachophoresis 210–211, 214
free-flow zone electrophoresis 211, 214

gel concentration 327
genome mapping 345
geometric radius 9, 11
glass fiber paper 71
$\alpha_1$-globulins 81
$\alpha_2$-globulins 82
$\beta$-globulins 78, 82–83
$\gamma$-globulins 84
electroosmosis 52
glomerular proteinuria 98
L-glutamic acid 452
glutardialdehyde 290, 303, 453
glycinate ion 35, 147
glycine 453
glycylglycinate ion 34
glycylglycine 453
glyoxal 17
Gouy 30
gradient gel 135, 141
gradient gels 46–47, 131, 135, 138–140
gradient maker 55, 57–58
gradient polyacrylamide electrophoresis 136
gradient polyacrylamide gels 134

hemoglobin electrophoresis 89, 93
hemoglobins 3, 82–83, 89–90, 93, 102, 104

hemopexin 83
Henry's function 11, 30, 50
HEPES 453
HEPESate ion 34
HEPPS 453
high-strength agarose 41
L-histidine 453
Hjerten 40, 42
horizontal electrophoresis 46, 61, 131
horizontal-type elution 243–244
horseradish peroxidase (HRP 390
Hückel 29–31
Hückel equation 29
Huntington disease 353
hydrochloric acid 453
hydrogen phosphate ion 34
hydrogen phosphate buffer 22
hydronium ion 34
hydroxide ion 35
hyperhidrosis 401
hyperkeratosis 402
hyperlipoproteinemias 87

IgA 84–85, 95, 98
IgE 84
IgG 84–85, 94–95, 98
IgM 84–85, 98
imidazole 453
imidazolium ion 34
immobilines 65, 195–197, 204, v
immobilized pH gradients 65
immune complexes 226
immunoblotting 277, 285
immunodiffusion electrophoresis according to
    Grabar and Williams 110
immunoelectrophoresis 56, 63, 109, 262
immunofixation 63, 110–112, 115
immunoglobulin subtypes 225
immunoglobulins 109
immunoprecipitation 113, 116
immunoprinting 110–111
India ink 291, 293
indirect coupling 390
inhibitors 46
initialization 319
initiator-catalyst systems 43
injection 222
intermediate-density lipoproteins 86
2-iodoacetamide 453

ionic atmosphere 5–8, 9, 10, 11, 28–31
ionic boundary (function) of Kohlrausch 64
ionic product of water 21
ionic strength 50–51
ionic strength of buffer 326
iontophoresis 64, 397–402
iontophoresis device 399
IPG-Dalt 233
isobutanol 74–75
ISO-DALT 233
isoelectric focusing 61, 62, 65–66, 187, 189,
     195, 199, 201
isoelectric point 8
isoenzyme 226
Isotachophoresis 64, 147, 262

Joule heating 53

Kohlrausch regulating function 147, 149

lactate dehydrogenase 96
lactate ion 35
lactic acid 454
lectins 119
lipoproteins 80–83, 85–87, 98, 100, 102
liver ALP isoenzyme 120–121
low-density lipoproteins 86
low-melting agarose 40

macro ALP isoenzyme 121
MALDI-based techniques 266
mass spectroscopy 265
Maxam–Gilbert sequencing method 332
McLellan buffers 133
MES 454
MESate ion 34
messenger RNA (mRNA) 332, 339
methanol 74–75, 290, 292–293, 298,
     303–304, 454
methylene blue 2
methylmercury hydroxide 17
methylparaben 454
microchip 253, 256–257, 259, 261–265
microelectrodes 67
microfluidic chip 253, 265
microfluidic electrophoresis 371
microwave bonding 260
Middle-Eastern blotting 378–379, 381
mitochondrial DNA 330

mitochondrial DNA 346
mitochondrial genome 330
mixing chamber 57
mobility shift electrophoresis 126
model of Gouy–Chapman 6
model of Helmholtz 6
model of Stern 8
MOPS 454
moving ionic boundary 52
multiple sclerosis 84, 94
muscle spasm 402
myasthenia gravis 401

native disc-electrophoresis 65
native DNA 16–17
native nucleic acids 16
native proteins 15
native zone electrophoresis 357, 359
nitric acid 454
nitrocellulose membrane 272, 274, 276, 279,
     283–284
non-covalent binding of proteins 264
non-cross-linked (flowable) gels 351
Northern blotting 378–381
Northwestern blotting 278–279
nuclease protection assay 332, 337–338
nucleates 16–17
nucleic acid electrophoresis 317
nucleic acids 13, 15–17, 27, 325–327, 332
nylon 272

Onsager 29, 31–33

paper electrophoresis 71
parameter of Debye–Hückel 7, 50
parametric equation 29, 33
paternity or maternity testing 330
PC 254–255
PDMS 254–255, 257, 260
permanent capillary coating 220
permanent wall coating 259
persistent function 147
pH 14–15, 17
pherograms 2
phosphodiester bond 15
phosphoric acid 454
physicochemical factors 398
PIPES 455
placental ALP isoenzyme 121

planar warts 402
plasminogen 83
Plexiglas 56
PMMA 254–255, 257–260, 263, 265
poly(methyl methacrylate) 255
poly(N,N-dimethylacrylamide) (pDMA) 351
polyacrylamide 39, 41–42, 45–47
polyacrylamide gel electrophoresis 64–65, 131, 139, 142
polycarbonate 56, 255
polydimethylsiloxane 254
polydimethylsiloxane (PDMS) membrane 372
polyethylene oxide (PEO) 351
polyionic mobility 27, 29, 31, 49–51
polyions 5, 231, 236
polymerase chain reaction (PCR) 318
polymerization bonding 260
poly-N-hydroxyethylacrylamide (PHEA) 350
polynucleotide kinase 332, 339
polyvinylidene difluoride membrane 272
polyvinylpyrrolidone (PVP) 351
Ponceau S 74–75
Ponceau S staining of proteins 291
potassium dihydrogen phosphate 455
potassium hydroxide 455
potassium ion 34
ζ-potential 8–10
power supply 55–56, 59, 74
prealbumin 224
precast agarose gels 309, 395
precast polyacrylamide gels 309–310, 395
precipitation 14
preparation of polyacrylamide gels 358
preparative electrofocusing 247
preparative electrophoresis 241
primer extension assay 339, 373
protection 350
proteins 13–16, 27
pulsed-field capillary electrophoresis 353
pulsed-field gel electrophoresis 63, 343–346
pyridine 455
pyronin 455

QPNC-PAGE 241, 248–249
quadratic equation 29, 33

radioactive detection 277
recycling isoelectric focusing 248
regulating function of Kohlrausch 53

rehydratable polyacrylamide gels (CleanGels) 311
relative (di)electric permittivity 49, 51
relative molecular mass 13, 15
reptation model 326
reservoir 57
restriction endonuclease 332
restriction fragment length polymorphism (RFLP) 328
reverse iontophoresis 397
reverse transcriptase 339
RNA separation 338
rocket immunoelectrophoresis 110
rotating gel electrophoresis 345

sample injection module 350
Sanger sequencing 333, 340
scanner 55–56
scanners 57
scanning 299, 302
Schlieren-scanning system 62
scleroproteins 13
SDS 153, 161–172, 178–181, 183, 455
SDS disc-electrophoresis 65–66
SDS electrophoresis 65, 234–235, 238
SDS-proteins 46
second direction 231
semidry blotting 273–274
semidry transfer 377–378
sensitizing 294, 386
separation 222–226
separation medium 1–2
serum lipoproteins 224
serum proteins 78, 80–83, 98, 100, 102–104
short tandem repeats 345–346
sickle cell disease 91, 353
silver methods 3
silver nitrate 455
silver staining 294, 385
single-strand conformation polymorphism 334
slab gels 56
slipping plane 8
small intestine ALP isoenzyme 120–121
Smoluchowski 29–31, 49
sodium acetate 456
sodium carbonate 456
sodium dihydrogen phosphate 456
sodium dodecyl sulfate 162–163

sodium hydroxide  456
sodium ion  34
sodium thiosulfate  456
solvent bonding  260
D-sorbitol  456
Southern blotting  378–380, 382
Southwestern blotting  278–279
specific oligonucleotide primer  334
stacking gel  152, 160–161, 168–169, 179–180,
    182–183
staining method  171
staining of enzymes  296
staining of glycoproteins  295
staining of lipoproteins  296, 305
staining of proteins  276, 284, 290
starch (gel) electrophoresis  71
stationary ionic boundary  53
steady-state stacking elution  243, 245
stopping  294–295
submarine electrophoresis  327
sulfate ion  35
5-sulfosalicylic acid  290, 456
sulfuric acid  457
support films  40, 47
supported molecular matrix
    electrophoresis  119, 123
supported molecular matrix
    electrophoresis  123
surface electrophoresis  321–322
SYPRO Ruby staining  294, 298, 305

tank blotting  273
tank transfer  378
taurinate ion  34
taurine  457
temperature control  350
TES  457
TESate ion  34
thalassemias  91
thermal bonding  260
thermostat  55, 59
Tiselius electrophoresis  62
TMEDA  457
total monomeric concentration T  42
transcorneal iontophoresis  402
transdermal iontophoresis  400

transfer  271–274, 276, 279–282, 285–286,
    375–379, 381–382
transferrin  83
transthyretin  80
transverse alternating field
    gelelectrophoresis  344–345
triacylglycerols  81–82, 85–86, 88
trichloroacetic acid  74, 290, 302, 457
TRICINE  457
TRICINEate ion  34
triethanolamine  457
TRIS  22–23, 25, 457
TRIS ion  34, 36
TRIS-borate buffer  2, 36
TRIS-borate ion  36
TRIS-borate-EDTA buffer  23, 373
TRIS-buffered saline  116
TRIS-formate buffer  22
TRIS-formate-taurinate buffer system
    361–362
TRIS-glycinate buffer  74
TRIS-histidinate buffer  2
TRIS-taurinate buffer  74, 78, 99, 101–102,
    104, 106
TRIS-taurinate-EDTA buffer  23
Triton X-100  458
tubular proteinuria  98
Tween 20  458
two-dimensional electrophoresis  66, 231

ultrathin gels  46
urea  17, 458

vacuum transfer  375–376
variable number tandem repeat (VNTR)  329
vertical electrophoresis  61, 131
vertical-type elution  243
viral infections  401

Western blotting  278–279

xylene cyanol  2

Y chromosome  346

zone electrophoresis  61–65, 131–132, 134–136

www.ingramcontent.com/pod-product-compliance
Lightning Source LLC
Chambersburg PA
CBHW060957210326
41598CB00031B/4852